AUSTRALIA'S MEGAFIRES

Biodiversity Impacts and Lessons from 2019–2020

Editors: Libby Rumpff, Sarah M. Legge, Stephen van Leeuwen, Brendan A. Wintle and John C. Z. Woinarski

CSIRO PUBLISHING

ISBN: 9781486316649 (pbk)
ISBN: 9781486316656 (epdf)
ISBN: 9781486316663 (epub)

How to cite:
Rumpff L, Legge SM, van Leeuwen S, Wintle BA, Woinarski JCZ (Eds) (2023) *Australia's Megafires: Biodiversity Impacts and Lessons from 2019–2020*. CSIRO Publishing, Melbourne.

Published by:

CSIRO Publishing
Private Bag 10
Clayton South VIC 3169
Australia

Telephone: +61 3 9545 8400
Email: publishing.sales@csiro.au
Website: www.publish.csiro.au
Sign up to our email alerts: publish.csiro.au/earlyalert

Front cover: (Top) Fires rolling over the ridges of Tidbinbilla Nature Reserve and Namadgi National Park (photo by Markus Dirnberger); (Bottom, left to right) NSW National Parks and Wildlife Service staff in post-fire conditions (photo by Alex Pike/Saving our Species/DPE), eucalypt regrowth after fire (photo by Dan Rogers), an echidna in extensively burnt landscape (photo by WWF-Australia/Douglas Thron)

Edited by Joy Window (Living Language)
Cover design by Cath Pirret
Typeset by Envisage Information Technology
Index by Max McMaster
Printed in Australia by Ligare Book Printers

CSIRO acknowledges the Traditional Owners of the lands that we live and work on across Australia and pays its respect to Elders past and present. CSIRO recognises that Aboriginal and Torres Strait Islander peoples have made and will continue to make extraordinary contributions to all aspects of Australian life including culture, economy and science. CSIRO is committed to reconciliation and demonstrating respect for Indigenous knowledge and science. The use of Western science in this publication should not be interpreted as diminishing the knowledge of plants, animals and environment from Indigenous ecological knowledge systems.

FSC
www.fsc.org
MIX
Paper from responsible sources
FSC® C011613

The paper this book is printed on is in accordance with the standards of the Forest Stewardship Council®. The FSC® promotes environmentally responsible, socially beneficial and economically viable management of the world's forests.

Oct22_01

Contents

Acknowledgements

We would first like to acknowledge the Traditional Owners of Australia, their knowledge and continuing connection to the land, sea and waters of Australia. We pay our deep respects to Elders past and present.

This book represents a collaboration involving many researchers, government agencies and conservation NGOs, reflective of the extraordinary number and diversity of those who responded to the Australian wildfires of 2019–20. Nearly 200 people from more than 70 government agencies, universities, conservation non-government organisations, Indigenous organisations and other groups co-authored chapters, and this book is the better for the many perspectives thus offered. We thank all of these contributors for their wisdom, words, friendship and commitment.

The book was inspired by the concerns shown by so many Australians about the impacts that these wildfires had on our nature, and the extraordinary responses made by so many organisations and individuals to help recover species, environments and landscapes. In this book, we wanted to celebrate, pay tribute to and document that spirit, and to provide a resource that would help our community prepare better for next time.

A substantial part of this book springs from research conducted in the aftermath of the wildfires by the National Environmental Science Program's Threatened Species Recovery Hub, supported by the Australian Government. All book editors were affiliated with this program, however, we note that the contributors to this book had many other affiliations. Within the Threatened Species Recovery Hub, we acknowledge particularly Heather Christensen, Chris Fenwick, Rachel Morgain and Jaana Dielenberg for their expertise and ready guidance.

Part of our involvement in the 2019–20 wildfires stemmed from membership (LR, SL, SvL and JW) of the Wildlife and Threatened Species Bushfire Recovery Expert Panel, established by the Australian Minister for the Environment, while the fires were still burning; and we acknowledge the Chair of that panel, the then Threatened Species Commissioner Dr Sally Box, for helping to coordinate such an extraordinary response.

For their reviews of chapter drafts we thank the generosity and expert guidance of Michael Braby, Elinor Breman, Allan Burbidge, Gareth Catt, Melissa Danks, Kelly de Bie, Chris Dickman, Tim Doherty, Megan Good, Conrad Hoskins, Lindy Lumsden, Dan Lunney, Michael Mahony, Kristina Macdonald, Peter Mawson, Anna Richards, Kevin Tolhurst, Rob Whelan, Dick Williams and Mike Wouters. We thank also the many people who willingly contributed photographs to this book.

From CSIRO Publishing, we thank Briana Melideo, Mark Hamilton and Tracey Kudis for their expert guidance and care, and we thank Joy Window for copy editing.

Finally, we would like to dedicate this book to our children, and their future.

Libby Rumpff, Sarah Legge, Stephen van Leeuwen, Brendan Wintle and John Woinarski

List of contributors

Rachael Alderman
Department of Natural Resources and Environment Tasmania, 5/134 Macquarie Street Hobart, Tasmania 7000, Australia. (Rachael.Alderman@nre.tas.gov.au)

Tim Allard
Australian Wildlife Conservancy, 280 Hay Street, Subiaco East, Western Australia 6008, Australia. (tim.allard@australianwildlife.org)

Tony D. Auld
Science, Economics and Insights, NSW Department of Planning, Industry and Environment, Parramatta, New South Wales 2150, Australia; *and* Centre for Ecosystem Science, School of Biological, Earth and Environmental Science, University of NSW, Kensington, New South Wales 2052, Australia; *and* School of Earth, Atmospheric and Life Sciences, University of Wollongong, Wollongong, New South Wales 2522, Australia. (Tony.Auld@environment.nsw.gov.au)

David Bain
Ecosystems and Threatened Species, South East, Biodiversity and Conservation, NSW Department of Planning, Industry and Environment, Queanbeyan, New South Wales 2620, Australia. (david.bain@environment.nsw.gov.au)

Payal Bal
School of Ecosystem and Forest Sciences, University of Melbourne, Parkville, Victoria 3010, Australia. (payal.bal@unimelb.edu.au)

Sarah Barrett
Parks and Wildlife Service, Department of Biodiversity, Conservation and Attractions, Western Australian Government, Albany, Western Australia 6330, Australia. (sarah.barrett@dbca.wa.gov.au)

Stephen A. J. Bell
School of Environmental and Life Sciences, University of Newcastle, Callaghan, New South Wales 2308, Australia. (stephen.bell@newcastle.edu.au)

Karleah K. Berris
Kangaroo Island Landscape Board, 35 Dauncey Street, Kingscote South Australia 5223, Australia. (karleah.berris@sa.gov.au)

James R. Biggs
Director, Conservation & Population Management, Zoo and Aquarium Association Australasia, Mosman, New South Wales 2088, Australia; *and* School of Life and Environmental Sciences, University of Sydney, New South Wales 2066, Australia. (james@zooaquarium.org.au)

Leigh Blackmore
Murray Local Land Services, 931 Garland Ave, Albury, New South Wales 2640, Australia. (leigh.blackmore@lls.nsw.gov.au)

Glenn Block
Principal Director Bushfire Recovery Branch, Australian Government Department of Climate Change, Energy, the Environment and Water, John Gorton Building, King Edward Terrace, Parkes, Australian Capital Territory 2600, Australia. (glenn.block@environment.gov.au)

Lucas Bluff
Victorian Department of Environment, Land, Water and Planning, 574 Main St, Bairnsdale, Victoria 3875, Australia. (lucas.bluff@delwp.vic.gov.au)

Rebecca L. Boulton
Department of Ecology and Evolutionary Biology, University of Adelaide, Adelaide, South Australia 5005, Australia. (rlboulton@gmail.com)

Elle Bowd
Fenner School of Environment and Society, Australian National University, Canberra, Australian Capital Territory 2601, Australia. (elle.bowd@anu.edu.edu)

Anne Buchan
Biodiversity Division, Victorian Department of Environment, Land, Water and Planning, 8 Nicholson St, East Melbourne, Victoria 3002, Australia. (Anne.buchan@delwp.vic.gov.au)

Allan H. Burbidge
Biodiversity and Conservation Science, Department of Biodiversity, Conservation and Attractions, Western Australian Government, Kensington, Western Australia 6151, Australia. (allan.burbidge@dbca.wa.gov.au)

Amelia Caddy
Bush Heritage Australia, PO Box 329, Flinders Lane, Melbourne, Victoria 3009, Australia. (amelia.caddy@bushheritage.org.au; a.caddy@icloud.com)

Michelle Campbell-Ward
Veterinary Department, Taronga Western Plains Zoo, Dubbo, New South Wales 2830, Australia. (shellzoovet@googlemail.com)

David E. A. Catcheside
College of Science and Engineering, Flinders University, Adelaide, South Australia 5042, Australia. (david.catcheside@flinders.edu.au)

Pamela S. Catcheside
Botanic Gardens and State Herbarium, Adelaide, South Australia 5000, Australia; *and* College of Science and Engineering, Flinders University, Adelaide, South Australia 5042, Australia. (pam.catcheside@sa.gov.au)

Renee Catullo
School of Biological Sciences, University of Western Australia, 35 Stirling Highway, Crawley, Western Australia 6009, Australia. (renee.catullo@uwa.edu.au)

Tracey Churchill
Queensland Parks and Wildlife Service & Partnerships, Department of Environment and Science, 55 Priors Pocket Rd, Moggill, Queensland 4070, Australia; *and* Wildlife and

Threatened Species Operations, Queensland Parks and Wildlife Service and Partnerships, Department of Environment and Science, Brisbane, Queensland 4000, Australia. (Tracey.Churchill@des.qld.gov.au)

Hamish Clarke
NSW Bushfire Risk Management Research Hub, Wollongong, New South Wales 2522, Australia; *and* Hawkesbury Institute for the Environment, Western Sydney University, Locked Bag 1797, Penrith, New South Wales 2751, Australia. (hamishc@uow.edu.au)

David Coates
Biodiversity and Conservation Science, Department of Biodiversity, Conservation and Attractions, 17 Dick Perry Avenue, Kensington, Western Australia 6151, Australia; *and* School of Biological Sciences, University of Western Australia, 35 Stirling Hwy, Crawley, Western Australia 6009, Australia. (dave.coates@dbca.wa.gov.au)

Luke Collins
Pacific Forestry Centre, Canadian Forest Service, Natural Resources Canada, 506 West Burnside Road, Victoria BC V8Z 1M5, Canada; *and* Department of Ecology, Environment & Evolution, La Trobe University, Bundoora, Victoria 3086, Australia. (L.Collins3@latrobe.edu.au)

Sarah Comer
Parks and Wildlife Service, Department of Biodiversity, Conservation and Attractions, Western Australian Government, Albany, Western Australia 6330, Australia. (sarah.comer@dbca.wa.gov.au)

Lachlan M. Copeland
Eco Logical Australia, Coffs Harbour, New South Wales 2450, Australia. (LachlanC@ecoaus.com.au)

Oliver Costello
Bundjalung Jagun, Rosebank, New South Wales 2480, Australia. (oliverycostello@gmail.com)

Rebecca L. Cramp
School of Biological Sciences, University of Queensland, Brisbane, Queensland 4072, Australia. (r.cramp@uq.edu.au)

Jemma Cripps
Arthur Rylah Institute for Environmental Research, Department of Environment, Land, Water and Planning, 123 Brown Street, Heidelberg, Victoria 3084, Australia. (jemma.cripps@delwp.vic.gov.au)

Kaye Currey
Charles Darwin University, Casuarina, Northern Territory 0800, Australia; *and* School of Ecosystem of Forest Sciences, University of Melbourne, Parkville, Victoria 3010, Australia. (kaye.currey@unimelb.edu.au)

Naveed Davoodian
Royal Botanic Gardens Victoria, Melbourne, Victoria 3004, Australia. (naveed.davoodian@rbg.vic.gov.au)

Kelly de Bie
School of Ecosystem of Forest Sciences, University of Melbourne, Parkville, Victoria 3010, Australia. (kelly.debie@unimelb.edu.au)

Christopher R. Dickman
Desert Ecology Research Group, School of Life and Environmental Sciences, University of Sydney, New South Wales 2006, Australia. (chris.dickman@sydney.edu.au)

Rebecca Dillon
Rebecca Dillon, Department of Biodiversity, Conservation and Attractions, Albany, Western Australia 6330, Australia. (rebecca.dillon@dbca.wa.gov.au)

Jingyi Ding
Centre for Ecosystem Sciences, School of Biological, Earth and Environmental Sciences, University of NSW, Sydney, New South Wales 2052, Australia. (jyd@mail.bnu.edu.cn)

Paul Donatiu
Healthy Land and Water, Level 19, 160 Ann Street, Brisbane, Queensland 4000, Australia. (paul.do@hlw.org.au)

Don Driscoll
School of Life and Environmental Sciences, Deakin University, Melbourne Burwood Campus, 221 Burwood Highway, Burwood, Victoria 3125, Australia. (d.driscoll@deakin.edu.au)

Brenda Duffy
Bush Heritage Australia, PO Box 329, Flinders Lane, Melbourne, Victoria 3009, Australia. (Brenda.Duffy@bushheritage.org.au)

David H. Duncan
School of Ecosystem and Forest Sciences, University of Melbourne, Parkville, Victoria 3010, Australia. (david.duncan@unimelb.edu.au)

Louise Durkin
Arthur Rylah Institute for Environmental Research, Department of Environment, Land, Water and Planning, 123 Brown Street, Heidelberg, Victoria 3084, Australia. (louise.durkin@delwp.vic.gov.au)

Dan Duval
South Australian Seed Conservation Centre, Botanic Gardens and State Herbarium of South Australia, Adelaide, South Australia 5000, Australia. (daniel.duval@sa.gov.au)

David J. Eldridge
Centre for Ecosystem Sciences, School of Biological, Earth and Environmental Sciences, University of NSW, Sydney, New South Wales 2052, Australia. (d.eldridge@unsw.edu.au)

Michelle J. Ensbey
Research Institute of the Environment and Livelihoods, Charles Darwin University, Casuarina, Northern Territory 0800, Australia. (mikijensbey@gmail.com)

Pamela Fallow
BirdLife Australia, Suite 2-05, 60 Leicester Street, Carlton, Victoria 3053, Australia. (pamela.fallow@birdlife.org.au)

Adam Fawcett
Northern Inland Branch, NSW National Parks and Wildlife Service, Department of Planning and Environment, Locked Bag 5022, Parramatta, New South Wales 2124, Australia. (adam.fawcett@environment.nsw.gov.au)

David M. Forsyth
Vertebrate Pest Research Unit, NSW Department of Primary Industries, Orange, New South Wales 2800, Australia. (dave.forsyth@dpi.nsw.gov.au)

Craig E. Franklin
School of Biological Sciences, University of Queensland, Brisbane, Queensland 4072, Australia. (c.franklin@uq.edu.au)

Fiona Fraser
Acting Threatened Species Commissioner, Australian Government Department of Climate Change, Energy, the Environment and Water, John Gorton Building, King Edward Terrace, Parkes, Australian Capital Territory 2600, Australia. (fiona.fraser@awe.gov.au)

Hannah Fraser
School of Ecosystem and Forest Sciences, University of Melbourne, Parkville, Victoria 3010, Australia. (hannahsfraser@gmail.com)

Rosemary Gales
Department of Natural Resources and Environment Tasmania, 5/134 Macquarie Street Hobart, Tasmania 7000. (Rosemary.Gales@nre.tas.gov.au)

Rachael V. Gallagher
Hawkesbury Institute for the Environment, Western Sydney University, Richmond, New South Wales 2753, Australia. (Rachael.Gallagher@westernsydney.edu.au)

Stephen T. Garnett
Research Institute of the Environment and Livelihoods, Charles Darwin University, Casuarina, Northern Territory 0800, Australia. (stephen.garnett@cdu.edu.au)

Jody A. Gates
National Parks and Wildlife Service South Australia, Department for Environment and Water, 81-95 Waymouth St, Adelaide, South Australia 5000, Australia. (jody.gates@sa.gov.au)

William Geary
Biodiversity Division, Department of Environment, Land, Water and Planning, Biodiversity Strategy and Knowledge, East Melbourne, Victoria 3002, Australia. (billy.geary@delwp.vic.gov.au)

Heloise Gibb
School of Life Sciences, La Trobe University, Melbourne, Victoria 3086, Australia. (h.gibb@latrobe.edu.au)

Rebecca K. Gibson
Science, Economics and Insights Division, Department of Planning Industry and Environment, Alstonville, New South Wales 2477, Australia. (Rebecca.Gibson@environment.nsw.gov.au)

Kate Giljohann
School of Ecosystem and Forest Sciences, University of Melbourne, Parkville, Victoria 3010, Australia. (kmgi@unimelb.edu.au)

Carl R. Gosper
Biodiversity and Conservation Science, Department of Biodiversity, Conservation and Attractions, Western Australian Government, Kensington, Western Australia 6151, Australia. (carl.gosper@dbca.wa.gov.au)

Jenny Gray
Zoos Victoria, Elliot Avenue, Parkville, Victoria 3052, Australia. (jgray@zoo.org.au)

Matthew Greenlees
School of Biological Sciences, Monash University, Melbourne, Victoria 3800, Australia; *and* Department of Biological Sciences, Macquarie University, Sydney, New South Wales 2109, Australia; *and* Australian Museum Research institute, The Australian Museum, 1 William Street, Sydney, New South Wales 2010, Australia. (matthew.greenlees@mq.edu.au)

Aaron Greenville
School of Life and Environmental Sciences, University of Sydney, New South Wales 2006, Australia. (aaron.greenville@sydney.edu.au)

Heidi Groffen
Kangaroo Island Land for Wildlife, PO Box 966, Kingscote, South Australia 5223, Australia. (heidi@terrainecology.com.au)

Darren Grover
WWF-Australia, 60 Leicester Street, Carlton, Victoria 3053, Australia. (Dgrover@wwf.org.au)

Jenny Guerin
South Australian Seed Conservation Centre, Botanic Gardens and State Herbarium of South Australia, Adelaide, South Australia 5000, Australia. (jenny.guerin@sa.gov.au)

Fern Hames
Arthur Rylah Institute for Environmental Research, Department of Environment, Land, Water and Planning, 123 Brown St, Heidelberg, Victoria 3084, Australia. (fern.hames@delwp.vic.gov.au)

Tianxiao Hao
School of Ecosystem and Forest Sciences, University of Melbourne, Victoria 3010, Australia. (tianxiaoh@student.unimelb.edu.au)

Mark S. Harvey
Collections & Research, Western Australian Museum, 49 Kew Street, Welshpool, Western Australia 6106, Australia. (mark.harvey@museum.wa.gov.au)

Anna-Lisa Hayes
Australian Government Department of Climate Change, Energy, the Environment and Water, John Gorton Building, King Edward Terrace, Parkes, Australian Capital Territory 2600, Australia. (anna-lisa.hayes@environment.gov.au)

Jason Higham
National Parks & Wildlife Service, Department for Environment and Water, South Australian Government, GPO Box 1047, 81–95 Waymouth Street, Adelaide, South Australia 5000, Australia. (jason.higham@sa.gov.au)

Harry B. Hines
Department of National Parks, Sport and Racing, GPO Box 2454, Brisbane, Queensland 4001, Australia. (Harry.Hines@des.qld.gov.au)

Pat Hodgens
Kangaroo Island Land for Wildlife, PO Box 966, Kingscote, South Australia 5223, Australia. (pat@terrainecology.com.au)

Katja Hogendoorn
School of Agriculture, Food and Wine, University of Adelaide, Adelaide, South Australia 5005, Australia. (katja.hogendoorn@adelaide.edu.au)

Rosemary Hohnen
Research Institute for the Environment and Livelihoods, Charles Darwin University, Darwin, Northern Territory 0810, Australia. (rosie.hohnen@gmail.com)

Bronwyn Hradsky
School of Ecosystem and Forest Sciences, University of Melbourne, Victoria 3010, Australia. (bronwyn.hradsky@unimelb.edu.au)

Paul Jennings
Kangaroo Island Landscape Board, 35 Dauncey Street, Kingscote, South Australia 5223, Australia. (Paul.Jennings@sa.gov.au)

Chris Jolly
Gulbali Institute, Charles Sturt University, Elizabeth Mitchell Drive, PO Box 789, Albury, New South Wales 2640, Australia; *and* Department of Biological Sciences, Macquarie University, Sydney, New South Wales 2109, Australia. (chris.jolly@mq.edu.au)

Liana Joseph
Australian Wildlife Conservancy, 280 Hay Street, Subiaco East, Western Australia 6008, Australia. (Liana.Joseph@australianwildlife.org)

Greg Keighery
Biodiversity and Conservation Science, Department of Biodiversity, Conservation and Attractions, 17 Dick Perry Avenue, Kensington, Western Australia 6151, Australia. (Greg.Keighery@dpird.wa.gov.au)

David A. Keith
Science, Economics and Insights, NSW Department of Planning, Industry and Environment, Parramatta, New South Wales 2150, Australia; *and* Centre for Ecosystem Science, School of Biological, Earth and Environmental Science, University of NSW, Kensington, New South Wales 2052, Australia. (david.keith@unsw.edu.au)

Luke Kelly
School of Ecosystem & Forest Sciences, University of Melbourne, Victoria 3010, Australia. (ltkelly@unimelb.edu.au)

Mark J. Kennard
Australian Rivers Institute, Griffith University, Nathan, Queensland 4111, Australia. (m.kennard@griffith.edu.au)

Robert M. Kooyman
Department of Biological Sciences, Macquarie University, Sydney; *and* National Herbarium of NSW, Royal Botanic Gardens, Sydney; *and* Missouri Botanical Garden (MBG) (robert.kooyman@mq.edu.au)

Tanya Latty
School of Life and Environmental Sciences, University of Sydney, Sydney, New South Wales 2006, Australia. (tanya.latty@sydney.edu.au)

Jenny Lau
BirdLife Australia, Suite 2-05, 60 Leicester Street, Carlton, Victoria 3053, Australia. (jenny.lau@birdlife.org.au)

Brad Law
Forest Science, NSW Primary Industries, Parramatta, New South Wales 2124, Australia. (brad.law@dpi.nsw.gov.au)

Tom D. Le Breton
Centre for Ecosystem Science, School of Biological, Earth and Environmental Science, University of NSW, Kensington, New South Wales 2052, Australia. (t.lebreton@unsw.edu.au)

Joshua Lee
Centre for Ecosystem Science, UNSW Sydney, Sydney, New South Wales 2052, Australia. (josh.lee2415@gmail.com)

Sarah M. Legge
Fenner School of Environment and Society, Australian National University, Canberra, Australian Capital Territory 2601, Australia; *and* Research Institute for the Environment and Livelihoods, Charles Darwin University, Darwin, Northern Territory 0810, Australia; *and* Centre for Biodiversity Conservation Science, University of Queensland, St Lucia, Queensland 4072, Australia. (SarahMariaLegge@gmail.com)

Frank Lemckert
Eco Logical Australia Pty Ltd, PO Box 12, Sutherland, New South Wales 2232, Australia; *and* Forest Corporation NSW, Hardwood Forests, PO Box 702, Eden, New South Wales 2551, Australia. (Frank.Lemckert@ecoaus.com.au)

Viyanna Leo
Australian Wildlife Conservancy, 280 Hay Street, Subiaco East, Western Australia 6008, Australia. (viyanna.leo@australianwildlife.org)

Steve Leonard
Environment, Heritage and Land Division, Department of Natural Resources and Environment Tasmania, 5/134 Macquarie Street Hobart, Tasmania 7000. (Steve.Leonard@nre.tas.gov.au)

David B. Lindenmayer
Fenner School of Environment and Society, Australian National University, Canberra, Australian Capital Territory 2601, Australia. (david.lindenmayer@anu.edu.au)

Mark Lintermans
Centre for Applied Water Science, University of Canberra, Australian Capital Territory 2617, Australia. (mark.lintermans@canberra.edu.au)

Hannah Lloyd
Biodiversity and Conservation Division | Department of Planning and Environment, 4 Parramatta Square, 12 Darcy St Parramatta, New South Wales 2150 (Locked Bag 5022, Parramatta, New South Wales 2124). (Hannah.Lloyd@environment.nsw.gov.au)

Lisien Loan
National Parks & Wildlife Service, Department for Environment and Water, South Australian Government, GPO Box 1047, 81–95 Waymouth Street, Adelaide, South Australia 5000, Australia. (lisien.loan@sa.gov.au)

Michele Lockwood
Research Institute for the Environment and Livelihoods, Charles Darwin University, Darwin, Northern Territory 0810, Australia. (m.lockwood.nrm@gmail.com)

Ella Loeffler
Biodiversity Division, Victorian Department of Environment, Land, Water and Planning, Level 2, 8 Nicholson St, East Melbourne VIC 3002, PO Box 500, East Melbourne, Victoria 3002. (ella.loeffler@delwp.vic.gov.au)

Rachel Lowry
WWF-Australia, 60 Leicester Street, Carlton, Victoria 3053, Australia. (RLowry@wwf.org.au)

Phoebe V. Macak
Arthur Rylah Institute for Environmental Research, Department of Environment, Land, Water and Planning, 123 Brown Street, Heidelberg, Victoria 3084, Australia. (phoebe.macak@delwp.vic.gov.au)

Berin D. E. Mackenzie
Science Economics and Insights Division, NSW Department of Planning, Industry and Environment, Parramatta, New South Wales 2150, Australia; *and* Centre for Ecosystem Science, School of Biological, Earth and Environmental Science, University of NSW, Kensington, New South Wales 2052, Australia. (berin.mackenzie@environment.nsw.gov.au)

Nina Maclean
Environment, Heritage & Water Division, Environment, Planning and Sustainable Development Directorate, ACT Government, Canberra, Australian Capital Territory 2601, Australia. (Nina.McLean@act.gov.au)

Michael J. Mahony
School of Environmental and Life Sciences, University of Newcastle, New South Wales 2308, Australia. (michael.mahony@newcastle.edu.au)

Robert O. Makinson
Bob Makinson Consulting, 19 Fern St, Pymble, New South Wales 2073, Australia. (Robert.Makinson@environment.nsw.gov.au)

Justin Mallee
North Coast Branch, NSW National Parks and Wildlife Service, Alstonville, New South Wales 2477, Australia (justin.mallee@environment.nsw.gov.au)

George Malolakis
South Coast Branch, NSW National Parks and Wildlife Service, Merimbula, New South Wales 2548, Australia. (george.malolakis@environment.nsw.gov.au)

Helene Marsh
College of Science and Engineering, Bebegu Yumba Campus, James Cook University, Townsville, Queensland 4811, Australia. (helene.marsh@jcu.edu.au)

Jess R. Marsh
Harry Butler Research Institute, Murdoch University, Murdoch, Western Australia 6150, Australia; *and* Earth and Biological Sciences, South Australian Museum, Adelaide, South Australia 5000, Australia. (jess.marsh@murdoch.edu.au)

Golo Maurer
BirdLife Australia, Suite 2-05, 60 Leicester Street, Carlton, Victoria 3053, Australia; *and* College of Marine & Environmental Sciences and Australian Tropical Herbarium, James Cook University, Nguma-bada Campus, McGregor Road, Smithfield, Queensland 4878, Australia. (Golo.Maurer@birdlife.org.au)

Tom W. May
Royal Botanic Gardens Victoria, Melbourne, Victoria 3004, Australia. (tom.may@rbg.vic.gov.au)

Rob McCormack
Australian Aquatic Biological Pty Ltd, Karuah, New South Wales 2324, Australia. (rob@aabio.com.au)

David McCreery
South Coast Branch, NSW National Parks and Wildlife Service, Merimbula, New South Wales 2548, Australia. (david.mccreery@environment.nsw.gov.au)

Michael S. McFadden
Australian Fauna Precinct, Taronga Conservation Society Australia, Mosman, New South Wales 2088, Australia. (mmcfadden@zoo.nsw.gov.au)

Jane Melville
Museums Victoria, GPO Box 666, Melbourne, Vic 3001, Australia; *and* School of Biological Sciences, Monash University, Melbourne, Victoria 3800, Australia. (jmelv@museum.vic.gov.au)

Andre Messina
Royal Botanic Gardens Victoria, Birdwood Avenue, Melbourne, Victoria 3004, Australia. (andre.messina@rbg.vic.gov.au)

Damian Michael
Gulbali Institute, School of Agriculture, Environment and Veterinary Science, Charles Sturt University, Elizabeth Mitchell Drive, PO Box 789, Albury, New South Wales 2640, Australia. (DMichael@csu.edu.au)

Kim A. Miller
Healesville Sanctuary, Zoos Victoria, Healesville, Victoria 3777, Australia. (kmiller@zoo.org.au)

Christian Miller-Sabbioni
School of Molecular and Life Sciences, Curtin University, Bentley, Western Australia 6102, Australia. (christian.miller-sabbioni@curtin.edu.au)

Melinda L. Moir
Department of Primary Industries and Regional Development, South Perth, Western Australia 6151, Australia. (melinda.moir@dpird.wa.gov.au)

Rachel Morgain
NRM Regions Australia, 48 Linnaeus Way, Acton, Australian Capital Territory 2601, Australia. (rachelm@nrmregionsaustralia.com.au)

Eridani Mulder
Australian Wildlife Conservancy, Wongalara Wildlife Sanctuary, PMB 162, Katherine, Northern Territory 0852, Australia. (Eridani.Mulder@australianwildlife.org)

Angela Muscatello
Department of Environment, Land, Water and Planning, East Melbourne, Victoria 3002, Australia. (angela.muscatello@delwp.vic.gov.au)

V. John Neldner
Queensland Herbarium, Department of Environment and Science, Toowong, Queensland 4066, Australia. (john.neldner@des.qld.gov.au)

Jenny Nelson
Arthur Rylah Institute for Environmental Research, Department of Environment, Land, Water and Planning, 123 Brown Street, Heidelberg, Victoria 3084, Australia. (jenny.nelson@delwp.vic.gov.au)

David Newell
Faculty of Science and Engineering Southern Cross University Lismore, New South Wales 2480, Australia. (david.newell@scu.edu.au)

Michael R. Ngugi
Queensland Herbarium, Department of Environment and Science, Toowong, Queensland 4066, Australia. (michael.ngugi@des.qld.gov.au)

Dale Nimmo
Gulbali Institute, Charles Sturt University, School of Agriculture, Environment and Veterinary Science, Charles Sturt University, Elizabeth Mitchell Drive, PO Box 789, Albury, New South Wales 2640, Australia. (dnimmo@csu.edu.au)

Craig Nitschke
School of Ecosystem and Forest Sciences, University of Melbourne, 500 Yarra Blvd, Richmond, Victoria 3121, Australia. (craign@unimelb.edu.au)

Rachael H. Nolan
NSW Bushfire Risk Management Research Hub, Wollongong, New South Wales 2522, Australia; *and* Hawkesbury Institute for the Environment, Western Sydney University, Locked Bag 1797, Penrith, New South Wales 2751, Australia. (Rachael.Nolan@westernsydney.edu.au)

Catherine A. Offord
The Australian PlantBank, Australian Institute of Botanical Science, Australian Botanic Garden, Mount Annan, New South Wales 2567, Australia. (Cathy.Offord@botanicgardens.nsw.gov.au)

Damon L. Oliver
Ecosystems and Threatened Species, South East, Biodiversity and Conservation, New South Wales Department of Planning, Industry and Environment, Queanbeyan, New South Wales 2620, Australia. (Damon.Oliver@environment.nsw.gov.au)

Mark K. J. Ooi
Centre for Ecosystem Science, School of Biological, Earth and Environmental Science, University of NSW, Kensington, New South Wales 2052, Australia. (m.ooi@unsw.edu.au)

Manda Page
Wildlife and Threatened Species Operations, Queensland Parks and Wildlife Service and Partnerships, Department of Environment and Science, Brisbane, Queensland 4000. (Manda.page@des.qld.gov.au)

Marissa L. Parrott
Wildlife Conservation & Science, Zoos Victoria, Parkville, Victoria 3052, Australia. (mparrott@zoo.org.au)

Bruce Pascoe
Black Duck Foods, P.O. Box 27, Genoa, Victoria 3891, Australia. (brucepascoe@westnet.com.au)

Luke Pearce
New South Wales Department of Primary Industries (Fisheries), Albury, New South Wales 2640, Australia. (luke.pearce@dpi.nsw.gov.au)

Trent D. Penman
FLARE wildfire research, School of Ecosystem and Forest Sciences, University of Melbourne, Parkville, Victoria 3010, Australia. (trent.penman@unimelb.edu.au)

Petina L. Pert
CSIRO Land and Water, Townsville, Queensland 4814, Australia. (petina.pert@csiro.au)

Owen Price
Centre for Environmental Risk Management of Bushfire, School of Earth, Atmosphere and Life Sciences, University of Wollongong, New South Wales 2522, Australia. (oprice@uow.edu.au)

Ella Plumanns Pouton
School of Ecosystem & Forest Sciences, University of Melbourne, Victoria 3010, Australia. (ellap@student.unimelb.edu.au)

Tarmo A. Raadik
Arthur Rylah Institute for Environmental Research, Heidelberg, Victoria 3084, Australia. (tarmo.raadik@delwp.vic.gov.gov.au)

Tania Reid
Biodiversity and Conservation Division, Department of Planning and Environment, 4 Parramatta Square, 12 Darcy St, Parramatta, New South Wales 2150 (Locked Bag 5022, Parramatta, New South Wales 2124). (Tania.Reid@environment.nsw.gov.au)

J. Dale Roberts
School of Biological Sciences, University of Western Australia, PO Box 5771, Albany, Western Australia 6332, Australia. (dale.roberts@uwa.edu.au)

Cathy J. Robinson
CSIRO Land and Water, Dutton Park, Qld 4001, Australia; *and* Research Institute of the Environment and Livelihoods, Charles Darwin University, Casuarina, Northern Territory 0909, Australia. (catherine.robinson@csiro.au)

Daniel J. Rogers
Department for Environment and Water, South Australian Government, GPO Box 1047, 81-95 Waymouth Street, Adelaide, South Australia 5000, Australia; *and* School of Biological Sciences, University of Adelaide, Adelaide, South Australia 5005, Australia. (daniel.rogers@sa.gov.au)

Simon Rowe
OceanWatch Australia, Cnr Bank St and Pyrmont Bridge Rd, Pyrmont, New South Wales 2009, Australia. (simon@oceanwatch.org.au)

Jodi J. L. Rowley
Australian Museum Research Institute, Australian Museum, 1 William St, Sydney 2010, Australia; *and* Centre for Ecosystem Science, School of Biological, Earth and Environmental Sciences, University of NSW, Sydney, New South Wales 2052, Australia. (Jodi.Rowley@austmus.gov.au)

Libby Rumpff
School of Ecosystem and Forest Sciences, University of Melbourne, Parkville, Victoria 3010, Australia. (lrumpff@unimelb.edu.au)

Claudia Santori
OceanWatch Australia, Cnr Bank St and Pyrmont Bridge Rd, Pyrmont, New South Wales 2009, Australia. (claudia@oceanwatch.org.au)

Ben C. Scheele
ARC Centre of Excellence for Environmental Decisions, National Environmental Research Program Environmental Decisions Hub, Fenner School of Environment and Society, Forestry Building (48), Australian National University, Canberra, Australian Capital Territory 0200 Australia. (Ben.scheele@anu.edu.au)

Julian Seddon
Environment, Heritage & Water Division, Environment, Planning and Sustainable Development Directorate, ACT Government, Canberra, Australian Capital Territory 0200 Australia. (julian.seddon@act.gov.au)

Sally L. Sherwen
Wildlife Conservation and Science, Zoos Victoria, Melbourne, Victoria 3010, Australia. (ssherwen@zoo.org.au)

Chris Smyth
Chris Smyth Consulting, Queenscliff, Victoria 3225, Australia. (chris.smyth36@gmail.com)

Darren Southwell
Quantitative and Applied Ecology Group, School of Ecosystem and Forest Sciences, University of Melbourne, Parkville, Victoria 3010, Australia. (darren.southwell@unimelb.edu.au)

Emma Spencer
WWF-Australia, Level 1, 1-3 Smail Street, Ultimo, New South Wales 2007, Australia. (espencer@wwf.org.au)

Rebecca Spindler
Bush Heritage Australia, PO Box 329, Flinders Lane, Melbourne, Victoria 3009, Australia. (rebecca.spindler@bushheritage.org.au)

Elizabeth M. Tasker
Science, Economics and Insights, NSW Department of Planning, Industry and Environment, Parramatta, New South Wales 2150, Australia. (liz.tasker@environment.nsw.gov.au)

Chris Taylor
Fenner School of Environment and Society, Australian National University, Canberra, Australian Capital Territory 2601, Australia. (Christopher.Taylor@anu.edu.au)

Thai Te
South Australian Seed Conservation Centre, Botanic Gardens and State Herbarium of South Australia, Adelaide, South Australia 5000, Australia. (Thai.te@sa.gov.au)

Daniella Teixeira
Queensland University of Technology, Gardens Point, Brisbane 4000, Australia. (d.teixeira@uq.edu.au); *and* Bush Heritage Australia, PO Box 329, Flinders Lane, Melbourne, Victoria 3009, Australia. (daniella.teixeira@bushheritage.org.au)

Stephanie Todd
BirdLife Australia, Suite 2-05, 60 Leicester Street, Carlton, Victoria 3053, Australia; *and* College of Marine & Environmental Sciences and Australian Tropical Herbarium, James Cook University, Nguma-bada Campus, McGregor Road, Smithfield, Queensland 4878, Australia. (stephanie.todd@birdlife.org.au)

Arn Tolsma
Arthur Rylah Institute for Environmental Research, Department of Environment, Land, Water and Planning, 123 Brown St, Heidelberg, Victoria (arn.tolsma@delwp.vic.gov.au)

Mark Tozer
Science, Economics and Insights, NSW Department of Planning, Industry and Environment, Parramatta, New South Wales 2150, Australia; *and* Centre for Ecosystem Science, School of Biological, Earth and Environmental Science, University of NSW, Kensington, New South Wales 2052, Australia. (Mark.Tozer@environment.nsw.gov.au)

Samantha K. Travers
Centre for Ecosystem Sciences, School of Biological, Earth and Environmental Sciences, University of NSW, Sydney, New South Wales 2052, Australia. (samantha.k.travers@gmail.com)

Camille Truong
Royal Botanic Gardens Victoria, Melbourne, Victoria 3004, Australia. (camille.truong@rbg.vic.gov.au)

Ayesha Tulloch
School of Life and Environmental Sciences, University of Sydney, Camperdown, New South Wales 2006, Australia. (ayesha.tulloch@sydney.edu.au)

Kate Umbers
School of Science & Hawkesbury Institute for the Environment, Western Sydney University, Locked Bag 1797 Penrith, New South Wales 2751, Australia. (k.umbers@westernsydney.edu.au)

Lily M. van Eeden
Arthur Rylah Institute for Environmental Research, Department of Environment, Land, Water and Planning, 123 Brown Street, Heidelberg, Victoria 3084, Australia. (lily.vaneeden@delwp.vic.gov.au)

Stephen van Leeuwen
School of Molecular & Life Sciences, Faculty of Science & Engineering, Curtin University, GPO Box U1987, Perth, Western Australia 6845, Australia. (stephen.vanleeuwen@curtin.edu.au)

Monique Van Sluys
Welfare, Conservation & Science, Taronga Conservation Society Australia, Mosman, New South Wales 2088, Australia. (mvansluys@zoo.nsw.gov.au)

Ian S. Walker
Environment, Planning and Sustainable Development Directorate, ACT Government, Naas Neighbourhood, Level 2, 480 Northbourne Avenue, Dickson (GPO Box 158 Canberra Australian Capital Territory 2602). (IanS.Walker@act.gov.au)

Neville Walsh
Royal Botanic Gardens Victoria, Birdwood Avenue, Melbourne, Victoria 3004, Australia. (Neville.Walsh@rbg.vic.gov.au)

Michelle Ward
Centre for Biodiversity and Conservation Science, University of Queensland, St Lucia, Queensland 4072, Australia; *and* School of Earth and Environmental Sciences, The University of Queensland, Brisbane, Queensland 4072, Australia; *and* WWF – Australia, Brisbane, Queensland 4000, Australia. (m.ward@uq.edu.au)

James E. M. Watson
Centre for Biodiversity and Conservation Science, University of Queensland, St Lucia, Queensland 4072, Australia; *and* School of Earth and Environmental Sciences, The University of Queensland, Brisbane, Queensland 4072, Australia. (james.watson@uq.edu.au)

Jonathan Webb
School of Life Sciences, University of Technology Sydney, Broadway, New South Wales, 2007 Australia. (jonathan.webb@uts.edu.au)

Noel Webster (Uncle Nook)
Sussex Inlet, New South Wales 2540, Australia. (noel.webster@dpie.nsw.gov.au)

Judy West
Executive Director, Australian National Botanic Gardens, Acton, Australian Capital Territory 2601, Australia. (Judy.West@awe.gov.au)

Matt West
School of BioSciences, University of Melbourne, Parkville 3010, Victoria, Australia. (m.west@unimelb.edu.au)

Cherie White
Riverina Local Land Services, 64 Fitzroy Street, Tumut, New South Wales 2720, Australia. (cherie.white@lls.nsw.gov.au)

Jacob White
C/– Level 19, 160 Ann Street, Brisbane, Queensland 4000, Australia. (jcb.white@gmail.com)

Nick S. Whiterod
Aquasave-Nature Glenelg Trust, Victor Harbor, South Australia 5211, Australia. (nick.whiterod@aquasave.com.au)

Dick Williams
Centre for Applied Alpine Ecology, La Trobe University, Bundoora, Australia. (dickwilliams1955@gmail.com)

Brendan A. Wintle
School of Ecosystem and Forest Sciences, University of Melbourne, Parkville, Victoria 3010, Australia. (b.wintle@unimelb.edu.au)

John C. Z. Woinarski
Research Institute of the Environment and Livelihoods, Charles Darwin University, Casuarina, Northern Territory 0909, Australia; *and* School of Ecosystem and Forest Sciences, University of Melbourne, Parkville, Victoria 3010, Australia. (john.woinarski@cdu.edu.au)

Rupert Woods
Wildlife Health Australia, Mosman, New South Wales 2088, Australia. (rwoods@wildlifehealthaustralia.com.au)

Damian Wrigley
National Coordinator, Australian Seed Bank Partnership, Acton, Australian Capital Territory 2601, Australia. (coordinator@seedpartnership.org.au)

Colin J. Yates
Department of Biodiversity, Conservation and Attractions, Kensington, Western Australia 6151, Australia. (colin.yates@dbca.wa.gov.au)

Heidi C. Zimmer
Centre for Australian National Biodiversity Research, Centre for Australian National Biodiversity Research, Canberra, Australian Capital Territory 2061, Australia. (Heidi.Zimmer@csiro.au)

Sylvia Zukowski
Aquasave-Nature Glenelg Trust, Victor Harbor, South Australia 5211, Australia. (sylvia.zukowski@aquasave.com.au)

Phil Zylstra
School of Molecular and Life Sciences, Curtin University, Bentley, Western Australia 6102, Australia. (pzylstra@uow.edu.au)

1

Introduction

Libby Rumpff, Sarah M. Legge, Stephen van Leeuwen, Brendan A. Wintle and John C. Z. Woinarski

Fire is an integral part of the Australian landscape. Over a long history, including tens of thousands of years under Indigenous use and management, it has shaped most of Australia's terrestrial environments and moulded its plant and animal assemblages (Bowman 1998; Bowman *et al.* 2009).

Yet in 2019–20, much of eastern and southern Australia was burnt in 'megafires' at high severity, across an unprecedented extent and over an unusually long duration (Collins *et al.* 2021). The term and incidence of megafire is relatively new: reflecting a global increase in fire sizes, it is recognised as 'an emerging concept' (Linley *et al.* in press). Here we apply it to fires that are more than 100 000 ha in extent, although we note that others have advocated applying the term megafire to fires > 10 000 ha and have coined the terms *gigafire* for fires > 100 000 ha, and *terafire* for fires > 1 000 000 ha (Linley *et al.* in press).

This season of fires was a landmark event in the country's post-colonial history, a year of exceptional loss. For many, these megafires provided a stark demonstration that the consequences of global climate change are occurring now. This was a frightening glimpse of a dystopian future, of a world beyond our control (Fig. 1.1).

The 2019–20 wildfires exacted a considerable toll on people and the economy. They killed 33 people directly and in fire-fighting operations, led to the death of approximately another 430 people through smoke pollution, imposed a $2 billion impost on the Australian health budget (Johnston *et al.* 2021), destroyed many homes and livelihoods and caused significant damage to the national and regional economies. However, in this book our focus is on the environmental impact of these fires.

Australians, and the global community more broadly, watched on as images of burnt koalas and kangaroos haunted the nightly news footage, and experts estimated that billions of animals may have been killed (van Eeden *et al.* 2020). But the dead or injured animals represented only the most conspicuous margins of the impacts of these fires on wildlife. Places of outstanding biodiversity significance, such as World Heritage areas, were damaged. Some environments intolerant of fire, such as rainforests, burnt for the first time in living memory (Kooyman *et al.* 2020). A preliminary analysis estimated that, over the period of 6–8 months, the fires burnt the habitat or range of 107 of Australia's 436 threatened terrestrial and freshwater vertebrate species. The fires rapidly pushed many already imperilled species further to the brink, and caused hundreds of other species formerly considered secure to now be threatened (Ward *et al.* 2020). Scientists have begun

Fig. 1.1. The 2019–20 wildfires were exceptional in scale and duration, burning large areas at high severity. Here, flames from the fires rolling over the ridges of Tidbinbilla Nature Reserve and Namadgi National Park. (Photo: Markus Dirnberger)

documenting likely extinctions as a result of the fires (Moir 2021) – we will learn more in the coming years. However, because of decades of chronic underfunding of survey and monitoring efforts, the actual number of such extinctions may never be known (Scheele *et al.* 2019; Wintle *et al.* 2019).

The fires galvanised an extraordinary response by the Australian community, governments, Indigenous groups and conservation organisations. Record sums of money were contributed to wildlife rescue and recovery efforts. The Australian Government established a $200 million wildlife recovery fund, for urgent and short-term actions designed to limit biodiversity losses and support post-fire recovery – this one-off funding initiative was almost twice that of the normal annual expenditure by the Australian Government on management of all threatened species (Wintle *et al.* 2019). Rapid responses – often heroically undertaken in dangerous situations and with great urgency – helped save some species and constrain some losses (Morton 2020). These responses showed how much the Australian community valued our nature, and were concerned and moved by its potential loss.

But while many of these responses were admirable and achieved some successes, these fires also highlighted the *ad hoc* status, and inadequacies, of planning, management, policy and legal settings for the protection of biodiversity. Emergency management for biodiversity was not previously given prominence by government agencies, and these fires exposed the limited capacity to deal with situations that were beyond what was previously thought to be the extreme limit. Established soon after the fires, a series of government inquiries probed these revealed weaknesses, and proposed some major changes to attempt to reduce the likelihood of future comparable events and to improve the effectiveness of fire planning and operational control (Chapter 30).

Of course, there is a history of severe and destructive wildfires; these have been a recurring and scarring feature of Australian life, especially in temperate Australia. Although exceptional, the 2019–20 wildfires had precursors that also had major impacts on biodiversity. For example, although little of Tasmania was burnt in the 2019–20 season, extensive fires there in the previous season, and in 2016, imposed significant impacts on the Tasmanian Wilderness World Heritage Area and on species and ecological communities that are highly susceptible to wildfire (Foulkes *et al.* 2021). But no previous Australian wildfires have caused such loss of biodiversity as the 2019–20 wildfires. And our starting point in 2019–20 was influenced and contextualised by a changing fire regime – the complex spatial and temporal patterning of fires has been shifting. Some species and environments may recover from the impacts of the 2019–20 fires, but recovery may not occur for many decades, and recovery for others may be stymied if another comparable fire occurs in the near future. The precarious prospects for these species and environments depend now on our willingness and capability to better manage this country.

Why we wrote this book

We write as scientists, respectful of the objectivity of data and fascinated by the analysis and interpretation of it. But here we also write for the benefit of our children, and for their descendants, and for the world that they will live in. We write at a critical time for the health of our planet, for the wellbeing of humanity, for the future of nature. These are at increasing risk from cascading episodes of catastrophic events such as the 2019–20 Australian wildfires, from the factors that made those fires happen, from our demonstrable ineffectiveness in managing them, and from an insufficient provision of supporting resources and inadequate protection (Samuel 2020). We write because we don't want to bequeath to our children a tarnished world in which such fires – and the devastating and compounding losses of nature that they cause – become the new normal. Such a world would become inhospitable for humanity and for nature (Fig. 1.2).

So, in this account we want to document and communicate as accurately as possible the impacts of this single year of fire across the breadth of Australia's species and natural environments; we also want to propose that such losses are unacceptable. Furthermore, we want to distil the lessons hard won from these fires, to recommend responses that will make it less likely that comparable fires will blight our future, so that we can respond more effectively (before, during and after fire) if they do happen. We want to use the 2019–20 fires as a pivot in learning to live in this country and for better safeguarding of our natural heritage.

Accounting for biodiversity loss in these fires is not straightforward, because shortcomings in Australia's biodiversity monitoring (Legge *et al.* 2018) mean that there is no robust pre-fire benchmark of the population status of most species, at most sites, against which to measure loss. Furthermore, post-fire sampling, of population sizes or the condition of ecosystems, is a time-consuming process, and especially so across the gamut of species and ecological communities affected by these fires. And while it is important for accounting purposes to document the immediate population reductions due to these fires, a more critical conservation focus should be on the longer-term impact and the pathway to and likelihood of recovery (the return to pre-fire status): how long such

Fig. 1.2. Landscape in the aftermath of the 2019–20 wildfires, Kangaroo Island. (Photo: Nicolas Rakotopare/Threatened Species Recovery Hub)

recovery may take; what factors may help support it; and what factors may constrain or prevent it. We can document the wounds to nature relatively easily, but more critical is the process of healing, and our capability and willingness to contribute to such healing. We are at the edge of the unknown. There are few precedents for chronicling recovery from environmental catastrophes of the scale of the 2019–20 fires, and such recovery will be tenuous in a future world if comparable fires recur, and highly unlikely without a significant increase in focus on recovery and increasing the resilience of our biodiversity and landscapes.

The shape of this book

This account is focused on those regions of southern and eastern Australia – mostly forested areas – in which wildfires in 2019–20 were exceptional (Fig. 1.3). This area was circumscribed by the Australian Department of Agriculture, Water and the Environment and designated the 'Preliminary Analysis Area' (PAA), and has been used as the basis of many analyses of impacts (e.g. Ward *et al.* 2020; Legge *et al.* 2022). The total area burnt in this region by fires in 2019–20 was 10.3 million ha. There were many fires in 2019–20 elsewhere in Australia, but those fires were much more within normal bounds.

The set of 2019–20 Australian fires has been labelled the Black Summer, an evocative description of the sense of doom felt by many people in fire-affected areas, and by those in nearby capital cities blanketed for many weeks by thick smoke. However, that epithet overlooks one of the key distinctive features of these fires – that they started unseasonably

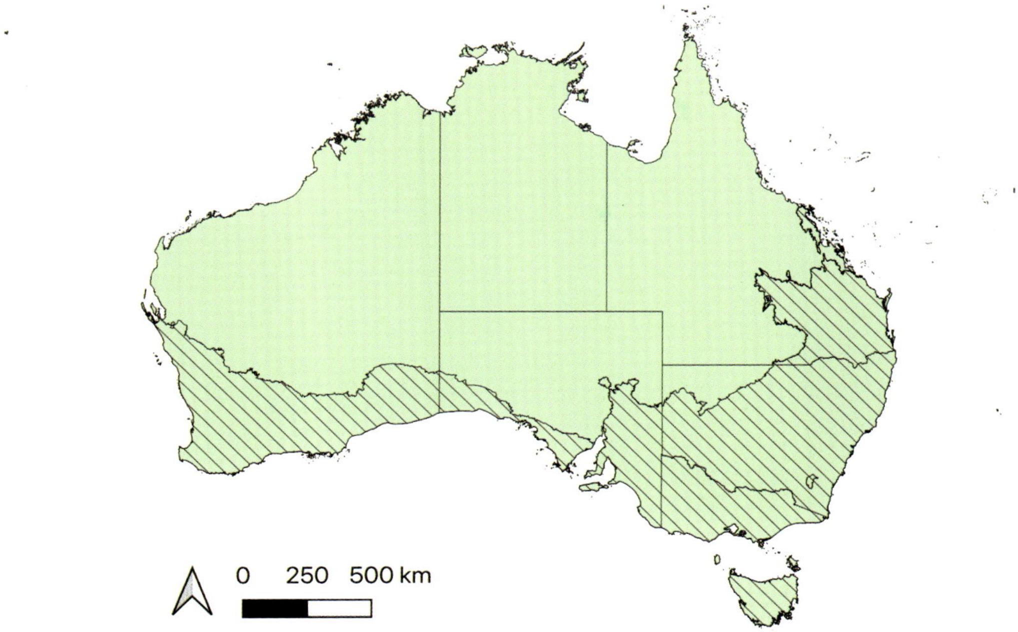

Fig. 1.3. The geographical scope of this book is focused on those regions of Australia (hatched in this figure) in which the 2019–20 fires were exceptional: the 'Preliminary Analysis Area' (PAA).

early, in south-eastern Queensland in early September 2019 (Inspector-General Emergency Management 2020; Smith *et al.* 2021).

In the following chapter, we contextualise and characterise the 2019–20 fires, and the factors that catalysed them. We then describe (in Chapter 3) the impacts of these fires on Indigenous cultural values and Country. In Chapters 4 to 16, we document the immediate and short-term impacts of these fires across different components of Australia's biodiversity. As noted above, the longer-term impacts cannot yet be known. In Chapters 17 to 19, we also recognise that many species and ecosystems faced threats (such as introduced animals, weeds and plant disease, and forestry) before the 2019–20 wildfires, and that these factors may continue to operate after fire, compounding the impacts from fire alone. For post-fire recovery to occur, such interactions need to be understood and the interacting factors effectively managed. We acknowledge that other threats may also compound the impacts of fire, including habitat loss and fragmentation. Chapter 20 provides a distillation across the different components of biodiversity of the direct and compounding impacts of the 2019–20 wildfires, and summarises key results and characteristics of this impact.

Chapter 21 provides an account of prescribed (fuel reduction) burning, an integral component of the management of fires in Australia. It notes that such pre-emptive burning to some extent reduced risks associated with the 2019–20 fires, and that prescribed burning itself may have varied impacts on biodiversity.

In the following section (Chapters 22 to 28), we describe the responses made to help support the recovery of biodiversity following the 2019–20 wildfires. A feature of this response was that it involved so many groups and individuals, but that there was generally a high degree of collaboration across this network of respondents. Biodiversity conservation in Australia is a shared responsibility, and this was very evident in the response to these wildfires. We include in this section chapters describing the contributions made by the Australian Government, state and territory governments, some conservation non-government organisations (NGOs) and natural resource management (NRM) groups. We acknowledge here that many other conservation groups, animal welfare groups, community groups, Indigenous groups, landholder organisations, local councils, corporations and individuals also made significant contributions to this recovery effort – not all of this extraordinary effort could be reported here. This section also considers two issues that are important generally in the recovery effort, and specifically in the case of the 2019–20 fires: animal welfare and *ex situ* conservation.

In the final section (Chapters 29 to 36), we describe the planning and management context that forms the basis for dealing with fire in Australia. In this final set of chapters, we seek to identify what improvements need to be made in the way biodiversity is managed in relation to fire in Australia. In particular, we:

- consider how adequate the knowledge base is for species' response to and needs for fire, and what the key knowledge gaps are;
- assess whether biodiversity assets in fire-prone regions are clearly prioritised, defined and circumscribed, such that they can be most readily protected during fire operations;
- evaluate the extent to which biodiversity is adequately represented in fire planning and whether such inclusion has sufficient clout during fire operations and management;
- assess the extent to which conservation of biodiversity assets during fire operations is sufficiently grounded in policy and law;
- describe the current extent of biodiversity monitoring, and its role in benchmarking fire impacts and in assessment of recovery and management effectiveness;
- assess the successes and failures for biodiversity outcomes of other actions taken before, during and after fire;
- recognise the need for a more pivotal role for Indigenous land and fire management; and
- consider what actions may be taken before fire events to improve environmental resilience, to reduce the likelihood of severe fire, and to spread and reduce environmental risks from fire.

Recognising that the impacts of these fires on biodiversity also affect our sense of the value of nature and our connections to country, we also include a chapter (Chapter 34) that offers a set of personal perspectives from some individuals involved in the fire management and recovery process.

We use the information reported in all chapters to derive recommendations for planning, policy, law and management that would serve to reduce the likelihood of future catastrophic fires and their impacts on biodiversity, and that would help develop and implement manuals or strategies for enhancing the recovery of biodiversity following fire.

Note that some examples feature in several chapters. These include some of the species and ecosystems with highest profile impacts or conservation efforts in these wildfires: Wollemi pine (*Wollemia nobilis*), Kangaroo Island glossy black-cockatoo (*Calyptorhynchus lathami halmaturinus*), Kangaroo Island dunnart (*Sminthopsis griseoventer aitkeni*), Gondwanan rainforests, the heathlands of the Stirling Range, and montane ash forests. These examples recur partly because different chapters describe impacts and responses, because we wanted to describe how different agencies and groups contributed in different ways to their rescue or recovery, and because the impacts of fire on these species and communities were also influenced by other management contexts.

The 2019–20 wildfires showed that our community cared deeply about our natural environments and wildlife, and particularly about the extent to which our nature may be lost in catastrophic events. For example, as at January 2020, 53% of Australians had donated to an appeal to help recovery from the 2019–20 wildfires, with a median donation of $50, and an attitudinal survey concluded that 'The bushfire crisis brought out the best in us' (morestrategic 2020). If we can learn the lessons evident from the fires themselves, their impacts and our responses, these fires will provide us with the opportunity to care better for our country and to plan for a more sustainable future.

As we write this chapter, exceptional flood events are causing loss of human life, infrastructure and biodiversity in northern New South Wales and south-eastern Queensland. Writing about wildfire in such settings seems perverse, but fire and flood are part of the same fabric of escalating environmental catastrophes due to climate change that collectively will winnow biodiversity and challenge our future. Furthermore, although this book specifically documents the impacts on biodiversity of the 2019–20 Australian wildfires, the responses made, and the lessons that can be learnt, much of what we have written can be applied more generically to any environmental disaster. We cannot continue to lurch blindly from crisis to crisis, treating each as an unexpected exceptionality, learning little from each, and each prompting a patched emergency response. That pathway is a descending spiral. We need instead to try to reduce the likelihood of such disasters, to build capability to respond to them, and to better manage our country such that the status of biodiversity is not so precarious.

References

Bowman DMJS (1998) Tansley Review No. 101: The impact of Aboriginal landscape burning on the Australian biota. *New Phytologist* **140**, 385–410. doi:10.1111/j.1469-8137.1998.00289.x

Bowman DMJS, Balch JK, Artaxo P, Bond WJ, Carlson JM, *et al.* (2009) Fire in the Earth system. *Science* **324**, 481–484. doi:10.1126/science.1163886

Collins L, Bradstock RA, Clarke H, Clarke MF, Nolan RH, *et al.* (2021) The 2019/2020 mega-fires exposed Australian ecosystems to an unprecedented extent of high-severity fire. *Environmental Research Letters* **16**, 044029. doi:10.1088/1748-9326/abeb9e

Foulkes JA, Prior LD, Leonard SWJ, Bowman DMJS (2021) Demographic effects of severe fire in montane shrubland on Tasmania's central plateau. *Fire (Basel, Switzerland)* **4**, 32. doi:10.3390/fire4030032

Inspector-General Emergency Management (2020) 'Queensland Bushfires Review. Report 2: 2019–2020'. Inspector-General Emergency Management, Brisbane.

Johnston FH, Borchers-Arriagada N, Morgan GG, Jalaludin B, Palmer AJ, *et al.* (2021) Unprecedented health costs of smoke-related PM 2.5 from the 2019–20 Australian megafires. *Nature Sustainability* **4**, 42–47. doi:10.1038/s41893-020-00610-5

Kooyman RM, Watson J, Wilf P (2020) Protect Australia's Gondwana rainforests. *Science* **367**, 1083. doi:10.1126/science.abb2046

Legge S, Lindenmayer DB, Robinson NM, Scheele BC, Southwell DM, *et al.* (Eds) (2018) *Monitoring Threatened Species and Ecological Communities.* CSIRO Publishing, Melbourne.

Legge S, Woinarski J, Scheele B, Garnett ST, Lintermans M, *et al.* (2022) Rapid assessment of the biodiversity impacts of a megafire to guide urgent management intervention and recovery. *Diversity & Distributions* **28**, 571–591. doi:10.1111/ddi.13428

Linley GD, Jolly CJ, Doherty TS, Geary WL, Armenteras D, *et al.* (in press) What do you mean,'megafire'? *Global Ecology and Biogeography.* doi:10.1111/geb.13499

Moir ML (2021) Coextinction of *Pseudococcus markharveyi* (Hemiptera: Pseudococcidae): a case study in the modern insect extinction crisis. *Austral Entomology* **60**, 89–97. doi:10.1111/aen.12506

morestrategic (2020) 'Public attitudes to bushfire fundraising: insights from national opinion poll 16–18 January 2020'. Fundraising Institute Australia, Sydney.

Morton A (2020) 'Dinosaur trees': firefighters save endangered Wollemi pines from NSW bushfires. *The Guardian Australia*, 15 January (online).

Samuel G (2020) 'Independent review of the EPBC Act - Final report'. Department of Agriculture Water and the Environment, Canberra.

Scheele BC, Legge S, Blanchard W, Garnett ST, Geyle HM, *et al.* (2019) Continental-scale assessment reveals inadequate monitoring for threatened vertebrates in a megadiverse country. *Biological Conservation* **235**, 273–278. doi:10.1016/j.biocon.2019.04.023

Smith I, Velasquez E, Pickering C (2021) Quantifying potential effect of 2019 fires on national parks and vegetation in South-East Queensland. *Ecological Management & Restoration* **22**, 160–170. doi:10.1111/emr.12479

van Eeden L, Nimmo D, Mahoney M, Herman K, Ehnke G, *et al.* (2020) *Australia's 2019–2020 Bushfires: The Wildlife Toll.* World Wide Fund for Nature Australia, Melbourne.

Ward M, Tulloch AIT, Radford JQ, Williams BA, Reside AE, *et al.* (2020) Impact of 2019–2020 mega-fires on Australian fauna habitat. *Nature Ecology & Evolution* **4**, 1321–1326. doi:10.1038/s41559-020-1251-1

Wintle BA, Cadenhead NCR, Morgain RA, Legge SM, Bekessy SA, *et al.* (2019) Spending to save: what will it cost to halt Australia's extinction crisis? *Conservation Letters* **12**, e12682. doi:10.1111/conl.12682

2

The 2019–20 Australian wildfires: precursors, characteristics and implications for the future

Trent D. Penman, Hamish Clarke, Rebecca K. Gibson,
Luke Collins and Rachael H. Nolan

Summary

- Extensive severe wildfires are occurring more frequently across much of the world, driven by changes in land use and management, and especially by climate change.
- There were exceptional precursors to the 2019–20 Australian wildfires, notably a 3-year drought. The calendar year of 2019 was the hottest and driest year on record for the country, resulting in exceptionally low moisture levels in forest fine fuels.
- Extending from August 2019 to March 2020, the wildfires in southern and eastern Australia burnt 10.3 million ha, mostly of temperate forests.
- The 2019–20 Australian fires were globally exceptional. Since 2000, no forest biome anywhere on Earth has had a higher percentage of its area burnt in a single season than the 21% reported for forest biome burnt in the 2019–20 Australian fires.
- The 2019–20 wildfires also impacted many vegetation communities that are typically too wet to burn, such as peatlands and Gondwanan rainforest.
- A feature of many of the megafires of the 2019–20 season was their unusually long duration, with one fire burning for 108 days.
- Extreme fire activity was largely driven by complex dynamics of fire coupling with atmospheric conditions, resulting in an unusually high incidence of violent pyrocumulonimbus storms.
- The 2019–20 fires set Australian records for area burnt at high severity, houses lost and the largest individual fire originating from a single ignition.
- Estimates of economic impacts from the fires are more than $10 billion.

Introduction

The size and frequency of wildfires has increased across many forested regions globally (Pausas and Fernández-Muñoz 2012; Kelly *et al.* 2013; Bradstock *et al.* 2014; Hanes *et al.* 2019; Williams *et al.* 2019). Increased extent of wildfire is broadly caused by two key factors – changing human practices increasing the structure and load of fuels, and climate change altering the occurrence of weather conducive to fire spread (Abatzoglou and Williams 2016). In many countries, a cessation of Indigenous burning has altered vegetation communities favouring systems with high fuel levels (Mariani *et al.* 2022). Similarly, the abandonment of agriculture in many Mediterranean countries has resulted in grass-dominated agricultural areas being replaced with shrubby systems that carry higher-intensity fires (Pausas and Fernández-Muñoz 2012). Contemporary fire management practices such as fire suppression aim to reduce the impact of wildfires, but policies of wildfire exclusion have increased fuel levels, leading to large high intensity fires (Brotons *et al.* 2013; Calkin *et al.* 2015). The most consistent cause for shifting fire regimes is the effect of global climatic change. Increases in weather conducive to fire spread have been observed in fire-prone systems around the world (Westerling and Bryant 2008; Moritz *et al.* 2012; Abram *et al.* 2021). Further, the incidence of severe drought is increasing across forests globally under climate change (Miralles *et al.* 2019; Brodribb *et al.* 2020), leading to increased forest flammability through changes to fuel structure and declines in fuel moisture content (Ruthrof *et al.* 2016; Abatzoglou *et al.* 2018).

Major fire events must be taken as a time of learning and reassessment of current knowledge. Each major fire prompts new calls for a re-evaluation of policy and practice, attitudes, understanding and expectations of future fire regimes (Moritz *et al.* 2014; Moreira *et al.* 2020). The Black Summer fire season in Australia began in August 2019 and finished in March 2020. The area burnt by these fires was the largest on record for south-eastern Australia (Boer *et al.* 2020; Bowman *et al.* 2020a): the total area burnt in the study area (see Fig. 2.2) of eastern and southern Australia was 10.3 million ha. The devastating impacts of these fires will be felt for years to come by those that lived through them and by the environment. It is therefore vital that we learn to reduce the impacts of future extreme fire seasons. In this chapter, we look at the environment before Black Summer and the weather during the season, and describe the extent and behaviour within each of the fires and the impact of the fires. Finally, we compare these data with preceding fire seasons and extreme fire seasons across the globe to place the fires in both a global and climate change context.

Environmental precursors

Wildfire occurrence is strongly determined by the biomass and moisture content of fuels. The temperate forests of Australia typically have enough fuel biomass, but it is not until periods of drought that sufficient quantities of fuels become dry enough to facilitate large wildfires (Bradstock 2010). In the winter of 2019, much of eastern Australia was in the grip of a third consecutive year of drought (Bureau of Meteorology 2019a). In the latter half of 2019, many regions were subject to a flash drought (Nguyen *et al.* 2021), which is the rapid intensification of an extreme drought event (Pendergrass *et al.* 2020). As the season progressed, there were multiple heatwaves, with record-breaking temperatures recorded across Australia (Bureau of Meteorology 2020). The calendar year of 2019 was the hottest and driest year on record for the country (Fig. 2.1) (Bureau of Meteorology 2019a, 2020). These conditions triggered widespread canopy die-back in many forests across eastern

Australia (Fig. 2.1) (De Kauwe *et al.* 2020; Nolan *et al.* 2021b) and primed the landscape for fire.

A key metric of forest fire danger is the moisture content of fine fuels (plant biomass < 6 mm diameter, such as leaves and twigs) (Hines *et al.* 2010). Estimates of fine fuel

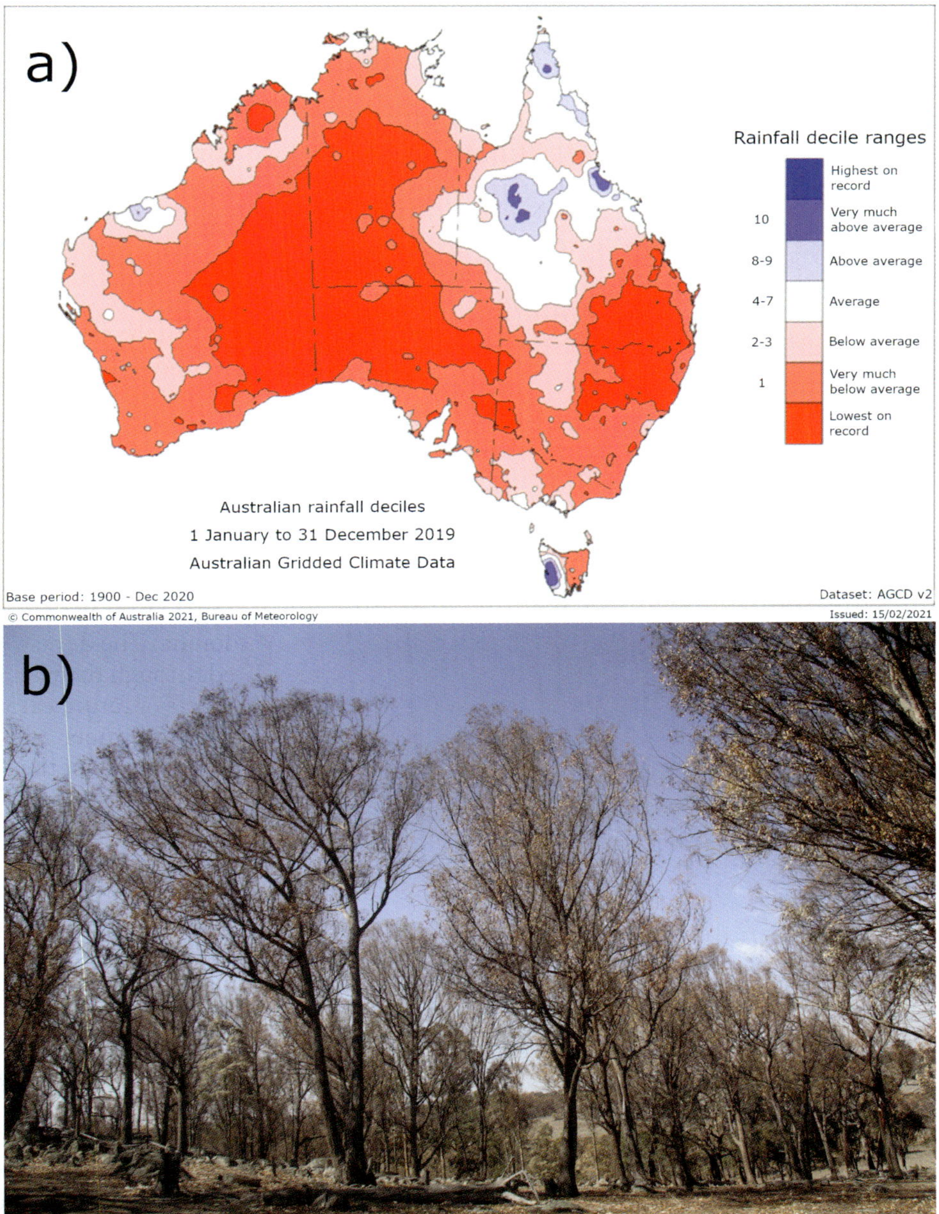

Fig. 2.1. a) 12-month rainfall deciles for Australia in 2019 (source: https://www.bom.gov.au); and b) an example of a drought-stressed forest experiencing canopy dieback in eastern Australia in late 2019. (Photo: Rachael Nolan)

moisture content over the 2019–20 fire season indicated that values were very much below average, or the lowest on record, across much of eastern and southern Australia (Nolan et al. 2020; Abram *et al.* 2021). The canopy die-back in eucalypt forest triggered by the drought also probably resulted in an increase in the ratio of dead fuels to live fuels, which would have further amplified the flammability of the temperate forest estate. The combination of an influx of dead fuels to the forest floor and the opening of the forest canopy, as the forests responded to severe drought stress, may have contributed to increased rates of fire spread (Ruthrof *et al.* 2016).

The biomass of fine fuels in a forest is another key predictor of fire behaviour (Gould *et al.* 2011; McCaw *et al.* 2012). Analyses of fine fuel load and structure in the lead-up to the 2019–20 fires in New South Wales indicated that, on average, fine fuel load and structure were no higher than historical values (Nolan *et al.* 2021a). This is not to say that fuel load and structure were not important drivers of the fires at particular locations, but the overriding influence on the 2019–20 fire season was climate, not fuel load and structure.

Fire weather during the fire season

Fire weather conditions during the 2019–20 season were unquestionably extreme and by many (but not all) measures unprecedented. The McArthur Forest Fire Danger Index (FFDI; Luke and McArthur 1978) is computed from standard meteorological observations or model predictions of precipitation, relative humidity, wind speed and temperature (Noble *et al.* 1980). Mean monthly FFDI was the highest ever recorded (since records began in 1950) over parts of northern New South Wales and south-east Queensland in September 2019 (Bureau of Meteorology 2019a; Clarke *et al.* 2020). These conditions continued through spring, with New South Wales, Queensland, Northern Territory, Western Australia and Tasmania all setting new records for mean spring FFDI (Bureau of Meteorology 2019b). December 2019 saw the highest ever average FFDI for Australia as a whole, with the majority of the country setting new record highs (Bureau of Meteorology 2020). Many locations in New South Wales set new records for highest minimum monthly FFDI values, indicating a lack of reprieve from elevated conditions and likely contributing to the severity and extent of the overall season.

During December 2019, FFDI values exceeding 100 ('Catastrophic') were observed in every mainland state and the Northern Territory (Bureau of Meteorology 2020). Much of eastern Tasmania recorded its highest ever December FFDI on 30 December. Australian Capital Territory set new records for maximum daily temperature in December, January and February (Bureau of Meteorology 2020). In South Australia there were days of FFDI exceeding 50 in November, December and January, including 3 days where multiple regions recorded FFDI values > 100 (Government of South Australia 2020). Southern Western Australia recorded near Catastrophic conditions on multiple days in December (Bureau of Meteorology 2020). Multiple regions in eastern New South Wales set records for the number of days exceeding FFDI of 25 (Clarke 2020). These same regions recorded high but not record-setting numbers of days with FFDI greater than 50.

Fire weather conditions experienced during the 2019–20 season were extreme but, in many cases, fitted within a long-term increasing trend of FFDI values (Clarke *et al.* 2013; Dowdy 2018; Abram *et al.* 2021) and atmospheric instability (Dowdy and Pepler 2018). While interannual variability is expected to persist, unmitigated climate change is expected to lead to further, substantial increases in FFDI and atmospheric instability (Clarke and Evans 2019; Di Virgilio *et al.* 2019; Dowdy *et al.* 2019). This would imply an increase in fire extent and likelihood of extreme fire seasons.

Timing, ignitions, spread and severity

In Australia, fire seasons typically coincide with broad latitudinal bands, occurring in spring in southern Queensland and northern New South Wales, spring to summer in southern New South Wales and summer to early autumn in southern regions (i.e. Victoria, South Australia, south-west Western Australia and Tasmania) (http://www.bom.gov.au/weather-services/fire-weather-centre/bushfire-weather/index.shtml). The start of the 2019–20 fire season was earlier than average (Filkov *et al.* 2020). Ignition dates followed the typical pattern of commencing earlier in the north and progressing through the season towards the south (Fig. 2.2). Numerous different ignition sources were reported over the 2019–20 fire season, including lightning and anthropogenic ignitions (e.g. arson, power-lines). Fires originating from suspected lightning ignitions accounted for the greatest amount of area burnt (Nolan *et al.* 2021a). Most of the fires across southern Australia were extinguished or declared under control by the end of February 2020 (Fig. 2.2).

A notable aspect of the 2019–20 fire season was the duration of many of the larger fire events across south-eastern Australia. Numerous megafires (as defined in Chapter 1, fires > 100 000 ha) burnt through forested terrain for several months (Table 2.1), the longest

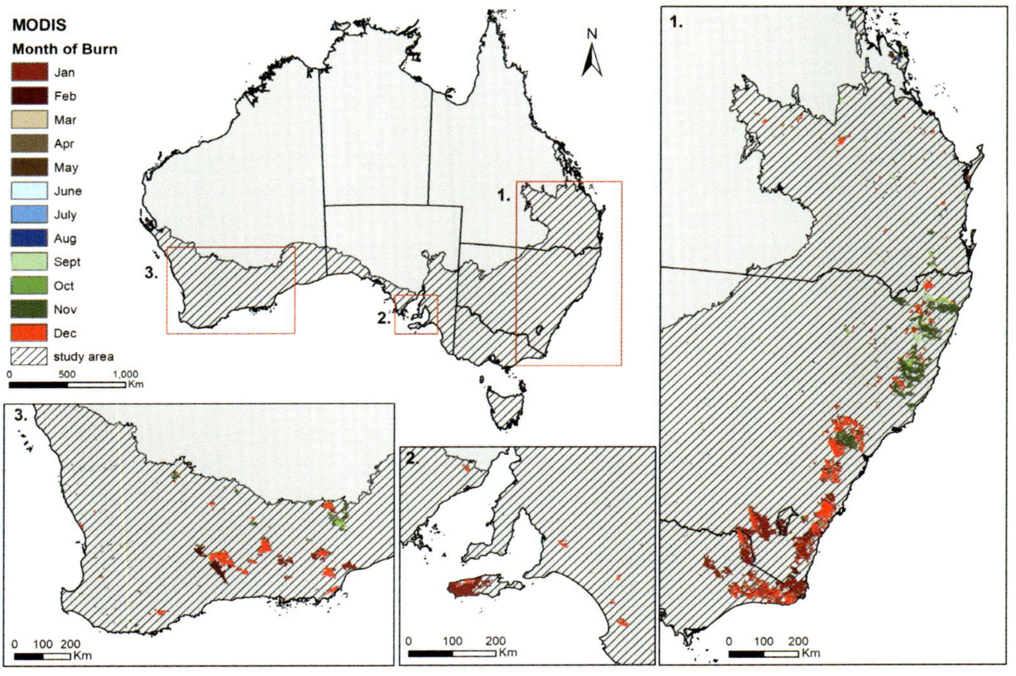

Fig. 2.2. The month of burn for the study area. Insets show a detailed view of areas of significant fire activity in (1) eastern Australia, (2) South Australia, and (3) Western Australia. Month of burn was derived from the Moderate Resolution Imaging Spectroradiometer (MODIS) day of burn product (https://modis.gsfc.nasa.gov/).

Table 2.1. Details of the start date, duration, size and the contribution (%) of each fire severity class for megafires (> 100 000 ha) occurring in south-eastern Australia.

The area burnt and severity statistics only consider native forest, woodland and shrublands. Fire severity classes are: unburnt canopy or low canopy scorch (Low); partial canopy scorch (Moderate); full canopy scorch with partial consumption (High); full canopy consumption (Extreme). The fire numbers align with those presented in Fig. 2.3. (Fires 1–9 occurred in New South Wales; fires 10–12 in Victoria)

Fire name	Start date	Duration (days)*	Area burnt (ha)	Severity statistics			
				Low	Moderate	High	Extreme
1. Liberation Trail	04/11/2019	51	218 995	16.3	32.0	27.0	24.7
2. Bees Nest	31/08/2019	75	108 551	6.1	34.5	44.9	14.5
3. Carrai East	17/10/2019	95	155 950	16.1	37.2	22.7	24.1
4. Gosper Mtn	26/10/2019	108	479 560	30.1	40.0	19.3	10.6
5. Green Wattle Ck	27/12/2019	46	282 279	19.0	34.8	24.4	21.7
6. Currowan	26/11/2019	75	318 792	16.7	30.1	22.5	30.7
7. Orroral	27/01/2020	32	92 313	10.7	16.9	31.2	41.2
8. Dunns Rd	28/12/2019	50	221 817	17.5	22.9	32.0	27.7
9. Badja Forest	27/12/2019	69	331 214	28.7	14.1	27.2	30.1
10. Upper Murray	29/12/2019	46	230 031	28.8	20.3	32.3	18.5
11. Ovens	30/12/2019	31	126 468	35.0	23.7	26.2	15.1
12. Gippsland	21/11/2019	91	1 070 341	25.1	23.9	33.2	17.8

* In the absence of published end dates, MODIS hotspots were used to calculate fire duration.

being the Gospers Mountain fire, which burnt for 108 days across complex terrain in largely inaccessible eucalypt forest west of Sydney. The duration of this and other fires in the Black Summer season exceeded that of notable megafire events occurring in south-eastern Australia in the 2003 (59 days) and 2007 (2 months) fire seasons (McCarthy *et al.* 2012; Tolhurst and McCarthy 2016).

Strong surface winds drove very rapid fire spread and long-range spotting in southern Queensland and northern New South Wales (Sharples 2020). In contrast, extreme fire activity in the southern regions was largely driven by complex dynamics of fire coupling with atmospheric conditions, resulting in violent pyrocumulonimbus (pyroCb) storms. PyroCB events generate extreme fire behaviour, including rapid rates of fire spread and long-range spotting, facilitating very rapid fire growth (Dowdy *et al.* 2017). An estimated 29 pyroCBs occurred during the 2019–20 season, which greatly increased the number of previously recorded pyroCB events for southern Australia (i.e. 60; Bowman *et al.* 2020b; Abram et al. 2021). The combination of extreme drought, frequent episodes of elevated fire danger and a high number of pyroCB events resulted in periods of rapid fire growth across eastern Australia. For example, 18 pyroCB storms occurring in southern New South Wales and Victoria over the period from 29 December 2019 to 4 January 2020 burnt an estimated 530 000 ha in a non-consecutive 51-hour period (Kablick *et al.* 2020; Peterson *et al.* 2021).

Satellite derived measures of fire severity have provided important insight into the impact of fire intensity on vegetation across the 2019–20 fires. In Australian forests, woodlands and shrublands, fire severity is quantified based on the degree of scorch and consumption of canopy and understorey foliage recorded in the days and months following fire (Collins *et al.* 2020; Gibson *et al.* 2020). Four fire severity classes are generally considered: (1) understorey burnt with low levels of canopy scorch (< 20% scorch; Low); (2) understorey burnt with moderate levels of canopy scorch (20–80% scorch; Moderate); (3) high levels of canopy scorch with some canopy consumption (> 80% scorch; High); and (4) complete canopy consumption (Extreme).

Fire severity patterns were highly heterogeneous across the extent of the 2019–20 fires (Table 2.1; Fig. 2.3), reflecting the complex interplay between antecedent drought, fire weather, topography, vegetation structure and past management (Bowman *et al.* 2021).

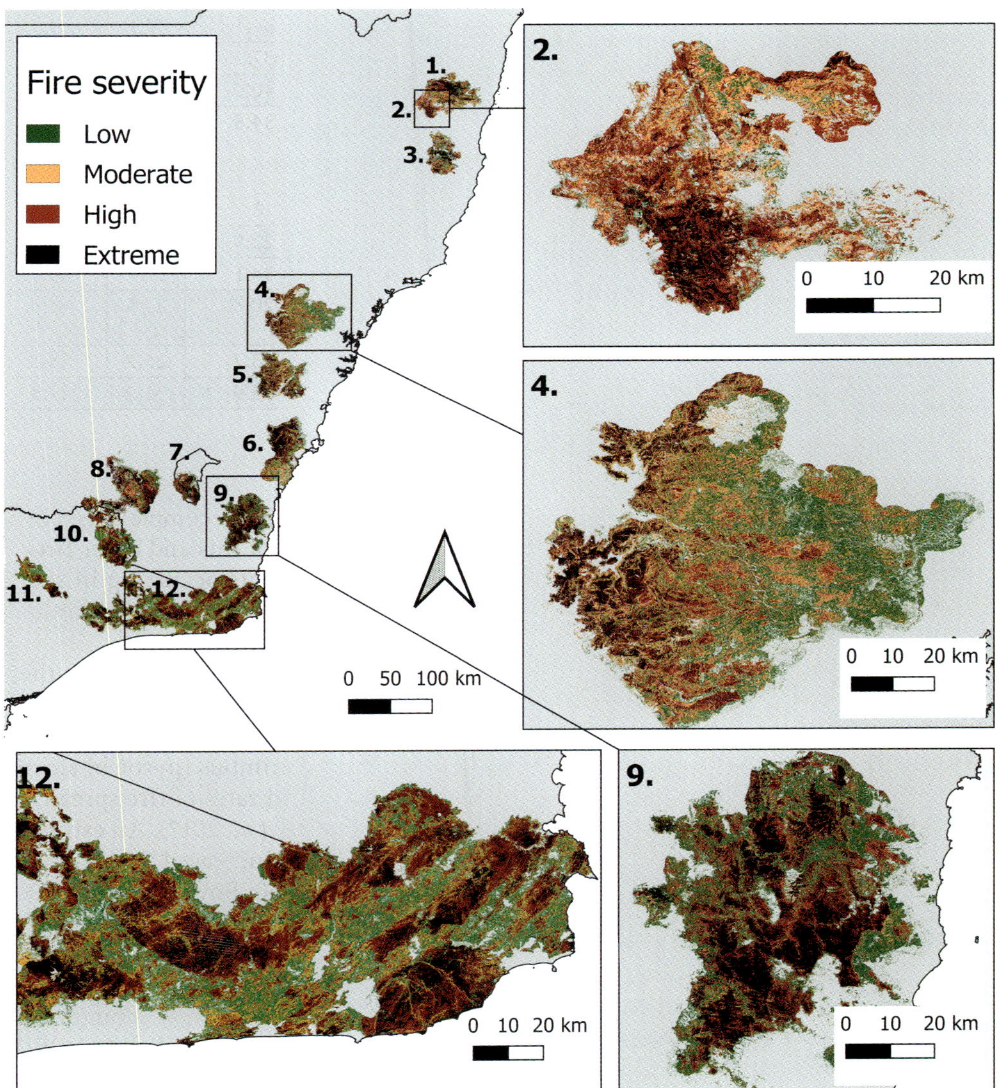

Fig. 2.3. Fire severity maps for 12 major fires in eastern Australia. Fire severity maps were produced using satellite imagery from either Landsat (Collins *et al.* 2020) or Sentinel 2 (Gibson *et al.* 2020). Insets show the severity maps for select fires in greater detail.

Analysis of fire severity patterns indicates that the proportional contribution of severity classes to fire extent typically fell within the historic range (1988–2018) for forests and shrublands in south-eastern Australia (Collins *et al.* 2021). Notable exceptions were rainforest and woodland communities, which experienced a proportionally greater extent of extreme severity fire (Collins *et al.* 2021). The 2019–20 fires impacted many vegetation communities that are typically considered too wet to burn, such as peatlands and Gondwanan rainforest (Nolan *et al.* 2020). Given the unprecedented size of the 2019–20 fires, the overall extent of high and extreme severity fire is likely to be a new record within a single fire season (Collins *et al.* 2021).

Fire severity that exceeded the resistance of canopy plants accounted for a considerable proportion (~44%) of the total extent of the temperate woody vegetation communities impacted by the 2019–20 fires (Bowman *et al.* 2021; Collins *et al.* 2021). This included many very large (> 10 000 ha) contiguous patches of high and extreme severity fire that were associated with extreme fire behaviour occurring during pyroCBs (Collins *et al.* 2021; Peterson *et al.* 2021). Unburnt vegetation and low severity fire were also common across the fire grounds (e.g. ~9% unburnt and 28% low severity) (Collins *et al.* 2021), with large (> 5000 ha) unburnt patches being observed across several megafires (Godfree *et al.* 2021). Areas that remain unburnt or burnt at low severity offer areas for survival, post-fire persistence and *in situ* recolonisation by plants and animals (e.g. Banks *et al.* 2011; Godfree *et al.* 2021).

Emissions

Climate change was a contributor to the magnitude of the 2019–20 wildfires, but these (and comparable future) fires may themselves contribute to ongoing climate change through the exceptional amounts of greenhouse gas emissions that they produce. Satellite measurements of trace gases were used to estimate that 715 Tg (million tonnes) (range 517–867) of carbon dioxide were emitted by these wildfires between November 2019 and January 2020 (van der Velde *et al.* 2021). As context, this vastly exceeds the (16-year) average of 9 Tg from fires in south-eastern Australia (van der Velde *et al.* 2021), and exceeds Australia's typical annual fossil fuel emissions (e.g. 531 Tg CO_{2e} in 2019: Commonwealth of Australia 2020a). In some environments, fire-derived emissions are expected to be compensated for when vegetation regrows, but such offsetting is unlikely to be realised in the future if comparable severe fires recur with shorter intervals (van der Velde *et al.* 2021).

Global comparisons

Placing the 2019–20 Australian fires in context is challenging as there is a continually growing list of internationally significant fire events. In the 2 years since the start of Black Summer, there have been extreme fire seasons across the globe including the United States of America, Canada, Greece and Russia. Although they may differ in properties like fire size, dominant fuel type, severity and impact, together they illustrate the depth of the worldwide wildfire problem (Gill *et al.* 2013).

The lengthy record of major Australian fires since European colonisation provides a natural point of comparison for the 2019–20 Australian fires. The 2019–20 fires set records for area of temperate forest burnt (Nolan et al. 2020), area burnt at high severity (Collins *et al.* 2021), houses lost (Filkov *et al.* 2020) and the largest individual fire originating from a single ignition (Boer *et al.* 2020). Several previous fire seasons have caused greater loss of life, including the Black Saturday fires of 2009 (173 lives, > 2000 houses). However, the Black Saturday fires burnt a much smaller area (~400 000 ha) and impacted primarily on a single day rather than months (Blanchi *et al.* 2014).

The 2019–20 Australian fires were globally remarkable. Since 2000, no forest biome anywhere on Earth has ever had a higher percentage of its area burnt in a single season than the 21% reported for forest biome burnt in the 2019–20 Australian fires (Boer *et al.* 2020). This figure also represents a massive departure from typical annual values of well below 5% in this and other forest biomes.

While comparisons between different major fires are commonplace, they are generally informal and limited to a few fires. Three exceptions are Filkov *et al.* (2020) (n = 14), Duane *et al.* (2021) (n = 18) and Williams *et al.* (2011) (n = 8). Across the three studies, the only fires of comparable size to the Australian 2019–20 season were the 1997–98 Indonesian peat fires (9.7 Mha), and the Siberian boreal forest fires of 2019 (13 Mha) and 2020 (14 Mha). In contrast, the number of people killed by the 2019–20 Australian fires was below many of the examples listed, which included 112 (Portugal 2017), 102 (Greece 2018) and 85 (USA 2018). Other fires that destroyed multiple thousands of houses were the 2017, 2018 and 2020 fire seasons in the USA (~10 000, ~19 000 and ~10 000 structures respectively). Clearly, these types of impacts are a complex function of the interaction between fire and human settlements, which can vary considerably in their size, density and other factors that influence their exposure and vulnerability to fire.

Human and economic costs

Thirty-three people lost their lives directly as a result of the fires (Filkov *et al.* 2020). Smoke exposure is estimated to have caused an additional 417 deaths and 3230 hospitalisations for cardiovascular and respiratory problems, and 1523 presentations to emergency departments due to asthma (Johnston *et al.* 2021). The cost of these health impacts from the fires are estimated at $1.95 billion (Johnston *et al.* 2021). The fires are also likely to have caused mental health issues associated with the trauma of evacuation and loss of family, friends, livestock, pets, properties and livelihoods (Zhang *et al.* 2020). Frontline workers, particularly firefighters, may have similarly experienced mental health impacts from exhaustion, emotional trauma, smoke exposure and, for volunteer firefighters, loss of income (Zhang *et al.* 2020). Further, the fires particularly affected many Indigenous communities in relation to their relationships with Country (Williamson *et al.* 2020). At a larger societal level, media coverage of the catastrophic nature of the fires, in combination with coverage of the COVID-19 pandemic shortly thereafter, may have affected mental health across Australia and beyond (Looi *et al.* 2020).

Over 3100 houses were destroyed by the fires (Filkov et al. 2020). In New South Wales, which was subject to the largest loss of houses, 2476 were destroyed and 1034 damaged (NSW Government 2020). In addition to house loss, 3 schools and 284 facilities were destroyed by fire in New South Wales, along with losses of critical agricultural and telecommunication infrastructure (NSW Government 2020). Fire damage of power poles, powerlines and the telecommunications network are estimated to have resulted in power outages to over 280 000 customers and over 800 telecommunications outages (Commonwealth of Australia 2020b). In New South Wales alone, over 88 000 km of agricultural fencing and 600 000 ha of pasture were damaged by the fires (NSW Government 2020).

The loss of lives and property was disproportionately borne by regional and socially disadvantaged communities (Nolan *et al.* 2021a).

Estimates of economic impacts from the fires are upwards of $10 billion (Commonwealth of Australia 2020b). Many of the fire-affected communities are reliant on tourism (Schweinsberg *et al.* 2020), which was heavily impacted by COVID-19 travel restrictions immediately following the Black Summer fires (Pham *et al.* 2021). The compound effect of loss of income from the fires and COVID-19 may delay social and economic recovery from the fires. Another important factor likely to affect recovery of fire-affected communities is the pace of rebuilding, which is often linked to insurance coverage (Eriksen and de Vet 2020). However, many homes in fire-affected areas are likely to have been underinsured, as was found following the 2013 fires in the Blue Mountains area of New South Wales (Eriksen and de Vet 2020). Thus, the socio-economic impacts of the fires and compounding effects of COVID-19 may take many years to become apparent.

Conclusions

The 2019–20 Black Summer set Australian records for a range of fire and weather statistics. However, of greatest concern is that the seasons fits within a global trend of worsening fire weather and fire impacts. Whatever records the 2019–20 Australian fire season set may not last long, given the continuing effects of climate change. This places fire and conservation managers in a difficult position as the main driver of change – rising global temperatures and increasing occurrence and intensity of drought – is beyond their control and subject to national and international decision making.

The heart of the worldwide wildfire problem is its complexity. Across diverse ecosystems and stakeholders, at multiple spatial and temporal scales, simple solutions are elusive (Gill *et al.* 2013). Improving the evidence base of national fire history datasets and associated impacts remains critical, regardless of policy directions (Syphard and Keeley 2019; Coops *et al.* 2020; Bowman *et al.* 2020a; Collins et al. 2021). Estimates of environmental impact have been hampered by limited baseline information from which to compare and the limitations created by the COVID 19 global pandemic. Bringing together the available data is important for scientists to interpret the Black Summer season, but is vital for predicting, measuring and evaluating future change across Australia's diverse ecosystems.

References

Abatzoglou JT, Williams AP (2016) Impact of anthropogenic climate change on wildfire across western US forests. *Proceedings of the National Academy of Sciences of the United States of America* **113**, 11770. doi:10.1073/pnas.1607171113

Abatzoglou JT, Williams AP, Boschetti L, Zubkova M, Kolden CA (2018) Global patterns of interannual climate–fire relationships. *Global Change Biology* **24**, 5164–5175. doi:10.1111/gcb.14405

Abram N, Henley B, Sen Gupta A, Lippmann TJR, Clarke H, *et al.* (2021) Connections of climate change and variability to large and extreme forest fires in southeast Australia. *Communications Earth & Environment* **2**, 8. doi:10.1038/s43247-020-00065-8

Banks SC, Dujardin M, McBurney L, Blair D, Barker M, *et al.* (2011) Starting points for small mammal population recovery after wildfire: recolonisation or residual populations? *Oikos* **120**, 26–37. doi:10.1111/j.1600-0706.2010.18765.x

Blanchi R, Leonard J, Haynes K, Opie K, James M, *et al.* (2014) Environmental circumstances surrounding bushfire fatalities in Australia 1901–2011. *Environmental Science & Policy* **37**, 192–203. doi:10.1016/j.envsci.2013.09.013

Boer MM, Resco de Dios V, Bradstock R (2020) Unprecedented burn area of Australian mega forest fires. *Nature Climate Change* **10**, 171–172. doi:10.1038/s41558-020-0716-1

Bowman DMJS, Williamson GJ, Yebra M, Lizundia-Loiola J, Lucrecia Pettinari M, *et al.* (2020a) Wildfires: Australia needs national monitoring agency. *Nature* **584**, 188–191. doi:10.1038/d41586-020-02306-4

Bowman DMJS, Kolden CA, Abatzoglou JT, Johnston FH, van der Werf GR, *et al.* (2020b) Vegetation fires in the Anthropocene. *Nature Reviews Earth & Environment* **1**, 500–515. doi:10.1038/s43017-020-0085-3

Bowman DMJS, Williamson GJ, Gibson RK, Bradstock RA, Keenan RJ (2021) The severity and extent of the Australia 2019–20 Eucalyptus forest fires are not the legacy of forest management. *Nature Ecology & Evolution* **5**, 1003–1010. doi:10.1038/s41559-021-01464-6

Bradstock RA (2010) A biogeographic model of fire regimes in Australia: current and future implications. *Global Ecology and Biogeography* **19**, 145–158. doi:10.1111/j.1466-8238.2009.00512.x

Bradstock R, Penman T, Boer M, Price O, Clarke H (2014) Divergent responses of fire to recent warming and drying across south-eastern Australia. *Global Change Biology* **20**, 1412–1428. doi:10.1111/gcb.12449

Brodribb TJ, Powers J, Cochard H, Choat B (2020) Hanging by a thread? Forests and drought. *Science* **368**, 261–266. doi:10.1126/science.aat7631

Brotons L, Aquilué N, de Cáceres M, Fortin M-J, Fall A (2013) How fire history, fire suppression practices and climate change affect wildfire regimes in Mediterranean landscapes. *PLoS One* **8**, e62392. doi:10.1371/journal.pone.0062392

Bureau of Meteorology (2019a) *Special Climate Statement 70 – Drought Conditions in Australia and Impact on Water Resources in the Murray–Darling Basin*' Commonwealth of Australia, Canberra.

Bureau of Meteorology (2019b) *Special Climate Statement 72 – Dangerous Bushfire Weather in Spring 2019*. Commonwealth of Australia, Canberra.

Bureau of Meteorology (2020) *Special Climate Statement 73 – Extreme Heat and Fire Weather in December 2019 and January 2020*. Commonwealth of Australia, Canberra.

Calkin DE, Thompson MP, Finney MA (2015) Negative consequences of positive feedbacks in US wildfire management. *Forest Ecosystems* **2**, 9. doi:10.1186/s40663-015-0033-8

Clarke H (2020) 'How will the 2019–20 NSW bushfires influence near-future risk? Report to the NSW Bushfire Inquiry'. NSW Bushfire Risk Research Hub, University of Wollongong, Wollongong.

Clarke H, Evans JP (2019) Exploring the future change space for fire weather in southeast Australia. *Theoretical and Applied Climatology* **136**, 513–527. doi:10.1007/s00704-018-2507-4

Clarke H, Lucas C, Smith P (2013) Changes in Australian fire weather between 1973 and 2010. *International Journal of Climatology* **33**, 931–944. doi:10.1002/joc.3480

Clarke H, Cirulis B, Penman T (2020) 'Historical and seasonal context analyses – fire weather. Report prepared by the NSW Bushfire Mangement Research Hub for the NSW Bushfire Inquiry.' University of Wollongong, Wollongong.

Collins L, McCarthy G, Mellor A, Newell G, Smith L (2020) Training data requirements for fire severity mapping using Landsat imagery and random forest. *Remote Sensing of Environment* **245**, 111839. doi:10.1016/j.rse.2020.111839

Collins L, Bradstock RA, Clarke H, Clarke MF, Nolan RH, *et al.* (2021) The 2019/2020 mega-fires exposed Australian ecosystems to an unprecedented extent of high-severity fire. *Environmental Research Letters* **16**, 044029. doi:10.1088/1748-9326/abeb9e

Commonwealth of Australia (2020a) *Quarterly Update of Australia's National Greenhouse Gas Inventory: December 2019*. Commonwealth of Australia, Canberra.

Commonwealth of Australia (2020b) 'Royal Commission into National Natural Disaster Arrangements – Report'. Commonwealth of Australia, Canberra.

Coops NC, Shang C, Wulder MA, White JC, Hermosilla T (2020) Change in forest condition: Characterizing non-stand replacing disturbances using time series satellite imagery. *Forest Ecology and Management* **474**, 118370. doi:10.1016/j.foreco.2020.118370

De Kauwe MG, Medlyn BE, Ukkola AM, Mu M, Sabot MEB, *et al.* (2020) Identifying areas at risk of drought-induced tree mortality across South-Eastern Australia. *Global Change Biology* **26**, 5716–5733. doi:10.1111/gcb.15215

Di Virgilio G, Evans JP, Blake SAP, Armstrong M, Dowdy AJ, *et al.* (2019) Climate change increases the potential for extreme wildfires. *Geophysical Research Letters* **46**, 8517–8526. doi:10.1029/2019GL083699

Dowdy AJ (2018) Climatological variability of fire weather in Australia. *Journal of Applied Meteorology and Climatology* **57**, 221–234. doi:10.1175/JAMC-D-17-0167.1

Dowdy AJ, Pepler A (2018) Pyroconvection risk in Australia: climatological changes in atmospheric stability and surface fire weather conditions. *Geophysical Research Letters* **45**, 2005–2013. doi:10.1002/2017GL076654

Dowdy AJ, Fromm MD, McCarthy N (2017) Pyrocumulonimbus lightning and fire ignition on Black Saturday in southeast Australia. *Journal of Geophysical Research Atmospheres* **122**, 7342–7354. doi:10.1002/2017JD026577

Dowdy AJ, Ye H, Pepler A, Thatcher M, Osbrough SL, *et al.* (2019) Future changes in extreme weather and pyroconvection risk factors for Australian wildfires. *Scientific Reports* **9**, 10073. doi:10.1038/s41598-019-46362-x

Duane A, Castellnou M, Brotons L (2021) Towards a comprehensive look at global drivers of novel extreme wildfire events. *Climatic Change* **165**, 43. doi:10.1007/s10584-021-03066-4

Eriksen C, de Vet E (2020) Untangling insurance, rebuilding, and wellbeing in bushfire recovery. *Geographical Research* **59**, 228–241.

Filkov AI, Ngo T, Matthews S, Telfer S, Penman TD (2020) Impact of Australia's catastrophic 2019/20 bushfire season on communities and environment. Retrospective analysis and current trends. *Journal of Safety Science and Resilience* **1**, 44–56. doi:10.1016/j.jnlssr.2020.06.009

Gibson R, Danaher T, Hehir W, Collins L (2020) A remote sensing approach to mapping fire severity in south-eastern Australia using Sentinel 2 and random forest. *Remote Sensing of Environment* **240**, 111702. doi:10.1016/j.rse.2020.111702

Gill AM, Stephens SL, Cary GJ (2013) The worldwide 'wildfire' problem. *Ecological Applications* **23**, 438–454. doi:10.1890/10-2213.1

Godfree RC, Knerr N, Encinas-Viso F, Albrecht D, Bush D, *et al.* (2021) Implications of the 2019–2020 megafires for the biogeography and conservation of Australian vegetation. *Nature Communications* **12**, 1023. doi:10.1038/s41467-021-21266-5

Gould JS, McCaw WL, Cheney NP (2011) Quantifying fine fuel dynamics and structure in dry eucalypt forest (*Eucalyptus marginata*) in Western Australia for fire management. *Forest Ecology and Management* **262**, 531–546. doi:10.1016/j.foreco.2011.04.022

Government NSW (2020) 'Final report of the NSW Bushfire Inquiry'. Government of NSW, Sydney.

Government of South Australia (2020) 'Independent review into South Australia's Bushfire Season.' Government of South Australia, Adelaide.

Hanes CC, Wang X, Jain P, Parisien M-A, Little JM, *et al.* (2019) Fire-regime changes in Canada over the last half century. *Canadian Journal of Forest Research* **49**, 256–269. doi:10.1139/cjfr-2018-0293

Hines F, Tolhurst KG, Wilson AAG, McCarthy GJ (2010) 'Overall fuel hazard assessment guide, 4th edition. Fire and Adaptive Management Report no. 82'. Department of Sustainability and Environment, East Melbourne.

Johnston FH, Borchers-Arriagada N, Morgan GG, Jalaludin B, Palmer AJ, *et al.* (2021) Unprecedented health costs of smoke-related PM2.5 from the 2019–20 Australian megafires. *Nature Sustainability* **4**, 42–47. doi:10.1038/s41893-020-00610-5

Kablick III GP, Allen DR, Fromm MD, Nedoluha GE (2020) Australian pyroCb smoke generates synoptic-scale stratospheric anticyclones. *Geophysical Research Letters* **47**, e2020GL088101. doi:10.1029/2020GL088101

Kelly R, Chipman ML, Higuera PE, Stefanova I, Brubaker LB, *et al.* (2013) Recent burning of boreal forests exceeds fire regime limits of the past 10,000 years. *Proceedings of the National Academy of Sciences of the United States of America* **110**, 13055. doi:10.1073/pnas.1305069110

Looi JCL, Allison S, Bastiampillai T, Maguire P (2020) Fire, disease and fear: Effects of the media coverage of 2019–2020 Australian bushfires and novel coronavirus 2019 on population mental health. *The Australian and New Zealand Journal of Psychiatry* **54**, 938–939. doi:10.1177/0004867420931163

Luke RH, McArthur AG (1978) *Bushfire in Australia.* Australian Government Publishing Service, Canberra.

Mariani M, Connor SE, Theurkauf M, Herbert A, Kunes P, *et al.* (2022) Disruption of cultural burning promotes shrub encroachment and unprecedented wildfires. *Frontiers in Ecology and the Environment* **20**, 292–300. doi:10.1002/fee.2395

McCarthy GJ, Plucinski MP, Gould JS (2012) Analysis of the resourcing and containment of multiple remote fires: The Great Divide Complex of fires, Victoria, December 2006. *Australian Forestry* **75**, 54–63. doi:10.1080/00049158.2012.10676385

McCaw WL, Gould JS, Cheney NP, Ellis PFM, Anderson WR (2012) Changes in behaviour of fire in dry eucalypt forest as fuel increases with age. *Forest Ecology and Management* **271**, 170–181. doi:10.1016/j.foreco.2012.02.003

Miralles DG, Gentine P, Seneviratne SI, Teuling AJ (2019) Land-atmospheric feedbacks during droughts and heatwaves: state of the science and current challenges. *Annals of the New York Academy of Sciences* **1436**, 19–35. doi:10.1111/nyas.13912

Moreira F, Ascoli D, Safford H, Adams MA, Moreno JM, *et al.* (2020) Wildfire management in Mediterranean-type regions: paradigm change needed. *Environmental Research Letters* **15**, 011001. doi:10.1088/1748-9326/ab541e

Moritz MA, Parisien M-A, Batllori E, Krawchuk MA, Van Dorn J, *et al.* (2012) Climate change and disruptions to global fire activity. *Ecosphere* **3**, art49. doi:10.1890/ES11-00345.1

Moritz MA, Batllori E, Bradstock RA, Gill AM, Handmer J, *et al.* (2014) Learning to coexist with wildfire. *Nature* **515**, 58–66. doi:10.1038/nature13946

Nguyen H, Wheeler MC, Hendon HH, Lim EP, Otkin JA (2021) The 2019 flash droughts in subtropical eastern Australia and their association with large-scale climate drivers. *Weather and Climate Extremes* **32**, 100321. doi:10.1016/j.wace.2021.100321

Noble IR, Bary GAV, Gill AM (1980) Mcarthur's fire danger meters expressed as equations. *Australian Journal of Ecology* **5**, 201–203. doi:10.1111/j.1442-9993.1980.tb01243.x

Nolan RH, Boer MM, Collins L, Resco de Dios V, Clarke H, *et al.* (2020) Causes and consequences of eastern Australia's 2019–20 season of mega-fires. *Global Change Biology* **26**, 1039–1041. doi:10.1111/gcb.14987

Nolan RH, Bowman DMJS, *et al.* (2021a) What do the Australian Black Summer fires signify for the global fire crisis? *Fire (Basel, Switzerland)* **4**, 97. doi:10.3390/fire4040097

Nolan RH, Gauthey A, Losso A, Medlyn BE, Smith R, *et al.* (2021b) Hydraulic failure and tree size linked with canopy die-back in eucalypt forest during extreme drought. *New Phytologist* **230**, 1354–1365. doi:10.1111/nph.17298

Pausas JG, Fernández-Muñoz S (2012) Fire regime changes in the Western Mediterranean Basin: from fuel-limited to drought-driven fire regime. *Climatic Change* **110**, 215–226. doi:10.1007/s10584-011-0060-6

Pendergrass AG, Meehl GA, Pulwarty M, Hobbins M, Hoell A, *et al.* (2020) Flash droughts present a new challenge for subseasonal-to-seasonal prediction. *Nature Climate Change* **10**, 191–199. doi:10.1038/s41558-020-0709-0

Peterson DA, Fromm MD, McRae RHD, Campbell JR, Hyer EJ, *et al.* (2021) Australia's Black Summer pyrocumulonimbus super outbreak reveals potential for increasingly extreme stratospheric smoke events. *npj Climate and Atmospheric Science* **4**, 38. doi:10.1038/s41612-021-00192-9

Pham TD, Dwyer L, Su J-J, Ngo T (2021) COVID-19 impacts of inbound tourism on Australian economy. *Annals of Tourism Research* **88**, 103179. doi:10.1016/j.annals.2021.103179

Ruthrof KX, Fontaine JB, Matusick G, Breshears DD, Law DJ, *et al.* (2016) How drought-induced forest die-off alters microclimate and increases fuel loadings and fire potentials. *International Journal of Wildland Fire* **25**, 819–830. doi:10.1071/WF15028

Schweinsberg S, Darcy S, Beirman D (2020) 'Climate crisis' and 'bushfire disaster': Implications for tourism from the involvement of social media in the 2019–2020 Australian bushfires. *Journal of Hospitality and Tourism Management* **43**, 294–297. doi:10.1016/j.jhtm.2020.03.006

Sharples JJ (2020) Submission to the Senate Select Committee into lessons to be learned in relation to the Australian bushfire season 2019–20, <https://www.aph.gov.au/Parliamentary_Business/Committees/Senate/Finance_and_Public_Administration/Bushfirerecovery/Submissions>.

Syphard AD, Keeley JE (2019) Factors associated with structure loss in the 2013–2018 California wildfires. *Fire (Basel, Switzerland)* **2**, 49. doi:10.3390/fire2030049

Tolhurst KG, McCarthy G (2016) Effect of prescribed burning on wildfire severity: a landscape-scale case study from the 2003 fires in Victoria. *Australian Forestry* **79**, 1–14. doi:10.1080/00049158.2015.1127197

van der Velde IR, van der Werf GR, Houweling S, Maasakkers JD, Borsdorff T, *et al.* (2021) Vast CO_2 release from Australian fires in 2019–2020 constrained by satellite. *Nature* **597**, 366–369. doi:10.1038/s41586-021-03712-y

Westerling A, Bryant B (2008) Climate change and wildfire in California. *Climatic Change* **87**, 231–249. doi:10.1007/s10584-007-9363-z

Williams R, Bradstock R, Matthews S, Price O, Sullivan A, *et al.* (2011) 'Climate change, fire regimes and risk in Australian landscapes: lessons for adaptation'. Report to the Department of Climate Change and Energy Efficiency, Canberra.

Williams AP, Abatzoglou JT, Gershunov A, Guzman-Morales J, Bishop DA, *et al.* (2019) Observed impacts of anthropogenic climate change on wildfire in California. *Earth's Future* **7**, 892–910. doi:10.1029/2019EF001210

Williamson B, Weir J, Cavanagh V (2020) Strength from perpetual grief: how Aboriginal people experience the bushfire crisis. *The Conversation*, 10 January (online).

Zhang Y, Beggs PJ, *et al.* (2020) The 2020 special report of the MJA-Lancet Countdown on health and climate change: lessons learnt from Australia's 'Black Summer'. *The Medical Journal of Australia* **213**, 490–492.e10. doi:10.5694/mja2.50869

3

Impacts of wildfire on Indigenous cultural values

Stephen van Leeuwen and Christian Miller-Sabbioni

Summary

- Indigenous Australians have an indisputable right to be consulted when decisions that impact upon them or their Country are made. As the owners of a vast and growing estate they also have the right to lead the decision making and implementation process for the management of their Country.
- Fire was and still is an efficient tool used with great skill and effect by Indigenous Australian land management practitioners; it is embedded in their stewardship of Country and entwined from time immemorial with normative customary traditions, cultural values, myths, rituals and lore.
- The impacts of the 2019–20 Black Summer wildfires on Indigenous cultural heritage values were undoubtedly immense; however, no systematic quantification of the impacts have been undertaken.
- Deficiencies in understanding the notion of cultural significance from an Indigenous Australian's perspective and in identifying, managing and protecting Indigenous cultural heritage values from the impacts of wildfire are being recognised by authorities who are implementing risk minimisation strategies in partnership and with the consent of Traditional Owners.
- The impacts of the wildfires were not all bad, with some Indigenous cultural heritage values being further exposed to enable greater documentation and better protection from future fires and other threats.
- The upswelling of interest in cultural burning practices as a mitigation strategy to reduce the impacts of wildfire is also a positive outcome, especially as Indigenous Australians are being recognised for the stewardship of Country, their traditional knowledge is being celebrated and they are being encouraged to revitalise traditional practices.

Introduction

The misconception of fire, its purpose, its cultural value, its ecological value and its inherent resourcefulness is distorted by a European-style axiology and imposed colonial preconceptions from the past. Fire has been and continues to be a quintessential element in the sustainability of the economy and spirit of Indigenous Australian communities, and is inextricably linked to the wellbeing of individuals, kinship systems, customs and cultural practices.

Indigenous Australians recognise fire as a living thing and have an innate understanding that Country is fire-prone. The stewardship of Country by Indigenous Australians for millennia has shaped Australia 'as the Aborigines made it' (Hallam 2014). This stewardship, conferred through the resourceful utilisation of fire and burning practices, is valued as part of a broader suite of responsibilities inherited from ancestors and ancestral beings, and passed down through the generations: care for Mother Earth and she will reciprocate by caring for you (Prober *et al.* 2011; Maclean *et al.* 2018; Neale *et al.* 2019).

In precolonial Australia fire was utilised to rejuvenate ecosystems, for hunting game, to promote and prepare bush foods (both plant and animal), for communication between communities and in rituals and ceremony (Lake and Christianson 2019). Essentially, fire was, and still is, a systematic tool used with great skill and effect to buttress the subsistence of Aboriginal society while maintaining the ecology (composition, structure, function) and ecosystems services that permitted Aboriginal society to endure for millennia.

An important although often neglected understanding of fire from an Indigenous Australian perspective is its spiritual and ceremonial significance. These intangible quantities have been embedded in Aboriginal myth and lore from time immemorial and are entwined with normative cultural values, procedures and protocols. For example, on Noongar boodja in the south-west corner of Australia the lighting of fires in burial pits/graves was a cultural and spiritual procedure to promote the transfer of life from the earth to the sky – into the sun-mother, which metaphorically held the womb of life and enabled life to continue after death in the spiritual sense (Hallam 2014).

In 2019–20 the Black Summer wildfires affected large parts of the Australian landscape, in particular Country on the eastern seaboard and throughout the Stirling Range and Great Western Woodlands in the west (Davey and Sarre 2020). Indigenous Australians who reside on Country impacted by the wildfires were severely affected as the event not only decimated community infrastructure but also destroyed sites and objects of cultural heritage significance and impacted on environmental values of cultural significance.

Key findings

Impacts on Indigenous cultural heritage

Indigenous cultural heritage, both tangible and intangible, includes sites, artefacts, practices and biological entities (species, ecological communities, landscapes) that are inseparable from the concept of Country (Brumann 2015; Weir *et al.* 2020). Consequently, Indigenous cultural heritage values may be impacted by fire and other extreme events along with the management practices, such as prescribed burning, deployed to mitigate threats.

The entirety of the area burnt in the 2019–20 season is Country belonging to Aboriginal people according to Indigenous law (Williamson *et al.* 2020). According to the laws of Western civil society conferred through legislative instruments such as native title determinations, land rights and heritage Acts, Indigenous land use agreements and through the ownership of freehold land, Aboriginal people hold significant legal rights and interests over large areas of the land affected by the 2019–20 fires.

Indigenous cultural heritage values such as scar trees, rock art and stone arrangements, along with objects of nature such as culturally significant entities, were all impacted by the

2019–20 fires. Quantification of the impact is challenging as the values ascribed to Indigenous cultural heritage objects of significance to Indigenous Australians vary with community and Country. The absence of a formal assessment of the impacts to Indigenous cultural heritage values across the fire ground also hinders any quantification of the impacts. Nevertheless, the impacts can be assumed to be significant given the large population of Indigenous Australians living in areas that were fire affected and the extent of the Indigenous estate impacted. For example, Williamson *et al.* (2020) estimated that over 96 000 Indigenous people were directly affected in New South Wales, Victoria, the Australian Capital Territory and the Jervis Bay Territory, while archaeologists and Aboriginal Elders (Pickrell 2020) suggest that tens of thousands of historic Aboriginal sites and artefacts, many of which had not been officially recorded, may have been damaged.

The impacts of the 2019–20 fires on the Indigenous land estate (Jacobsen *et al.* 2020) have also not been qualified; however, where they have occurred, they have been extensive. For example, the majority of the 5500 ha Budj Bim National Park, a World Heritage Cultural Landscape property, was razed, while 81% of the Greater Blue Mountains Area and 54% of the Gondwana Rainforests, both World Heritage properties, were impacted. These localities harbour Indigenous cultural heritage sites and objects that contribute to the Outstanding Universal Values of the properties.

From a viewpoint focused just on objects of nature, the numbers estimated to have been impacted, both at the species level (e.g. 9092–17197 plant species (Gallagher 2020)) and total number of individuals (~3 billion vertebrates (van Eeden *et al.* 2020)), suggest that many totemic and customary-use (e.g. bush foods, bush medicines, tool-making materials) culturally significant species, such as the koala (*Phascolarctos cinereus*) and black bean (*Castanospermum australe*), would have been impacted.

The mode of impact of fire on Indigenous cultural heritage values is complex, and varies temporally and spatially with reference to the fires and the type of cultural value under threat. For example, tangible objects such as scar trees can be impacted by direct consumption during the fire, or through subsequent operations to make the fire ground safe and secure when dangerous trees are intentionally felled. Similarly, petroglyphs and art on rock faces and overhangs can be damaged directly by radiant heat, and the same heat may also promote accelerated weathering through fracturing, exfoliation and spalling (Gunn 2011; Buckman *et al.* 2021). These types of values can also be deleteriously impacted by the deposition on rock surfaces of soot that may interact chemically with pigments or the rock itself to promote image deterioration (Black *et al.* 2017). Changes in local hydrology as a consequence of reduced vegetative cover, including soil/substrate moisture and the flow pathways of water during storm events (Chapters 5 and 6) may also impact Indigenous cultural heritage values through mechanisms such as increased surface weathering (Mol and Grenfell 2022), chemical compositional changes to substrates, and accelerated erosion of floor deposits and artefact scatters (Gunn 2011). For some values that appear resistant to fire – such as stone arrangements, tool quarry sites and shell middens – fire suppression activities such as clearing mineral earth breaks, or the violent deluge of water/retardant during aerial bombing campaigns, can noticeably dislodge objects and compromise the integrity of sites.

Although not a consequence of the 2019–20 fire event, one of the best-documented examples of the deleterious impacts of fire on Indigenous cultural heritage comes from the destruction of rock art dated at around 19 500 years old in Baloon Cave in Carnarvon Gorge, Queensland. In December 2018 a wildfire ignited the viewing platform and walkway at Baloon Cave, resulting in an extremely hot fire with heat so intense that nearby

Fig. 3.1. Baloon Cave Carnarvon Gorge, Queensland after wildfire destroyed the recycled plastic viewing platform. The heat was so intense much of the rock shelter surface exfoliated and collapsed, destroying or severely damaging rock art. Black soot covered remaining parts of the shelter wall and ceiling. (Photos: Paul S. C. Taçon, Griffith University, Queensland)

rock surfaces suffered massive exfoliation, destroying or severely damaging rock art, including rare, hafted stone axe stencils, as well as coating rock surfaces with soot derived from recycled plastic products (Taçon 2019; Taçon and Harding 2021) (Fig. 3.1). The Bidjara-Garingbal Aboriginal community on whose Country the Baloon Cave is located were extremely upset with what occurred. The community has subsequently been warning other Aboriginal communities not to use recycled plastic or other flammable material at rock art sites as it can put the art at significant risk.

The 2019–20 fire event does have an upside with respect to Indigenous cultural heritage in that across the fire ground the fires cleared away dense vegetation, permitting access to previously inaccessible areas and exposing previously hidden Indigenous cultural values. The best example of this positive outcome comes from the 7000 ha razed by fires on land around Lake Condah and in the Budj Bim Cultural Landscape World Heritage area in Victoria. As a consequence of the fires, previously unknown sections of channels and pools carved out of the rocky volcanic pavement by Gunditjmara People over 6600 years ago to harvest eels, part of the Muldoon trap complex, were revealed (Johnson 2020). The additional structures identified as a consequence of vegetation removal also include 27 stone houses (Indigenous Mapping Workshop 2020). Another positive outcome of the fires at Budj Bim were the low-impact control techniques employed to suppress the fires, avoiding the need to use heavy machinery to clear mineral earth breaks and thereby preventing any damage to stone structures or arrangements. These fire suppression approaches were co-designed and jointly implemented with the Gunditjmara People (Country Fire Authority 2020). Coincidentally, another positive outcome from the Budj Bim fires was the removal of 1100 feral animals (feral pigs and deer) during aerial shooting operations, a result achieved due to the significant reduction in vegetative cover following the fires (Miles 2021).

Response from the Aboriginal community

> *... I've watched in anguish and horror as fire lays waste to precious Yuin land, taking everything with it – lives, homes, animals, trees – but for First Nations people it is also burning up our memories, our sacred places, all the things that make us who we are. (Allam 2020)*

The 2019–20 fires compounded the perpetual trauma and helplessness that many Aboriginal people experience as a result of historic dispossession, institutional racism, and the contemporary disconnection and exclusion from Country and their obligations to care for Country (Williamson *et al.* 2020). This predicament is reinforced by the notion that the stewardship of Country as practised before colonisation protected sacred, culturally significant and spiritual values that are at the crux of Aboriginal identity and wellbeing, and are now those cultural values most at risk of being lost to future generations. For Aboriginal people, moving on, healing people, is very much a process that entails Country, healing Country, because Country is a living entity that intrinsically informs Aboriginal culture.

In concert with understanding and quantifying the impacts of the fires and suppression activities on cultural heritage values, many Aboriginal people were concerned about the ongoing protection post-fire of Country and the heritage values within. These concerns included protection of newly exposed and vulnerable heritage values, reconnecting with Country to support recovery, and subsequent mitigation of the likelihood of further and future impacts (Australian Broadcasting Commission 2020; Pickrell 2020). Ultimately Aboriginal people are seeking to prioritise their interests and draw upon their strengths and values to collaborate in the ongoing stewardship of their Country.

In the immediate aftermath of the fires, much attention was drawn to determining the extent to which, and how badly, Indigenous cultural heritage values had been impacted, in conjunction with the discovery of previously undocumented heritage values and how such values would now be protected. As happened in the wash-up for the 2003 Victorian Alps fires (Freslov 2004; Freslov *et al.* 2004), Aboriginal people and archaeologists sought a formal assessment and thus called on government to launch a thorough investigation to gauge the fire's effect on culturally significant sites and artefacts.

A greater recognition and better protection of Indigenous cultural heritage values susceptible to fire impacts, or associated suppression and management activities, is happening through collaborations between Traditional Owners, government institutions and other land managers. After the Baloon Cave tragedy, for example, the Queensland Environment Minister ordered an audit of the infrastructure at cultural heritage site and subsequently the removal of all recycled plastic walkways and viewing platforms (Australian Broadcasting Commission 2019). It is worth noting that the Baloon Cave tragedy was not the first occasion that recycled plastic infrastructure has caused the unsalvageable destruction of rock art. An identical outcome was the consequence of the ignition of a boardwalk during a hazard reduction burn in 2008 at a site in Keep River, Northern Territory (Lambert and Welsh 2011). Clearly, communicating the risk such infrastructure affords to Indigenous cultural heritage site managers, including Traditional Owners, is critical, a challenge the Bijara-Garingbal community is actively addressing (Taçon 2019).

In the Australian Capital Territory (ACT) on Ngunnawal Country during and following the fires, a range of strategies was undertaken in collaboration with Traditional Owners to safeguard Indigenous cultural heritage values. These strategies included the retrieval and temporary storage of a 700–1000-year-old culturally significant, women's business object from a remote area of the Namadgi National Park. This outcome was achieved through a collaboration between the Dhawura Ngunnawal Community, the Environment, Planning and Sustainable Development Directorate, the ACT Emergency Services Agency and the Australian Defence Force (Weaver 2020). Post-fire the ACT government has also reinforced its efforts to provide greater protection for Indigenous cultural heritage sites and artefacts during emergencies by implementing a range of safeguards (ACT

Environment, Planning and Sustainable Development Directorate 2020; Freeman *et al.* 2021). These safeguards include:

- low-fuel buffer zones, bare earth breaks and sprinklers around rock art sites;
- removing of timber viewing boardwalk at rock art sites;
- bare earth breaks to protect scar trees;
- 'no water bombing zones' to protect stone arrangements and burial sites; and
- the embedding of an environmental and cultural 'values officer' in the Incident Management Team to assist in protecting Country during emergencies.

Similar strategies to protect Indigenous cultural heritage values have also been executed in other jurisdictions (Forestry Corporation of NSW 2021; NSW Rural Fire Service 2021).

A transformational consequence of the fires driven by politicians, bureaucrats, land management practitioners, and welcomed by Aboriginal people, has been the drive to unlock and enable Indigenous land management practices, in particular cultural burning. It is perceived that this holistic approach to landscape management, which is based on a person's 'sense of place', their connection to Country and the anthology of traditional knowledge amassed over millennia, can confer significant social, cultural and environmental benefits that deliver security for Indigenous cultural heritage and promote healthy people and healthy Country (Burgess *et al.* 2005; Barber and Jackson 2017; Steffenson 2020; Taylor-Bragge *et al.* 2021).

While many Indigenous Australians and land management practitioners enthusiastically embrace Indigenous land management programs and cultural burning, it will take time, resources and considerable effort to make the transformation to this new paradigm, a new way of working that is Indigenous-led and Indigenous-owned. This is particularly the case in southern Australia where the revival of cultural burning faces many challenges to its widespread adoption, including land tenure and management regime intricacies, burdensome administrative requirements over small spatial scales and altered environmental setting from past pre-European conditions (e.g. weeds, feral animals, climate change) (Baker *et al.* 2001). Further compounding this revival is the under-resourcing of Indigenous communities who are still endeavouring to regain access to Country and advocating for their traditional cultural management practices to be recognised as legitimate evidence-based practices equivalent in legitimacy to those deployed by European-style land management practitioners and not simply dismissed as historical practices of the past (Williamson *et al.* 2020).

Busting paradigms that devalue traditional knowledge systems and practices, and decolonising Indigenous land management, underscores discussions and initiatives to reinvigorate and revive cultural burning (Fletcher *et al.* 2021; Plange 2021). For areas impacted by the 2019–20 events, the Commonwealth Government's provision of $2 million in grant funding, to support Indigenous-led knowledge exchange of Indigenous fire and land management practices, goes some way to paradigm busting. Thirteen Traditional Owner and Indigenous enterprises were able to participate in the grant program that facilitated the sharing of cultural burning knowledge among community members and with partner organisations while re-engaging community with their Country and culture.

Recommendations

The summer wildfires of 2019–20 had a detrimental effect on the entirety of Indigenous communities in areas that were impacted, although the consequences of this extreme event on Indigenous cultural heritage remain unquantified.

The contemporary Australian perception of fire as an evil and dangerous element or hazard is as prevalent as ever (Fletcher *et al.* 2021). This attitude has led to the deliberate suspension of traditional land management practices, including cultural burning, in which fire was skilfully used by Indigenous stewards as a tool to care for Country and in so doing care for their community and themselves. Ironically, as noted by Fletcher *et al.* (2021), fire is now perceived as a Promethean threat to Australia's natural environment, its unique species and ecological communities, and its ancient Indigenous heritage.

This prevailing attitude, if allowed to persist, will devalue the wellbeing of all Australians and our unique environments. Consequently, limiting the impacts and mitigating the risk of future wildfires and similar extreme events on Indigenous cultural heritage values will be effectively improved through:

- enhanced knowledge of the spatial distribution and cultural significance to Indigenous Australians of places and objects/entities of significance and their vulnerability to fire;
- greater legislative and regulatory protection from destruction and desecration for tangible and intangible cultural heritage values;
- development of risk minimisation strategies, for routine land management and for emergency response, that do not deliberately or unintentionally ignore Indigenous cultural heritage values, but rather help to secure such values in perpetuity; and
- improved comprehension of the viewpoint of Indigenous Australians on fire to find common ground or a common understanding in relation to fire and the stewardship of Country that recognises the rights of Indigenous Australia to participate in decisions about themselves and their Country.

References

ACT Environment, Planning and Sustainable Development Directorate (2020) Protecting Aboriginal sites and artefacts in emergencies, <https://www.cmtedd.act.gov.au/open_government/inform/act_government_media_releases/esdd/2020/protecting-aboriginal-sites-and-artefacts-in-emergencies>.

Allam L (2020) For First Nations people the bushfires bring a particular grief, burning what makes us who we are. *The Guardian,* 6 January, <https://www.theguardian.com/commentisfree/2020/jan/06/for-first-nations-people-the-bushfires-bring-a-particular-grief-burning-what-makes-us-who-we-are>.

Australian Broadcasting Commission (2019) Statewide audit after ancient rock art destroyed by Carnarvon Gorge bushfire. *South Queensland Drive*, 9 December, <https://www.abc.net.au/radio/programs/south-queensland-drive/leanne-enoch-moves-to-prevent-new-baloon-cave-disaster/11850060>.

Australian Broadcasting Commission (2020) The summer bushfires exposed Indigenous artefacts protected by bushland. The race is on to save them from neglect. *ABC News*, 7 June, <https://www.abc.net.au/news/2020-06-07/bushfire-exposed-indigenous-art-we-must-save-from-neglect/12322258?utm_campaign=abc_news_web&utm_content=link&utm_medium=content_shared&utm_source=abc_news_web>.

Baker R, Davies J, Young E (2001) Managing country: an overview of the prime issues. In *Working on Country: Contemporary Indigenous Land Management of Australia's Lands and Coastal Regions*. (Eds R Baker, J Davies and E Young) pp. 1– 23. Oxford University Press, Melbourne.

Barber M, Jackson S (2017) Identifying and categorizing co-benefits in state-supported Australian indigenous environmental management programs: international research implications. *Ecology and Society* **22**, 11. doi:10.5751/ES-09114-220211

Black JL, MacLeod ID, Smith BW (2017) Theoretical effects of industrial emissions on colour change at rock art sites on Burrup Peninsula, Western Australia. *Journal of Archaeological Science, Reports* **12**, 457–462. doi:10.1016/j.jasrep.2017.02.026

Brumann C (2015) Cultural heritage. In *International Encyclopedia of the Social and Behavioural Sciences.* (Ed. JD Wright) pp. 414–419. Elsevier, Oxford.

Buckman S, Morris RH, Bourman RP (2021) Fire-induced rock spalling as a mechanism of weathering responsible for flared slope and inselberg development. *Nature Communications* **12**, 1–14.

Burgess CP, Johnston FH, Bowman D, Whitehead PJ (2005) Healthy country: healthy people? Exploring the health benefits of Indigenous natural resource management. *Australian and New Zealand Journal of Public Health* **29**, 117–122. doi:10.1111/j.1467-842X.2005.tb00060.x

Country Fire Authority (2020) Fighting fires on cultural land in partnership with traditional owners. Victoria, <https://news.cfa.vic.gov.au/news/fighting-fires-on-cultural-land-in-partnership-with-traditional-owners>.

Davey SM, Sarre A (2020) Editorial: the 2019/20 Black Summer bushfires. *Australian Forestry* **83**, 47–51. doi:10.1080/00049158.2020.1769899

Fletcher MS, Romano A, Connor S, Mariani M, Maezumi SY (2021) Catastrophic bushfires, indigenous fire knowledge and reframing science in southeast Australia. *Fire (Basel, Switzerland)* **4**, 61. doi:10.3390/fire4030061

Forestry Corporation of NSW (2021) Operational guideline for Aboriginal cultural heritage management – a due diligence code of practice, <https://www.forestrycorporation.com.au/__data/assets/pdf_file/0018/503442/operational-guidelines-aboriginal-cultural-heritage.pdf>.

Freeman D, Williamson B, Weir J (2021) Cultural burning and public sector practice in the Australian Capital Territory. *The Australian Geographer* **52**, 111–129. doi:10.1080/00049182.2021.1917133

Freslov J (2004) 'Post Wildfire Indigenous Heritage Survey Volume 2: Management of Impacts from Wildfire and Suppression Activities. A Report to Parks Victoria, the Department of Sustainability and Environment and Aboriginal Affairs, Victoria'. Perspectives Heritage Solutions Pty Ltd, Norman Park, Queensland.

Freslov J, Hughes P, Mullett R (2004) 'Post Wildfire Indigenous Heritage Survey Volume 1: Background, Survey, Results and Recommended Management Options. A Report to Parks Victoria'. AAV Project No. 2833. Perspectives Heritage Solutions Pty Ltd, Norman Park, Queensland.

Gallagher RV (2020) Final national prioritisation of Australian plants affected by the 2019–2020 bushfire season. Commonwealth Department of Agriculture, Water and Environment, Canberra. <https://www.dcceew.gov.au/sites/default/files/env/pages/289205b6-83c5-480c-9a7d-3fdf3cde2f68/files/final-national-prioritisation-australian-plants-affected-2019-2020-bushfire-season.pdf>.

Gunn RG (2011) The impact of bushfires and fuel reduction burning on the preservation of shelter rock art. *Rock Art Research* **28**, 53.

Hallam SJ (2014) *Fire and Hearth: Karla Yoorda – A Study of Aboriginal Usage and European Usurpation in South-western Australia.* University of Western Australia Publishing, Nedlands.

Indigenous Mapping Workshop (2020) Ancient aquaculture sites in the Budj Bim Cultural Landscape. Winyama Digital Solutions, Perth. <https://www.winyama.com.au/news-room/ancient-aquaculture-sites-in-the-budj-bim-cultural-landscape>.

Jacobsen R, Howell C, Read S (2020) 'Australia's Indigenous land and forest estate: separate reporting of Indigenous ownership, management and other special rights'. ABARES Research Report 20.15, <https://www.agriculture.gov.au/abares/forestsaustralia/publications/indigenous-estate-report>.

Johnson S (2020) Budj Bim Cultural Landscape fire reveals new sections of ancient aquatic system. Australian Broadcasting Commission, South West Victoria, posted 19 January 2020, <https://www.abc.net.au/news/2020-01-19/fire-reveals-further-parts-of-6600-year-old-aquatic-system/11876228?utm_campaign=abc_news_web&utm_content=link&utm_medium=content_shared&utm_source=abc_news_web>.

Lake FK, Christianson AC (2019) Indigenous fire stewardship. In *Encyclopedia of Wildfires and Wildland-urban Interface (WUI) Fires.* (Ed. SL Manzello) pp. 1–9. Springer, New York. doi:10.1007/978-3-319-51727-8_225-1

Lambert D, Welsh B (2011) Fire and rock art. *Rock Art Research: The Journal of the Australian Rock Art Research Association* **28**, 45–48[AURA].

Maclean K, Robinson CJ, Costello O (2018) 'A national framework to report on the benefits of Indigenous cultural fire management. Final Report'. CSIRO, Canberra. https://publications.csiro.au/rpr/download?pid=csiro:EP188803&dsid=DS1.

Miles D (2021) Feral pigs and deer killed in aerial cull to protect Budj Bim Cultural Landscape. Australian Broadcasting Commission, South West Victoria, <https://www.abc.net.au/news/2021-05-14/aerial-shoot-kills-feral-pigs-and-deer-in-world-heritage-site-/100136532?utm_campaign=abc_news_web&utm_content=link&utm_medium=content_shared&utm_source=abc_news_web>.

Mol L, Grenfell M (2022) Influence of landscape moisture sources and topography on rock weathering patterns associated with wildfire. *Earth Surface Processes and Landforms* doi:10.1002/esp.5345.

Neale T, Carter R, Nelson T, Bourke M (2019) Walking together: a decolonising experiment in bushfire management on Dja Dja Wurrung country. *Cultural Geographies* **26**, 341–359. doi:10.1177/1474474018821419

NSW Rural Fire Service (2021) Bush fire environmental assessment code – supporting document: conditions for hazard reduction and Aboriginal heritage, <https://www.rfs.nsw.gov.au/__data/assets/pdf_file/0015/24333/CMR1502-BFEAC-Conditions-for-HR-and-Aboriginal-Heritage-2021-final.pdf>.

Pickrell J (2020) Thousands of ancient Aboriginal sites probably damaged in Australian fires. *Nature* doi:10.1038/d41586-020-00164-8.

Plange NAL (2021) Decolonising fire: recognition justice and Aboriginal fire knowledge in the 2019–2020 Australian bushfire news narrative. PhD thesis. The University of Sydney, Sydney, <https://hdl.handle.net/2123/25697>.

Prober SM, Connor MHO, Walsh FJ (2011) Australian Aboriginal peoples' seasonal knowledge: a potential basis for shared understanding in environmental management. *Ecology and Society* **16**, 12. doi:10.5751/ES-04023-160212

Steffenson V (2020) *Fire Country: How Indigenous Fire Management could Help save Australia.* Hardie Grant Travel, Melbourne.

Taçon P (2019) The aftermath of fire damage to important rock art at the Baloon Cave tourist destination, Carnarvon Gorge, Queensland. Bradshaw Foundation, <https://www.bradshawfoundation.com/rockartnetwork/baloon_cave.php>

Taçon P, Harding D (2021) 'Baloon Cave visit and assessment, Carnarvon Gorge, Queensland Fieldwork Report, 7 February 2019'. Centre for Social and Cultural Research, Griffith University, Brisbane. <http://hdl.handle.net/10072/411211>.

Taylor-Bragge RL, Whyman T, Jobson L (2021) People needs country: the symbiotic effects of landcare and wellbeing for Aboriginal peoples and their countries. *Australian Psychologist* **56**, 458–471. doi:10.1080/00050067.2021.1983756

van Eeden LM, Nimmo D, Mahony M, Herman K, Ehmke G, *et al.* (2020) 'Impacts of the unprecedented 2019–2020 bushfires on Australian animals'. Report prepared for WWF-Australia, Sydney.

Weaver M (2020) Sacred Aboriginal sites and artefacts protected during bushfire emergency. *The Riotact*, <https://the-riotact.com/sacred-aboriginal-sites-and-artefacts-protected-during-bushfire-emergency/355843>.
Weir JK, Williamson B, Markham F (2020) Submission to the NSW independent expert inquiry into the 2019 – 2020 bushfire season, <https://files.igem.vic.gov.au/2021-03/WEIR%20Jessica%20Ors%20129S.pdf>.
Williamson B, Weir J, Cavanagh V (2020) 'Strength from perpetual grief: how Aboriginal people experience the bushfire crisis. Faculty of Social Sciences – Papers. 4709'. University of Wollongong, Wollongong, <https://ro.uow.edu.au/sspapers/4709>.

4

Consequences of the Australian 2019–20 wildfires for sites important for biodiversity and other World Heritage values

Michelle Ward, James E. M. Watson, William Geary, Aaron Greenville, Golo Maurer, Stephanie Todd and Ayesha Tulloch

This chapter is based on, and sections reproduced with kind permission from, Ward M, Watson JEM, Greenville A, Maurer G, Todd S, Geary W, Tulloch A. (2022) Consequences of the Australian 2019/20 wildfires for sites important for biodiversity and other world heritage values. *Australian Zoologist* **42**(2), 182–198. doi:10.7882/AZ.2022.034

Summary

- Australia's 2019–20 wildfire season was unprecedented in its extent and severity, yet the consequences for sites important for biodiversity conservation and other World Heritage values remain unknown.
- Here, we summarise the 2019–20 wildfire impacts on key sites set aside for, or identified as being important for, Australian biodiversity, with specific reference to nationally designated protected areas, World Heritage listings and Key Biodiversity Areas (KBAs). We also explore patterns between burn extent and severity, and how they differ when considering underlying landscape characteristics.
- Over 7 months, approximately 10 million ha of native vegetation burnt. Of these burnt landscapes, ~3.2 million ha (41%) were within the Australian protected area estate, with 834 protected areas impacted.
- Six Australian World Heritage listings were impacted by the 2019–20 wildfires, with the largest impact in the Greater Blue Mountains Area (680 000 ha or 67% was affected by the fires).
- The 2019–20 fires impacted over 2 million ha across 69 KBAs. Of these, 18 KBAs had > 15% of their area burnt.
- Of management importance, we show that the degree of overall forest 'integrity' affected the chances of fire severity, with the more degraded forests experiencing higher severity burns. This held true across both the protected area estate and those areas identified as KBAs.
- Recovery of many of these fire-affected areas will require targeted restoration efforts as well as resources to help reduce the likelihood of future megafires, as well as increase resilience in the face of other catastrophic climate events.

Introduction

The 2019–20 wildfire season had devastating impacts on many populations of Australia's species and was unprecedented in its extent and severity. An estimated 10 million ha (approximately the size of Iceland) burnt from July 2019 to the March 2020 (Chapter 2), affecting sites important for Australian biodiversity such as national parks, Indigenous Protected Areas, World Heritage listings and Key Biodiversity Areas (KBAs).

Australia has ~19% of its terrestrial extent protected in its National Reserve System, which includes more than 13 500 protected areas. Since Indigenous Protected Areas were first recognised in the early 1990s, there are now > 78 declared areas covering > 74 million ha across Australia making up 46% of the areal extent of Australia's network of protected areas (Commonwealth of Australia 2021b). Every Australian protected area is classified into an International Union for Conservation of Nature (IUCN) protected area management category according to its management objectives. IUCN categories I–IV make up 8.10% of the 19% of Australia's landmass under protection, while categories V–VI make up 11.62% (Table 4.1). Many of these conservation areas are important not only for biodiversity (Watson *et al.* 2010; Kearney *et al.* 2020), but also for cultural values (through Indigenous Protected Areas such as Dorodong Indigenous Protected Area) and ecosystem services. Such services provide Australian communities with purified water to drink and protection against natural hazards such as floods and soil erosion (Potts *et al.* 2014; Figgis *et al.* 2015; Ferraro *et al.* 2015). These ecosystem services within Australia's protected area estate have an estimated value of more than $1.3 trillion per year (Commonwealth of Australia 2021a).

Nominated UNESCO World Heritage sites are inscribed on the list only after they have been carefully assessed as representing the best examples of the world's cultural and natural heritage. Australia currently has 20 properties on the World Heritage List, including the Gondwana Rainforests of Australia, Greater Blue Mountains Area and K'gari (Fraser Island). The *Environment Protection and Biodiversity Conservation Act 1999* is Australia's primary legislative tool for protecting and managing World Heritage listings, as per its obligations under the World Heritage Convention.

KBAs are places identified as globally significant for the persistence of biodiversity and are recognised, by some groups, as focal areas for practical conservation action (IUCN 2016). There are currently 314 KBAs spread across the Australian continent, covering 4 468 900 ha. All 314 were first identified as Important Bird and Biodiversity Areas (Birdlife International 2021) or Alliance of Zero Extinction Sites (n = 34) (Alliance for Zero Extinction 2018). KBAs do not convey a legal conservation status in Australia.

Using a study region that includes 2.2 million km^2 across 43 temperate, Mediterranean, and subtropical bioregions, as defined in the Interim Biogeographic Regionalisation for Australia (IBRA) dataset (Commonwealth of Australia 2018; Roff 2020), we summarise the 2019–20 wildfire impacts on key sites set aside for, or identified as being important for, biodiversity, with specific reference to protected areas, World Heritage listings and KBAs. We investigate total area burnt and the severity of burns, and provide examples to further highlight the fine-scale impacts of this disaster. We also explore patterns between burn extent or severity and underlying landscape characteristics. To do this, we quantify the relationship between forest integrity and either the 2019–20 wildfire severity or proportion burnt for each protected area, World Heritage listing or KBA.

Table 4.1. Overview of terrestrial protected areas in Australia by IUCN Management Category (Commonwealth of Australia 2021a).

IUCN category	Description	Number	Area (ha)	Average size (ha)	% of Australia	Contribution to protected area network
Ia	A Strict Nature Reserve is set aside to protect biodiversity and also possibly geological/geomorphological features, where human visitation, use and impacts are strictly controlled and limited to ensure protection of the conservation values.	2619	15 981 368	6102	2.08%	10.52%
Ib	A Wilderness Area is usually large unmodified or slightly modified area, which retains its natural character and influence without permanent or significant human habitation and is protected and managed to preserve its natural condition.	66	3 847 645	58 298	0.50%	2.53%
II	A national park is a large natural or near natural area set aside to protect large-scale ecological processes, along with the complement of species and ecosystems characteristic of the area, which also provides a foundation for environmentally and culturally compatible, spiritual, scientific, educational, recreational, and visitor opportunities.	1096	38 190 170	34 845	4.97%	25.14%
III	A Natural Monument or Feature is set aside to protect a specific natural monument, which can be a landform, sea mount, submarine cavern, geological feature such as a cave or even a living feature such as an ancient grove.	2395	1 867 857	780	0.24%	1.23%
IV	A Habitat/Species Management Area aims to protect species or habitats and management reflects this priority.	4248	2 401 392	565	0.31%	1.58%
I–IV Total		**10 424**	**62 288 431**	**5975**	**8.10%**	**41.01%**

continued

Table 4.1. Continued

IUCN category	Description	Number	Area (ha)	Average size (ha)	% of Australia	Contribution to protected area network
V	A Protected Landscape/Seascape is managed for both conservation and recreation, where the interaction of people and nature has produced an area of distinct character with significant, ecological, biological, cultural, and scenic value, and where safeguarding the integrity of this interaction is vital to protecting and sustaining the area and its associated nature conservation and other values.	359	7 924 748	22 075	1.03%	5.22%
VI	A Managed Resource Protected Area is managed mainly for the sustainable use of natural ecosystems.	1320	81 450 986	61 705	10.59%	53.63%
V–VI Total		**1679**	**89 375 734**	**53 232**	**11.62%**	**58.85%**
Not applicable		8	2444	305	0.00%	0.00%
Not assigned		1377	214 860	156	0.03%	0.14%
Not reported		1	0.4	0.4	0.00%	0.00%
TOTAL		**13 488**	**151 881 469**	**11 260**	**19.75%**	**100.00%**

To calculate the area and intensity of the fire's impacts on iconic places, we intersected the Australian Google Earth Engine Burnt Area Map (AUS GEEBAM) (Commonwealth of Australia 2020a) with the extent of protected areas, World Heritage listings and KBAs. The AUS GEEBAM dataset is based on 40 m^2 Sentinel 2 satellite imagery from 1 July 2019 to 24 February 2020. AUS GEEBAM relies on a vegetation reflectance index (Relativised Normalized Burnt Ratio, RNBR) that calculates the difference in reflectance before and after the fire season for burnt areas and adjacent unburnt areas within the fire footprint. Combined with national vegetation mapping (Commonwealth of Australia 2020b) the result is a map of four fire severity classes that represent how severely vegetation was burnt during the 2019–20 wildfires. The classes presented have been designed for rapid response and were not trained with ground data, and so do not have confidence intervals or accuracy reports.

A wildfire may have very different impacts on biodiversity at different locations; these impacts depend on many factors, one of which is ecosystem condition before the fire. For example, forest degradation and fragmentation can interact with natural disturbance regimes such as fire to perturb forests beyond their capacity to regenerate (Lindenmayer *et al.* 2017), whereas forests with high ecologically integrity might provide refuge for forest species from increased fire frequencies in degraded landscapes (Watson *et al.* 2018). To explore patterns between forest integrity and wildfire severity for each protected area, World Heritage listing and KBA, we used ordinal response regression models (Guisan and Harrell 2000). To identify 'forest integrity' we used a global 'forest integrity' metric for forest ecosystems (Grantham *et al.* 2020). The forest integrity metric uses integrated data on forest extent, observed human pressures (e.g. infrastructure), other inferred human pressures (e.g. collection of forest materials and logging), and alterations in forest connectivity. Higher cell scores (at a 500 m^2 resolution) indicate higher landscape-level forest integrity for the beginning of 2019, whereas lower scores indicate lower integrity.

Key findings

Sites set aside for biodiversity: protected areas

Over 7 months, ~10 million ha of native vegetation burnt. Of these burnt landscapes ~3.2 million ha (41%) were within the protected area estate (see Box 4.1). More than half of these fires was classified as high to very high severity (1.8 million ha), with 1.4 million ha falling within the low to moderate severity classification (Fig. 4.1). In total, 834 protected areas burnt, the majority of which were IUCN classifications II (285) and Ia (228). Approximately 70% of all Australia's 1b IUCN protected areas were impacted by the 2019–20 wildfires (Table 4.2).

When we disaggregated protected area burning by IUCN category, the most impacted category by area burnt was II (national parks: 1 601 460 ha), closely followed by Ib (Wilderness Areas: 1 201 878 ha; Table 4.2). This ranking changed when we assessed proportion burnt, with IUCN category Ib most impacted (31% burnt, of which ~54% was high or very high severity).

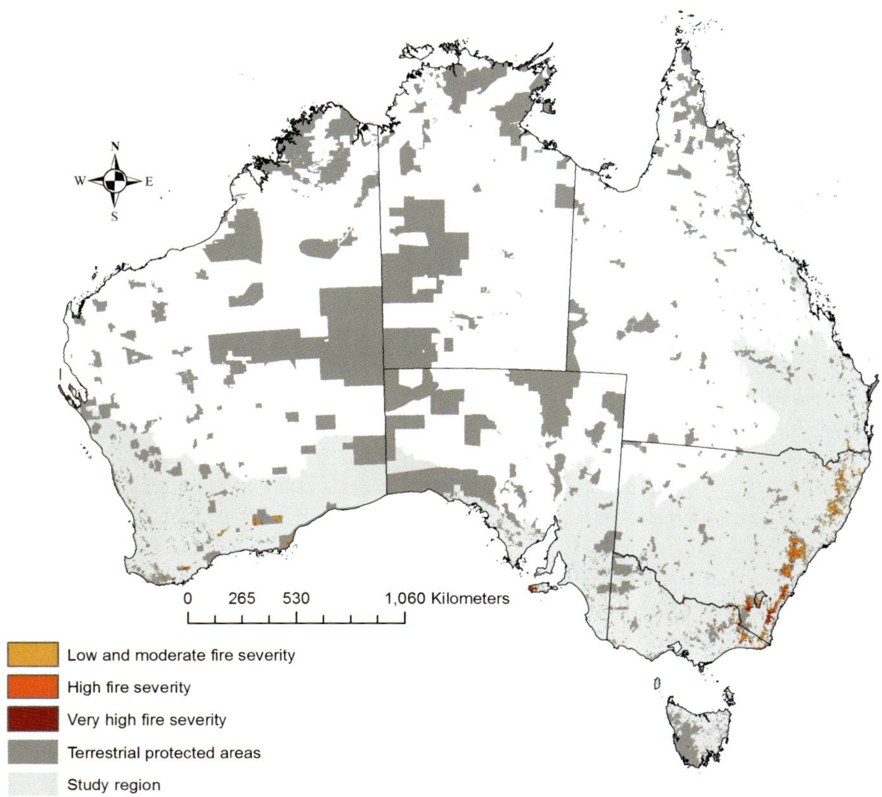

Fig. 4.1. Extent and severity of wildfires between July 2019 and February 2020 and impacts on protected areas. Yellow indicates low and moderate fire severity, orange indicates high fire severity, and red indicates very high fire severity. Dark grey shows all terrestrial protected areas, while light grey shows the study region.

Table 4.2. Overview of the proportion burnt in each burn severity category within each IUCN protected area listing.

IUCN listing	Number of protected areas burnt (% of all protected areas in this category)	Extent burnt – low and moderate fire severity (ha)	Extent burnt – high fire severity (ha)	Extent burnt – very high fire severity (ha)	Cumulative extent burnt (ha)	Proportion of total extent burnt (%)
Ia	228 (9%)	186 362	120 614	30 164	337 150	2%
Ib	46 (70%)	550 015	448 492	203 371	1 201 878	31%
II	285 (27%)	647 436	582 567	371 457	1 601 460	4%
III	126 (5%)	17 887	9754	14 901	42 542	2%
IV	88 (2%)	92 546	53 705	13 261	159 512	7%
V	10 (3%)	10 804	8678	5288	24 770	0.3%
VI	51 (4%)	12 370	4959	2431	19 760	0.02%
Total	834	1 517 420	1 228 769	640 873	3 387 072	

Box 4.1. Protected areas in Victoria: impacts and consequences for management

The 2019–20 wildfires burnt ~1.5 million ha across Victoria, including 30 national parks and nature conservation reserves that had over 80% of their area within the fire extent (DELWP 2020). This comprised some of Victoria's best-known national parks, including Croajingolong (87%), Snowy River (76%) and Errinundra (66%). National parks such as the Burrowa-Pine Mountain National Park and the Lind National Park had their entire extents burnt (DELWP 2020). Across Victoria's reserve system, over 280 000 ha were impacted by high-severity fire in the 2019–20 fire season (DELWP 2020).

The 2019–20 wildfires disproportionately affected areas of Victoria's highest biodiversity value (DELWP 2020), much of which occurs in Victoria's protected areas. Species such as spotted tree frog (*Litoria spenceri*) and eastern bristlebird (*Dasyornis brachpterus*) have important populations within Victoria's protected areas that were affected by the 2019–20 wildfires. These wildfires also compounded the effects of other recent large fires (e.g. 2003 and 2006–07), meaning some parts of Victoria's protected areas have been burnt multiple times in the last 20 years, severely impacting species and environments such as Victoria's alpine ash (*Eucalyptus delegatensis*) ecosystems (Fig. 4.2) (Geary *et al.* 2022). Since the fires, management of impacted species has focused on facilitating recovery through management of threats (e.g. invasive predator and herbivore control), as well as proactive actions that can help to reduce the risks of future extreme fires to these species such as emergency extraction (Selwood *et al.* 2021), translocation to new areas and captive breeding.

Fig. 4.2. Long unburnt (1939 regrowth) alpine ash forest in Victoria's Central Highlands (A), and alpine ash forest burnt by successive fires between 2007 and 2019–20 leading to population collapse (B). (Photos: Tom Fairman, University of Melbourne)

Sites set aside for biodiversity: World Heritage listings

Of the 18 terrestrial Australian World Heritage listings, six were impacted by the 2019–20 wildfires (Fig. 4.3; Table 4.3). The World Heritage listing with the largest impact was the Greater Blue Mountains Area, with ~680 000 ha burnt (of which 362 000 ha was high and very high severity and 321 000 ha was low and moderate severity) (see Box 4.2). This represents more than 65% of the entire Greater Blue Mountains Area. The Budj Bim Cultural Landscape, located in the traditional Country of the Gunditjmara people in south-eastern Australia, was also greatly impacted with ~54% of its extent affected by the 2019–20

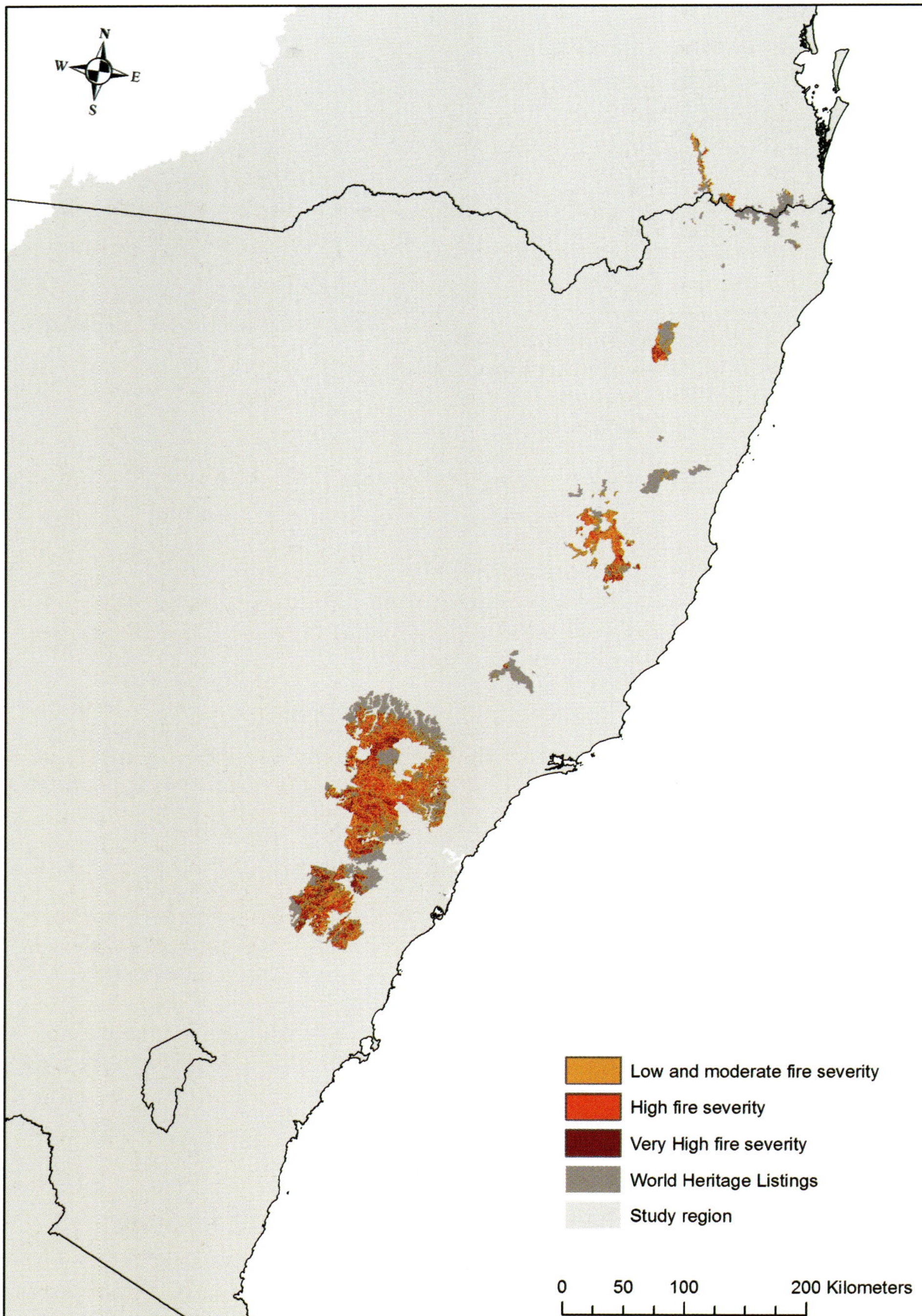

Fig. 4.3. Extent and severity of wildfires between July 2019 and February 2020 and impacts on World Heritage listings. Yellow indicates low and moderate fire severity, orange indicates high fire severity, and red indicates very high fire severity. Dark grey shows all terrestrial World Heritage listings, while light grey shows the study region.

Table 4.3. Overview of World Heritage listings and the extent and severity of 2019–20 wildfires.

Note: Some World Heritage listings spatially overlap with protected areas.

World Heritage listing	Total area (ha)	Extent burnt – low and moderate fire severity (ha)	Extent burnt – high fire severity (ha)	Extent burnt – very high fire severity (ha)	Extent burnt – high and very high fire severity (ha)	Cumulative extent burnt (ha)	Proportion of total extent burnt (%)	Proportion of total extent burnt not overlapping with PA (%)
Greater Blue Mountains Area	1 043 825	320 903	292 176	69 333	361 509	682 412	65%	0%
Budj Bim Cultural Landscape	9935	5329	84	1	85	5414	54%	0%
Gondwana Rainforests of Australia	368 727	85 273	53 533	6989	60 522	145 795	40%	0%
Australian Convict Sites	389	122	23	0	23	145	37%	0%
Fraser Island	181 838	5073	2970	844	3814	8887	5%	0%
Tasmanian Wilderness	1 585 523	197	371	1782	2153	2350	0%	0%
TOTAL	3 190 237	416 897	349 157	78 950	428 106	845 003		

wildfires. The Gondwana Rainforests of Australia, an area that does not generally burn (Kooyman *et al.* 2020) and harbours the highest concentrations of threatened species in subtropical south-east Queensland and northern New South Wales (Kooyman *et al.* 2020), had 40% of its extent burnt.

Box 4.2. Greater Blue Mountains Area

The Greater Blue Mountains Area west of Sydney is one of the most fire-prone regions in the world (Bradstock *et al.* 2009). The region experiences wildfires during dry El Niño years that result in a major wildfire season once or twice a decade (Bradstock *et al.* 2009; Hammill and Tasker 2010). Following extreme drought in 2019–20, the moisture content of fine dead fuel was well below the critical threshold for large wildfires to occur within the Sydney (Nolan *et al.* 2020). Lightning from dry summer storms and one escaped hazard reduction burn provided the ignition points for seven major wildfires, with some joining up across the 'Black Summer'. The 2019–20 megafires burnt 65% of the Greater Blue Mountains World Heritage Area (682 412 ha), with 35% of the region recording either

Fig. 4.4. A very high fire severity site one month after the Gospers Mountain Fire was extinguished after a major rainfall event. (Photo: Aaron Greenville)

high or extreme fire severity. The first fire – Gospers Mountain Fire and the largest of the megafires (Nolan *et al.* 2021) – started in late October 2019 and was also the last fire to be extinguished after a major rainfall event in February 2020 (Fig. 4.4). The Gospers Mountain Fire is the largest single-ignition fire in modern records for the region – burning 512 000 ha over 107 days (NSW Government 2020; Nolan *et al.* 2021). As a result of the extensive nature and severity of the wildfires, the conservation outlook for the Greater Blue Mountains World Heritage Area was downgraded to 'Significant Concern' by the IUCN and the thresholds for optimal return-time of fire were exceeded, placing the region's vegetation into vulnerable–too frequently burnt ecological states (Osipova *et al.* 2020; Nolan *et al.* 2021).

Important sites for biodiversity: Key Biodiversity Areas

The 2019–20 wildfire impacted over 2 million ha across 69 KBAs (see Box 4.3). Of these, 18 KBAs had > 15% of their area burnt. Jerrawangala KBA – based on Jerrawangala National Park boundary – is located on the Great Eastern Escarpment of New South Wales. Approximately 99% of Jerrawangala KBA was impacted by the 2019–20 wildfires. The Gibraltar Range and Werrikimbe KBAs were also severely impacted the 2019–20 wildfire, with ~68% and 67% burnt, respectively (Fig. 4.5; Table 4.4).

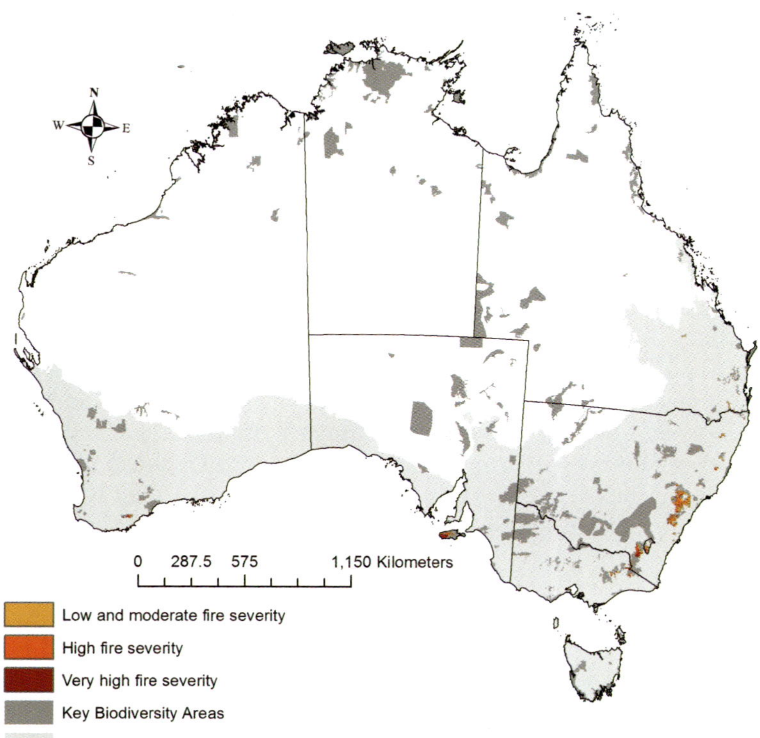

Fig. 4.5. Extent and severity of wildfires between July 2019 and February 2020 and impacts on KBAs. Yellow indicates low and moderate fire severity, orange indicates high fire severity, and red indicates very high fire severity. Dark grey shows all KBAs, while light grey shows the study region.

Table 4.4. Overview of the top 10 KBAs with the highest proportion of total extent and severity of the 2019–20 wildfires.

Name	Total area (ha)	Extent burnt – low and moderate fire severity (ha)	Extent burnt – high severity (ha)	Extent burnt – very high severity (ha)	Extent burnt – high and very high severity (ha)	Cumulative extent burnt (ha)	Proportion of total extent burnt (%)
Jerrawangala	4024	2100	1300	600	1900	4000	99
Gibraltar Range	36 563	10 000	12 600	2200	14 800	24 800	68
Werrikimbe	1 074 505	9000	1800	1700	14 500	23 500	67
Greater Blue Mountains	35 127	322 600	286 700	68 800	355 500	678 100	63
Nadgee to Mallacoota Inlet	38 171	11 200	5400	1300	6700	17 900	47
Fitzroy Falls and associated hydrobasin	12 395	1500	2000	1800	3800	5300	43
Palmgrove	25 188	9600	200	–	200	9800	39
Kangaroo Island	441 672	27 600	40 700	94 700	135 400	163 000	37
Stirling Range	112 580	7000	15 300	15 800	31 100	38 100	34
Ulladulla to Merimbula	138 357	34 400	22 200	6200	28 400	62 800	29

Box 4.3. Stirling Range KBA

Not only eastern states were affected by the 2019–20 wildfires – large areas of south-west Western Australia were also impacted. This is a region famous for its high levels of biodiversity and endemism (Beard *et al.* 2000). Stirling Range KBA is internationally recognised as important for the persistence of 15 bird species (DPaW 2014) and also contains 80 endemic plant species, 13 species of endemic trapdoor spiders, an isolated subpopulation of quokkas (*Setonix brachyurus*), and a threatened ecological community (Parks and Wildlife Service 2017) (Chapter 8). Over a third of Stirling Range KBA (37 826 ha) burnt in December 2019 during two blazes ignited by lightning strikes. Aided by drought and hot, windy weather, 37 826 ha of vegetation burnt with very high intensity, scorching the canopy and leaving few unburnt refuges (Fig. 4.6; Roff 2020). Impacts were compounded by a large fire having occurred the previous year, leaving a combined scar covering 43% of the KBA. While both Carnaby's (*Zanda latirostris*) and Baudin's black-cockatoo (*Z. baudinii*) feeding habitat in Stirling Range KBA were badly damaged by the 2019–20 wildfires, most of their known roosting and nesting habitat remained

Fig. 4.6. Stirling Range KBA 3 months post-fire (A) and one year post-fire (B) showing severely burnt areas are slow to recover. (Photos: Vicki Stokes)

intact (BirdLife Australia 2021), and breeding hollows were confirmed active in spring and summer 2020 (V. Stokes, *pers. comm.*). Other species did not get away so lightly. The fire sensitive wheatbelt western whipbird (*Psophodes nigrogularis oberon*), for which Stirling Range KBA is one of two remaining strongholds (McNee 1986; Birdlife International 2016), had ~30–40% of its habitat burnt in 2019, mostly with high severity. Record densities of calling whipbirds found in habitat adjacent to burnt areas suggests individuals immigrated from severely burnt areas, now uninhabitable (Stokes *et al.* 2021). Unfortunately, these refuges are unlikely to sustain whipbirds in the long term and burnt areas will take many years to regrow suitable whipbird breeding habitat.

Patterns of fire

We explored the patterns between forest integrity, measured using the global 'forest integrity' metric (Grantham *et al.* 2020) and 2019–20 wildfire severity for each protected area, World Heritage listing and KBA. The statistical analysis suggests that for every unit increase in forest integrity, the odds of an increase in fire severity decreased by 13.7% (CI: 13.5–13.8) in KBAs and 3.1% (CI: 3.0–3.2) in protected areas (Fig. 4.7A, C). More simply, the higher the forest integrity in a KBA or protected area, the less likely that area will experience very high severity fires.

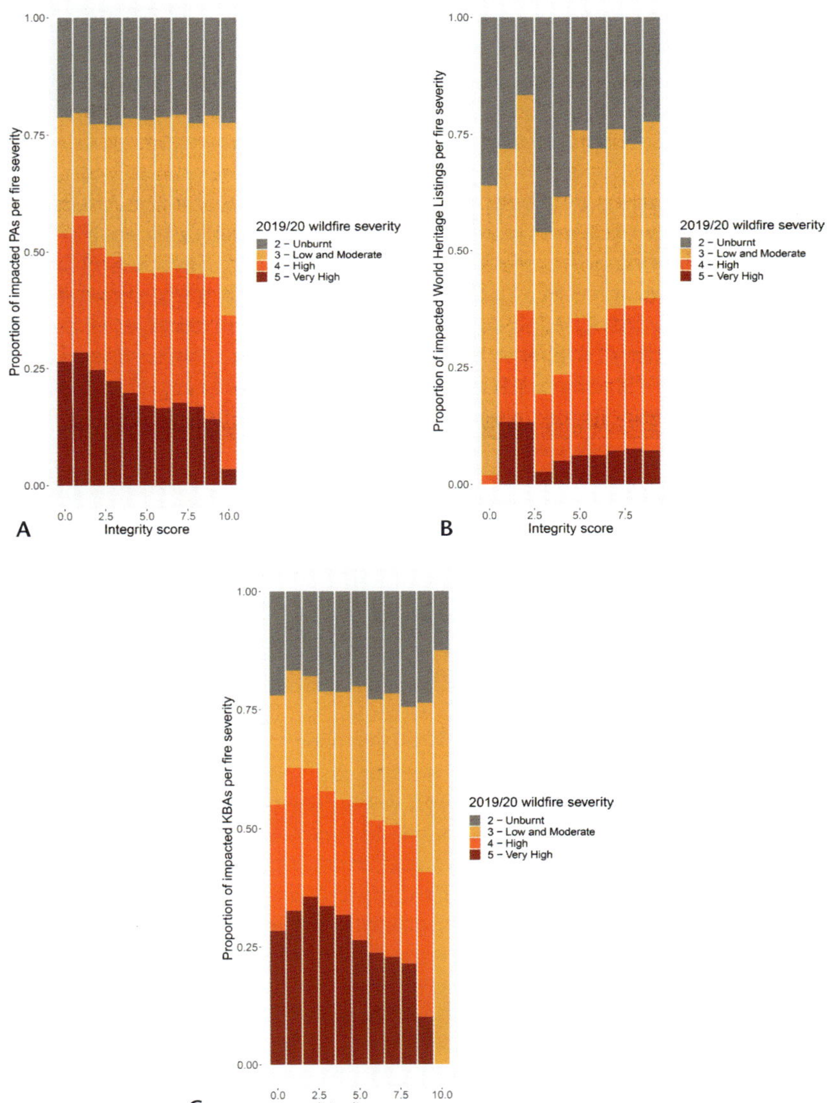

Fig. 4.7. Broad patterns between forest integrity and 2019–20 wildfire severity for each protected area (PA), World Heritage listing and KBA.

In contrast, for every unit increase in forest integrity within World Heritage listings the odds of an increase in fire severity increased by 8.4% (CI: 7.9–8.7) (Fig. 4.7B). A possible reason for the difference between protected area and KBA results, and World Heritage results is the small sample size of the World Heritage Listings relative to the size of the 2019–20 wildfires, meaning the results could be based on the particular characteristics of the six affected World Heritage areas. While many complex and interacting variables drive fire severity within the landscape, these initial exploratory patterns require further investigation, as they indicate that protected areas that are well managed and recovering from past degradation and fragmentation processes, and as a consequence have higher 'integrity', may be able to buffer megafire events like those experienced in 2019–20. In comparison, the fact that higher integrity places in World Heritage areas had higher odds of being more severely burnt raises concern for the resilience of these particular areas and the future value of such areas as ecological refugia from anthropogenic threats.

Conclusions

The 2019–20 wildfires had a devastating impact on many important Australian sites of high conservation value. Recovery for many of these places will require targeted effort and resources, which include actions to reduce the likelihood of future fire, as well as increase resilience in the face of other catastrophic climate events. However, as illustrated in the examples (Boxes 4.1 to 4.3), the achievement of such an objective will be challenging, given the large variety of threatening processes these sites face. The fire management objective is also unlikely to be successful unless the fundamental drivers of such fires, anthropogenic climate change and mismanagement of landscapes, are mitigated effectively. Without ambitious climate action and appropriate large-scale management, catastrophic fires will recur, and important conservation sites will continue to be diminished.

Recommendations

- Targeted management to support the recovery of all important fire-affected conservation sites must be resourced and implemented immediately, to prevent cumulative threats from worsening impacts.
- Long-term investment and commitment are needed to ensure full recovery of species, ecosystems and key ecological processes.
- The main objective of management should be to mitigate all known and likely threats to species and ecosystems within important conservation sites in a climate sensible way, to increase resilience in the face of future catastrophic climate events, as well as reduce the likelihood of future megafires.
- Investigation of the link between forest integrity and wildfire impacts, to test our preliminary evidence that higher integrity areas potentially buffer megafire events and act as species refugia under anthropogenic climate change, is of high importance. Such evidence would help allocate funding for new protected areas towards higher integrity areas that are refugial habitats of at-risk species. They also inform wider landscape scale forest management and forestry and agricultural policy when it comes to vegetation management.
- Long-term monitoring programs in important unburnt and fire-prone conservation sites that enable adaptive management are essential to ensure recovery of species and

ecosystems. These need to occur alongside protected area designation in those particularly important unprotected fire refugial habitats of at-risk species.
- Early warning systems and resourced response scenarios must include important conservation sites.
- Programs to ensure that species' habitat is available both on and off reserves for species that move over large areas are essential as this would ensure landscape-scale fire management is practised, rather than being restricted to local scale efforts.

References

Alliance for Zero Extinction (2018) Conserving the world's most threatened species, <https://zeroextinction.org/>.

Beard JS, Chapman AR, Gioia P (2000) Species richness and endemism in the Western Australian flora. *Journal of Biogeography* **27**, 1257–1268. doi:10.1046/j.1365-2699.2000.00509.x

BirdLife Australia (2021) KBA health check database, <https://portal.birdlife.org.au/kba-health-check/>.

Birdlife International (2016) *Psophodes nigrogularis*. doi:10.2305/IUCN.UK.2016-3.RLTS.T22705330A94013438.en.

Birdlife International (2021) Important bird and biodiversity areas: protecting birds where they live and migrate, <https://www.birdlife.org/projects/ibas-mapping-most-important-places/.

Bradstock RA, Cohn JS, Gill AM, Bedward M, Lucas C (2009) Prediction of the probability of large fires in the Sydney region of south-eastern Australia using fire weather. *International Journal of Wildland Fire* **18**, 932–943. doi:10.1071/WF08133

Commonwealth of Australia (2018) Australia's bioregions (IBRA), <https://www.environment.gov.au/land/nrs/science/ibra>.

Commonwealth of Australia (2020a) AUS GEEBAM Fire Severity NIAFED20200224. The Remote Sensing and Landscape Science Branch, Science Economics and Insights Division, New South Wales Department of Planning, Industry and Environment, <http://www.environment.gov.au/fed/catalog/search/resource/details.page?uuid=%7B8CE7D6BE-4A82-40D7-80BC-647CB1FE5C08%7D>.

Commonwealth of Australia (2020b) National Vegetation Information System (NVIS), <https://www.environment.gov.au/land/native-vegetation/national-vegetation-information-system>.

Commonwealth of Australia (2021a) Australia's protected areas. Department of Agriculture, Water and the Environment, <https://www.dcceew.gov.au/environment/land/nrs/about-nrs/australias-protected-areas>.

Commonwealth of Australia (2021b) Indigenous Protected Areas. Department of Agriculture, Water and the Environment, <https://www.dcceew.gov.au/environment/land/indigenous-protected-areas>.

DELWP (2020) 'Victoria's bushfire emergency: Biodiversity response and recovery (Preliminary report - Version 2)'. Melbourne, Victoria.

DPaW (2014) 'South Coast Threatened Birds Recovery Plan'. Department of Parks and Wildlife, Perth.

Ferraro PJ, Hanauer MM, Miteva DA, Nelson JL, Pattanayak SK, *et al.* (2015) Estimating the impacts of conservation on ecosystem services and poverty by integrating modeling and evaluation. *Proceedings of the National Academy of Sciences of the United States of America* **112**, 7420–7425. doi:10.1073/pnas.1406487112

Figgis P, Mackey B, Fitzsimons J, Irving J, Clarke P (2015) *Valuing Nature: Protected Areas and Ecosystem Services*. Australian Committee for IUCN, Sydney.

Geary WL, Buchan A, Allen T, Attard D, Bruce MJ, *et al.* (2022) Responding to the biodiversity impacts of a megafire: A case study from south-eastern Australia's Black Summer. *Diversity & Distributions* **28**, 463–478. doi:10.1111/ddi.13292

Grantham HS, Shapiro A, Bonfils D, Gond V, Goldman E, *et al.* (2020) Spatial priorities for conserving the most intact biodiverse forests within Central Africa. *Environmental Research Letters* **15**, 0940b5. doi:10.1088/1748-9326/ab9fae

Guisan A, Harrell FE (2000) Ordinal response regression models in ecology. *Journal of Vegetation Science* **11**, 617–626. doi:10.2307/3236568

Hammill K, Tasker EM (2010) *Vegetation, Fire and Climate Change in the Greater Blue Mountains World Heritage Area.* NSW Department of Environment, Climate Change and Water, Sydney.

IUCN (2016) '*A Global Standard for the Identification of Key Biodiversity Areas, Version 1.0. First edition*'. IUCN, Gland, Switzerland.

Kearney SG, Adams VM, Fuller RA, Possingham HP, Watson JEM (2020) Estimating the benefit of well-managed protected areas for threatened species conservation. *Oryx* **54**, 276–284. doi:10.1017/S0030605317001739

Kooyman RM, Watson J, Wilf P (2020) Gondwana World Heritage site burns. *Science* **367**, 1083. doi:10.1126/science.abb2046

Lindenmayer D, Thorn S, Banks S (2017) Please do not disturb ecosystems further. *Nature Ecology & Evolution* **1**, 31. doi:10.1038/s41559-016-0031

McNee S (1986) 'Surveys of the Western Whipbird and Western Bristlebird in Western Australia, 1985'. Royal Australasian Ornithologists Union (RAOU), Perth.

Nolan RH, Boer MM, Collins L, Resco de Dios V, Clarke H, *et al.* (2020) Causes and consequences of eastern Australia's 2019-20 season of mega-fires. *Global Change Biology* **26**, 1039–1041. doi:10.1111/gcb.14987

Nolan RH, Bowman DMJS, Clarke H, Haynes K, Ooi MKJ, *et al.* (2021) What do the Australian Black Summer fires signify for the Global Fire Crisis? *Fire (Basel, Switzerland)* **4**, 97. doi:10.3390/fire4040097

NSW Government (2020) 'Final Report of the NSW Bushfire Inquiry'. Government of NSW, Sydney.

Osipova E, Emslie-Smith M, Osti M, Murai M, Åberg U, *et al.* (2020) Greater Blue Mountains Area: 2020 conservation outlook assessment. In *IUCN World Heritage Outlook 3: A Conservation Assessment of all Natural World Heritage Sites.* (Ed. IUCN) Gland, Switzerland.

Parks and Wildlife Service (2017) Stirling Range National Park visitor guide, <https://parks.dpaw.wa.gov.au/sites/default/files/downloads/parks/20170430 StirlingRangeNP A3bro_Web.pdf>.

Potts T, Burdon D, Jackson E, Atkins J, Saunders J, *et al.* (2014) Do marine protected areas deliver flows of ecosystem services to support human welfare? *Marine Policy* **44**, 139–148. doi:10.1016/j.marpol.2013.08.011

Roff A (2020) *Australian Google Earth Engine Burnt Area Map: A Rapid, National Approach to Fire Severity.* doi:10.13140/RG.2.2.13434.52167

Selwood KE, Antos M, Bramwell M, Lee A, Lynch M, *et al.* (2021) Emergency conservation interventions during times of crisis: A case study for a threatened bird species in the Australian Black Summer bushfires. *Conservation Science and Practice* **4**, e606.

Stokes V, McNee S, Danks A, Comer S, Burbidge A (2021) 'Western Whipbird (Psophodes nigrogularis oberon) surveys in the Stirling Range region of WA 2020'. Report prepared for BirdLife Australia and the South Coast Threatened Birds Recovery Team, Perth.

Watson JEM, Evans MC, Carwardine J, Fuller RA, Joseph LN, *et al.* (2010) The capacity of Australia's protected-area system to represent threatened species. *Conservation Biology* **25**, 324–332. doi:10.1111/j.1523-1739.2010.01587.x

Watson JEM, Evans T, Venter O, Williams B, Tulloch A, *et al.* (2018) The exceptional value of intact forest ecosystems. *Nature Ecology & Evolution* **2**, 599–610. doi:10.1038/s41559-018-0490-x

5

Wildfire effects on soils and soil processes

David J. Eldridge, Jingyi Ding and Samantha K. Travers

Summary

- Fires influence soil physical, chemical and biological processes, but the long-term effects of intense, widespread fires such as those in 2019–20 on these soil processes is little known.
- Fires heat the soil surface, convert the protective layer of litter and vegetation to ash, and may make the surface water repellent. These processes have flow-on effects to soil chemical and biological processes, and run-off and erosion. The nature and magnitude of these effects will vary with soil type and moisture content, but also depend on fire intensity and landscape context.
- The magnitude of erosion varies with slope, soil type, geology, aspect, vegetation cover, fire intensity, fire connectivity and rainfall intensity. Erosion tends to be greater on ridgelines and in steep terrain, and average rates are in the order of 20–80 t sediment ha^{-1} or 47–77 t ha^{-1} for the 2019–20 fires. Recovery of eroded soils from megafires is likely to be very slow, given the extremely low rates of soil formation in Australia (~1 mm every century).
- The effects of large wildfires, such as the 2019–20 fire, on soil nutrient dynamics and subsequent plant growth are complex and uncertain. A single wildfire can substantially reduce soil carbon and nutrient pools, but plant and nutrient pools are likely to recover within a decade. Larger, high temperature wildfires, however, have more substantial effects. They deposit sizeable quantities of ash containing large amounts of phosphorus that have been stored in vegetation. Nitrogen (N) loss and added phosphorus (P) can reduce N:P ratios, with a risk of lasting nitrogen limitation. This retards long-term vegetation recovery, which may be offset by a pulse of recruitment of N-fixing plants such as *Acacia*. Full recovery of nutrient cycles and plant cover and diversity after the 2019–20 fires is likely to be slow in many severely burnt places.
- There is also strong feedback between the recovery of plants and soils and that of soil fauna such as ants and termites that promote water flow through the soil and help to reduce the run-off of water. The extent to which wildfires impact soil-dwelling microbes and arthropods, and then how such impacts on

invertebrate communities will influence soil recovery processes such as litter decomposition and nutrient mineralisation, is poorly resolved.

- There are critical knowledge gaps about the impacts of severe fires, and of fire regimes, on soils, and the broader ecological ramifications of such impacts; and such knowledge gaps severely constrain our ability to manage or mitigate such impacts. Accordingly, our recommendations highlight priority research needs.

Introduction

> *Fire, in result, and an example of such result is erosion, affects each one of us, whether we are of the towns or of the country. The common weal is most grievously threatened by erosion. (Stretton 1939)*

Wildfire is a major ecological process worldwide and an important driver of above-ground processes. Less well understood, however, is how fire affects soils and below-ground processes. Soils support a range of critical ecosystem services. They absorb water, store and release nutrients, filter pollutants, provide a substrate for plants and habitat for biota that drive critical nutrient processes, and are a source of sediment, organic matter and nutrients (DeBano *et al.* 1998).

The 2019–20 wildfires in eastern and southern Australia cast a light on the ecological effects of large-scale wildfires, with significant losses in biodiversity (see Chapters 4–16), plant production, forestry assets and property (see Chapter 2). While the immediate effects of fire on livelihoods and human capital have been devastating, fire is a natural event, and many plant species need fire to initiate germination (Hodges *et al.* 2019; Chapter 9). Yet it is known that megafires, such as the 2019–20 wildfires, are devastating for soils and soil processes, with huge losses of sediment, water, carbon and nutrients, and reductions in soil structural integrity. These changes have substantial flow-on effects to a range of biota, processes and functions at varying spatial scales.

In this chapter we discuss how wildfire affects those soil processes that are associated with the consumption of vegetation, the modification of soil surface structure, and changes in soil carbon and nutrient cycles. These factors alter the way that water moves through the landscape and increase the likelihood of catchment-wide sediment removal. We draw on Australian and international literature, with a general focus on large-scale fires to understand how the 2019–20 wildfires impacted Australian soils. Finally, we recognise critical knowledge gaps and make recommendations to remedy those deficiencies.

Physical processes: heating, ash deposition and repellancy

Fire heats the soil surface

The magnitude of soil surface heating by fire determines the extent to which the major soil components (e.g. nutrients, ions, organic matter) will change during fires. This effect of heating depends on the conductivity of the soil, bulk density (how densely packed the soil particles are) and the moisture content, as well as fuel loads and fire characteristics such as intensity and duration (DeBano *et al.* 1998). Overall, sandy soils are more susceptible to fire than clay soils, not only because they are more conductive (i.e. fire can penetrate deeper into sands), but also because they contain less organic matter, often occur on steep land and are more poorly structured, and therefore erode more easily.

Soil surface temperatures range from ~100°C during mild fire, to 800°C in moderate to high-intensity fires, to 1500°C during extreme fire events, such as many of the 2019–20 wildfires (DeBano *et al.* 1998). While a low to moderate fire will leave some litter and organic matter on the surface, high to intense fires can cause total canopy loss, and the complete combustion of litter and soil organic matter to several centimetres depth. These impacts have the potential to physically alter soil properties that influence infiltration, runoff and erodibility of soil (Fig. 5.1).

In general, organic matter loss commences at temperatures exceeding 100°C. The temperature at which soil constituents (e.g. nutrients and cations) are vaporised depends on their temperature tolerances. Carbon, sulphur and nitrate are volatilised by 200°C, and by ~450–500°C almost all organic matter has been combusted (De Bano *et al.* 1998). Substantial soil physical and chemical changes occur once the surface reaches temperatures of 400–600°C. Most temperature effects occur within a few centimetres of the surface, with values rarely exceeding 150°C at depths of 50 mm (DeBano *et al.* 1998).

Soil temperatures also affect biological processes. Low-intensity fires (e.g. hazard reduction burns) may fail to heat the surface above 40°C at 50 mm depth, thereby failing to break seed dormancy for the germination for some plants (Penman and Towerton 2008). High-intensity fires are therefore critical for the germination of many plant species (Chapter 9).

Ash deposition on the fire ground

Ash comprises charred, loose material < 1 cm in size and is formed from the complete oxidation of vegetation and soil organic matter. Ash loads increase with fire intensity, with values of 1 to 35 t ha^{-1} recorded for low to severe fires (respectively) in Australia

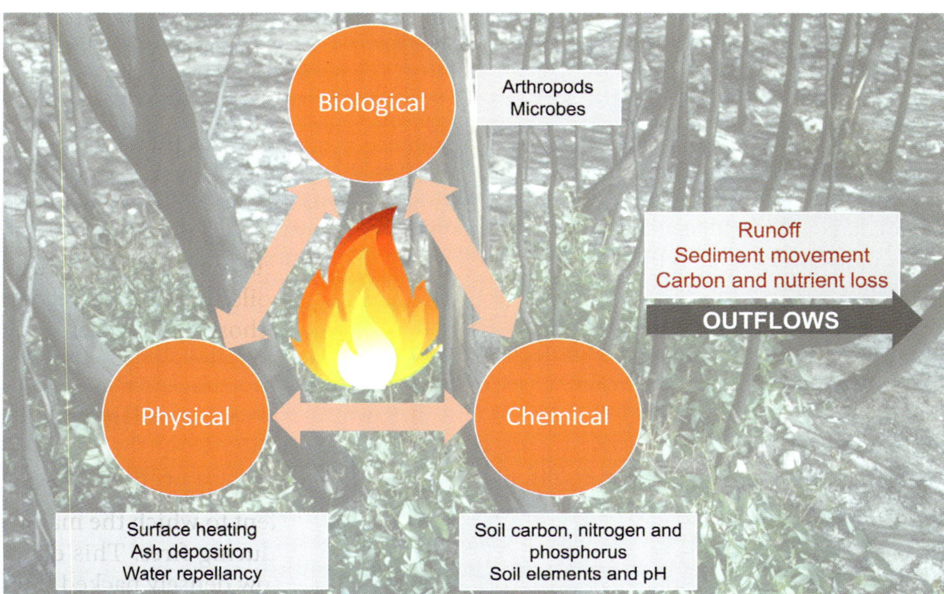

Fig. 5.1. Conceptual model of the direct effects of fire on biological, physical and chemical properties and processes, and resultant outflows.

(Santin *et al.* 2014). The chemical composition of ash also varies with fire severity. Low-intensity fires generate ash of finely combusted litter and vegetation, whereas extreme fires generate mineral-rich ash (Santin *et al.* 2014). Ash increases the bioavailability of nutrients, particularly nitrates and phosphorus ('ash bed effect', Chambers and Attiwill 1994), releases cations, increases soil pH, and can alter soil composition for several years after fire. Reacidification of burnt soils may take half a century to reach pre-burn soil pH levels (Khanna *et al.* 1996).

Soil surfaces may repel water

The soil beneath Australian vegetation, particularly eucalypts, is generally hydrophobic (non-wetting) due to high concentrations of phenols and oils from decomposing litter, and a proliferation of fungal hyphae (Howell *et al.* 2006). The effects of fire on the wettability of soils are complex; low-intensity fires tend to activate or induce hydrophobicity, which can persist for many months until invertebrates (ants, termites and worms) reactivate their burrows and alter the distribution of hydrophobic hyphae and oils. Hydrophobicity is likely to affect post-fire seedling establishment (Bailey *et al.* 2015), by affecting moisture around seeds. Under moderate or intense fires (above ~400°C) this repellency tends to be reduced (Cawson *et al.* 2016), promoting moisture accumulation and germination (Bailey *et al.* 2015).

Chemical processes: nutrients, carbon and elements

Fire alters soil biogeochemical cycles (Adams *et al.* 2020), and nutrient balances and functions via multiple pathways of losses and inputs (Fig. 5.1; Bowd *et al.* 2019). Fire can reduce soil nutrient pools through volatilization (e.g. conversion of urea to ammonia, which is lost from the soil), leaching (where soluble nutrients are removed from the soil by water), extensive plant uptake as vegetation recovers, and post-fire soil erosion (Bowd *et al.* 2019). Soil carbon and nutrient loss is greatest i) at the surface (Hobley *et al.* 2019), ii) from labile (available or organic) pools, which are susceptible to heating (Butler *et al.* 2018), and iii) at sites that are frequently burnt or under greater fire intensity (Toberman *et al.* 2014).

Elemental responses to fire are varied

Fire reduces elements such as Na, K, Mg, and S through volatilization and consumption of plant tissue (Ludwig *et al.* 1998) but increases others such as Cu, Mn, and Zn through ash deposition (Abraham *et al.* 2018). The greatest losses occur when soil organic carbon is converted to CO_2. For example, 27 t ha^{-1} of carbon were lost from the top 10 cm of the soil during the 2013 wildfire in the Warrumbungles National Park (Tulau *et al.* 2019). These rates are comparable to those values lost during the 2019–20 fires. These soil organic carbon losses may be compensated by short-term increases in ash loads, as shown by studies in eucalypt forests in eastern Australia (1.55 t ha^{-1} C loss cf. 1.67 t ha^{-1} in ash; Krishnaraj *et al.* 2016) and long-term increases through plant regeneration. However, repeated fires are likely to reduce recovery of soil organic carbon (Hobley *et al.* 2019).

Fire typically reduces soil nitrogen pools when ammonia is lost to the atmosphere. Rapid regrowth after fires also increases the demand for nitrogen as microbes take up (immobilise) most of the nitrogen to help them break down organic matter (Chambers and Attiwill 1994). Post-fire systems therefore tend to become nitrogen limited. However, fire releases phosphorus, which is derived from soil parent material, and released when litter and plants combust and are returned as ash (Butler *et al.* 2018). Levels of released phosphorus can be an order of magnitude greater than that in long-unburnt sites (Blake *et al.* 2020). With more frequent and/or intense fires, phosphorus increases, and nitrogen declines, leading to reductions in the N:P ratio and therefore nutrient imbalance. Reductions in the N:P ratio can have substantial effects on vegetation composition and diversity, for example, the easing of phosphorus limitations or increasing nitrogen limitations, as well as on microbial biomass and the abundance of some invertebrates (Butler *et al.* 2017).

Fire, however, can also stimulate the germination of seeds of plants such as *Acacia, Casuarina* and leguminous species that fix nitrogen (Adams and Attiwill 1984), leading to a recovery of nitrogen. The rapid recovery of *Acacias* after the intense Warrumbungle fires (Gordon *et al.* 2017) suggests that a similar response from N-fixing plants is likely within a few years of the 2019–20 wildfires.

Outflows: runoff and sediment

Though the impacts of fires on soil is complex, it is likely that the 2019–20 fires had a major impact on landscape stability. Estimates of sediment removal during the 2019–20 wildfires suggest erosion rates ranging from 47 to 77 t ha^{-1} across 5.37 million ha of burn area in New South Wales alone (DPIE 2020), which would represent a phenomenal displacement of 47 million t of soil, after accounting for background sediment movement. The magnitude of erosion depends on terrain (slope, soil type, geology, aspect), remaining vegetation cover, fire intensity and connectivity, and rainfall intensity (Yang *et al.* 2020), but tends to be greatest on steep-sided gullies and ridgelines where fire intensity is greatest (Nyman *et al.* 2014; Tulau *et al.* 2019; Blake *et al.* 2020). Apart from total erosion loads, event-based erosivity estimated using satellite and radar following these mega fires was estimated at ~5 t ha^{-1} $month^{-1}$, which is ~30 times higher than the pre-fire erosion rates, largely due to intense rainstorms following fire (Yang *et al.* 2020). Furthermore, the magnitude and extent of sediment loss may also increase with the intensity of mechanical disturbance after fire, which in itself creates considerable runoff (Smith *et al.* 2011).

These rates are similar to those from other comparable fire events across temperate areas in eastern Australia. For example, sediment removal following wildfire in the Warrumbungle National Park in 2013 was estimated to be ~25 t ha^{-1} across an area of ~56 000 ha, compared with background pre-fire levels of ~1 t ha^{-1} (Tulau *et al.* 2019; Fig. 5.2). These estimates are consistent with those resulting from subalpine woodland and forest fires in the Snowy Mountains (3–95 t ha^{-1}; Smith and Dragovich 2008), the 1983 Sydney fires (28–64 t ha^{-1}; Atkinson 2012), and 2013 Balmoral fires in Victoria (34 t ha^{-1}; Santin *et al.* 2015). Overall, erosion rates following wildfire can be three orders of magnitude greater than background rates (Langhans *et al.* 2016).

Effects on biodiversity

The impacts of fire on soils have many consequences for biodiversity, with these effects influenced by fire severity and environmental settings. Increased run-off and deposition of soils and ash in aquatic systems after fire leads to major impacts on the health and biodiversity of aquatic systems, that may extend many kilometres downstream of direct fire impacts (Chapter 6), and in some cases to estuarine and coastal marine systems (Chapter 7).

Less well considered are the impacts on the soil biological communities, and the subsequent feedback effects on recovery of ecological structure, function and composition

Fig. 5.2. Surface of the soil after the 2013 Warrumbungles fire showing the complete removal of the surface soil layers by fire and exposure of the B horizon. (Photo: Sally McInnes-Clarke)

(Chapter 11). For instance, soil arthropods and microbes process organic matter and release carbon and nitrogen. Like many faunal groups (Chapters 13–16), arthropods vary greatly in their susceptibility to fire, and the magnitude of impacts of fire on invertebrate communities is influenced by fire severity and fire regimes, and environmental settings. For example, collembola may take many years to return to pre-fire levels (Driessen and Greenslade 2004), affecting the ability of soils to process organic matter. Fire can suppress litter processing taxa such as cockroaches (Arnold *et al.* 2017) and alter predator–prey relationships among litter-dwelling invertebrates (Dawes-Gromadzki 2007).

Recovery of hydrological function and surface stability generally occurs 18–36 months after fire, depending on vegetation regrowth, reductions in soil repellency and reduced soil erodibility (Smith *et al.* 2011). However, recovery of infiltration capacity may depend on the reinstatement of animal-created pores and burrows of ants and spiders, and even the passages around plant roots (macropores; Holden *et al.* 2014), which drive hydraulic conductivity.

The effect of fire on invertebrates also varies widely among ecosystems, and with fire severity and fire regimes. For example, few differences were found in collembolan communities 12–24 years after fire in a Tasmanian buttongrass community (Driessen and Greenslade 2004) whereas ants were more abundant in burnt sites in arid mallee woodlands (Kwok and Eldridge 2015), suggesting a rapid response to fire in this ecosystem. In contrast, the impact of even a single wildfire may still be evident in some invertebrate communities for at least 50 years (Henry *et al.* 2022). Though the 2019–20 occurred predominantly across forested systems, there is a real opportunity to explore the impacts of fires on the invertebrate communities from different vegetation types, and attempt to understand the subsequent implications for recovery of these ecosystems.

The implications of the fires on the soil microbial community, and the implications for recovery of systems, remain poorly understood. Though soil type, moisture content and temperature determine the response of soil microbes to fire (Chambers and Attiwill 1994), fire reduces soil microbial biomass and therefore soil respiration, even at low temperatures < 120°C, and this effect is greatest close to the soil surface (DeBano *et al.* 1998). Soil heating and reduced soil acidity from ash can alter competition between microbial autotrophs and heterotrophs, thereby promoting soil nitrification (Bauhus *et al.* 1993). Microbial changes may limit vascular plant recovery as fire can destroy specialist microbes that affect seed germination (Senior *et al.* 2018). However, microbial biomass can recover relatively quickly within a few years of fire (Liu *et al.* 2015), but not under frequent or high-intensity fires, which can suppress soil respiration and reduce soil organic carbon pools (Liu *et al.* 2015).

Conclusions

In the report from the Royal Commission into the 1939 fires, Stretton wrote at length on the serious consequences of resultant soil erosion in relation to infrastructure, water quality and livelihoods: 'The damage to river banks causes encroachments upon and destruction of areas otherwise usable for agricultural and pastoral purposes. Erosion generally is a constant crippling enemy to water supply, whether for the big cities or for the farmer who draws upon an irrigation scheme for the nourishment of his crops and pastures.' Ecological impacts of such severe fires are also likely to be profound. Soil is the foundation of myriad ecological processes, and hence the influence of wildfires on soils is likely to have many long-lasting ramifications across ecosystems. However, in terms of biodiversity and ecosystem health, knowledge of impacts, and feasible and effective pre- and post-fire management solutions, is limited, and there is a pressing need to address critical knowledge gaps (see 'Recommendations' below).

The 2019–20 wildfires will have long-lasting and substantial legacy effects on soils, and hence on the many other environmental components influenced by or associated with soils. Aquatic systems may take many years to recover from the high loads of sediment deposited from burnt areas. Source areas where soil biological activity is low and where soils are shallow will take many thousands of years to recover. This is because natural rates of soil formation in Australia are extremely low (1 mm per century; Stockmann *et al.* 2014). This recovery is likely to become increasingly tenuous with more frequent and intense fires predicted over the next century.

Invertebrates and microbes are critical to ecosystem function, and the feedbacks between fire, soils and ecosystem recovery remain poorly understood (see Chapters 10 and 11). It is critical to try to better understand not only the impacts and outcomes of high severity fires such as those of 2019–20, but also the impacts of multiple fires, and compounding threats such as drought. Only then perhaps can we expect to have a clear picture of the plant and animal communities that we might expect to dominate, or persist in, our landscape into the future.

Recommendations

Our review indicates that there will be many detrimental and diverse impacts of the 2019–20 wildfires on soils, and hence on the health, productivity and composition of environments.

However, this review also recognises that many critical knowledge gaps constrain our ability to evaluate the magnitude and duration of such impacts and, even more so, on how such impacts can be managed or mitigated. Such knowledge will become increasingly important for a future in which such megafires are likely to become more prevalent. Accordingly, our recommendations focus mostly on the need to resolve these key knowledge gaps in order to better understand and manage the impacts of megafires such as the 2019–20 wildfires, and fire regimes, on structure, function and composition of future landscapes:

- Remote sensing has provided good understanding of the extent of the 2019–20 fires (e.g. DPIE 2020, Yang *et al.* 2020), but we have poor understanding of the extent and magnitude of impacts of large-scale fires on soil structure, function and composition. Most of our understanding of ecological effects of fire is based on more localised or lower intensity events. There is priority need to build the evidence base from the 2019–20 fires, including to identify those soil and landscape attributes that most influence the levels of erosion and sediment deposition.
- There is a priority need to better understand the responses of soils to fire regimes. Legacy effects of previous fires are likely to drive future fire impacts on soils, but the interactions between previous and future fires is poorly known.
- More research is needed to improve knowledge of the linkages between fire effects on soils, and flow-on effects to other parts of the ecosystem, such as the changes in structure, function and composition of plants and animal communities in terrestrial, freshwater and marine systems.
- Given their importance for the recovery of 'soil health', more research is needed to document and monitor the rate and characteristics of post-fire changes in soil fauna communities.
- Relatively little is known about how we can improve the ecological resilience of soils to cope with the eventual next wave of high-intensity wildfire. Current evidence suggests that recovery of soil function following high-intensity wildfire will be protracted, but techniques to mitigate the effects of fires on soils are not known. Hence, there is a priority need to enhance the evidence base needed to underpin the design and implementation of such mitigation and management techniques.

References

Abraham J, Dowling K, Florentine S (2018) Influence of controlled burning on the mobility and temporal variations of potentially toxic metals (PTMs) in the soils of a legacy gold mine site in Central Victoria, Australia. *Geoderma* **331**, 1–14. doi:10.1016/j.geoderma.2018.06.010

Adams MA, Attiwill PM (1984) Role of *Acacia* spp. in nutrient balance and cycling in regenerating *Eucalyptus regnans* F. Muell. forests. I. Temporal changes in biomass and nutrient content. *Australian Journal of Botany* **32**, 205–215. doi:10.1071/BT9840205

Adams MA, Shadmanroodposhti M, Neumann M (2020) Causes and consequences of eastern Australia's 2019–20 season of mega-fires: a broader perspective. *Global Change Biology* **26**, 3756–3758. doi:10.1111/gcb.15125

Arnold KT, Murphy NP, Gibb H (2017) Post-fire recovery of litter detritivores is limited by distance from burn edge. *Austral Ecology* **42**, 94–102. doi:10.1111/aec.12404

Atkinson GA (2012) Soil erosion following wildfire in Royal National Park, NSW. *Proceedings of the Linnean Society of New South Wales* **134**, 1–13.

Bailey TG, Davidson NJ, Close DC (2015) Patterns of soil water repellency in response to coarse woody debris and fire: implications for eucalypt regeneration in dry forests. *Plant and Soil* **397**, 93–102. doi:10.1007/s11104-015-2605-3

Bauhus J, Khanna PK, Raison RJ (1993) The effect of fire on carbon and nitrogen mineralization and nitrification in an Australian forest soil. *Australian Journal of Soil Research* **31**, 621–639. doi:10.1071/SR9930621

Blake D, Nyman P, Nice H, D'Souza FML, Kavazos CRJ, *et al.* (2020) Assessment of post-wildfire erosion risk and effects on water quality in south-western Australia. *International Journal of Wildland Fire* **29**, 240–257. doi:10.1071/WF18123

Bowd EJ, Banks SC, Strong CL, Lindenmayer DB (2019) Long-term impacts of wildfire and logging on forest soils. *Nature Geoscience* **12**, 113–118. doi:10.1038/s41561-018-0294-2

Butler OM, Lewis T, Chen CR (2017) Fire alters soil labile stoichiometry and litter nutrients in Australian eucalypt forests. *International Journal of Wildland Fire* **26**, 783–788. doi:10.1071/WF17072

Butler OM, Ekser JJ, Lewis T, Mackey B, Chen C (2018) The phosphorus-rich signature of fire in the soil–plant system: a global meta-analysis. *Ecology Letters* **21**, 335–344. doi:10.1111/ele.12896

Cawson JG, Nyman P, Smith HG, Lane PNJ, Sheridan GJ (2016) How soil temperatures during prescribed burning affect soil water repellency, infiltration and erosion. *Geoderma* **278**, 12–22. doi:10.1016/j.geoderma.2016.05.002

Chambers DP, Attiwill PM (1994) The ash-bed effect in *Eucalyptus regnans* forest: chemical, physical and microbiological changes in soil after heating and partial sterilization. *Australian Journal of Botany* **42**, 739–749. doi:10.1071/BT9940739

Dawes-Gromadzki TZ (2007) Short-term effects of low intensity fire on soil macroinvertebrate assemblages in different vegetation patch types in an Australian tropical savanna. *Austral Ecology* **32**, 663–668. doi:10.1111/j.1442-9993.2007.01752.x

DeBano LF, Neary DG, Ffolliott PF (1998) *Fire's Effects on Ecosystems.* John Wiley & Sons, New York.

DPIE (2020) *NSW Fire and the Environment 2019–20 Summary. Biodiversity and Landscape Data and Analyses to Understand the Effects of the Fire Events.* NSW Department of Planning, Industry and Environment, Sydney.

Driessen MM, Greenslade P (2004) Effect of season, location and fire on Collembola communities in buttongrass moorlands, Tasmania. *Pedobiologia* **48**, 631–642. doi:10.1016/j.pedobi.2004.08.002

Gordon CE, Price OF, Tasker EM, Denham AJ (2017) Acacia shrubs respond positively to high severity wildfire: Implications for conservation and fuel hazard management. *The Science of the Total Environment* **575**, 858–868. doi:10.1016/j.scitotenv.2016.09.129

Henry SC, Kirkpatrick JB, McQuillan PB (2022) The half century impact of fire on invertebrates in fire-sensitive vegetation. *Austral Ecology* **47**, 590–602. doi:10.1111/aec.13141

Hobley EU, Zoor LC, Shrestha HR, Bennett LT, Weston CJ, *et al.* (2019) Prescribed fire affects the concentration and aromaticity of soluble soil organic matter in forest soils. *Geoderma* **341**, 138–147. doi:10.1016/j.geoderma.2019.01.035

Hodges JA, Price JN, Nimmo DG, Guja LK (2019) Evidence for direct effects of fire-cues on germination of some perennial forbs common in grassy ecosystems. *Austral Ecology* **44**, 1271–1284. doi:10.1111/aec.12806

Holden J, Wearing C, Palmer S, Jackson B, Johnston K, *et al.* (2014) Fire decreases near-surface hydraulic conductivity and macropore flow in blanket peat. *Hydrological Processes* **28**, 2868–2876. doi:10.1002/hyp.9875

Howell J, Humphreys GS, Mitchell PB (2006) Changes in soil water repellence and its distribution in relation to surface microtopographic units after a low severity fire in eucalypt woodland, Sydney, Australia. *Australian Journal of Soil Research* **44**, 205–217. doi:10.1071/SR04176

Khanna PK, Ludwig B, Raison RJ (1996) Comparing modelled and observed effects of ash additions on chemistry of a highly acid soil. *Australian Journal of Soil Research* **34**, 999–1013. doi:10.1071/SR9960999

Krishnaraj SJ, Baker TG, Polglase PJ, Volkova L, Weston CJ (2016) Prescribed fire increases pyrogenic carbon in litter and surface soil in lowland *Eucalyptus* forests of south-eastern Australia. *Forest Ecology and Management* **366**, 98–105. doi:10.1016/j.foreco.2016.01.038

Kwok ABC, Eldridge DJ (2015) Does fire affect the ground-dwelling arthropod community through changes to fine-scale resource patches? *International Journal of Wildland Fire* **24**, 550–559. doi:10.1071/WF14088

Langhans C, Smith HG, Chong DMO, Nyman P, Lane PNJ, *et al.* (2016) A model for assessing water quality risk in catchments prone to wildfire. *Journal of Hydrology* **534**, 407–426. doi:10.1016/j.jhydrol.2015.12.048

Liu X, Chen CR, Wang WJ, Hughes JM, Lewis T, *et al.* (2015) Vertical distribution of soil denitrifying communities in a wet sclerophyll forest under long-term repeated burning. *Microbial Ecology* **70**, 993–1003. doi:10.1007/s00248-015-0639-y

Ludwig B, Khanna PK, Raison RJ, Jacobsen KL (1998) Modelling cation composition of soil extracts under ashbeds following an intense slashfire in a eucalypt forest. *Forest Ecology and Management* **103**, 9–20. doi:10.1016/S0378-1127(97)00173-4

Nyman P, Sheridan GJ, Smith HG, Lane PNJ (2014) Modeling the effects of surface storage, macropore flow and water repellancy on infiltration after wildfire. *Journal of Hydrology* **513**, 301–313. doi:10.1016/j.jhydrol.2014.02.044

Penman TD, Towerton AL (2008) Soil temperatures during autumn prescribed burning: implications for the germination of fire responsive species? *International Journal of Wildland Fire* **17**, 572–578. doi:10.1071/WF07092

Santin C, Doerr SH, Chafer CJ (2014) Ash deposition during wildfire and its threat to water quality. In *Advances in Forest Fire Research*. (Ed. DX Viegas) pp. 1658–1660. Imprensa da Universidade de Coimbra, Coimbra, Portugal.

Santin C, Doerr SH, Otero XL, Chafer CJ (2015) Quantity, composition and water contamination potential of ash produced under different wildfire severities. *Environmental Research* **142**, 297–308. doi:10.1016/j.envres.2015.06.041

Senior JK, O'Reilly-Wapstra JM, Schweitzer JA, Bailey JK, Potts BM (2018) Forest fire may disrupt plant-microbial feedbacks. *Plant Ecology* **219**, 497–504. doi:10.1007/s11258-018-0811-9

Smith HG, Dragovich D (2008) Post-fire hillslope erosion response in a sub-alpine environment, south-eastern Australia. *Catena* **73**, 274–285. doi:10.1016/j.catena.2007.11.003

Smith HG, Sheridan GJ, Lane PNJ, Bren LJ (2011) Wildfire and salvage harvesting effects on runoff generation and sediment exports from radiata pine and eucalypt forest catchments, south-eastern Australia. *Forest Ecology and Management* **261**, 570–581. doi:10.1016/j.foreco.2010.11.009

Stockmann U, Minasny B, McBratney AB (2014) How fast does soil grow? *Geoderma* **216**, 48–61. doi:10.1016/j.geoderma.2013.10.007

Stretton LEB (1939) 'Report of the Royal Commission into the Causes of and Measures Taken to Prevent the Bush Fires of January, 1939'. Government Printer, Melbourne.

Toberman H, Chen CR, Lewis T, Elser JJ (2014) High-frequency fire alters C:N:P stoichiometry in forest litter. *Global Change Biology* **20**, 2321–2331. doi:10.1111/gcb.12432

Tulau MJ, McInnes-Clarke SK, Yang XH, McAlpine RA, Karunaratne SB, *et al.* (2019) The Warrumbungle Post-Fire Recovery Project – raising the profile of soils. *Soil Use and Management* **35**, 63–74. doi:10.1111/sum.12467

Yang XH, Zhang MX, Oliveira L, Ollivier QR, Faulkner S, *et al.* (2020) Rapid assessment of hillslope erosion risk after the 2019–2020 wildfires and storm events in Sydney drinking water catchment. *Remote Sensing* **12**, 3805. doi:10.3390/rs12223805

6

The impact of the 2019–20 Australian wildfires on aquatic systems

Nick S. Whiterod, Mark Lintermans, Rebecca L. Cramp, Craig E. Franklin, Mark J. Kennard, Rob McCormack, Luke Pearce, Tarmo A. Raadik, Michelle Ward and Sylvia Zukowski

Summary

Context and challenges

- Wildfires pose profound, but often underappreciated, threats to aquatic ecosystems and species.
- The complex and dynamic impacts are predominantly realised with run-off events that operate at greater temporal (i.e. months to years) and spatial (i.e. tens of kilometres) scale than that imposed by the wildfire footprint.
- Freshwater fish and spiny crayfish (and other aquatic species) are typically understudied and receive little management attention, hindering the ability to understand impacts and recovery.

Main findings

- The studies summarised in this chapter reveal mortality of individuals (tens to thousands) from at least 32 aquatic species, with these impacts as far as 70 km downstream of fire-affected areas
- Spatial analyses predicted that fire-related impacts overlapped substantially with the distributions of 21 freshwater fish species and 21 spiny crayfish species; the entire known range of nine species was estimated to be affected.
- The population status of impacted species was often predicted to be worse 1-year post-fire than 1-week post-fire, because fire impacts can occur well after the fire event.
- For many species, recovery over 10 years or three generations was predicted to be slow, depending on pre-fire status and intrinsic traits of species.
- Emergency conservation efforts were initiated for at least 24 aquatic species across three jurisdictions and involved rescue, *ex situ* captive maintenance and later return to wild.

- Actions initiated and knowledge gained in response to the 2019–20 wildfires helped alleviate some impacts; future actions should attempt to build resistance/resilience, improve knowledge, escalate legislative protection and coordinate management in the face of more frequent and intense wildfires.

Introduction

While the impacts of wildfires on terrestrial ecosystems are obvious, the impacts to aquatic ecosystems are less visible but can be profound and more enduring. Impacts to aquatic ecosystems can extend beyond the fire footprint, tens of kilometres downstream. Wildfires can cause mass mortality and the loss of populations but also sublethal changes to individuals, and long-lasting alteration of habitats.

Conditions leading up to the 2019–20 Australian wildfires included a prolonged 3-year drought (see Chapter 2), which stressed many aquatic ecosystems and exacerbated the subsequent impacts of wildfires. The fire footprint overlapped with many aquatic ecosystems, including upland streams and bogs, subtropical rainforest streams, lowland rivers, and streams, lakes and swamps in the sandy coastal heathlands (wallum) (Silva *et al.* 2020; Shelley *et al.* 2021). Post-fire, some burnt areas experienced higher than average rainfall, or localised but high-intensity convective storm cells, leading to severe run-off, erosion, and high streamflow, magnifying the impacts of the wildfires on these aquatic ecosystems. The consequences for aquatic fauna can be profound; direct impacts were documented for some species, while ongoing impacts are predicted for many more. The full extent and severity of these impacts, which continue to be experienced, may never be fully appreciated.

In this chapter, we summarise the effects of the 2019–20 Australian wildfires on aquatic fauna. We focus on freshwater fish and, as prominent and threatened representatives of Australian aquatic ecosystems (Furse and Coughran 2011), crayfish of the Australian endemic genus *Euastacus* ('spiny crayfish'). Other invertebrates and frogs are covered elsewhere in this book (Chapters 11 and 13 respectively). We first describe what is generally known about wildfire impacts on aquatic ecosystems, freshwater fish and spiny crayfish; we then outline the impacts of the 2019–20 wildfires, including assessments to identify species at most risk; and we document the emergency conservation efforts that took place in response to these wildfires. Finally, we summarise key lessons for improving responses to future wildfires, and improving conservation management of aquatic ecosystems and fauna.

The content presented here consolidates the efforts of many people, whose work is captured in published and unpublished sources, and aims to provide up-to-date information to understand and guide future responses when wildfires affect aquatic fauna.

How do fires impact aquatic ecosystems and species?

Wildfires are complex disturbances for aquatic ecosystems and species (Fig. 6.1). The nature and severity of impacts are context-dependent, varying with the wildfire severity as well as the characteristics and health of the aquatic ecosystem. In rivers and streams, catchment topography (i.e. slope), geology, soil porosity and degree of water repellence, vegetation condition and cover, and the frequency and intensity of subsequent storm run-off events, all influence the effect of wildfires (Legge *et al.* in press). In lake and wetland systems, the characteristics and extent of riparian and emergent vegetation cover, as well as waterbody depth, influence the severity of fire impacts on resident aquatic

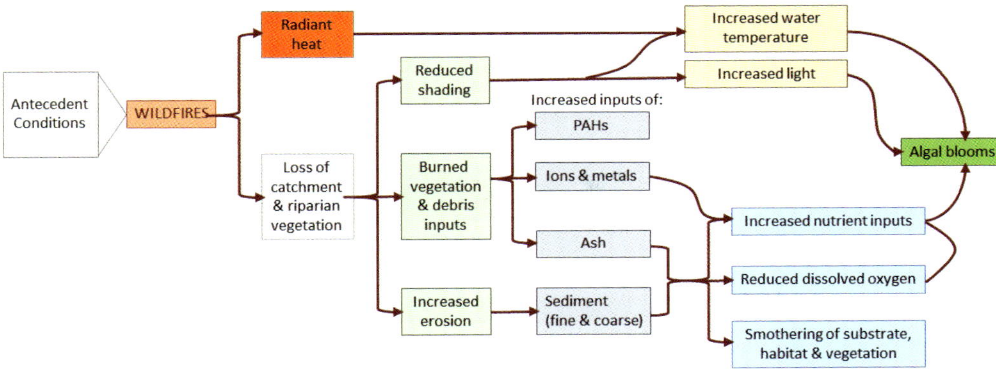

Fig. 6.1. Drivers of, and responses to, wildfires in aquatic ecosystems.

species. Antecedent conditions, such as the drought that preceded the 2019–20 wildfires, and other stressors, including habitat destruction and flow regulation, will also influence wildfire impacts. The nature of impacts varies between ecosystem types, from headwaters to lowland streams, wetlands, and estuary and marine ecosystems (Chapter 7). Importantly, post-fire run-off can create impacts that extend many kilometres downstream of the fire footprint (Lyon and O'Connor 2008; Reale *et al.* 2015; Shelley *et al.* 2021). Impacts may endure for years to decades, well beyond the actual period of the wildfire (Box 6.1). Despite this variability, some generalities are evident in the mechanisms and processes that lead to the impacts (Gomez Isaza *et al.* 2022).

Box 6.1. Physical impacts to aquatic ecosystems

The ACT Orroral Valley wildfire ignited on 27 January 2020 and burnt 88 000 ha including 80% of Namadgi National Park. The fire was extinguished on 27 February 2020 by widespread, intense rainfall (~158 mm between 10 and 14 February 2020: ACT Government 2020). The Orroral Valley wildfire was one of a series that burnt much of the upper Murrumbidgee catchment in New South Wales, and was more damaging than the extensive areas previously burnt in the 2003 Canberra fires. The rainfall that extinguished the fire caused significant sedimentation of the upper Cotter River. Mass erosion formed gullies > 2 m deep and deposited boulders, cobble, sand, silt, and ash into the stream, altering the stream course, and eroding banks which deposited more sediment. By mid-March 2020, pools at one site that were previously > 1.5 m deep had filled with sand and silt, with depth reduced to ~0.3 m; extensive sediment slugs moved down the Cotter River from upstream fire-affected areas by mid to late April 2020. Monitoring of one representative deep pool over time revealed accumulated sediment of 0.9–1.5 m depth in mid-March 2020, with maximum sediment depth slowly decreasing over 2 years to 0.95 m (Fig. 6.2) with a concomitant increase in water depth. However, recovery to pre-fire levels is expected to be slow. After the 2003 fires, pool habitats in the lower Cotter had not recovered 10 years later. Stream substrates in the 2019–20 fire-impacted reach have changed from rocky-bottomed (predominantly

Fig. 6.2. Fire-impacted pool on the upper Cotter River over time: (A) boulder and cobble from erosion gully (March 2020); (B) fine silt smothers the rocky substrate (March 2020); (C) sand persisting in stream (April 2021); and (D) sedimentation still evident 2 years post-fire (February 2022). (Photos: M. Lintermans)

cobble and boulder with some bedrock) to pebble, gravel, sand and silt. Instream aquatic vegetation is now almost absent in many areas, with previously abundant beds of water milfoil (*Myriophyllum* sp.) now smothered by sediment. The previously locally abundant population of two-spined blackfish (*Gadopsis bispinosus*) declined by ~80% immediately following the fire and sedimentation. This species is listed as Vulnerable in the Australian Capital Territory, but persists in well-forested upland streams where it relies on rocky substrates for spawning, refuge and feeding sites.

Key message: Over 2 years post-fire, aquatic habitats are still severely impacted.

During wildfire, radiant heating of the water and the immediate loss of vegetative cover can alter aquatic thermal regimes (Cooper *et al.* 2015). Extreme water temperatures, up to 55°C, can occur during wildfires (Victorian DSE 2009) with elevated temperatures (7–15°C above normal) persisting in some systems for decades until vegetation cover returns (Koontz *et al.* 2018). The rapid reduction in shading also contributes to altered primary productivity and exposes aquatic animals to greater predation threat (e.g. Rosenberger *et al.* 2015).

The largest impacts to aquatic ecosystems often occur well after the wildfire. High volume, intensity or frequency rainfall events that cause run-off across burnt, hydrophobic soils, mobilise and deposit ash-laden sediments and timber debris into aquatic ecosystems. This can alter the physical nature of aquatic ecosystems by introducing large volumes of sediment (silt, sand, gravel, pebble, cobble and boulder), by creating log jams, and altering streamflow. The ash and soil from the fireground may contain a complex milieu of nutrients (e.g. phosphorus, nitrate), ions (e.g. sodium, chloride), metals (e.g. magnesium, iron, copper) and polycyclic aromatic hydrocarbons (PAHs), which can be toxic to aquatic fauna (Cramp *et al.* 2021 and references therein). The vegetation loss reduces rainfall interception, increasing erosivity of hydrophobic burnt soils. The loss of timber debris on the forest floor reduces the ability of the riparian zone to intercept run-off and filter sediment (Reale *et al.* 2015). The influx of run-off-derived inputs into streams can adversely impact water quality, often through rapid reduction in dissolved oxygen concentrations (Lyon and O'Connor 2008). Increased sedimentation can also alter habitat suitability (through smothering), and the mobilisation of nutrients (and increased light) can promote excessive growth of aquatic plants and cause algal blooms.

How aquatic species cope with the changes to water and instream habitat quality following wildfires depends on their ability to physiologically tolerate or behaviourally adjust to these changes (Box 6.2). Aquatic organisms have a narrow range of water quality tolerances; fluctuations outside these tolerances can be lethal or impair physiological performance and fitness. For freshwater fish and spiny crayfish, abrupt or substantial changes in water temperature or oxygen concentrations affect physiological performance (Cramp *et al.* 2021) and can lead to increased predation risk, increased incidence of disease, reduced fecundity, or even kill animals. In addition, ash and sediment leachates can pose significant health problems for fish and crayfish (Gonino *et al.* 2019); however, the exact nature and composition of wildfire leachates, and therefore their potential for impact, is often highly site-specific (Harper *et al.* 2019). Ash and sediment leachates can raise water pH to > 10, compromising the ionic balance and disrupting the excretory pathways and capacities of aquatic animals. Wildfire run-off often contains ammonia, nitrate, metals (e.g. Mg, Fe, Cu, Cr, Zn, As, Pb, Hg) and polycyclic aromatic compounds, which can damage gas exchange surfaces, impair ion regulation, and impede growth and development in fish (Gomez Isaza *et al.* 2020). Furthermore, ash and sediment inputs can promote aquatic microbial activity which rapidly strips waterbodies of oxygen, suffocating fish and causing crayfish to leave the water with chronic sublethal exposure, which affects tolerance thresholds, physiology (e.g. damages respiratory structures) and behaviour (Cramp *et al.* 2021 and references therein).

Box 6.2. Physiological tolerances of aquatic species

Changes in an animal's physiological performance are useful indicators of environmental stressors and can inform biologically relevant 'tolerance thresholds' to guide management decision making (e.g. define water quality management guidelines and identify emergency triggers such as post-fire run-off mitigation or population extractions/relocations). Tolerance thresholds can also inform habitat suitability models and monitoring of species recovery. However, currently we understand little about the water quality tolerance limits of most Australian fish and crayfish, nor the potential for ash and sediment exposures to influence physiological tolerances and aerobic performance longer-term.

In response to the 2019–20 wildfires, we used laboratory-based experimental approaches to examine thermal and oxygen tolerance limits for Australian aquatic species (Cramp *et al.* 2021). We also examined the impacts of medium-term (1–2 weeks) exposure to wildfire ash leachates and fine suspended sediments on temperature and oxygen, and on the aerobic performance of aquatic species (Cramp *et al.* 2021). We found that fish and spiny crayfish from cooler high-altitude climates (e.g. mountain galaxias (*Galaxias olidus*), strong spiny crayfish (*Euastacus valentulus*), Conondale spiny crayfish (*Euastacus hystricosus*)) are likely to be at greater risk from wildfire-induced changes to aquatic thermal and oxygen regimes because their thermal and oxygen tolerance limits are up to 20% lower than those of more broadly distributed and/or lowland species. In addition, we found that chronic exposure to even very low levels of suspended ash and sediment can damage respiratory surfaces and depress thermal and oxygen tolerance thresholds by as much as 10%. This can reduce aerobic performance even in comparatively tolerant species such as olive perchlet (*Ambassis agassizzi*) and redclaw crayfish (*Cherax quadricarinatus*). These results suggest that ash and sediment may compound the negative impacts of wildfire-induced changes in water temperature and oxygen on some fish and crayfish species. Importantly, the species studied differed quite substantially in their relative sensitivity to ash and sediment-induced physiological disruptions, suggesting that post-wildfire conditions may favour species with more resilient physiologies, leading to significant changes in community composition. The multifactorial nature of wildfire impacts supports the need for informed, species-specific and adaptive management in fire-affected systems, particularly for systems containing threatened or vulnerable aquatic species.

Key message: Fish and crayfish with existing restricted, upland or temperate distributions are likely to be disproportionately affected by wildfire-induced changes to aquatic environments.

Heavier, coarse material in sediment (sand, gravel, pebble, cobble and boulder) settles out of water columns rapidly, smothering substrates and increasing stream bedload volumes, and may persist for decades. After wildfires, fine sediments in suspension can rapidly increase the total suspended solids to > 57 000 mg L^{-1} (or 5.7% solids), and turbidity levels to > 6000 nephelometric turbidity units (NTU); these sediments may remain in suspension for weeks or months, and become repeatedly re-suspended for years (Lyon and O'Connor 2008; Shelley *et al.* 2021). Suspended sediments, both coarse and fine, can clog gills, reduce feeding behaviour, and increase stress biomarkers in fish and crayfish (Kemp *et al.* 2011; Rosewarne *et al.* 2014). Ash can also alter fish behaviour by suppressing activity levels, by disrupting foraging behaviours and shoal cohesion (Gonino *et al.* 2019). If wildfire-induced run-off impacts coincide with important life-cycle processes, spawning and movement cues may be disrupted, spawning sites smothered or the survival of critical life stages, such as eggs or larvae, reduced.

Fire-fighting retardants and foams can also impact aquatic fauna. Most chemical fire suppressants are predominantly water-based, but often contain high levels of inorganic salts (e.g. diammonium phosphates and ammonium polyphosphate salts) that can liberate large quantities of ammonia if delivered directly into waterbodies; this amount of ammonia can kill aquatic fauna and exacerbate eutrophication (Boulton *et al.* 2003).

Impacts from the 2019–20 wildfires

Mortality events

Mortality of aquatic animals following the 2019–20 wildfires was widespread, documented across 26 locations in 20 waterways, over four states and territories, from headwaters to coastal areas (Silva *et al.* 2020; Shelley *et al.* 2021). Mortality was documented in 32 aquatic species, including large-bodied Murray cod (*Maccullochella peelii*) and trout cod (*M. macquariensis*), small-bodied Australian smelt (*Retropinna semoni*), non-native common carp (*Cyprinus carpio*), redfin perch (*Perca fluviatilis*) and eastern gambusia (*Gambusia holbrooki*), and estuarine species such as black bream (*Acanthopagrus butcheri*) and dusky flathead (*Platycephalus fuscus*). Mortality of the Murray spiny crayfish (*Euastacus armatus*) was observed in four waterways, and platypus (*Ornithorhynchus anatinus*) deaths were also documented. While mortality was mostly observed within areas burnt by the wildfires, freshwater fish and spiny crayfish mortality was recorded as far as 100 km downstream. Mortality events involved tens to thousands of individuals. Most mortality certainly went undetected. The mechanistic causes of observed mortality events are unknown, as are the implications of the observed (and unobserved) mortalities on the impacted species (Bino *et al.* 2021).

Predicting the species at most risk

During disturbances such as wildfire, identifying the species most at risk is critical to guide resource allocation. Following the 2019–20 wildfires, freshwater fish and spiny crayfish were prioritised through a rapid preliminary assessment (in January 2020, revised in March 2020) by the Wildlife and Threatened Species Bushfire Recovery Expert Panel, before a National Environmental Science Program project allowed for a more considered refinement of the assessment during 2021 (Legge *et al.* in press; Ward *et al.* unpublished).

For fish, the initial assessment focused on threatened species listed either internationally (International Union for Conservation of Nature (IUCN) Red List of Threatened Species), nationally (*Environment Protection and Biodiversity Conservation Act 1999* (EPBC Act)); Australian Society for Fish Biology Threatened Fishes List), or at State/Territory level, and included select undescribed species. For crayfish the focus was the highly threatened genus of spiny crayfish (*Euastacus* spp.). Use of multiple threatened listings was necessary because aquatic species are under-represented in Australian legislation (Furse and Coughran 2011; Lintermans *et al.* 2020).

A key difference between the preliminary (2020) and subsequent (2021) assessments was the inclusion of spatial variation in fire severity in the later assessment. This refinement was important because distributional overlap with fire is an imprecise indicator of the immediate fire impact. For example, a low-intensity fire may not raise water temperature and inhabitants of deep pools may be little affected. Furthermore, fire severity mapping was incorporated into an aquatic sedimentation risk model, which was created by adapting an existing erosion model to include information on fire extent and severity, and post-fire rainfall events (Legge *et al.* in press; Ward *et al.* unpublished). In contrast, the preliminary assessment relied on extending the fire extent 80 km downstream to estimate sediment pulses. This aquatic sedimentation risk model allowed for a more sophisticated characterisation of the potential aquatic impacts given fire severity and prevailing conditions (e.g. soil type, slope steepness, rainfall, and run-off) (Ward *et al.* unpublished).

Both assessments utilised the best available knowledge to define the distribution of each species, although information on the distribution and status of populations is incomplete. For freshwater fish, distributions were based on observations collated during the 2019 IUCN assessment of Australian freshwater fish (Lintermans 2019). For spiny crayfish, species distribution polygons were compiled from previous distribution data and new observations collected in 2021 during a national conservation status assessment project (Whiterod *et al.* 2022). The assessments focused on species for which the 2019–20 wildfires overlapped their range by at least 10% if listed as threatened, and at least 25% if not listed as threatened. The list of spiny crayfish species used here was refined (relative to the lists used in the assessments) to reflect taxonomic revisions and new wildfire impact knowledge. Both assessments utilised knowledge of species' behavioural, ecological and life history traits to infer susceptibility to fire impacts. However, the refined assessment also used structured expert elicitation to estimate the immediate population impact and the likely population trajectories following the wildfires (Legge *et al.* in press). We note that experts were often uncertain about fire impacts, because empirical data are few and our understanding of basic ecology and biology is incomplete for some species.

The preliminary assessment prioritised 21 threatened freshwater fish and 22 spiny crayfish taxa for management intervention (Figs 6.3, 6.4) (Legge *et al.* 2022). These 'priority' species were distributed from Victoria, throughout New South Wales and into coastal southern Queensland. Ward *et al.* (unpublished) estimated that the entire known ranges of five freshwater fish and five spiny crayfish were vulnerable to sedimentation risk following the wildfires (Fig. 6.3). Priority species included small-bodied, non-migratory, habitat specialists with restricted ranges, all traits that are linked to greater risk of extinction in aquatic species (cf. Lintermans *et al.* 2020) although several large-bodied species from the genera *Maccullochella*, *Macquaria* and *Gadopsis* were also estimated to be impacted. Impact was also predicted to be high for some broadly distributed species, such as Sutton's crayfish (*Euastacus suttoni*), with over half its range overlapping with wildfires.

The structured expert elicitation of the refined assessment highlighted the potential delayed impact from sedimentation events, as experts consistently forecast that the status of aquatic species would be worse 1 year than 1 week post-fire. Moreover, it was forecast that only partial (but not full) recovery may be realised 10 years post-fire, in part reflecting pressure from existing threats, such as drought and habitat destruction, driving prior and ongoing declining population trends of many species (Legge *et al.* in press).

The assessments relied on spatial analyses and expert elicitation, informed by empirical evidence which in some cases was limited. Regardless, it represents the best approach for using available data to rapidly guide management responses during the complex and dynamic chaos of an extended wildfire season. For most species the immediate impacts could not be measured, as access to the fire ground was impossible while fires were still active, with surveys only occurring many weeks after the fire. The 2019–20 wildfires likely pushed some aquatic species closer to extinction, including eight that were already considered to have > 50% probability of extinction in the next 20 years (Lintermans *et al.* 2020). Conservation actions to reduce extinction risk for these species is an urgent challenge. Given the predictions for increased fire frequency and severity, we need to improve our understanding of the impacts of wildfire on aquatic species, and our knowledge of the fundamental ecology of these species, to better inform similar assessments during future wildfires (and other disturbances).

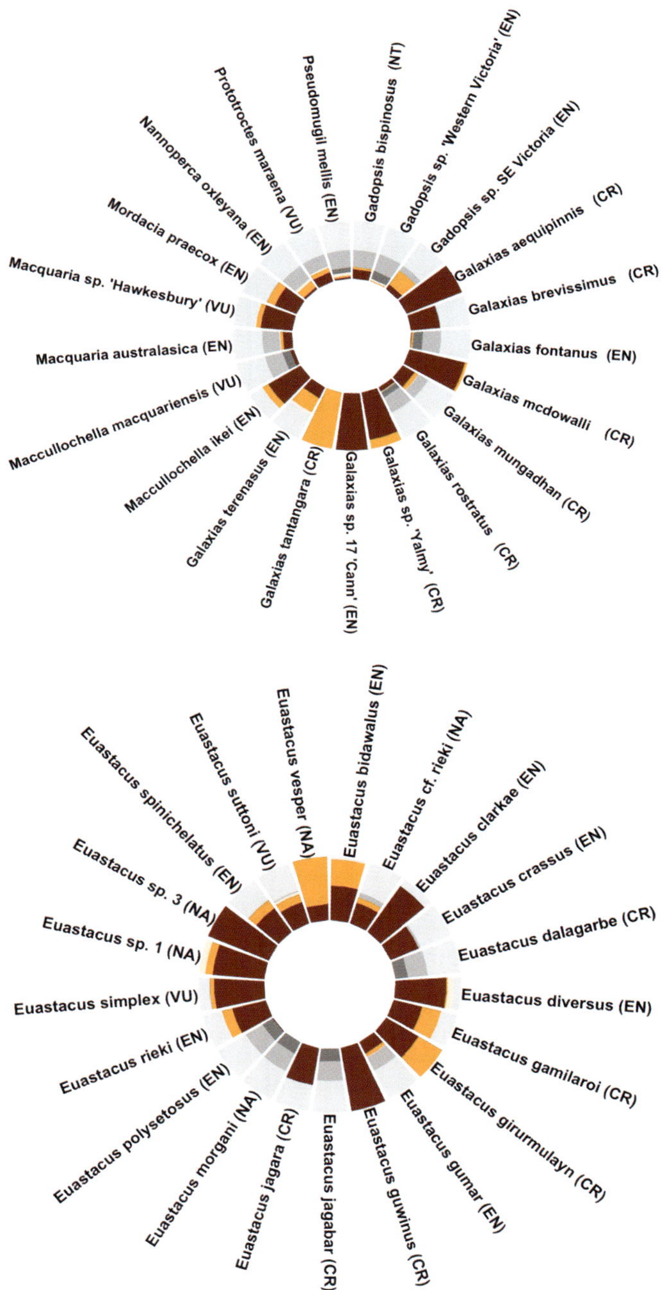

Fig. 6.3. Proportion of range at risk from sedimentation for each taxon assessed. Upper figure represents fish; lower figure is for spiny crayfish. Relative levels of risk are colour-coded as relatively mild risk (cream), relatively high risk (orange), relatively severe risk (brown) and, with background thresholds to aid visual comparisons of 30% (dark grey), 50% (medium grey) and 100% (light grey). Each taxon name includes information on their current IUCN conservation status, which can be Critically Endangered (CR), Endangered (EN), Vulnerable (VU), Near Threatened (NT), Least Concern (LC), or not assessed (NA) (modified from Ward *et al.* unpublished).

Fig. 6.4. Examples of priority species impacted by the 2019–20 wildfires: Clark's crayfish (*Euastacus clarkae*) (A); Tianjara crayfish (*Euastacus guwinus*) (B); short-tail galaxias (*Galaxias brevissimus*) (C). (Photos: (A, B) Rob McCormack, (C) Tarmo A. Raadik).

Emergency conservation efforts

While the wildfires were still burning, it was clear that urgent actions were needed to prevent the loss of populations, or even the extinction, of some aquatic species. In some cases, this involved sediment control, but mostly these actions involved emergency rescue of individuals of highly restricted species from threatened streams. Animals were translocated to unimpacted waters or to *ex situ* temporary holding aquaria or aquaculture facilities until water and habitat conditions improved and they could be returned (e.g. Raadik et al. 2021). Most rescues occurred in February 2020, following substantial rainfall in late January–early February 2020 that caused widespread sediment flows. These salvage efforts were stressful to the species affected, and to the rescuers themselves, and time critical. Pragmatically the scale of these efforts was limited, but without them entire populations would have been lost, and some threatened species may have faced extinction.

Emergency actions were initiated for 24 aquatic species across three jurisdictions (Table 6.1). For five species of *Galaxias* whose entire distributions were burnt, it is possible that emergency rescue averted species extinction. Two such species, Yalmy galaxias (*Galaxias* sp. nov. 'Yalmy') and Cann galaxias (Galaxias sp. nov. 'Cann'), faced the risk of extinction before being formally described (Raadik *et al.* 2021; Stoessel *et al.* 2021). In New South Wales, emergency rescues were implemented for two other range-restricted galaxiids (stocky

Table 6.1. Species subject to salvage efforts following the 2019–20 wildfires.

Conservation status based on state listing (for the state that the work was done), national listing (*Environment Protection and Biodiversity Conservation Act 1999* (EPBC Act)) and international listing (International Union for Conservation of Nature; IUCN Red List). Conservation statuses are Critically Endangered (CR), Endangered (EN) and Vulnerable (VU). Table adapted from Shelley *et al.* (2021).

Common name	Scientific name	Conservation status			Jurisdiction where emergency rescues occurred
		State listing	EPBC Act	IUCN Red List	
Freshwater fish					
Cann galaxias	*Galaxias* sp. 'Cann'		Under assessment		Vic.
Dargo galaxias	*Galaxias mungadhan*	CR	Under assessment	CR	Vic.
East Gippsland galaxias	*Galaxias aequipinnis*	CR	Under assessment	CR	Vic.
Eastern freshwater cod	*Maccullochella ikei*	EN	EN	EN	NSW
Gippsland blackfish	*Gadopsis* sp. SEV				Vic.
Macquarie perch	*Macquaria australasica*	EN	EN	VU	NSW, Vic.
McDowall's galaxias	*Galaxias mcdowalli*	CR	Under assessment	CR	Vic.
Mountain galaxias	*Galaxias olidus*			LC	Qld
Murray cod	*Maccullochella peelii*		VU	LC	Vic.
Olive perchlet	*Ambassis agassizii*	EN*		LC	NSW
Oxleyan pygmy perch	*Nannoperca oxleyana*	EN	Under assessment	EN	NSW
River blackfish	*Gadopsis marmoratus*			LC	Qld
Short-tail galaxias	*Galaxias brevissimus*	Under assessment	Under assessment	CR	NSW
Southern purple-spotted gudgeon	*Mogurnda adspersa*	EN	Under assessment*	LC	NSW
Stocky galaxias	*Galaxias tantangara*	CR	CR	CR	NSW
Yalmy galaxias	*Galaxias* sp. nov. 'Yalmy'	Under assessment	Under assessment	CR	Vic.
Freshwater crayfish					
East Gippsland spiny crayfish	*Euastacus bidawalus*		Under assessment	EN	Vic.
Mountain crayfish	*Euastacus sulcatus*		Under assessment	VU	Qld
Orbost spiny crayfish	*Euastacus diversus*	EN	Under assessment	EN	Vic.
Variable spiny crayfish	*Euastacus yanga*	EN		LC	Vic.
Freshwater mussel					
Austral river mussel	*Hyridella australis*				Vic.
Depressed river mussel	*Hyridella drapeta*				Vic.
Glenelg freshwater mussel	*Hyridella glenelgensis*	CR	CR	CR	Vic.

* Murray–Darling Basin population

galaxias (*Galaxias tantangara*), short-tail galaxias (*Galaxias brevissimus*)) as well as large-bodied species such as eastern freshwater cod (*Maccullochella ikei*) and Macquarie perch (*Macquaria australasica*) (Box 6.3). In addition, the severe drought and wildfires prompted NSW Department of Primary Industries (Fisheries) in December 2019 to rescue 252 Oxleyan pygmy perch (*Nannoperca oxleyana*) from coastal wallum wetlands in Broadwater National Park. Other populations of Macquarie perch were rescued in Victoria and either translocated to nearby unimpacted rivers or transferred to a temporary captive facility. Mountain galaxias (*Galaxias olidus*) and river blackfish (*Gadopsis marmoratus*) were the focus of emergency rescues from Spring Creek in the upper Condamine River Catchment in Queensland. Emergency rescue was implemented for five spiny crayfish species in Queensland and Victoria.

Box 6.3. Rescuing Macquarie perch in Mannus Creek

Mannus Creek has one of only four remaining natural populations of threatened Macquarie perch (*Macquaria australasica*) in the New South Wales section of the Murray–Darling Basin. The creek was devastated by the 2019–20 wildfires, which burnt the entire known distribution for Macquarie perch in the creek (Fig. 6.5). The creek was first accessed 3 days post-fire, when the landscape was blackened and denuded, and the creek was green from an algal bloom. Despite this utter devastation, sampling revealed that some Macquarie perch had survived the blaze, but the real danger was yet to come, with forecast rain that could cause toxic run-off across the very steep and unstable catchment. Given this imminent threat, a small team worked over the next week to rescue as many Macquarie perch as possible. Despite tireless efforts, only 10 Macquarie perch were salvaged and, before extra rescue teams could arrive, a large rainstorm in the

Fig. 6.5. The impact of wildfire on Mannus Creek (A, C); a rescued Macquarie perch (B). (Photos: Luke Pearce)

catchment turned the creek into a flowing river of black sludge. Dissolved oxygen in the creek plummeted to near zero and a fish kill followed.

Surveys over the following months helped to quantify the impacts on the habitat and the fish community. Habitat mapping revealed major changes to the creek, with reduced pool depth, large amounts of burnt debris, sediment and ash deposition and loss of riparian vegetation. Initial post-fire fish surveys failed to detect any Macquarie perch, and as the rescued fish were too few to re-establish a population, they were used to supplement the genetic diversity of another remnant population. However, over time, some hope emerged with fish surveys detecting small numbers of Macquarie perch in Mannus Creek. The focus is now to reinforce this population with captive-bred fish sourced from other populations. The recovery of Macquarie perch in Mannus Creek is a long and difficult challenge that those working on the species are committed to.

Key message: Early assessment and intervention is critical, but difficult, and should be planned and resourced in advance.

Emergency conservation efforts are complex operations. In response to the 2019–20 wildfires, a chain of responsibility originating in state government agencies provided over-arching guidance for those conducting the on-ground activities. Across jurisdictions, many stakeholders facilitated the emergency efforts (Shelley *et al.* 2021). On-ground rescues required several permits (including permission to access firegrounds), consideration of safety and welfare issues, and coordination with incident management teams (e.g. planning officers and operations officers), who play a critical role in providing access and logistical support (Shelley *et al.* 2021). Some species could not be assessed or extracted when roads and access tracks became impassable due to burnt trees, burnt bridges, landslides and erosion. For fish species in small, shallow streams that could be sampled easily and quickly, rescue was relatively straightforward; however, other species occupying deep habitats presented more problems. As rescued individuals were probably already stressed, transportation protocols were adapted to minimise further stress (Ebner *et al.* 2020; Shelley *et al.* 2021). Captive maintenance of animals involved strict protocols to address potential stress, disease and parasites, aggression and cannibalism, and health issues; and to ensure water quality, habitat and feeding were tailored for each species (Stoessel *et al.* 2021).

The scale of the emergency conservation effort was unprecedented, helping to avoid the loss of 24 aquatic species and populations. Releasing rescued individuals or their progeny may assist recovery. For some species, the lack of knowledge on captive husbandry and availability of suitable holding facilities made maintaining the rescued individuals problematic. Overall, the experience gained from the rescues has been invaluable (Shelley *et al.* 2021) and an important legacy could be achieving a proactive, flexible and appropriately resourced emergency rescue strategy for aquatic species. With more frequent and severe wildfires anticipated, the importance of such emergency conservation efforts will be critical for maintaining aquatic species in fire-prone regions.

Improving management of aquatic ecosystems in fire-prone regions

Several opportunities emerge – relating to prevention, preparedness, response and recovery – to better manage aquatic ecosystems and species across fire-prone regions.

Prevention: identifying and mitigating drivers

The single greatest challenge to managing wildfire impacts on aquatic ecosystems is preventing increases in the frequency and severity of wildfires, by identifying and mitigating drivers of wildfires as well as identifying areas that are more prone to the impacts of wildfires.

Preparedness: improving protection, building resilience, minimising other stressors and gaining knowledge

Some species of freshwater fish (Lintermans *et al.* 2020) and spiny crayfish are among the most threatened of all species in Australia. The prioritisation assessment (see predicting the species at most risk) drew attention to the under-representation of these taxonomic groups under national environmental legislation. Currently, only one-quarter (43) of freshwater fish and spiny crayfish species that are recognised as being of conservation concern on international, and state threatened lists, are listed under the EPBC Act (Furse and Coughran 2011). Of the wildfire-impacted species, only 10 priority fish and none of the 22 priority spiny crayfish were listed under the EPBC Act. This under-representation requires urgent rectification as it hampers the conservation of some of the most at-risk species in Australia. For instance, the current national Threatened Species Strategy (DAWE 2021) only considers EPBC-listed taxa in scope; thus 75% of threatened freshwater fish and crayfish are excluded. Fortunately, two Commonwealth funded projects are currently seeking to escalate the listing assessments for freshwater fish and spiny crayfish; this should increase attention for these species. The number of as-yet-undescribed freshwater species is also high, for example as many as 27 species of spiny crayfish (Austin *et al.* 2022). Consequently, adequate resources are needed to formally describe these candidate species, as instigating conservation management and growing community appreciation of these imperilled species is difficult without a scientific name and knowledge of distribution.

Preparing for a future with more frequent and severe wildfires will challenge the conservation of many aquatic species in fire-prone regions. Ecosystems and species at most risk need to be identified, with strategic fire management designed and implemented to lessen the impact of future wildfires (Srivastava *et al.* 2021; Legge *et al.* 2022). Resilience could be enhanced by establishing additional and/or larger populations of species to reduce extinction risk and improve evolutionary potential. Declining aquatic habitat availability and quality may be managed with actions such as restricting water extraction from refuge pools during extended dry periods, allocating more environmental flows and restoring habitat (including providing appropriate instream and overhead vegetation, rocks and logs; reducing sedimentation levels; ensuring access to burrows; and mitigating deforestation). Invasive species, both terrestrial (e.g. horses, pigs, deer, foxes, stock) and aquatic (trout, redfin, translocated native fish), should be removed and excluded from sites. Improved knowledge of the factors driving post-fire changes to aquatic ecosystems and physiological tolerances of aquatic species will provide a more holistic picture of the likely consequences of increasing wildfire frequency and severity on aquatic ecosystems.

In the wake of the 2019–20 wildfires, funding was directed towards knowledge generation for aquatic species (Box 6.4); this needs to be sustained. This includes establishing long-term monitoring programs to allow robust assessment of the impacts of, and recovery following, wildfires (Southwell *et al.* 2022). In addition to emergency rescue, facilities for short- and long-term captive maintenance and breeding along with translocation may have merit for high-risk species. These management interventions should be implemented in accordance with best practice (e.g. IUCN/SSC 2013) to ensure conservation responses are effective, ethical, socially acceptable and minimise ecological and environmental risks.

Box 6.4. Collaborative project to inform the recovery of spiny crayfish

'Saving the Spinys' was a multifaceted collaborative project, involving crayfish experts from eight organisations, which aimed to inform the recovery of wildfire-impacted spiny crayfish. Field surveys helped to resolve the contemporary range and identify key populations and threats for all 22 priority spiny crayfish species. The project also facilitated the most comprehensive molecular taxonomic analyses ever conducted on spiny crayfish (Austin *et al.* 2022), and the application of environmental DNA (eDNA) to define the range of species enable tracking of population trajectories over time. Zukowski *et al.* (2021) identified conservation translocations as an important management tool for spiny crayfish, although more species-specific investigation is warranted. The knowledge gained through the project enabled formal description of new species, guided assessment of priority species under national EPBC Act legislation and allowed for an action plan to be developed. By redressing long-standing knowledge gaps, Saving the Spinys will establish a platform for the conservation of spiny crayfish for decades to come.

The small crayfish (*Euastacus spinichelatus*) showcases how the project informed conservation of a wildfire-susceptible species (Fig. 6.6). Formally described in 1997,

Fig. 6.6. The small crayfish (*Euastacus spinichelatus*) sampled during post-fire surveys. (Photo: Rob McCormack)

the molecular taxonomic analyses of Austin *et al.* (2022) confirmed its validity and identified two divergent and geographically disjunct lineages. Field surveys revealed the loss of the species from several sites (including the type locality) following the 2019–20 wildfires. Two years on, the species has not naturally recolonised these sites despite the forest regenerating and streamflows normalising; reintroductions may be possible (Zukowski *et al.* 2021). The small crayfish is forecast to be listed as Endangered under the EPBC Act (Whiterod *et al.* 2022), with the projected increase in wildfire frequency and severity a key threat. While many knowledge gaps have been filled, the new focus is investigation and description of the two divergent lineages.

Key message: Proactive strategic assessments for poorly known taxonomic groups will help fill knowledge gaps and establish priorities for future conservation responses.

Response: emergency management to mitigate

The 2019–20 wildfire experience led to several recommendations for future improvement, including creating a dedicated emergency fund in each jurisdiction to allow for time-critical, coordinated conservation responses (Shelley *et al.* 2021). Stakeholders need access to relevant information and training to be 'emergency-ready' and permits and approvals need to be processed rapidly. The capacity of existing fish hatcheries and aquariums to support and assist emergency captive management programs should be enhanced, or purpose-built facilities should be established. Long-term collaborative relationships between conservation managers and emergency agencies are required to ensure future responses are feasible. The prioritisation of resources will increasingly be necessary as the 2019–20 wildfires highlighted the fact that not all populations can be saved.

Recovery: overcoming the impacts to ecosystems and species

Helping aquatic species to recover from wildfires will require targeted implementation of threat abatement, habitat protection and enhancement, and *ex situ* conservation (and translocations) actions (Geary *et al.* 2022; Legge *et al.* 2022). These actions should occur in both burnt and unburnt areas, which are refuges and sources of dispersal to assist recovery. The recovery outlook for freshwater fish and spiny crayfish is considered to be poorer than for other taxonomic groups (Legge *et al.* in press). Improving our ecological knowledge of these species should help our ability to manage them. Better identifying high-risk areas and species likely to be present will be necessary to refine the blueprint for emergency responses (Legge *et al.* 2022).

Conclusions

The impacts of the 2019–20 wildfires will endure in aquatic ecosystems for decades. The consequences may be irreversible for some taxa. While the conservation responses to the wildfires may have supported recovery in some aquatic fauna, recovery is generally anticipated to be slow and future wildfires will create additional risks. Sustained investment and attention are needed to measure and support recovery, and to increase preparedness for future events.

Recommendations

While these recommendations have been developed with freshwater fish and spiny crayfish in mind, they are equally relevant to other wildfire-prone freshwater species.

- Build the resilience of freshwater species to future wildfires by reducing other stressors, protecting and enhancing habitat, expanding distributions and increasing population sizes.
- Establish conservation facilities to house insurance populations and facilitate emergency rescues.
- Undertake statutory listing assessments of threatened freshwater fish and spiny crayfish (including those impacted by wildfires); encourage greater incorporation of EPBC Act-listed species into national and regional strategies.
- Pending the comprehensive updating of aquatic species in EPBC listings, adopt the approach used in the initial fire impact prioritisation process of using all threatened species lists when considering priority species for management/conservation.
- Resource and implement long-term monitoring and research programs for fire-impacted ecosystems and species to address knowledge gaps.
- Streamline the ability to rapidly react by establishing a centrally coordinated and appropriately funded response framework to enable time-critical and strategic conservation responses to catastrophic disturbances such as wildfires (as well as floods and droughts).

Addressing these recommendations will help to improve the future management of aquatic fauna in response to wildfires, and improve the conservation of these 'underwater, out of sight' Australian aquatic fauna.

Acknowledgements

The authors thank the funding sources and numerous people who contributed to the work described in this chapter. The chapter was improved through constructive comment by Sarah Legge (Australian National University/University of Queensland), Ashley Sparrow (Arthur Rylah Institute) and an anonymous reviewer.

References

ACT Government (2020) *ACT Bushfire and Flood Recovery Plan*. ACT Government, Canberra. <https://www.environment.act.gov.au/data/assets/pdf_file/0011/1622855/act-bushfire-and-flood-recovery-plan-2020.pdf>.

Austin CM, Whiterod NS, McCormack R, Raadik TA, Ahyong ST, *et al.* (2022) *Molecular Taxonomy of Australia's Endemic Freshwater Crayfish Genus* Euastacus *(Parastacidae), with Reference to Priority 2019–20 Bushfire-impacted Species – 2022 Update*. Deakin University and Aquasave-NGT, Victor Harbor.

Bino G, Hawke T, Kingsford RT (2021) Synergistic effects of a severe drought and fire on platypuses. *Science of the Total Environment* **777**, 146137. doi:10.1016/j.scitotenv.2021.146137

Boulton AJ, Moss G, Smithyman D (2003) Short-term effects of aerially-applied fire-suppressant foams on water chemistry and macroinvertebrates in streams after natural wild-fire on Kangaroo Island, South Australia. *Hydrobiologia* **498**, 177–189. doi:10.1023/A:1026213301871

Cooper SD, Page HM, Wiseman SW, Klose K, Bennett D, *et al.* (2015) Physicochemical and biological responses of streams to wildfire severity in riparian zones. *Freshwater Biology* **60**, 2600–2619. doi:10.1111/fwb.12523

Cramp R, Mulvey C, Cameron J, Wintour M, Gomez Isaza D, *et al.* (2021) 'Impacts of post-fire ash and runoff sediment on the physiological tolerances of Australian freshwater aquatic fauna'. NESP Threatened Species Recovery Hub Project 8.3.7 report, Brisbane.

DAWE (2021) *The Australian Government's Threatened Species Strategy 2021–2031*. Department of Agriculture, Water and the Environment, Canberra.

Ebner BC, Lobegeiger J, Coe J, Balcombe S, Latimer D, *et al.* (2020) 'Killarney fish rescue'. Centre for Tropical Water & Aquatic Ecosystem Research (TropWATER) Publication Report No. 20/43, James Cook University, Cairns.

Furse JM, Coughran J (2011) An assessment of the distribution, biology, threatening processes and conservation status of the freshwater crayfish, genus *Euastacus* (Decapoda: Parastacidae) in continental Australia. II. Threats, conservation assessments and key findings. *Crustaceana Monographs: New Frontiers in Crustacean Biology* **15**, 253–263.

Geary WL, Buchan A, Allen T, Attard D, Bruce MJ, *et al.* (2022) Responding to the biodiversity impacts of a megafire: a case study from south-eastern Australia's Black Summer. *Diversity & Distributions* **28**, 463–478 doi:10.1111/ddi.13292.

Gomez Isaza DF, Cramp RL, Franklin CE (2020) Living in polluted waters: a meta-analysis of the effects of nitrate and interactions with other environmental stressors on freshwater taxa. *Environmental Pollution* **261**, 114091. doi:10.1016/j.envpol.2020.114091

Gomez Isaza DF, Cramp RL, Franklin CE (2022) Fire and rain: a systematic review of the impacts of wildfire and associated runoff on aquatic fauna. *Global Change Biology* **8**, 2578–2595. doi:10.1111/gcb.16088

Gonino GM, Figueiredo BR, Manetta GI, Alves GHZ, Benedito E (2019) Fire increases the productivity of sugarcane, but it also generates ashes that negatively affect native fish species in aquatic systems. *Science of the Total Environment* **664**, 215–221. doi:10.1016/j.scitotenv.2019.02.022

Harper AR, Santin C, Doerr SH, Froyd CA, Albini D, *et al.* (2019) Chemical composition of wildfire ash produced in contrasting ecosystems and its toxicity to *Daphnia magna*. *International Journal of Wildland Fire* **28**, 726–737. doi:10.1071/WF18200

IUCN/SSC (2013) *Guidelines for Reintroductions and Other Conservation Translocations, Version 1.0*. International Union for Conservation of Nature (IUCN) Species Survival Commission, Gland, Switzerland.

Kemp P, Sear D, Collins A, Naden P, Jones I (2011) The impacts of fine sediment on riverine fish. *Hydrological Processes* **25**, 1800–1821. doi:10.1002/hyp.7940

Koontz ED, Steel EA, Olden JD (2018) Stream thermal responses to wildfire in the Pacific Northwest. *Freshwater Science* **37**, 731–746. doi:10.1086/700403

Legge S, Woinarski J, Scheele B, Garnett ST, Lintermans M, *et al.* (2022) Rapid assessment of the biodiversity impacts of the 2019–20 Australian megafires to guide urgent management intervention and recovery, and lessons for other regions. *Diversity & Distributions* **28**, 571–591. doi:10.1111/ddi.13428

Legge S, Rumpff L, Woinarski JCZ, Whiterod NS, Ward M, *et al.* (in press) The conservation impacts of ecological disturbance: Time-bound estimates of population loss and recovery for fauna affected by the 2019–2020 Australian megafires. *Global Ecology and Biogeography* doi:10.1111/geb.13473.

Lintermans M (2019) IUCN Red List. *Lateral Lines* **49**, 163.

Lintermans M, Geyle HM, Beatty S, Brown C, Ebner BC, *et al.* (2020) Big trouble for little fish: identifying Australian freshwater fishes in imminent risk of extinction. *Pacific Conservation Biology* **26**, 365–377. doi:10.1071/PC19053

Lyon JP, O'Connor JP (2008) Smoke on the water: can riverine fish populations recover following a catastrophic fire-related sediment slug? *Austral Ecology* **33**, 794–806. doi:10.1111/j.1442-9993.2008.01851.x

Raadik TA, Tonkin Z, Johnson G (2021) Fish, crayfish and mussels. In *Biodiversity Bushfire Response Supplementary Report: Bushfire Impacts on Species in Victoria*. Department of Environment, Land, Water and Planning, Melbourne.

Reale JK, Van Horn DJ, Condon KE, Dahm CN (2015) The effects of catastrophic wildfire on water quality along a river continuum. *Freshwater Science* **34**, 1426–1442. doi:10.1086/684001

Rosenberger AE, Dunham JB, Neuswanger JR, Railsback SF (2015) Legacy effects of wildfire on stream thermal regimes and rainbow trout ecology: an integrated analysis of observation and individual-based models. *Freshwater Science* **34**, 1571–1584. doi:10.1086/683338

Rosewarne PJ, Svendsen JC, Mortimer RJ, Dunn AM (2014) Muddied waters: suspended sediment impacts on gill structure and aerobic scope in an endangered native and an invasive freshwater crayfish. *Hydrobiologia* **722**, 61–74. doi:10.1007/s10750-013-1675-6

Shelley JJ, Raadik TA, Lintermans M (2021) 'Summary of the 2019/20 bushfire impacts on freshwater fish and emergency conservation response in south-eastern Australia'. NESP Threatened Species Recovery Hub, Brisbane.

Silva LG, Doyle KE, Duffy D, Humphries P, Horta A, *et al.* (2020) Mortality events resulting from Australia's catastrophic fires threaten aquatic biota. *Global Change Biology* **26**, 5345–5350. doi:10.1111/gcb.15282

Southwell D, Legge S, Woinarski J, Lindenmayer D, Lavery T, *et al.* (2022) Design considerations for rapid biodiversity reconnaissance surveys and long-term monitoring to assess the impact of wildfire. *Diversity & Distributions* **28**, 559-570. doi:10.1111/ddi.13427.

Srivastava SK, Lewis T, Behrendorff L, Phinn S (2021) Spatial databases and techniques to assist with prescribed fire management in the south-east Queensland bioregion. *International Journal of Wildland Fire* **30**, 90–111. doi:10.1071/WF19105

Stoessel D, Raadik TA, Shelley JJ (2021) *Emergency Extraction, Temporary Housing and Return of Small-bodied Aquatic Species following the 2019/20 Megafire*. Arthur Rylah Institute for Environmental Research, Heidelberg.

Victorian DSE (2009) *Murrindindi Complex North Fire Emergency Stabilization and Rehabilitation Plan*. Department of Sustainability and Environment, Melbourne.

Ward M, Southwell D, Gallagher R, Raadik TA, Whiterod NS, *et al.* (unpublished) Modelling post-fire sedimentation risk to estimate the impacts of fire on waterways and vulnerability to aquatic species.

Whiterod NS, Furse JF, Lutz M, Lintermans M, McCormack R, *et al.* (2022) *The 2022 Action Plan for Priority 2019–20 Bushfire-impacted Species from Australia's Endemic Freshwater Crayfish Genus* Euastacus *(Parastacidae)*. Aquasave–Nature Glenelg Trust, Victor Harbor.

Zukowski S, Whiterod NS, Furse JF, McCormack R, Chara S, *et al.* (2021) *Assessing the Feasibility of Conservation Translocations for Australia's Endemic Freshwater Crayfish Genus* Euastacus*, with Reference to Priority 2019–20 Bushfire-impacted Species*. Aquasave-NGT, Victor Harbor.

7

The impacts of the 2019–20 wildfires on marine species and ecosystems

Claudia Santori, Simon Rowe and Chris Smyth

Summary

Context and challenges

- There has been little previous appreciation of, or research on, the impacts of wildfire on marine systems.
- Due to the limited data, this chapter considers Australian and international research (before and after the fires), and anecdotal observations in media reports and by coastal communities who contributed to an OceanWatch Australia survey.
- We consider a broad definition of marine ecosystems to include nearshore (subtidal and surf zone) and offshore waters (pelagic, water column and benthic components), estuaries, intertidal zones (beaches, mangroves, and saltmarsh), sand dunes and the mouths of coastal streams.

Main findings

- Anecdotal evidence and some recently published data indicate that the 2019–20 wildfires impacted marine species and ecosystems. However, the extent and duration of that impact are unknown.
- The impacts of the wildfires could continue as wildfire-derived sediments slowly flow down waterways and reach estuaries long after the 2019–20 wildfire season.
- Australia lacks a comprehensive and systematic monitoring of impacts on marine species and ecosystems generally, and by wildfires specifically.
- There is inadequate funding of marine research generally, and more specifically inclusive of wildfire impacts.
- The 2019–20 wildfires showed the strong connections between terrestrial, freshwater and marine ecosystems, with terrestrial and freshwater impacts having consequences for the marine environment.

- Marine, terrestrial and freshwater research often provides only a snapshot in time with no longitudinal data collection, and this limits the assessment of wildfire impacts. This is largely a feature of short-term funding provided by government programs.
- Marine research is often focused on commercial and recreational target species or large charismatic species (e.g. cetaceans, turtles, sharks) rather than building a broader knowledge of how marine ecosystems work and how marine species interact within those ecosystems. This is very limiting for building an understanding on how marine ecosystems are affected by wildfires.

Introduction

The scale and intensity of the 2019–20 wildfires have been linked to climate change and a prolonged drought (Abram *et al.* 2021). Their impacts were exacerbated by subsequent storms and rainfall, which led to widespread severe flooding with ash, debris, nutrients, metals and chemicals flowing into streams and down into estuaries, onto beaches and out to sea. However, the nature and full extent of the wildfire impacts on marine species and ecosystems is largely unknown. The main reasons for this are:

1. During and in the immediate aftermath of the wildfires, access roads were closed and entry to fire grounds was restricted due to safety concerns. Post-fire storms and floods and then the emergence of the COVID-19 pandemic also delayed and then limited the scope of fieldwork in marine environments.
2. The majority of government, academic and citizen scientists work on terrestrial and freshwater species and ecosystems, with research funding directed accordingly.
3. Australia lacks a comprehensive and systematic monitoring framework for marine species and ecosystems that could have been used to establish baselines and report on change in response to the wildfires.
4. The focus of wildfire recovery projects were animals and plants on threatened species lists, which skew towards terrestrial species, in large part due to point 2.
5. There is a long-held belief that dynamic oceans dilute pollution and any impacts from wildfire ash, debris, nutrients, metals and chemicals will be short-lived, of little concern and beyond human intervention.

Even so, anecdotal observations, coincidental research, water-quality monitoring in fire-impacted estuaries, and Australian and international research into past fires can provide some insight into their impacts on marine species and ecosystems (Fig. 7.1). Freshwater research can also aid understanding due to some similarities in species and ecological processes (see Chapter 6). However, determining impacts of fire may be complex, in part due to compounding factors. One of the future research challenges will be to separate the impacts of wildfires from those caused by drought, climate change and floods, as well as from coastal development, pollution, over-harvesting and other stressors.

In this chapter, we describe three mechanisms for wildfire impacts on marine species and ecosystems: death and injury from incineration, smoke inhalation and radiant heat; run-off laden with ash, debris, nutrients, metals and chemicals; and wildfire smoke settling in estuarine and offshore waters. We also report on two case studies (Boxes 7.2 and 7.3) and one survey that compiled anecdotal observations on marine impacts of the 2019–20 wildfires (Box 7.1).

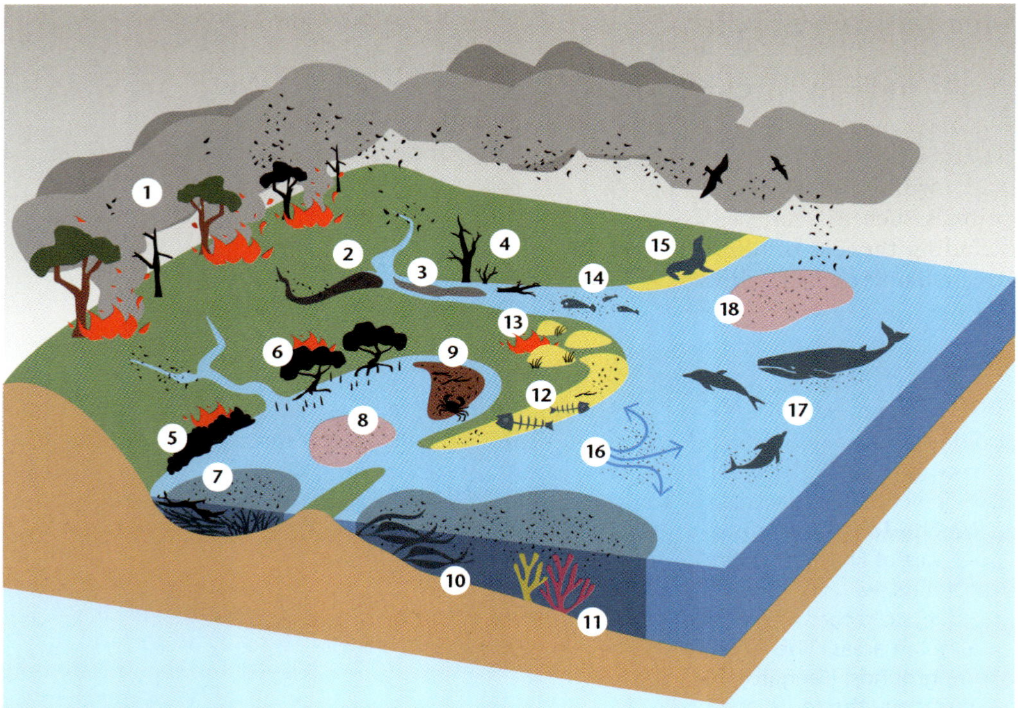

Key

1 Smoke containing ash, nutrients, metals and chemicals released to atmosphere, impacting coastal birdlife.

2 Ash and debris including nutrients, metals and chemicals run off into waterways after rains.

3 Sediment slug moves downstream, killing fish and reducing invertebrate habitats.

4 Riparian vegetation burnt, removing habitat and reducing shade – water temperatures increase.

5 Saltmarsh burnt, removing habitat and reducing shoreline protection.

6 Mangroves burnt, removing habitat and reducing shoreline protection.

7 Seagrass smothered by ash and debris. Turbid water reduces light and impacts seagrass photosynthesis.

8 Algal bloom leads to deoxygenation and death of fish.

9 Mudflats covered in ash and debris, killing mud crabs and removing food source for shorebirds.

10 Subtidal reefs and kelp beds smothered by ash and debris.

11 Coral reef smothered by ash and debris. Turbid water reduces photosynthesis.

12 Beaches covered by ash, debris and dead fish.

13 Sand dunes burnt, leading to erosion.

14 Fish that migrate upstream during life cycle killed by ash and debris.

15 Seals accumulate toxic PFAS from fire retardants.

16 Tides, currents and winds disperse ash and debris and spread the impact in surf zone and nearshore waters.

17 Whales and dolphins affected by inhaling wildfire smoke.

18 Smoke particles fall out of smoke plume, settling on water and causing algal blooms.

Fig. 7.1. Summary of possible wildfire impacts to marine ecosystems. (Figure created by Chris Tsernjavski, with permission from University of Melbourne)

Mechanisms of wildfire impacts on marine systems

Death and injury from incineration, smoke inhalation and radiant heat

Although evidence is limited, the OceanWatch Australia survey (see Box 7.1) received reports of burnt mangroves, wetlands, saltmarshes and sand dunes (Santori and Rowe 2021), which probably led to the immediate injury or death of plants and less-mobile animals such as molluscs and crabs. The loss of vegetation cover has flow-on effects, including the loss of nursery habitat, reduced carbon storage and the destabilisation of stream banks and shorelines.

Box 7.1. OceanWatch Australia survey of wildfire impacts on marine wildlife, habitats and water quality, and socio-economic consequences

Methodology

As part of an investigation into the impacts of the 2019–20 fires on marine wildlife, habitats, water quality and industry, OceanWatch Australia conducted an online public survey to record anecdotal observations from people who lived or worked near the affected areas. The focus of the survey (and its advertising for participants) was on six fire grounds: Kangaroo Island (South Australia), East Gippsland (Victoria), northern Tasmania, the south and north coasts of New South Wales, and south-east Queensland. The survey was built on the ESRI Survey123 platform and ran from December 2020 to June 2021 (Santori and Rowe 2021).

Results

A total of 45 surveys that included 82 observations in the marine environment were submitted to the survey either by the public or by OceanWatch Australia staff who had interviewed stakeholders. Observations that the respondents were only 'slightly confident' about or 'not confident at all', and observations related to terrestrial or freshwater systems, were excluded from this analysis.

Observations from the survey respondents were not equally distributed among the six focus regions (Fig. 7.2):

- north and south coasts of New South Wales: 67 wildfire impact observations, including two 'no impact' observations);
- south-eastern Queensland: six, including three 'no impact' observations;
- East Gippsland: one (the burning of Mallacoota's abalone processing plant);
- Kangaroo Island: four, and two 'no impact' observations;
- coastal Western Australia: two.

Most impact observations (number of observations in brackets) were of impacts to water quality (29), followed by socio-economic impacts (22), habitat (15) and marine wildlife (9). Seven surveys were returned reporting no impacts observed. The impacts were mostly observed from November 2019 to February 2021. Most impacts were reported as 'not recovered' (23), followed by 'partially back to pre-fire conditions' (20), while 20% (15) of the impacts reported were 'fully back to pre-fire conditions' at the time of submission.

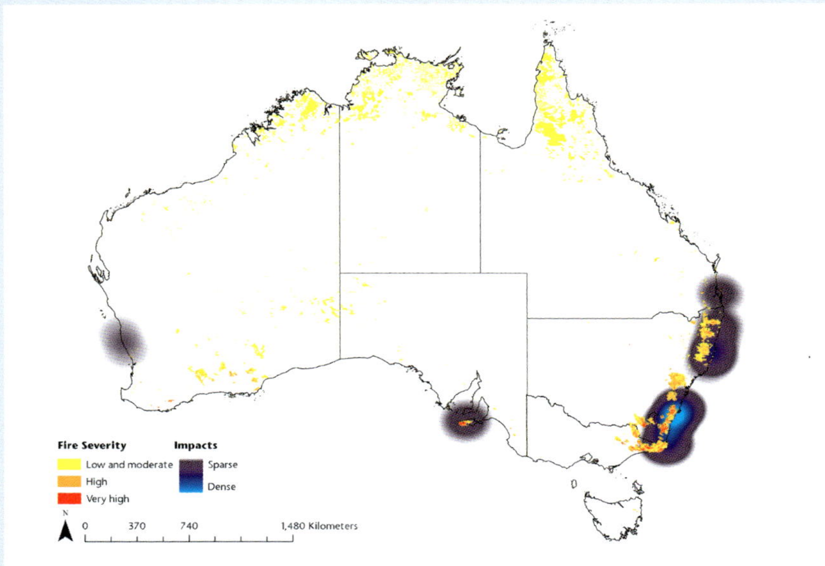

Fig. 7.2. Density of wildfire impact observations around mainland Australia (Australian Bureau of Statistics 2011) collected by OceanWatch Australia, relative to fire severity (the classes 'unburnt' and 'no data' are not displayed; New South Wales Department of Planning, Industry and Environment 2020).

Eight respondents reported observing a negative wildfire impact on marine wildlife, while one reported a positive impact (increase in numbers of big eye tuna and prawns, 6–7 months post wildfires). Five reports were about dead oysters; however, other dead animals observed were mullet, abalone, grouper, crabs, lobsters and 'fish'. Respondents reported that they knew or were uncertain that no action had been taken to address eight out of the nine animal impacts.

Water quality impacts dominated the observations, with most respondents reporting ash in the water (25) among the causal factors, followed by debris (18). Fifty-nine per cent of respondents (17) reported that water quality had not recovered or had only partially recovered from the wildfire impacts, over the survey period. Respondents reported that they either knew or were uncertain that no action had been taken to treat 21 out of the 29 water quality impacts, while actions such as estuary openings and clean-ups were reported for six of the impacts, with mixed success.

Most observations of habitat impacts included habitat destruction (10) followed by habitat degradation (5). The habitat most affected was mangroves (8) followed by wetland (5), saltmarsh (5) and seagrass (4). The main causes identified were fire itself, and ash and debris. Eight respondents stated that the habitat did not recover and six stated that it had only partially recovered. Moreover, respondents reported that to their knowledge no measures had been implemented to treat the impact in 60% (9) of 15 habitat impact observations. However, one respondent reported that coir logs were used to reduce sedimentation of waterways to safeguard seagrass.

Socio-economic impacts observed by the survey respondents covered businesses reliant on marine environments (e.g. aquaculture, fishing), marine-based tourism (e.g. loss of aesthetic appeal and visitor numbers), local use of marine environments (e.g. impeded access), or wellbeing of coastal communities. The most reported impact was impeded access to water resources (9), followed by loss of stock (8) and damage or loss of infrastructure (7). Twenty-three per cent (5) of the socio-economic impact observations reported were classified as not recovered, while 32% (7) were reported as fully back to pre-fire conditions. Respondents reported that to their knowledge no measures had been implemented to treat the socio-economic impact in 11 cases, while for eight of the impacts, respondents were unsure.

Survey limitations

One of the major limitations of this survey is the time that passed between the wildfires and the survey launch, which in many cases was more than 1 year. Moreover, access to online surveys can be difficult for communities that are remote or limited in their digital literacy. Nevertheless, given the paucity of Australian-based scientific information on the effects of wildfires on aquatic environments, collecting anecdotal observations from people who live along the Australian coast and work in Australian waters has built an understanding of post-wildfire impacts. However, it is important to consider that wildfire impacts can be subtle, difficult to attribute to a specific cause, or could be inaccurately reported due to the time passed between the fires and the survey. For example, the impact of the wildfires was exacerbated by the drought that preceded them, as well as the extraordinary summer heat. The 2019–20 wildfires were then followed by significant flooding in many of the affected areas. The impacts reported by survey participants might have been attributed to wildfires when in fact they could have been the result of the combined impacts of these events. As a result, these observations should be treated as hypotheses to be tested, and a first attempt to understand the impacts that wildfires have on marine environments at a regional and national scale.

Where riparian vegetation is burnt, the radiant heat and the loss of shade (Water Quality Australia 2018) can increase water temperature and lead to the death of fish and macroinvertebrates in impacted coastal streams. For example, the richness, composition, function and resilience of aquatic macroinvertebrates, which include midges, black flies, stoneflies and caddisflies, declined after the 2013 fires in northern Victoria (Verkaik *et al.* 2015).

In Mallacoota, hundreds of birds from 25 bird species were found dead on Tip Beach during the 2019–20 wildfires, including lorikeets, cockatoos, honeyeaters and robins, possibly due to smoke inhalation and exhaustion as they tried to escape out to sea (Butt and McManus 2020).

Not all marine and coastal animals in the path of the fires and smoke will die. However, injured animals and those that have lost their habitat may eventually succumb to disease, predation and competition with other surviving wildlife. Moreover, marine habitats altered by wildfires take time to recover and might be unable to support returning fauna.

Post-fire changes to floral composition, age class and density are common and were observed during the 2019–20 wildfires. During a tour of fire grounds in February 2021, two of the authors observed that the dominant foredune canopy species of banksia were killed and not regenerating at Noosa Heads, in south-eastern Queensland, with little evidence of seedling propagation. In the South Coast of New South Wales (NSW), along the Clyde River, Mogendoura Creek and Wonboyn Lake, grey and river mangroves, which

Fig. 7.3. Dead grey mangroves (*Avicennia marina*) along the Clyde River, likely killed by radiant heat due to the 2019–20 wildfires. (Photo: OceanWatch Australia)

are rarely affected by wildfires, were likely killed by radiant heat along hundreds of metres of shoreline (Fig. 7.3; Santori and Rowe 2021).

Run-off laden with ash, debris, nutrients, metals and chemicals

Wildfires do not just burn the 'wild'. They can also engulf houses, factories, furniture, machinery and vehicles, creating a toxic cocktail of ash, carbon dioxide, debris, nutrients (phosphorus, nitrogen and potassium), metals (copper, zinc, lead and mercury), synthetics and fire-retardant chemicals that can be washed into waterways and down to estuaries and river mouths. Wildfires can mobilise industrial and leaded petrol depositions (Wu *et al.* 2017) as well as soil mercury, which can move up the food chain from macroinvertebrates to top-order predators (Turetsky *et al.* 2006; Struzik 2018). Metals in high concentrations can be toxic to marine life and affect reproduction and growth rates (Sydney Institute of Marine Science 2020).

Seagrasses meadows, kelp beds and mangrove forests provide very important ecosystem services in Australia's estuaries, including nutrient cycling, sediment stabilisation, water quality improvement and shoreline protection. They also provide refuge and nursery habitats for the flathead, whiting, bream, mulloway, prawns and shellfish targeted by commercial and recreational fishers. However, they are highly vulnerable to poor water quality, which is worsened by wildfire debris washed down coastal rivers and streams.

Mangroves rely on pneumatophores or breathing roots to extract oxygen from the air to survive in muddy shorelines and can be impacted by the smothering of their roots by sediments, algal blooms and seagrass wrack (OzCoasts 2007), all of which could increase after wildfires in the catchment.

After the 2019–20 wildfires, fire-affected waterways were described as fluid mud, sludge, inky black or resembling cake mix. Wave, wind and tidal action eventually cleared blackened beaches of ash, some of it floating into nearshore waters and the surf zone, before breaking down or sinking to the seabed and subtidal reefs. Ash was also observed underwater at North Solitary Island off the New South Wales coast near Coffs Harbour.

Along the coast and in estuaries between the Hawkesbury River and Botany Bay (NSW), monitoring by the Abyss Project showed that climate change and drought, in combination with the wildfires and subsequent severe storms and floods, had led to significant marine impacts to depths of eight metres (The Abyss Project 2020) and could lead to ecological change.

The Abyss Project had been monitoring the area for 12 years and found that changes in water quality (and ocean chemistry) following the 2019–20 wildfires had impacted soldier crabs, sea urchins, kelp, turban snails, soft sponges and coral-like bryozoan invertebrates. The post-fire recovery of some species was faster than others, and invasive species were spreading into impacted areas where they could out-compete indigenous species.

NSW Government scientists monitored dissolved oxygen levels in seven estuaries during the 2019–20 wildfire season. In the Wonboyn River estuary, this monitoring showed that dissolved oxygen declined from 4.3 mg/L to 0 mg/L in 2 weeks during February 2020, the result of the excess nutrients and accompanying phytoplankton blooms (Manjalay 2020). Lowered water levels during drought had also prevented the estuary entrance from opening, concentrating the ash and debris from catchment run-off and increasing estuary vulnerability.

Excess phosphorus, nitrogen and potassium flowing into Australia's estuaries, which are generally low in nutrients, can lead to algal blooms that, when decomposing, reduce oxygen in the water and kill fish and other marine animals. Along with the immediate wildfire impacts, there is likely to be a long-term and gradual drip-feed of nutrients into estuaries over months and years; these nutrients may accumulate and lead to algal blooms well after the fire events.

In a Before-After-Control-Impact study, Barros *et al.* (2022) demonstrated substantial impacts of the 2019–20 wildfires on estuarine habitat. Where fires were of high intensity and burnt to the estuary edge (i.e. with no protective buffer), sedimentation from burnt areas led to higher silt content and increased concentrations of nutrients and metals in estuarine benthic habitats.

Unprecedented algal blooms that developed in the Gippsland Lakes (Victoria) after fires in 2006–07 were followed by floods that led to large increases in loads of suspended sediment, nitrogen, and phosphorus above background levels (Cook *et al.* 2008). Heavy rains and thunderstorms following the 2019–20 wildfires in late January 2020 washed large volumes of ash and debris into the Tambo River in Gippsland, killing eels, commonly regarded as the hardiest of animals (McNaughton 2020). The Tambo River flows into the Gippsland Lakes, where a scientific audit is now underway to assess the impacts of the fires on the Ramsar-listed coastal lagoons (Ley and Chester 2021).

The water column and seabed can be filled or covered by ash and debris, smothering habitats and clogging the gills of fish and the feeding structures of filter feeders. Increased turbidity can reduce sunlight and hinder the ability of aquatic plants, including seagrasses, to photosynthesise. The OceanWatch Australia survey (see Box 7.1) received reports of dead luderick, flathead, mullet, bream, abalone, rock oysters, rock lobsters and mud crabs on beaches and in estuaries. The NSW Department of Primary Industries reported that at the Tilba Lake estuary there were ~2000 dead flathead and bream, 100 mullet and tens of eels and crabs (see forum post by Barnzey 2020).

Sedimentation in fire-affected streams fills habitat spaces needed by invertebrates, kills fish and impacts their recruitment (Stoessel *et al.* 2012). Ash and debris can also form sediment slugs that slowly move downstream over what can be years, impacting animals and plants along the way and potentially entering a river mouth or estuary (see Box 7.2). A sediment slug that formed in an alpine waterway after Victoria's 2003 fires travelled 200 km downstream, and reduced fish abundance by 95–100%. Signs of recovery did not appear until 2 years later (Lyon and O'Connor 2008). Avoidance behaviour, if possible, could take animals away from their preferred feeding and breeding grounds into areas where they are outcompeted by the existing wildlife.

Fire suppression activities also have the potential to harm marine life. Fire retardants contain fertilisers, corrosion inhibitors, ammonia, and chemicals such as per- and polyfluoroalkyl substances (PFAS). Fish have been shown to avoid areas of fire retardants (Giménez *et al.* 2004), which can also lead to eutrophication and fish kills if applied incorrectly (Kalabokidis 2000). PFAS have been found in Australian fur seals (*Arctocephalus pusillus doriferus*) on Phillip Island, Victoria, and could be responsible for a significant decline in local seal numbers (Taylor *et al.* 2021). Researchers suggest that seal pups likely received the chemicals in gestation or in their mother's milk, while juvenile and adult seals could have accumulated them from the fish, crustaceans, octopus and squid they eat (Taylor *et al.* 2021).

Box 7.2. Potential long-term threats from wildfires in Kangaroo Island waters

Kangaroo Island, 10 km off the South Australian coast, was devastated by the 2019–20 wildfires. More than 200 000 ha or 46% of the island was burnt between December 2019 and February 2020 (National Parks & Wildlife Service South Australia 2020).

Heavy rain began to fall on 1 February 2020, with 55 mm recorded in 1 day, which was the daily record for 2020 (Bureau of Meteorology 2021). Although the rain helped firefighting efforts, it had the potential to severely impact marine ecosystems. Many of the island's estuary systems are not permanently open to the sea, so the water could remain stagnant while being contaminated by sediments and chemicals. Then the forced opening of estuaries as wildfire run-off accumulated within them could cause nearshore turbid plumes and affect estuarine and coastal fish communities.

As soon as it was safe, University of Adelaide scientists began collecting data on water quality, metal contaminants, stable isotopes and fish. Their aims were to:

- estimate the area burnt per catchment;
- evaluate the spatial extent of turbid river plumes washing into the island's coastal areas;
- assess water quality changes due to local fluxes of wildfire-mobilised solutes and sediment run-off; and
- investigate the effects on the composition and abundance of fish species over time (Gillanders and Reis-Santos 2020).

At time of writing, data analysis was continuing. However, detrimental impacts on biota and changes in water quality have been observed, with nutrient enrichment and low dissolved oxygen in several estuarine sites, along with fish die-offs (tens to hundreds of individuals) in the aftermath of the wildfires at Middle River, Harriet River and Stun'Sail Boom, and localised mortality during 2020 (P. Reis-Santos, *pers. comm*).

Like many studies on the marine impacts of wildfires, the Kangaroo Island research was constrained by the lack of pre-fire baseline data on water quality and fish community composition. This makes it difficult to determine if/when water conditions and fish communities will return to pre-fire conditions. Nevertheless, the ability to link sediment run-off, loads and impacts on water quality will be key, and ongoing monitoring critical, in appraising the long-term marine impacts of the 2019–20 wildfires on estuaries and coastal waters. The risk of degraded water remains as wildfire-related sediments are likely to remobilise or enter estuarine systems in future run-off (P. Reis-Santos, *pers. comm.*).

Wildfire smoke settling in estuarine and offshore waters

The smoke generated by wildfires contains ash, nutrients, metals and chemicals that are released into the atmosphere and eventually settle on land, inland waterways, estuaries and distant oceans (Fig. 7.4). For example, smoke from the 2019–20 wildfires was recorded in New Zealand where it had increased atmospheric particles to 150 parts per billion in early December 2019 (normal levels are 60 parts per billion; National Institute of Water and Atmospheric Research 2019).

Haze from wildfires can decrease sunlight and undermine photosynthesis in coral reefs, mangroves and seagrasses (Jaafar and Loh 2014) and lead to a die-off of phytoplankton, a reduction in oxygen and the death of marine life (Murray and Ajani 2020). Moreover, inhaling smoke may be harmful for air-breathing marine animals such as dolphins and whales, which could suffer from lung diseases and a compromised immune system as a result (Basu 2020).

Where nutrients are at naturally high levels, there may be little response from marine life when the particles in wildfire smoke settle in seawater. In contrast, nutrient-deficient waters that receive a significant influx of nutrients can experience algal blooms (Sundarambal *et al.* 2010). This occurred in 1997 off the island of Sumatra, Indonesia, and smothered coral reefs over an area one-quarter the size of the Great Barrier Reef (Abram *et al.* 2003). Offshore algal blooms were also caused by the iron-rich smoke particles from Australia's 2019–20 wildfires, as the next case study (Box 7.3) outlines.

Box 7.3. Wildfire smoke particles cause algal bloom in the South Pacific

While coastal impacts were more visible and relatively better understood, the 2019–20 wildfires also affected marine ecosystems as far as the South Pacific Ocean (Li *et al.* 2021) by releasing iron-rich aerosol plumes into the atmosphere.

The waters of the South Pacific Ocean are limited in their iron content. This restricts the growth of phytoplankton, which require iron for photosynthesis (Boyd *et al.* 2007). The iron-rich smoke particles eventually settled on these iron-limited waters, triggering a strong response by the phytoplankton (Tang *et al.* 2021). An algal bloom appeared around October 2019 and lasted for more than four months (Tang *et al.* 2021).

The bloom in the South Pacific Ocean exceeded the size of the Australian continent, reaching the highest concentrations ever observed by the European Space Agency's Ocean Colour Climate Change Initiative, although its consequences remain unclear (Tang *et al.* 2021). It acted as a sink for the CO_2 emitted by the wildfires, absorbing 680 million t of carbon dioxide: most of the 715 million t released by the wildfires from November 2019 to January 2020 (Tang *et al.* 2021; Lu 2021). However, it could have significantly impacted fish populations by reducing dissolved oxygen in the water as the bloom decomposed, and the carbon sequestration may be only temporary (Lu 2021).

Fig. 7.4. Thick plumes of smoke rise over Batemans Bay on the southern coast of New South Wales, Australia, on 31 December 2019. (Image credit to the European Space Agency. © Contains modified Copernicus Sentinel data (2019), processed by ESA, CC BY-SA 3.0 IGO. Image was cropped from the original: https://www.esa.int/var/esa/storage/images/esa_multimedia/images/2020/01/smoke_and_flames_in_australia/21788537-1-eng-GB/Smoke_and_flames_in_Australia_pillars.jpg)

Conclusions

There are many limitations to the analysis presented in this chapter, mostly due to the very limited information available. As a result, it has sought to synthesise the available knowledge on wildfire impacts on marine species and ecosystems and summarise the post-wildfire issues. Australia's wildfire season is becoming longer, the wildfires are increasing in severity, intensity and frequency, and the impacts on marine species and ecosystems illustrated in this chapter will occur again. Hopefully, the immediate government, scientific and community responses to the wildfires and their impacts will encourage long-term improvement in the way we monitor, manage and conserve marine species and ecosystems. Acting upon the following recommendations will contribute towards mitigating negative impacts to marine ecosystems of future wildfire events.

Recommendations

- Support the establishment and operation of a collaborative, integrated, comprehensive and consistent research and monitoring program for marine environments that include components that build an understanding of wildfire impacts and their mitigation. This should engage government agencies at all levels throughout Australia, citizen scientists and research institutions in a nationally funded program to enable accurate and consistent measurement of environmental change in marine environments.
- Ensure that monitoring programs gathering data on ecosystems and marine species in fire-prone regions support management to include a focus on planning to intervene if wildfire damage to high value marine habitats can be mitigated.
- Establish protected areas that can act as refuges for surviving wildlife in fire-prone locations.
- Ensure that agencies and the natural resource management sector charged with preparing for, assessing and responding to the fire impacts have knowledge of and are sufficiently resourced to include marine ecosystems in their assessments and responses.

Acknowledgements

The authors wish to thank the scientists who advised on the preparation of this chapter, especially Dr Patrick Reis Santos, who is currently analysing the impacts of the 2019–20 wildfires on Kangaroo Island. We also thank respondents to the OceanWatch Australia survey for contributing their valuable time and observations. The 'Spatial Thinking – Mitigation of Wildfire Impacts on Our Marine Environment' project, of which the survey was an integral part, was supported by the Australian Government's investment in wildfire recovery for wildlife and their habitats.

References

Abram NJ, Gagan MK, McCulloch MT, Chappell J, Hantoro WS (2003) Coral reef death during the 1997 Indian Ocean Dipole linked to Indonesian wildfires. *Science* **301**, 952–955. doi:10.1126/science.1083841

Abram NJ, Henley BJ, Sen Gupta A, Lippmann TJR, Clarke H, *et al.* (2021) Connections of climate change and variability to large and extreme forest fires in southeast Australia. *Communications Earth and Environment* **2**, 8. doi:10.1038/s43247-020-00065-8

Australian Bureau of Statistics (2011) ABS boundaries 2011. Bioregional assessment source dataset, <http://data.bioregionalassessments.gov.au/dataset/8b65c3a4-7010-4a79-8eaa-5621b750347f>.

Barnzey (2020) 'Tilba lake had some significant kills last week'. [online forum post], *Fish Raider*, accessed 13 January 2020, <https://www.fishraider.com.au/topic/90445-dead-fish/>.

Barros TL, Bracewell SA, Mayer-Pinto M, Dafforn KA, Simpson SL, *et al.* (2022) Wildfires cause rapid changes to estuarine benthic habitat. *Environmental Pollution* **308**, 119571.

Basu B (2020) Wildfire smoke may harm whales and dolphins: here's what we know. *National Geographic*,<https://www.nationalgeographic.com/animals/article/whales-dolphins-harmed-wildfires-smoke-west>.

Boyd PW, Jickells T, Law CS, Blain S, Boyle EA, *et al.* (2007) Mesoscale iron enrichment experiments 1993–2005: synthesis and future directions. *Science* **315**, 612. doi:10.1126/science.1131669

Bureau of Meteorology (2021) Parndana (Turkey Lane) Station, <http://www.bom.gov.au/jsp/ncc/cdio/weatherData/av?p_nccObsCode=136&p_display_type=dailyDataFile&p_stn_num=022835&p_startYear=2020&p_c=-104446814>.

Butt C, McManus J (2020) 'It's a sorry sight': dead birds wash up on Mallacoota's beaches. *The Age*, Melbourne, 7 January.

Cook P, Holland D, Longmore A (2008) 'Interactions between phytoplankton dynamics, nutrient loads and the biogeochemistry of the Gippsland Lakes'. A report prepared for the Gippsland Lakes Taskforce, Water Studies Centre, Monash University, Clayton.

Gillanders BM, Reis-Santos P (2020) '*Effects of the January 2020 Bushfires on Estuarine Systems of Kangaroo Island Post Event Data Collection.*' Bushfire and Natural Hazards CRC, Melbourne.

Giménez A, Pastor E, Zarate L, Planas E, Arnaldos J (2004) Long-term forest fire retardants: a review of quality, effectiveness, application and environmental considerations. *International Journal of Wildland Fire* **13**, 1–15. doi:10.1071/WF03001

Jaafar Z, Loh T (2014) Linking land, air and sea: potential impacts of biomass burning and the resultant haze on marine ecosystems of Southeast Asia. *Global Change Biology* **20**, 2701–2707. doi:10.1111/gcb.12539

Kalabokidis KD (2000) Effects of wildfire suppression chemicals on people and the environment – a review. *Global NEST Journal* **2**, 129–137.

Ley S, Chester D (2021) Joint media release: scientists to assess Gippsland Wetlands, <https://minister.awe.gov.au/ley/media-releases/scientists-assess-gippsland-wetlands>.

Li M, Shen F, Sun X (2021) 2019-2020 Australian bushfire air particulate pollution and impact on the South Pacific Ocean. *Scientific Reports* **11**, 12288. doi:10.1038/s41598-021-91547-y

Lu D (2021) Australian wildfire smoke caused massive phytoplankton bloom in Southern Ocean. *The Guardian Australia*, 16 September (online).

Lyon JP, O'Connor JP (2008) Smoke on the water: can riverine fish populations recover following a catastrophic fire-related sediment slug? *Austral Ecology* **33**, 794–806. doi:10.1111/j.1442-9993.2008.01851.x

Manjalay B (2020) Our estuaries change after bushfires. Sydney, <https://data.nsw.gov.au/blog/our-estuaries-change-after-bushfires>.

McNaughton J (2020) Bushfire debris turns Gippsland's Tambo River to sludge, suffocating eels. *ABC Gippsland*, Gippsland, 23 January.

Murray S, Ajani P (2020) Ah shucks, how bushfires can harm and even kill our delicious oysters. *The Conversation*, 21 August (online).

National Institute of Water and Atmospheric Research (2019) Unprecedented carbon monoxide readings measured in Central Otago, <https://niwa.co.nz/news/unprecedented-carbon-monoxide-readings-measured-in-central-otago>.

National Parks & Wildlife Service South Australia (2020) Kangaroo Island bushfire 2019–20, <https://sagov-env.maps.arcgis.com/apps/Cascade/index.html?appid=ef29c850d8e34e98a8307dbdfe0aaaab>.

New South Wales Department of Planning, Industry and Environment (2020) AUS GEEBAM, <http://www.environment.gov.au/fed/catalog/search/resource/details.page?uuid=%7B8CE7D6BE-4A82-40D7-80BC-647CB1FE5C08%7D>

OzCoasts (2007) Changes in mangrove areas, <https://ozcoasts.org.au/indicators/biophysical-indicators/mangrove_areas/>.

Santori C, Rowe S (2021) Fish habitat protection protocol against bushfires. OceanWatch Australia, <https://www.oceanwatch.org.au/bushfire-recovery/>.

Stoessel D, Ayres R, Raadik TA (2012) *Improving Spawning Success for Barred Galaxias (*Galaxias fuscus*) in Streams affected by Bushfire – An Aid to Recovery, Black Saturday Victoria 2009*. Natural Values Fire Recovery Program, Department of Sustainability and Environment, Heidelberg.

Struzik E (2018) How wildfires are polluting rivers and threatening water supplies. *Yale e360*, <https://e360.yale.edu/features/howwildfires-are-polluting-rivers-and-threatening-water-supplies>.

Sundarambal P, Tkalich P, Balasubramanian R (2010) Impact of biomass burning on ocean water quality in Southeast Asia through atmospheric deposition: eutrophication modelling. *Atmospheric Chemistry and Physics* **10**, 11337–11357. doi:10.5194/acp-10-11337-2010

Sydney Institute of Marine Science (2020) Potential impacts of bushfires on our marine environment: Contaminants and toxic effects of ash to marine ecosystems, <https://www.sims.org.au/news/94/bushfire-impact-statement>.

Tang W, Llort J, Weis J, Peron MMG, Basart S, *et al.* (2021) Widespread phytoplankton blooms triggered by 2019–2020 Australian wildfires. *Nature* **597**, 370–375. doi:10.1038/s41586-021-03805-8

Taylor S, Terkildsen M, Stevenson G, de Araujo J, Yu C, *et al.* (2021) Per and polyfluoroalkyl substances (PFAS) at high concentrations in neonatal Australian pinnipeds. *The Science of the Total Environment* **786**, 147446. doi:10.1016/j.scitotenv.2021.147446

The Abyss Project (2020) New report highlights delayed, inconsistent recovery for marine life devastated by natural disasters, <https://www.scoop.co.nz/stories/WO2007/S00040/new-report-highlights-delayed-inconsistent-recovery-for-marine-life-devastated-by-natural-disasters.htm>.

Turetsky MR, Harden JW, Friedli HR, Flannigan MD, Payne N, *et al.* (2006) Wildfires threaten mercury stocks in northern soils. *Geophysical Research Letters* **33**, L16403. doi:10.1029/2005GL025595

Verkaik I, Vila-Escalé M, Rieradevall M, Baxter CV, Lake PS, *et al.* (2015) Stream macroinvertebrate community responses to fire: Are they the same in different fire-prone biogeographic regions? *Freshwater Science* **34**, 1527–1541. doi:10.1086/683370

Water Quality Australia (2018) Bushfires and water quality, <https://www.waterquality.gov.au/issues/bushfires>.

Wu L, Taylor MP, Handley HK (2017) Remobilisation of industrial lead depositions in ash during Australian wildfires. *The Science of the Total Environment* **599–600**, 1233–1240. doi:10.1016/j.scitotenv.2017.05.044

8

Ecological communities: impact and response

David A. Keith, Sarah Barrett, Rebecca Dillon, Carl R. Gosper, Robert M. Kooyman, Nina Maclean, Justin Mallee, Craig Nitschke, Libby Rumpff, Julian Seddon, Arn Tolsma, Dick Williams, Colin J. Yates and John C. Z. Woinarski

Summary

Context and challenges

- Many ecological communities (ECs) are in decline due to the impacts of inappropriate fire regimes and other threatening processes, many of which compound the impacts of fire.
- Although Australian legislation, and that of most states/territories, provides for listing of ECs as threatened (TECs), national assessment of the impacts of the 2019–20 wildfires on ECs is constrained by inconsistent EC definition, classification and mapping across jurisdictions.
- Of TECs that were nationally listed as threatened prior to the 2019–20 wildfires, three had > 40% of their distributions burnt, and another three had 20–40% of their extent burnt. Of a broader group of 92 fire-impacted ECs, including many listed as threatened by state jurisdictions, five had more than 90% of their extent burnt and 26 had more than half of their extent burnt.
- The impacts of fire on ECs are diverse, but not always negative, and consequences can include changes in species composition (including reduced abundance of foundational species), vegetation structure, ecosystem functions, resource availability and incidence and extent of other threats.
- The interval between successive fires is critical to community persistence, with irreversible changes ('ecosystem collapse') to disequilibrium states likely in some ECs, especially if inappropriate fire regimes disrupt life-cycle processes or transform habitats of key species.

Main findings

- Priority management responses needed to recover fire-affected ECs, and reduce the likelihood of further loss in future fires, include post-fire control of introduced herbivores; control of disease; protection of unburnt refuges from fire; protection of burnt areas from further fire; and protection of hydrological integrity.

- Major knowledge gaps (including the mechanisms of threats, their interactions and effective control methods) constrain effective management needed for recovery.
- Underlying these threats is a more pervasive need to reduce anthropogenic climate change. If this cannot be achieved, there will be ongoing diminution at least of rainforest, montane ash, peatland and montane heathland ecosystems.

Introduction

The 2019–20 wildfires of southern and eastern Australia burnt extensive areas of eucalypt open forests, and a wide range of environments including rainforests, eucalypt woodlands, alpine and coastal heathlands, and peatlands, with impacts extending also to aquatic and marine environments (Table 8.1; Chapters 6 and 7).

Ecological communities (ECs) are typically described at a finer scale than major vegetation groups, and are the main focus of this chapter. As defined broadly under the *Environment Protection and Biodiversity Conservation Act 1999* (EPBC Act), ECs are assemblages of native species that inhabit a particular area in nature. They may be characterised by species composition, structure, habitat features, distribution, biological interactions and ecosystem function (Keith 2009); and include a biotic component as well as the physical environment supporting that biota. ECs can be listed as threatened (TECs) under the EPBC Act, and ~90 are so listed (as of 2021). TECs are also listed in most states and territories; and ecosystems can also be listed as threatened at a global level through the IUCN Red List of Ecosystems (Keith *et al.* 2013). The listing and protection of TECs provides a

Table 8.1. Major Australian vegetation groups (as defined by the National Vegetation Information System, NVIS version 5) and their proportion burnt in the 2019–20 wildfires (from Gallagher *et al.* 2021).

Note: All other vegetation groups had < 1% of their national extent burnt.

Major vegetation group (MVG)	Total area burnt (km^2)	% of total area of MVG burnt
Eucalypt Tall Open Forests	12 016	33.8
Eucalypt Open Forests	35 695	15.7
Heathlands	1894	12.2
Cleared, modified, aquatic, unclassified	11 909	11.6
Rainforests and Vine Thickets	2562	7.1
Eucalypt Low Open Forests	746	6.7
Eucalypt Woodlands	23 847	2.8
Mallee Woodlands and Shrublands	5644	2.7
Other Shrublands	1670	1.4
Low Closed Forests and Tall Closed Shrublands	225	1.2
Other Forests and Woodlands	522	1.2
Callitris Forests and Woodlands	369	1.1
Casuarina Forests and Woodlands	156	1.0

complementary and more inclusive conservation focus to listing of individual threatened species, as it allows more opportunity to consider and manage ecological processes, and recognises the critical interactions among species. Furthermore, listing of TECs also provides a mechanism to safeguard many species, including poorly known species, that may not be individually eligible for listing because of critical knowledge gaps.

Many Australian ECs are in decline, with declines driven by multiple threats and their interactions. Fire has a complex impact on ECs. Individual fires – and the fire regimes in which they are embedded – may sustain, threaten or have negligible effect on the diversity and function of ECs. Such variable responses are associated with the inherent characteristics and pre-fire state of the EC, and of features of the fire, or fire regime, itself, as well as effects of fires on other threatening processes (Keith *et al.* 2021). In some ECs, such as rainforests, fire is an exceptional event, capable of causing long-lasting impacts, whereas it is an integral and necessary part of the ecological fabric for other ECs.

Changed or inappropriate fire regimes are causing major detrimental impacts for some ECs. Such impacts occur especially where fires recur at intervals that are too short to allow for continuation of critical life-cycle processes, such as development of habitat resources, maturation and accumulation of seed banks and development of fire-resistant organs – in such cases of 'interval squeeze', the EC may be transformed irretrievably (Enright *et al.* 2015). In some such cases, the ecological community may collapse irrevocably to a degraded state in which it has lost key defining features (including foundational species) and functions (Bergstrom *et al.* 2021). In some cases, this change is self-reinforcing as the degraded state has structural and fuel load characteristics that render future fire more likely or more severe (Camac *et al.* 2017). Individually, and within the context of their increasing frequency, recent severe wildfires, including the 2019–20 wildfires, are driving or contributing to such collapse for some Australian ecological communities including montane ash forests (Burns *et al.* 2015), conifer forests (Holz *et al.* 2020) and some heathlands (Barrett and Yates 2015).

In this chapter, we summarise the outcomes of a post-fire vulnerability assessment for ECs (Keith *et al.* 2021), providing case studies to highlight the impacts of the fires and interacting threats on some highly vulnerable TECs. The assessment built on an initial spatial assessment of nationally threatened ECs (DAWE 2020), by combining fire severity mapping with a broader range of distributional data for listed and unlisted ECs, as well as consideration of threats that interact with fire (Keith *et al.* 2021). The objective was to identify fire-affected ECs that were consequently priorities for national listing and management to support recovery. The approach was further developed by Keith *et al.* (in press) to consider a suite of 15 interactive threat types, categorised into four major groups (Table 8.2). These threat types mediate the pre-fire susceptibility of an EC to impacts from wildfire, the likelihood of compounding impacts of that fire with other threats, and the factors likely to constrain recovery. Consideration of these threats and their interactions is necessary to guide recovery efforts.

There are some noteworthy caveats on the assessments. First, their ambit has been restricted to terrestrial ECs, and no comparable assessment has yet been completed for freshwater ECs (but see Chapter 6). Second the current state of listing of Australian ECs is suboptimal, as many ECs likely to be eligible for national listing have not yet been assessed and, as of 2017, less than one-third of listed TECs were systematically monitored (Keith *et al.* 2018). Within Australia, the legislation of some jurisdictions (South Australia, Northern Territory) does not yet provide for listing TECs; there is inconsistency across jurisdictions in TEC definition and classification; and mapping of many ECs (including TECs) may be imprecise and inaccurate to varying degrees (Nicholson *et al.* 2015). As

Table 8.2. A categorisation of 15 fire-related threats to ecological communities, and the ecological mechanisms by which they may affect community structure, function and composition (modified from Keith *et al.* in press).

Threat type	Mechanism
A. *Fire regime components*	
A1. High fire frequency	Disrupts life-cycle processes for some organisms, such as seed bank accumulation, development of habitat structures and resources
A2. High fire severity	Causes long-term disruption to ecosystem structure and function; alteration to habitat suitability by limiting resources; or causes long-term population declines in species that cannot compensate high mortality with reproduction or immigration
A3. Out-of-season fires	Disrupt phenological processes, limit population turnover or expose organisms to risks of mortality
A4. Substrate fires	Consume organic substrates and cause long-term alteration to ecosystem structure and function, destroy regenerative organs and propagules, and alter habitat suitability, given long timeframes required for peat accumulation
B. *Fire-environmental interactions*	
B1. Fire–drought interactions	Drought before or after fires imposes additional stress on plants and animals, by depleting resources or health or limiting resources to support post-fire recovery.
B2. Fire–hydrological change	Fire may increase impacts of hydrological change by accelerating ecosystem adjustments to new stable states under altered water availability, with consequent changes to ecosystem structure and function.
B3. Post-fire erosion, sedimentation or pollution	Fire may increase risks of erosion and sedimentation by intense rainfall or wind events in the early post-fire period, causing long-term changes to ecosystem structure and function.
B4. Fire–climate change	Climate change may drive many of the above mechanisms, and may also alter bioclimatic habitat suitability for invasive species.
C. *Fire-biotic interactions*	
C1. Post-fire interactions with invasive herbivores	Herbivores may concentrate in post-fire regrowth, limiting survival and growth of post-fire seedlings and resprouts, and may degrade ecosystem function by reducing plant biomass and litter, and disrupting soil structure.
C2. Post-fire interactions with invasive predators	Predators may concentrate and hunt more efficiently in burnt areas, causing selective declines in faunal components of ecosystems.
C3. Fire–disease interactions	Fires accelerate or amplify impacts of disease, by increasing the invasiveness and infectiousness of the disease or by increasing susceptibility of affected organisms; this interaction may have severe impacts on foundational plant species that contribute to ecosystem structure and function.
C4: Fire–invasive plant interactions	Fires may promote invasion by introduced plants by increasing availability of nutrients, light and soil water; weed invasions may then competitively exclude some native plant species and reduce habitat suitability for some animals.
D. *Fire–human disturbance interactions*	
D1. Fire–logging interactions	Legacies of past timber extraction and post-fire disturbance (e.g. salvage logging) may disrupt post-fire regenerative processes.
D2. Fire–fragmentation interactions	Fragmentation decreases effective population sizes, and post-fire recolonisation, reducing diversity and function.
D3. Fire–human access interactions	Legacies and post-fire disturbance associated with human use may disrupt post-fire regenerative processes, with associated declines in ecosystem structure and function.

such, a comprehensive assessment of the impacts of the 2019–20 wildfires was constrained and charting post-fire recovery similarly challenged. Last, given the timeframe required to support emergency recovery decisions, the vulnerability assessments of most ECs were desktop based (see also Chapter 9). Some on-ground assessments have now been conducted, and here we present some initial findings from the case studies. With further monitoring, we expect more clarity on the impacts of this fire event.

Findings

Summary of impacts of the 2019–20 wildfires

Of TECs that were nationally listed before the 2019–20 wildfires, Keith *et al.* (in press) reported that three had > 50% of their distributions burnt (Upland Basalt Eucalypt Forests of the Sydney Basin Bioregion; Temperate Highland Peat Swamps on Sandstone; Eastern Stirling Range Montane Heath and Thicket), and another four had 20–40% within the fire footprint (New England Peppermint (*Eucalyptus nova-anglica*) Grassy Woodlands, Aquatic Root Mat Community in Caves of the Swan Coastal Plain, Alpine Sphagnum Bogs and Associated Fens, and Natural Temperate Grassland of the South Eastern Highlands). Two of these seven TECs that are currently listed as Endangered are now likely to be eligible for up-listing to Critically Endangered status. The highest priority for formal re-assessment of conservation status is for the Eastern Stirling Range Montane Heath and Thicket. An additional six nationally listed TECs, including Lowland rainforest of sub-tropical Australia, had more than 10% of their distributions burnt and major susceptibility to at least one fire-related threat (Keith *et al.* 2021).

These TECs formed part of a larger group of 92 ECs recognised by state, territory or Commonwealth authorities that occurred within the 2019–20 fire footprint. Five of these had more than 90% of their extent burnt in the 2019–20 fires, 25 had more than half of their extent burnt and 55 had more than 10% of their extent burnt (Keith *et al.* in press). Very few of these fire-affected ECs are currently listed as threatened at national level: however, as at February 2022, 16 fire-affected ECs are under formal assessment for such listing. The most affected ECs ranged from rainforests to peatlands, and included some, such as heathlands and sclerophyllous eucalypt forests, which are traditionally regarded as fire-prone and fire-adapted.

All of the ECs affected by the 2019–20 wildfires face a mix of individual and interactive threats, many of which compound, and are compounded by, the impacts of fire (Table 8.1; see also Chapters 17–19). Three fire-related threat types posed the most severe threats to large numbers of ECs: high-frequency fire (short fire interval); pre-fire drought; and post-fire invasive predator activity (Keith *et al.* in press). Given their variety and complex interactions, and the likely variable response of the constituent species in any community to threats and associated actions, management needs to address these threats in a strategic, integrated manner (Keith *et al.* 2021).

Temperate Highland Peat Swamps on Sandstone

Four peatland ECs are currently listed as threatened under the EPBC Act. In addition to fire, peatlands are subject to a range of threats, including changing climate, introduced herbivores, erosion and sedimentation, weed invasion and longwall mining.

The nationally listed Temperate Highland Peat Swamps on Sandstone TEC includes three state-listed communities, of which two were extensively burnt: Newnes Plateau shrub

Fig. 8.1. Collapse of a Newnes plateau shrub swamp peatland ecosystem caused by the combined effect of fire and hydrological disruption (due to underground longwall mining). (Photo: David Keith)

swamps (99% burnt) and Blue Mountains swamps (56%) (Tozer and Keith 2021). Approximately 20–30% of the burnt area of the TEC experienced high-severity fire (Keith *et al.* in press), as widespread drought resulted in low fuel moisture before ignition (Nolan *et al.* 2020). The responses to fire were notably affected by interactions with other threats. The hydrology of many peatlands has been disrupted by underground long-wall coal mining, leading to draining of the perched aquifer necessary to sustain the peatland community. Sampling after the 2019–20 wildfires revealed marked disparities in the impact and extent of recovery among burnt swamps, depending on whether prior underground mining had affected hydrology. Whereas many swamps recovered vigorously post-fire, in swamps where underground mining had led to substantial reduction in water retention, the fires consumed peat, killed plant regenerative organs and destroyed seed banks (Fig. 8.1). At such sites, recovery was largely subverted and the EC had collapsed, with almost no regrowth of the constituent species within a year after fire, but rather incursions of eucalypt seedlings into the drier soil, resulting in transformation from peatland EC to species-poor eucalypt thicket (Tozer and Keith 2021).

Upland bogs in the Australian Capital Territory

Nationally listed Alpine Sphagnum Bogs and Associated Fens include upland bogs in the Australian Capital Territory. They support distinct fauna and flora, including the Critically Endangered northern corroboree frog (*Pseudophryne pengilleyi*) and the Vulnerable broad-toothed rat (*Mastacomys fuscus*). Bogs intercept, filter and store large volumes of water that is slowly released into streams and rivers.

The bogs in this region are impacted by feral animals, weeds, pathogens, recreation, development, and climate change, as well as altered fire regimes, including an expected increase in peat fires. Post-fire recovery of bogs could be slowed or prevented if they are disturbed by introduced ungulates, burnt frequently over relatively short timeframes or affected by peat fires (Walsh and McDougall 2004). After being burnt, bogs can transition into grassland or heathland ecosystems (McMahon *et al.* 2015).

Upland bogs in the Australian Capital Territory are at the northern edge of the distribution of the nationally threatened EC, and have encountered three severe wildfires over the last 40 years. The most recent fire in 2020 burnt 86% of known bog sites in the territory, which completely removed all vegetation except for bleached *Sphagnum* moss and rare small unburnt patches (Fig. 8.2). Vegetation surveys conducted 9–13 months afterwards found that burnt sites on average had significantly more bare ground (43% burnt *v.* 0% unburnt) and less vegetation cover and complexity (summed cover index of 58% burnt *v.* 117% unburnt). This dramatic loss of vegetation is highly detrimental to bog fauna: sampling reported that broad-toothed rats were missing from five of six burnt survey sites, despite being found there before the fires. Further work is needed to distinguish long-term trends in these attributes from short-term fluctuations in the system as it responds to fire cycles.

After fires in 2003, Hope *et al.* (2009) inferred that re-wetting of degraded bogs was a critical priority to help ensure function and facilitate recovery. The main method used to

Fig. 8.2. Snowy Flats bog after fire showing mostly bare ground with burnt and bleached *Sphagnum* moss mounds. (Photo: Nathan Kay)

Fig. 8.3. Coir log damming the stream channel to slow and spread water throughout the peat soil, at Rotten Swamp in the Australian Capital Territory. (Photo: Nina McLean)

achieve this is the installation of leaky weirs (typically coir logs) in stream channels. Consequently, following the 2019–20 fires, the ACT Parks and Conservation Service installed leaky weirs in nine priority bogs (Fig. 8.3) with the aim to prevent and reduce erosion or incision in the event of surface disturbance (e.g. by ungulates) and heavy rains; increase peat wetness; and promote vegetation recovery. A long-term project has now been established to assess their effectiveness by tracking vegetation recovery, soil moisture, watertable depths and streambank stability at sites with and without constructed leaky weirs, with results from this monitoring contributing to potential improvements for future post-fire management.

Montane ash forests

Successive wildfires from 1998 to 2020 have occurred throughout the montane forests of Victoria. Of particular concern is that 87% of Victoria's alpine ash (*Eucalyptus delegatensis*) dominated forests (see Fig. 8.4) have fallen within these overlapping fire extents. In New South Wales, alpine ash forests have been similarly exposed to high-frequency fire, with large areas burnt in successive fires during 2003 and in 2019–20. Nationally, some 48% of mainland alpine ash forests has experienced at least two canopy fires in the past 50 years (Keith *et al.* in press). The impact of two or more high-severity fires in quick succession has led to concerns about the loss of alpine ash from these ecosystems (Bowman *et al.* 2014; Fairman *et al.* 2016) and the implementation of post-bushfire recovery and regeneration programs that sow alpine ash seeds into multiple burnt sites (Bassett *et al.* 2015). The 2019–20 wildfires led to many fire-impacted areas of *E. delegatensis* (and mountain ash (*E. regnans*)) forest across the Australian Alps being impacted by short-interval fires with ~11 500 ha reseeded to prevent the loss of these species. In Victoria, ~23% of the alpine ash has remained unburnt for many decades, 32% has burnt at low or moderate severity (on multiple occasions in some places), 39% has had a single high-severity fire while 7% has been impacted by two (5.7%), three (1.5%) or four (0.001%) fires. Given this complexity of

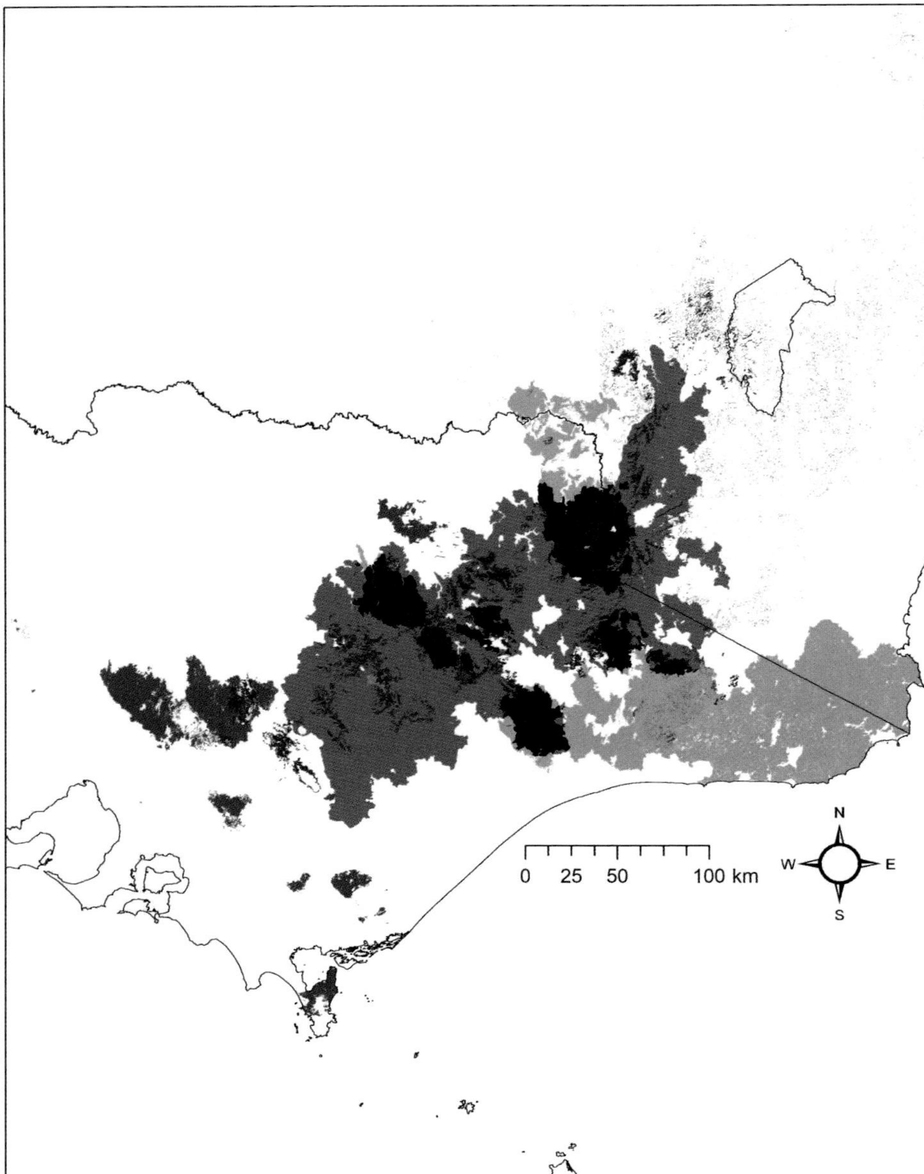

Fig. 8.4. Alpine ash extent in south-eastern Australia (dark blue) showing areas burnt by wildfires in the periods 1998 to 2018 (dark grey), 2019–2020 (light grey) and where the 2019–2020 fires overlapped with 1998 to 2018 fires (black).

fire occurrence, ~23% is at risk if a fire occurs in the near future and 7% has already been exposed to short-interval fires. This means that fires have occurred when the immaturity risk (i.e. no or limited seedbank) of these forests was high (McColl-Gausden *et al.* 2022) with ~23% of the alpine ash forest extent currently within this immaturity risk window following the 2020 bushfires. The high-risk window for complete loss of the structural

Fig. 8.5. Surviving alpine ash among standing and fallen fire-killed saplings at a site burnt by three high-severity fires (2007, 2014, 2019). Dead *Acacia obliquinervia* are prominent as is the amount of coarse wood debris from fallen ash killed in previous fires. Epicormically resprouting *E. dalrympleana* are in the background. (Photo: Craig Nitschke)

dominant of the community is likely to last for 15–20 years (McColl-Gausden *et al.* 2022). However, some alpine ash individuals have been found to produce viable seed at 8–9 years (Doherty *et al.* 2017). Nonetheless, substantial reductions in tree density and large structural changes can occur if forests are burnt in their early developmental stages.

Prior to the 2020 fires, multiple fire events had already impacted forests dominated by alpine ash and snow gum (*E. pauciflora*) and high elevation mixed eucalypt species forests. The impacts of short-interval fires have led to changes in forest structure in snow gum and mixed species forests (Fairman *et al.* 2017, 2019), a change in understorey composition in snow gum forests (Fairman *et al.* 2017) and changes in carbon stocks across all forest types (Bowman *et al.* 2014; Fairman *et al.* 2017). Assessments of the impacts of the 2019–20 wildfires in alpine ash forests have found a reduction in alpine ash abundance, reduction in standing carbon stocks, reduced sapwood area, reduced litter depth, increased coarse woody debris, increased flora species richness, and changes in species composition in areas impacted by multiple fires (Fairman *et al.* 2022). A decline in *Acacia* species in areas subjected to multiple fires has been documented, indicating that the hypothesised increase in *Acacia* dominance after multiple fires (Lindenmayer *et al.* 2011) may not be supported under scenarios involving multiple fires in montane forests. These changes may be shifting ecosystem functioning. However, there are some hopeful signs: reproductive ash as young as 7 years old have been found in burnt areas and seedlings from these trees (7–14 years old) found in recently burnt areas (Fig. 8.5). Another positive sign is the abundance of alpine ash in areas subjected to resowing. Notwithstanding these promising signs, climate change will increasingly place these montane landscapes at risk from more frequent fires. Species may persist in these landscapes but their dominance will shift; this will lead to changes in structure, composition, age distribution and availability of resources, such as tree hollows, that will influence ecosystem functioning at broader spatial scales over time.

Eastern Stirling Range Montane Heath and Thicket

The Eastern Stirling Range Montane Heath and Thicket EC occurs within the floristically diverse South-west Australian Floristic Region. It was assessed as Critically Endangered using IUCN Red List criteria for ecosystems (Barrett and Yates 2015) and is currently listed as Endangered under national legislation. The EC is geographically restricted and the impact of *Phytophthora* dieback in combination with recent wildfires has led to extensive changes in the EC (Department of Parks and Wildlife 2016). Of particular concern were wildfires in 1991 and 2000 that resulted in some 74% of the 322 ha EC experiencing a fire interval of 9 years, within the threshold for collapse for this EC (Barrett and Yates 2015).

A wildfire in autumn 2018 burnt up to 74% of the EC, and in December 2019 a second fire burnt 43% (Fig. 8.6). This habitat had been previously burnt in either 2000, 1991 or was long unburnt; therefore the shortest fire interval preceding the 2018–19 fires was 18–19 years. However, it is estimated that up to 15% of the EC burnt in both the 2018 and 2019 wildfires.

Monitoring commenced after the 2018 fire and was extended to areas burnt in 2019. There has been more abundant seedling emergence of some key *Phytophthora*-susceptible species compared with after the 2000 fire, plausibly associated with the longer fire interval. An initial analysis of long-term data suggests that species not susceptible to *Phytophthora* are relatively resilient to the recent fire regime. While the impact on the EC in areas that burnt in 2018 and 2019 is still being assessed, there has been a detrimental impact on seedlings of serotinous species such as the mountain dryandra (*Banksia montana*) that

Fig. 8.6. Eastern Stirling Range Montane Heath and Thicket TEC, autumn 2020. Habitat burnt in 2019 and 2018 shown on right and left sections of image, respectively. (Photo: Sarah Barrett)

germinated in 2018 and were burnt in 2019 (Chapter 9). Seedling densities of species with soil-stored seed such as the iconic mountain bells (*Darwinia* spp.) are also lower where the two fires overlapped. High levels of browsing of shrub species (Proteaceae, Fabaceae and Myrtaceae) has been documented in association with significant activity of the quokka (*Setonix brachyurus*), a threatened native macropod (Rathbone and Barrett 2017). Drought in the summer of 2019–20 was associated with elevated seedling mortality on shallow soils. The overall impact of *Phytophthora* dieback had been relatively low until spring 2021, when high disease impact on vulnerable juvenile plants became apparent after very wet conditions in winter and spring.

This EC is also considered in Chapters 11 and 34, in the context of the extinction of an invertebrate species narrowly endemic to the community. The example demonstrates that although ECs are often defined by plant species, ECs may also be characterised by strong interactions among constituent plant and animal species.

Gondwana rainforests

Rich in diversity and deep time paleo-history, parts of the Gondwana Rainforests of Australia World Heritage Area (GRAWHA) were severely impacted by the 2019–20 wildfires (Kooyman *et al.* 2020; Ward *et al.* 2020). The Gondwana rainforests of the south-east Queensland and northern New South Wales bioregion represent one of the three most biodiverse areas in Australia. These forests occur on a range of soil types and across elevations from near sea level to more than 1200 m. The wide range of environmental conditions produce mosaics of different rainforest types, including lowland warm subtropical, cooler upland subtropical, warm and cool temperate, and several dry rainforest types (Fig. 8.7). In combination these forests provide a diverse range of habitats for flora and fauna species that represent lineages with Gondwanan and Indo-Malesian origins (Kooyman *et al.* 2014, 2019).

For millions of years the locations that now form part of the GRAWHA reserves have acted as a refuge for ancient lineages (Kooyman *et al.* 2014; Rossetto and Kooyman 2021). In 2019–20, a combination of factors including catastrophic fire weather, a prolonged drought, and the likely influence of past land management including intensive logging of areas adjacent to rainforest, resulted in major impacts to some rainforest areas in the Nightcap Range in New South Wales (Kooyman *et al.* 2020; Lindenmayer *et al.* 2020). National Park agencies in Queensland and New South Wales commenced monitoring impacts immediately after the fires (Fig. 8.8). Targeted surveys were undertaken for threatened species such as Nightcap oak (*Eidothea hardeniana*) and long-nosed potoroo (*Potorous tridactylus*). The results have shown significant areas of important rainforest refugia survived the fires, but confirmed that many thin-barked rainforest trees are very susceptible, even to low-severity fire, with high mortality (up to 85% of standing tree stems), a feature of the most fire-affected rainforest areas. After 3 years of monitoring these sites, many dead tree stems are now falling, creating another round of forest disturbance and potentially adding to future fire risk associated with the accumulation of ground fuels and maintenance of a more open structure that promotes fuel drying. Some

Fig. 8.7. Border Ranges region with a transition from lowland to upland rainforest ECs. (Photo: Justin Mallee)

Fig. 8.8. Canopy image on permanent plot showing fire-impacted warm temperate rainforest (coachwood (*Ceratopetalum apetalum*)-dominated) with ca. 85% rainforest tree mortality. (Photo: Robert Kooyman)

rainforest species showed strong resprouting from the base despite the above-ground portion of the trees being killed, but this effect appeared to be related to the degree of basal smouldering. In these stands the impacts largely reflect changes in forest structure rather than major shifts in floristic composition. It is likely that re-establishment of canopy structure and the associated microclimate will take many decades, assuming future fires can be excluded. Regeneration trials show sometimes prolific natural rainforest regeneration from seed of a few species. However, areas of burnt warm-temperate rainforest formerly dominated by coachwood (*Ceratopetalum apetalum*) are now being dominated by the regeneration of tens of thousands of wattles (*Acacia* species), which could increase future fire risk and compete for space with the regenerating rainforest. Trials have now been implemented to both monitor and control the *Acacia* regeneration and investigate the effects of such control efforts. The interventions to protect priority (threatened) species rainforest habitats are ongoing and are a critical component of securing conservation outcomes for target taxa in the long term.

Detailed monitoring and ecological research are helping to guide decision making in these complex settings and inform management actions and fire planning to improve resilience and reduce future impacts.

Warm temperate rainforests in Victoria

Around one-third of Victoria's warm temperate rainforest and more than three-quarters of similar rainforests in south-eastern New South Wales were burnt at high severity during the 2019–20 fires (DELWP 2020); this was of great concern given the sensitivity of many rainforest species to fire and the strong influence of fire on rainforest distribution.

Post-fire sampling showed recovery of both canopy and understorey species was closely associated with local fire severity (Tolsma in press). Under low-severity fire, the rainforest canopy species lilly pilly (*Syzygium smithii*) was resprouting vigorously from buds on the trunk and branches. In contrast, under high-severity fire, resprouting was restricted to the very base of the trunk, or was absent altogether as trees were killed. The opening up of the canopy by fire has led to prolific germination of short-lived fire response species, such that the understorey is now often dominated by species that are substantially less common in unburnt rainforest, including kangaroo apple (*Solanum aviculare*), Indian weed (*Sigesbeckia orientalis*) and incense plant (*Calomeria amaranthoides*) (Tolsma in press; Fig. 8.9).

An immediate threat to burnt rainforest is invasion by eucalypt seedlings. Seedlings were establishing in rainforest stands to a distance approaching the height of the surrounding mature eucalypts, with higher seedling density closer to the original rainforest edge (Tolsma in press). Successful establishment and growth of eucalypts is likely to change the overstorey structure of severely burnt rainforest stands, especially small stands that may already be marginal. Eucalypts are predicted to increase with each successive fire in warm temperate rainforest (Chesterfield *et al.* 1990) until they eventually dominate through feedbacks in which fires promote eucalypts and eucalypts promote fires, a process known as a 'landscape trap' (Lindenmayer *et al.* 2011).

Climate change poses a longer-term threat to the extent and condition of rainforest, through its effects on direct drivers of species occupancy and its interactions with other

Fig. 8.9. Regrowth in the understorey of Warm Temperate Rainforest near Martin Creek, Victoria, currently dominated by Kangaroo Apple, a short-lived fire response species. (Photo: Arn Tolsma (DELWP 2020))

threats, particularly fire (Riddington 2014; Enright *et al.* 2015; Mariani *et al.* 2019). Warm temperate rainforests in Victoria might remain within the thresholds of their annual rainfall requirements and might be less impacted by climate change than cool temperate rainforest (Riddington 2014), but impacts will still be felt through increased frequency and intensity of fire. A self-perpetuating cycle of sclerophyll expansion and rainforest attrition will place severe limitations on the ability of rainforests to expand again to occupy their climatically- and edaphically-determined niches (Chesterfield *et al.* 1990), as those niches contract under a changing climate, and indeed limits their ability to persist in anywhere near their current (already depleted) extent.

Conclusions

The 2019–20 wildfires had major impacts on some nationally listed TECs and on many ECs not yet nationally listed. While the acute impacts of the 2019–20 Australian wildfires are significant conservation concerns, far-reaching ecological effects are best identified and understood by placing such single events in the context of fire regimes and interacting threats. In contrast, a disaster approach to conservation that has a more singular focus – on fire impacts and recovery from the fire event itself – runs the risk of focusing management responses on the treatment of those symptoms without addressing the full suite of their underlying causes.

Global climate change will drive an increase in the incidence and severity of wildfires, rendering some ECs likely to undergo continuing decline or, in some cases, collapse.

Recommendations

- Impacts of any fire event, even a catastrophic megafire, should be considered systematically within a broader context of interacting and compounding threats, with such information necessary to guide risk assessment, proactive management and post-fire recovery. Such a mechanistic understanding of fire-related risks to ecosystems needs to be built into planning and capacity building for future contingencies.
- For ECs affected by the 2019–20 wildfires, strategic and integrated management of significant threats (additional to fire) is required to support recovery and reduce susceptibility to future fires.
- For at least some ECs affected by the 2019–20 wildfires, more effective and targeted management is required to protect unburnt areas from future fire, to identify the locations and safeguard any refugial areas, and to seek to avoid the recurrence of fire until burnt areas have recovered sufficiently.
- National and state/territory processes should be undertaken to list (or up-list) as threatened the ECs for which the 2019–20 wildfires, and compounding factors, have resulted in losses and other impacts that meet assessment criteria. Moreover, more consistency among Australian jurisdictions in the classification, mapping and listing of ECs are required to enhance assessment of their national status and coordination of their conservation.
- Major knowledge gaps constrain assessments of the impacts of the 2019–20 wildfires, will hamper assessment of the conservation status of fire-affected ECs, hinder recovery efforts, and subvert preparedness for future wildfires. There is a need to undertake research to understand the fire regime requirements of different ECs, to complement spatial assessments of overlap with on-ground research and sampling to verify threats and impact, and to identify and refine priority management actions needed for recovery.

- A systematic monitoring program should be expanded to encompass all TECs – one that can provide a benchmark against which to measure impacts of future fires, help chart post-fire recovery, assess the effectiveness of threat management, and provide early warnings of undesirable trajectories.
- The ongoing decline of ECs that are highly susceptible to the increasing trend of more, and more severe, fires (e.g. rainforests) should be highlighted, and areas with these ECs should be recognised as critical heritage assets for protection in fire planning and operations.
- The most effective and immediate risk-reduction strategy for fire-impacted ECs is to reduce the extent and/or severity of future fires, where possible through reducing the rate of ignitions, early fire suppression and strategic fuel management within or in proximity to the ECs.
- The longer-term imperative is to constrain global climate change. If this cannot be done, the corollary of more frequent and severe droughts and fires will render Australia increasingly unsuitable for the persistence of, at least, rainforest and peatland ecosystems across large areas of their current range (Keith *et al.* 2021).

Acknowledgements

We thank Payal Bal for the analysis shown in Table 8.1.

References

Barrett S, Yates CJ (2015) Risks to a mountain summit ecosystem with endemic biota in southwestern Australia. *Austral Ecology* **40**, 423–432. doi:10.1111/aec.12199

Bassett OD, Prior LD, Slijkerman CM, Jamieson D, Bowman DMJS (2015) Aerial sowing stopped the loss of alpine ash (*Eucalyptus delegatensis*) forests burnt by three short-interval fires in the Alpine National Park, Victoria, Australia. *Forest Ecology and Management* **342**, 39–48. doi:10.1016/j.foreco.2015.01.008

Bergstrom DM, Wienecke BC, van den Hoff J, Hughes L, Lindenmayer DB, *et al.* (2021) Combating ecosystem collapse from the tropics to the Antarctic. *Global Change Biology* **27**, 1692–1703. doi:10.1111/gcb.15539

Bowman DMJS, Murphy BP, Neyland DLJ, Williamson GJ, Prior LD (2014) Abrupt fire regime change may cause landscape-wide loss of mature obligate seeder forests. *Global Change Biology* **20**, 1008–1015. doi:10.1111/gcb.12433

Burns EL, Lindenmayer DB, Stein J, Blanchard W, McBurney L, *et al.* (2015) Ecosystem assessment of mountain ash forest in the Central Highlands of Victoria, south-eastern Australia. *Austral Ecology* **40**, 386–399. doi:10.1111/aec.12200

Camac JS, Williams RJ, Wahren CH, Hoffmann AA, Vesk PA (2017) Climatic warming strengthens a positive feedback between alpine shrubs and fire. *Global Change Biology* **23**, 3249–3258. doi:10.1111/gcb.13614

Chesterfield EA, Taylor SJ, Molnar CD (1990) 'Recovery after wildfire: warm temperate rainforest at Jones Creek. Arthur Rylah Institute for Environmental Research, Technical Report Series No. 101'. Department of Conservation Forests and Lands, Melbourne.

DAWE (2020) *Wildlife and Habitat Bushfire Recovery Program 2019–20 to 2020–21. Australian Government's List of Priority Matters – Tranche 1: Threatened Ecological Communities.* Department of Agriculture, Water and Environment, Canberra.

DELWP (2020) *Victoria's Bushfire Emergency: Biodiversity Response and Recovery. Version 2.* Department of Environment Land Water and Planning, East Melbourne.

Department of Parks and Wildlife (2016) *Montane Heath and Thicket of the Southwest Botanical Province, Above Approximately 900 m above Sea Level (Eastern Stirling Range Montane Heath and Thicket Community). Interim Recovery Plan 2016–2021*. Department of Parks and Wildlife, Perth.

Doherty MD, Gill M, Cary J, Austin MP (2017) Seed viability of early maturing alpine ash (*Eucalyptus delegatensis* subsp. *delegatensis*) in the Australian Alps, south-eastern Australia, and its implications for management under changing fire regimes. *Australian Journal of Botany* **65**, 517–523. doi:10.1071/BT17068

Enright NJ, Fontaine JB, Bowman DMJS, Bradstock RA, Williams RJ (2015) Interval squeeze: altered fire regimes and demographic responses interact to threaten woody species persistence as climate changes. *Frontiers in Ecology and the Environment* **13**, 265–272. doi:10.1890/140231

Fairman TA, Nitschke CR, Bennett LT (2016) Too much, too soon? A review of the effects of increasing wildfire frequency on tree mortality and regeneration in temperate eucalypt forests. *International Journal of Wildland Fire* **25**, 831–848. doi:10.1071/WF15010

Fairman T, Bennett L, Tupper S, Nitschke CR (2017) Frequent wildfires erode tree persistence and alters stand structure and initial composition of a fire-tolerant subalpine forest. *Journal of Vegetation Science* **28**, 1151–1165. doi:10.1111/jvs.12575

Fairman T, Bennett L, Nitschke CR (2019) Short-interval wildfires increase likelihood of resprouting failure in fire-tolerant trees. *Journal of Environmental Management* **231**, 59–65. doi:10.1016/j.jenvman.2018.10.021

Fairman TA, Nitschke CR, Bennett LT (2022) Carbon stocks and stability are diminished by short-interval wildfires in fire-tolerant eucalypt forests. *Forest Ecology and Management* **505**, 119919. doi:10.1016/j.foreco.2021.119919

Gallagher RV, Allen S, Mackenzie BD, Yates CJ, Gosper CR, *et al.* (2021) High fire frequency and the impact of the 2019–2020 megafires on Australian plant diversity. *Diversity & Distributions* **27**, 1166–1179. doi:10.1111/ddi.13265

Holz A, Wood SW, Ward C, Veblen TT, Bowman DMJS (2020) Population collapse and retreat to fire refugia of the Tasmanian endemic conifer *Athrotaxis selaginoides* following the transition from Aboriginal to European fire management. *Global Change Biology* **26**, 3108–3121. doi:10.1111/gcb.15031

Hope G, Nanson R, Flett I (2009) 'The peat-forming mires of the Australian Capital Territory. Technical report 19'. Territory and Municipal Services, Canberra.

Keith DA (2009) The interpretation, assessment and conservation of ecological communities. *Ecological Management & Restoration* **10**, S3–S15. doi:10.1111/j.1442-8903.2009.00453.x

Keith DA, Rodríguez JP, Rodríguez-Clark KM, Nicholson E, Aapala K, *et al.* (2013) Scientific foundations for an IUCN Red List of Ecosystems. *PLoS One* **8**, e62111. doi:10.1371/journal.pone.0062111

Keith DA, Pellow BJ, Appleby M (2018) Monitoring threatened ecosystems and ecological communities. In *Monitoring Threatened Species and Ecological Communities.* (Eds S Legge, DB Lindenmayer, NM Robinson, BC Scheele, DM Southwell and BA Wintle) pp. 101–126. (CSIRO Publishing: Clayton)

Keith DA, Auld TD, Barrett S, English V, Gallagher R, *et al.* (2021) 'Terrestrial ecological communities in Australia: initial assessment and management after the 2019–20 bushfires. Report to the Wildlife and Habitat Bushfire Expert Panel and the Australian Department of Agriculture, Water and Environment'. Centre for Ecosystem Science, University of NSW, Sydney.

Keith DA, Allen SP, Gallagher R, Mackenzie BDE, Auld TD, *et al.* (in press) Fire-related threats and transformational change in Australian ecosystems. *Global Ecology and Biogeography*. doi:10.1111/geb.13500

Kooyman RM, Wilf P, Barreda VD, Carpenter RJ, Jordan GJ, *et al.* (2014) Paleo-Antarctic rainforest into the modern Old World Tropics: the rich past and threatened future of the 'Southern Wet Forest Survivors'. *American Journal of Botany* **101**, 2121–2135. doi:10.3732/ajb.1400340

Kooyman RM, Morley RJ, Crayn DM, Joyce EM, Rossetto M, *et al.* (2019) Origins and assembly of Malesian rainforests. *Annual Review of Ecology, Evolution, and Systematics* **50**, 119–143. doi:10.1146/annurev-ecolsys-110218-024737

Kooyman RM, Watson J, Wilf P (2020) Protect Australia's Gondwana rainforests. *Science* **367**, 1083. doi:10.1126/science.abb2046

Lindenmayer DB, Hobbs RJ, Likens GE, Krebs CJ, Banks SC (2011) Newly discovered landscape traps produce regime shifts in wet forests. *Proceedings of the National Academy of Sciences of the United States of America* **108**, 15887–15891. doi:10.1073/pnas.1110245108

Lindenmayer DB, Kooyman RM, Taylor C, Ward M, Watson JEM (2020) Recent Australian wildfires made worse by logging and associated forest management. *Nature Ecology & Evolution* **4**, 898–900. doi:10.1038/s41559-020-1195-5

Mariani M, Fletcher M-S, Haberle S, Chin H, Zawadzki A, *et al.* (2019) Climate change reduces resilience to fire in subalpine rainforests. *Global Change Biology* **25**, 2030–2042. doi:10.1111/gcb.14609

McColl-Gausden SC, Bennett LT, Ababei DA, Clarke HG, Penman TD (2022) Future fire regimes increase risks to obligate-seeder forests. *Diversity & Distributions* **28**, 542–558.

McMahon A, Tolsma A, McMahon J, Coates F, Lawrence R (2015) *Victorian Alpine Peatlands Spatial Action Plan: A Framework for Managing Victoria's Peatlands.* Ecology Australia, Fairfield, Australia.

Nicholson E, Regan TJ, Auld TD, Burns EL, Chisholm LA, *et al.* (2015) Towards consistency, rigour and compatibility of risk assessments for ecosystems and ecological communities. *Austral Ecology* **40**, 347–363. doi:10.1111/aec.12148

Nolan RH, Boer MM, Collins L, Resco de Dios V, Clarke H, *et al.* (2020) Causes and consequences of eastern Australia's 2019–20 season of mega-fires. *Global Change Biology* **26**, 1039–1041. doi:10.1111/gcb.14987

Rathbone DA, Barrett S (2017) Vertebrate browsing impacts in a threatened montane plant community and implications for management. *Ecological Management & Restoration* **18**, 164–171. doi:10.1111/emr.12259

Riddington M (2014) Victoria's rainforests and the potential impacts of a changing climate. *Victorian Naturalist* **131**, 209–218.

Rossetto M, Kooyman RM (2021) Conserving refugia – what are we protecting and why? *Diversity (Basel)* **13**, 67. doi:10.3390/d13020067

Tolsma A (in press) Rainforest surveys. In *Biodiversity Response and Recovery Supplementary Report: Bushfire Impacts on Species in Victoria*. pp. 211–225. Department of Environment, Land, Water and Planning, Melbourne.

Tozer MG, Keith DA (2021) *Fire-related Risks and Conservation Strategies for Ecological Communities.* NESP Threatened Species Recovery Hub, Brisbane.

Walsh NG, McDougall KL (2004) Progress in the recovery of the flora of treeless subalpine vegetation in Kosciuszko National Park after the 2003 fires. *Cunninghamia* **8**, 439–452.

Ward M, Tulloch AIT, Radford JQ, Williams BA, Reside AE, *et al.* (2020) Impact of 2019–2020 mega-fires on Australian fauna habitat. *Nature Ecology & Evolution* **4**, 1321–1326. doi:10.1038/s41559-020-1251-1

9

Blackened roots and green shoots: emerging trends in decline and recovery in Australian plant species after the 2019–20 wildfires

Rachael V. Gallagher, Sarah Barrett, Stephen A. J. Bell, Lachlan M. Copeland, Rebecca Dillon, Carl R. Gosper, David A. Keith, Tom D. Le Breton, Berin D. E. Mackenzie, Andre Messina, V. John Neldner, Michael R. Ngugi, Rachael H. Nolan, Mark K. J. Ooi, Elizabeth M. Tasker, Mark Tozer, Neville Walsh, Colin J. Yates and Tony D. Auld

Summary

- Australian plant diversity is maintained by variable fire regimes, and a single fire event – even one as large as the 2019–20 season – will have both positive and negative outcomes for species.
- We describe how impacts were assessed after the 2019–20 fires using vulnerability assessments and evidence from field-based observations.
- Many species are recovering well, and surprise findings of new populations are being reported across burnt areas.
- However, the 2019–20 fires have compounded threats already active across the range of many species, including Wollemi pine (*Wollemia nobilis*), Stirling Range dryandra (*Banksia montana*) and some terrestrial orchids.

Introduction

Previous analyses found that the 2019–20 wildfires of eastern and southern Australia burnt across suitable habitat for 17 197 plant taxa (hereafter 'species') – 69% of the currently described flora of Australia (Gallagher *et al.* 2021; Table 9.1). Collections made since 1950 confirmed the relatively recent presence of 9092 of these species across the fire grounds, including 587 plants listed as threatened under national legislation (44% of Australia's threatened plants). For many plant species, the 2019–20 fires brought welcome renewal to populations by promoting recruitment. However, the presence of several key pre-existing threats may jeopardise the long-term viability of many fire-affected populations and some species. Identifying which species may be most vulnerable to poor recovery involves (1) undertaking a rapid vulnerability

Table 9.1. Extent of area burnt in the 2019–20 fires across the ranges of 26 062 plant species.

All analyses were limited to fires within bioregions in the southern half of the continent. Plant geographical ranges were sourced from both cleaned occurrence records from the Australian Virtual Herbarium (https://avh.chah.org.au/) and species distribution modelling (see Andrew *et al.* 2021) or range mapping conducted for listing advice under the *Environment Protection and Biodiversity Conservation Act 1999* (EPBC Act) for threatened taxa. The spread of values provided reflects the use of these different range size estimates.

Species group	Count	Plant species with range data available	Plant species impacted by 2019–20 fires (%)	Plant species impacted by 2019–20 fires (count)	Plant species with range > 90% burnt (count (%))	Plant species with range > 50% burnt (count (%))	Plant species with range > 30% burnt (count (%))
NSW endemics	1320	1248	77–92%	956–1152	52–104 (4–8%)	306–311 (25%)	519–626 (42–50%)
WA endemics	8952	8578	32–68%	2754–5822	4–10 (< 1%)	40–45 (< 1%)	98–190 (1–2%)
Vic endemics	408	362	32–80%	113–278	12–13 (3–4%)	28–31 (8–9%)	43–57 (12–16%)
SA endemics	488	457	31–70%	160–319	3–6 (0–1%)	37–38 (8%)	56–58 (12–13%)
Qld endemics	3629	3436	11–54%	388–1865	1–3 (< 1%)	3–11 (< 1%)	19–30 (< 1%)
Tas endemics	543	525	9–77%	49–402	0 (0%)	0 (0%)	1–2 (< 1%)
ACT endemics	6	6	33–83%	2–5	0 (0%)	1 (17%)	1 (17%)
EPBC Act listed species	1335	1333	44%	585	35 (3%)	90 (7%)	148 (11%)
Listed as threatened in NSW	700	656	59–91%	386–598	32–34 (5%)	79–107 (12–16%)	139–199 (21–30%)
Listed as threatened in WA	436	434	13–51%	56–220	1–2 (< 1%)	4–12 (1–3%)	24–25 (6%)
Listed as threatened in Vic	1770	1617	60–93%	963–1499	17–18 (1%)	76–95 (5–6%)	229–249 (14–15%)
Listed as threatened in SA	807	769	53–87%	404–665	3–4 (< 1%)	19–21 (2–3%)	34–40 (4–5%)
Listed as threatened in Qld	935	899	15–43%	138–387	0–2 (< 1%)	3–14 (< 1%)	31–38 (3–4%)
Listed as threatened in Tas	460	434	59–79%	257–344	0–1 (< 1%)	2 (< 1%)	14–24 (3–6%)
Listed as threatened in ACT	13	13	31–85%	4–11	0–1 (0–8%)	0–1 (0–8%)	1–2 (8–15%)
All other taxa	7976	7969	40–65%	3203–5197	2–6 (< 1%)	23–71 (< 1%)	216–342 (3–4%)
Total	**26 062**	**25 052**	**36–69%**	**9092–17197**	**90–153 (< 1%)**	**517–593 (3%)**	**1319–1461 (5–6%)**

assessment, and (2) conducting field surveys to confirm, or revise, predictions about at-risk taxa.

To assess vulnerability to poor post-fire recovery, a framework that uses information on the extent of burning across a species' range, species fire-response traits and susceptibility to 11 key threats (Table 9.2; Auld *et al.* unpublished) was developed. From March to July 2020, this framework was tested using geographical range data (Andrew *et al.* 2021), trait information (Falster *et al.* 2021) and knowledge of the susceptibility of plant species to threats (Gallagher 2020). The primary goal of that desktop exercise was to identify the species in urgent need of surveys and conservation management interventions. In many instances ongoing monitoring remains essential to determine population recovery. This was the first prioritisation exercise across the entire breadth of the Australian flora, and it formed the evidence base for post-fire conservation prioritisation for plants by the Commonwealth and some state governments.

Of the 26 062 plant species assessed in the vulnerability analysis (Gallagher *et al.* 2021), 486 were prioritised as requiring immediate actions. These taxa had more than 80% of their range burnt or were listed as Endangered or Critically Endangered, or were considered as highly vulnerable to two or more threats. Threats potentially affecting the most species included pre-fire drought, high-severity fire, disease, and the cumulative impact of short fire intervals (Table 9.2). Overall, at least 831 species were ranked at high vulnerability to the interactive effects of existing threats and the impacts of the 2019–20 fires (Table 9.2).

While a desktop assessment could be completed rapidly to help inform decision making, the findings required confirmation. For instance, the outcomes of spatial analysis of vulnerability rely on the precision of datasets describing fire, as well as the quality of occurrence data and threat mapping. For many species, information on fire response remains unknown and therefore sensitivity to fire can only be assessed through field visits.

Table 9.2. Eleven criteria used to assess the vulnerability of plant species to recovery post-fire and their ranking after assessment.

Values in body of table are numbers of plant taxa. Criteria are described fully in Auld *et al.* (unpublished) and all methods for assessment are provided in Auld *et al.* (2020) and Gallagher (2020).

Criteria	High	Medium	Low	None	Data deficient
A. Interactive effects of fire and drought	235	340	10 375	14 102	1010
B. Short fire intervals (impacts of high fire frequency)	147	485	14 718	9498	1214
C. Post-fire herbivore impacts	97	47	4820	14 425	6673
D. Fire–disease interactions	186	444	3117	17 537	4778
E. High fire severity	258	474	16 080	8240	1010
F. Weed invasion	3	8	11 859	13 182	1010
G. Elevated winter temperatures or changed temperature regimes	9	19	235	25 654	145
H. Fire sensitivity	13	23	332	25 693	1
I. Post-fire erosion	4	17	5766	19 265	1010
J. Cumulative exposure to high fire frequency	291	593	1806	8543	14 829
K. Other plausible threats or expert-driven nominations	35	0	0	0	26 027

Below we use a set of examples to illustrate impacts and recovery of species and plant communities following the 2019–20 fires. We conclude with recommendations that aim to enhance plant conservation in a future likely to be marked by an increasing incidence of such wildfires.

Assessing impacts on iconic threatened species

Wollemi pine

Wollemi pine (*Wollemia nobilis*) is a Critically Endangered palaeoendemic conifer, restricted to a single population comprising just 46 mature individuals in Wollemi National Park, NSW (Mackenzie and Auld in press). It is a long-lived rainforest tree that reproduces sexually via seed and vegetatively via basal coppicing, both in the absence of, and in response to, disturbance (Mackenzie *et al.* 2021). Individual trees are usually multi-stemmed, each of which can live more than 400 years. There is no seed bank, but the population contains a bank of shade-tolerant, slow-growing juveniles (Zimmer *et al.* 2014).

The wild population of Wollemi pine came under threat during the 2019–20 fires and was the subject of an intensive, multifaceted, and much-lauded, multi-agency fire suppression effort (Hannam 2020; Fig. 9.1). The groves were still variously impacted by understorey fires; however, the intervention probably reduced the severity of impacts (Mackenzie and Auld).

Wollemi pine was assessed as high risk of poor recovery in post-fire vulnerability assessments (Auld *et al.* 2020; Gallagher 2020). The species' entire range was burnt and key risk drivers in these assessments included the prevalence of high fire frequency (Criterion B, Table 9.2), pathogens (Criterion D), and fire sensitivity (Criterion H) due to the species' longevity, slow growth and maturation, and vulnerability to basal collapse following prolonged charring (Auld *et al.* unpublished).

Post-fire inspections revealed that the majority of adult trees had survived the fires with their upper canopies intact; however, several large adults lost significant trunks via basal charring and impacts from falling rocks and trees (Mackenzie, Clarke and Auld, unpublished data). Delayed mortality of several individuals known to be infected with the pathogen *Phytophthora cinnamomi* was also noted post-fire and may be attributable to the interactive effects of disease and fire. Juvenile trees were disproportionately impacted by the fires with 95% of individuals < 2 m tall eliminated and ~60% of individuals > 2 m tall topkilled, only half of which had resprouted at 1.5 years post-fire (Mackenzie, Clarke and Auld, unpublished data). It is likely to take a decade or more to re-establish the juvenile cohort of plants. Three translocated populations established between 2012 and 2019 were also heavily impacted by the fires, although two of the populations have subsequently been re-established through additional planting (Mackenzie *et al.* 2021).

This case highlights the vulnerability of resprouters – especially long-lived species that are slow to develop fire-resistance and/or have low population turnover – to frequent or severe fire (although even fires of low severity may still cause significant damage, as was observed following the 2019–20 fires). A living collection and a seed bank of Wollemi pine maintained at the Australian Botanic Garden contains a broad range of genetic material from the wild population, which may enable the potential reintroduction of genetic diversity, if required (see also Chapter 27). Active fire management will be integral to the ongoing survival and conservation of the remaining wild populations to ensure appropriate fire-free intervals for post-fire recovery and recruitment (Mackenzie and Auld in press).

Fig. 9.1. Impacts of the 2019–20 fires on Wollemi pine and early stages of post-fire recovery. (Photos: (A) John Spencer/NPWS, (B) Phil Lamrock/NPWS, (C) Steve Clarke/DPE, (D, E) Berin Mackenzie/DPE)

Orchids

Orchids are considered iconic by many Australians and are subject to much public interest. Their response to fire is determined by the traits of their life cycle, as well as fire severity and timing. Metcalfe's greenhood (*Pterostylis metcalfei*) (Fig. 9.2), listed as Endangered in New South Wales, is a terrestrial orchid restricted to three populations, with many hundreds of plants in dozens of clusters, in north-eastern New South Wales (Metcalfe 2017). Most of these plants were burnt in a particularly intense wildfire in November 2019. Post-fire surveys (Eco Logical Australia 2021) documented a significant decline in the number of plants. Burnt areas originally known to contain hundreds of plants yielded only nine and 24 plants in 2020 and 2021 respectively. Most clusters of plants appear to have disappeared with no leaves or flowers visible in the two post-fire surveys. In contrast, a small adjacent, unburnt patch sampled in the same surveys yielded 97 and 89 plants respectively.

Fig. 9.2. (A) Metcalfe's greenhood; and (B) dense, post-fire infestation of invasive plants such as flax-leaf fleabane and spear thistle are compromising the recovery of Metcalfe's greenhood. (Photos: Lachlan Copeland)

The apparent decline in Metcalfe's greenhood in the severely burnt areas is probably a result of both direct mortality and habitat changes driven by the colonisation of weeds after fire (Criteria E and F respectively, Table 9.2). Death during or immediately after the fires was probably caused by burning that went deep into the leaf litter and killed the shallow tubers from which plants reshoot each year. Vegetation regrowth following the fires has also resulted in a very different physiognomic structure. Prior to the 2019 fire, the orchids grew in a tall moist open forest with a sparse shrub layer, a moderate density of *Poa sieberiana*, and relatively few weeds. The same area now has a very dense layer of introduced forbs such as flax-leaf fleabane (*Erigeron bonariensis*) and spear thistle (*Cirsium vulgare*). The decline of Metcalfe's greenhood in the burnt areas highlights how high fire severity and weed invasion can interact to create negative impacts on some ground orchids.

Stirling Range dryandra

The Critically Endangered Stirling Range dryandra (*Banksia montana*) is a fire-killed shrub with a canopy-held seed bank that is endemic to higher-elevation portions of the eastern Stirling Range in south-western Australia (Fig. 9.3). In common with many other threatened flora of the Eastern Stirling Range Montane Heath and Thicket ecological community (see Chapter 8), the Stirling Range dryandra is threatened by the combination of high susceptibility to *Phytophthora* dieback and exposure to short fire intervals (Criteria D and B respectively, Table 9.2) with a 9-year interval between fires in 1991 and 2000 being particularly damaging (Barrett and Yates 2015). Although known from six populations historically, two of these became extirpated by 2000 (Barrett *et al.* 2005).

In 2017, a full census of all populations recorded a total of 38 reproductively mature individuals, ranging in age from 17 years (recruited after the 2000 fire) to > 45 years, along with a handful of juveniles (Fig. 9.3). A fire in May 2018 killed all but eight of the mature individuals but led to the post-fire recruitment of 1354 seedlings – a much larger seedling cohort than seen after the 9-year interval in 2000. Another fire in December 2019 killed the remaining mature individuals, along with 36 of the seedlings that had germinated in 2018. The species was subsequently assessed as high risk in post-fire vulnerability assessments, and needing urgent intervention (Gallagher 2020).

Post-fire recruitment after the 2019 fire was more modest than in 2018, perhaps related to senescence in the > 45-year-old plants, and did not offset the mortality in the 2018

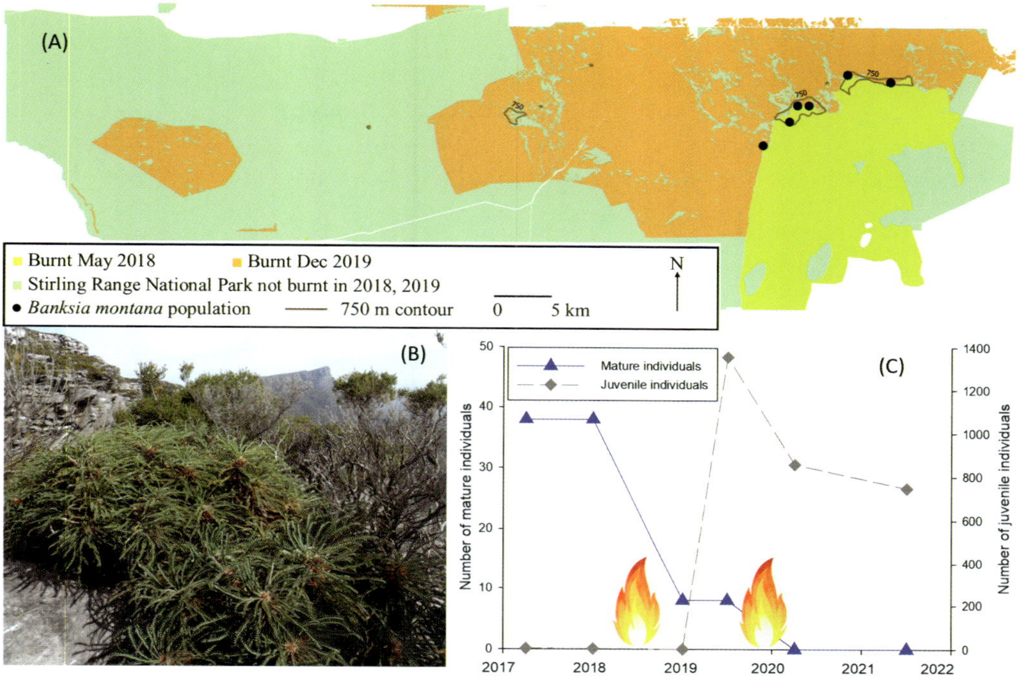

Fig. 9.3. Impacts of recent fires on Stirling Range dryandra: (A) Stirling Range National Park, showing the extent of fires in May 2018 and December 2019 and known occurrences of Stirling Range dryandra (note that the two most southerly populations became extinct by 2000); (B) mature plants pre-fire; and (C) temporal trends in the total number of reproductively mature and juvenile individuals over time, aggregated across all populations, in relation to the two recent fire events. Note that mature and juvenile individual axes have different scales. (Photo: (B): Sarah Barrett; Images: (A, C): Carl Gosper)

cohort. While the size of the juvenile cohort is still large, and has the potential to lead to a larger mature plant cohort than existed before the fires in 2018 and 2019, substantial decline in the number of seedlings over time is expected associated with *Phytophthora* dieback and mortality due to browsing, drought and competition.

Currently, there are no reproductively mature individuals in the wild, although small numbers occur in two conservation translocations outside the species' natural range. The combination of being a fire-killed species with a canopy-held seed bank, having a long juvenile period associated with a low-productivity ecosystem (Barrett and Yates 2015; Gosper *et al.* 2022), and now having populations solely consisting of juvenile plants, renders this species highly exposed to the risk of future fires (Criterion J, Table 9.2). The absence of fire for 20 years or more from areas with recruits will maximise the probability that sufficient seed accumulates to support self-replacement or population expansion after a future fire (Barrett and Yates 2015; Gosper *et al.* 2022).

New populations discovered as part of post-fire field surveys

The slender myoporum or weeping boobialla (*Myoporum floribundum*) (Fig. 9.4) is a shrub recognised as Endangered in Victoria. The species is known from only a handful of populations, most consisting of only a few plants, in the upper catchment of the Snowy River. It

Fig. 9.4. Slender myoporum, taken on the Ingeegoodbee Track, north-east of Suggan Buggan, in early 2022. (Photo: Andre Messina)

represents a species with many data gaps, so its vulnerability was difficult to assess after the fires (Gallagher 2020). A post-fire survey in the Snowy River Gorge area turned up a new, substantial population in a severely burnt area (Criterion E, Table 9.2), highlighting the critical importance of field surveys. The new population had ~100 plants, all regenerated from seed and all high above the flood level of the river, indicating that a population existed there pre-fire and had not recruited from flood-transported seed. Cuttings taken are now growing well at the Royal Botanic Gardens in Victoria.

Fire ephemerals emerge, increasing population size and extent

The Endangered sub-shrub *Androcalva rosea* is a short-lived species once thought to have a highly restricted distribution in the Hunter Valley, NSW. It was known only from four small populations in previously burnt scrubby heath on infertile sandstones, and it was postulated that this species may respond positively to fire (Bell and Copeland 2004).

New populations in their thousands were discovered 30 km away following a fire in 2013 in parts of northern Wollemi National Park that had remained unburnt for up to 70 years (Bell and Holzinger 2015). These plants had evidently germinated from soil-stored seed after decades of fire absence. After the 2019–20 fires, further new populations were uncovered, increasing again the species' known total population size and geographical extent (Bell and Woolley 2021). One new stand discovered on private land at Paynes Crossing proved particularly significant and established a new eastern limit. Although only several hundred plants were present here, this population lay 60 km from the nearest previously known occurrence of the species. Targeted surveys for *A. rosea* conducted around 18 months after the 2019–20 fires have resulted in a significant boost to the number of individuals and populations of this species, and dramatically increased its known geographical range.

The Endangered dwarf kerrawang (*Commersonia prostrata*) (Fig. 9.5) provides another example of how post-fire surveys can be important for locating populations of fire-ephemeral species. Dwarf kerrawang is an elusive species, generally only appearing for a few years following a disturbance such as fire or flood. Prior to the 2019–20 fires, in Victoria this species had only been recorded west of Bairnsdale in the Holey Plains and Providence Ponds areas. However, a small population of this ground-hugging shrub was observed on the banks of the Genoa River during post-fire surveys. This represents a range extension of some 200 km to the east of the known Victorian populations and links them to populations further north in New South Wales and Queensland.

The closely related genera *Androcalva* and *Commersonia* have demonstrated remarkable fire responses following the 2019–20 fires. As well as *A. rosea* and *C. prostrata* noted above, two extremely rare species in Victoria, *C. breviseta* and *C. pannosa*, known from one (1989) and three (1850, 1950 and 1952) herbarium collections respectively, were rediscovered during surveys in 2021 and 2022. Furthermore, a species never previously recorded

Fig. 9.5. A new population of dwarf kerrawang was found in Victoria in post-fire surveys. (Photo: Andre Messina)

in Victoria, *C. rugosa*, was collected in 2021, very near the occurrence of *C. prostrata*. These responses reflect deep 'hard-wired' relationships within this group, suggestive of their evolution in high fire frequency periods, probably millions of years ago.

Cyphanthera scabrella, a shrub species known from only five sites in the Blue Mountains in New South Wales, and largely unseen since 1979, was listed as a high priority after the 2019–20 wildfires (Gallagher 2020). It has been found since in two moderately burnt sites, with more than 600 individuals located. Curiously, a simultaneous survey located 20 mature adults at a site more than 30 km away which was not burnt in the 2019–20 fires and where the plant was last recorded in the 1970s. This discovery may be because they had been previously overlooked, because above-average rainfall in 2020–21 germinated some of the soil seed bank, or perhaps because the thick smoke that blanketed the entire Blue Mountains during the 2019–20 fires provided a germination cue for the buried seed (Commander et al. 2008), including in plants outside burnt areas (Ghebrehiwot *et al.* 2011).

Although the examples described above show the potential for substantial recruitment in post-fire landscapes, the interaction between fire and other threats can jeopardise these gains. For instance, early post-fire surveys of the Critically Endangered Megalong Valley bottlebrush (*Callistemon megalongensis*), known from just one location with ~2000 individuals, found that it was regenerating vigorously by both resprouts and seedlings. However, follow-up visits in March 2020 revealed severe herbivory (Criterion C, Table 9.2) and a substantially reduced population, and fencing was quickly installed to exclude grazing animals and feral pigs. Intense rainfall and subsequent flooding have since led to ongoing erosion adjacent to population that now poses additional threat (Criterion I, Table 9.2). This confirmed the high-risk status assigned after the fires, but, given the substantial data gaps for the species, also highlights the importance of a precautionary approach to post-fire vulnerability assessments and the importance of repeat visits.

For some species it is still too soon to say whether they will ultimately benefit or decline as a result of the fires. Some have had mass seedling germination post-fire, e.g. *Boronia imlayensis* (~160 000 seedlings) and *Kardomia prominens* (~24 000 seedlings), but, with almost all these populations now existing as juveniles, the plants remain extremely vulnerable to disease, herbivory and future fires. These species cannot be considered to have recovered until these recruits have reached maturity and are replenishing the seed bank.

Tales of unburnt species

The importance of investment in long-term monitoring of threatened species can be highlighted by four species that did not burn, even though they were (or were thought to be) in the fire footprint. *Euphrasia bella*, *Coronidium telfordii*, *Olearia* sp. Henrys Creek and *Grevillea bemboka* were initially assessed as having between 7% and 100% of their known occurrences burnt in the 2019–20 fires (Gallagher 2020). All are narrow range endemic species and typically occur in topographically complex environments such as cliff tops and gullies. Limitations inherent in mapping fire boundaries contributed to some unreliability in the initial assessment of fire overlap for these taxa, but assessment of fire overlap was also constrained by the limited availability of accurate location data for occurrences.

The first of these, Lamington eyebright (*Euphrasia bella*), is a perennial herb that occurs in the MacPherson Ranges on the New South Wales/Queensland border, where it is restricted to exposed sites on cliff edges near cool temperate rainforests and in low heathland on rock pavement. During the 2019–20 fires an estimated 68% of its known records were burnt, primarily in Queensland (Threatened Species Operations 2020). However, populations thought to be burnt in New South Wales were found to be incorrectly georeferenced and so not located in the fire footprint. Similarly, records in Lamington National Park were not burnt at all, and while fire mapping does indeed show locations with records in Mt Barney National Park that were burnt, the rainforest that covers the peaks where the species occurs meant that fire was only of low severity.

This is in some ways a good outcome for these species, as they were prioritised for survey based on available information. However, the implications for species that may have been impacted, but were not surveyed, is concerning. Inaccuracy is symptomatic of all species occurrence databases – and considerable ongoing resources are needed to maintain verified and reliable data.

Assessing multiple species by focussing on ecological communities

The Bendethera shrublands (NSW)

This ecosystem occurs on small limestone outcrops deep in the ranges of south-eastern New South Wales. The shrublands comprise an almost continuous thicket of Bendethera wattle (*Acacia covenyi*), with hakea (*Hakea eriantha*), bracelet honey-myrtle (*Melaleuca armillaris*) and blackthorn (*Bursaria spinosa*) occurring as subdominants. The ground cover is sparse and herbaceous.

The ecosystem's structural features are not conducive to the spread of low-intensity fire. In fact, the shrublands have remained unburnt since extensive wildfires in 1969 and several adjacent low-intensity fires have self-extinguished on the boundaries. There was therefore no risk of decline as a consequence of high fire frequency. Instead, the 2019–20 fire event was well timed to initiate regeneration of the shrublands which were in a state of senescence.

The area experienced severe and prolonged drought before the 2019–20 fires (Criterion A, Table 9.2). The high severity of the fire (Criterion E, Table 9.2) reinforced the dominance of the Bendethera wattle because serotinous seed banks of hakea were entirely consumed, while those of honey-myrtle survived only on individuals located within large rock outcrops. Conversely, high fire severity dramatically reduced populations of native and introduced herbivores (Criterion C, Table 9.2) and promoted extremely high rates of germination of Bendethera wattle, which caused a dramatic reduction in the size of the seed bank from an estimated 574 (± 190) seeds/m^2 to below the level of detection. The ecosystem is therefore vulnerable to decline if future fires occur before seed banks are replenished (Criterion J, Table 9.2).

In the immediate aftermath of the 2019–20 fires, these shrublands experienced rainfall estimated to be in the range of 100–200 mm over a 24-hour period. This initiated extensive erosion of topsoil (Criterion I, Table 9.2). On the steeper slopes, which comprise up to one-third of the extent, topsoil loss was associated with much lower densities of seedlings than in other areas.

K'gari (Fraser Island)

K'gari Great Sandy National Park in southern Queensland is an iconic World Heritage area. In November 2019, 12 600 ha at the southern end of the island burnt, and from October to December 2020 a wildfire burnt 85 000 ha in the north. A survey conducted in April 2021

(Neldner and Ngugi 2021; Fig. 9.6) compared the condition of the post-fire vegetation to benchmark values. Thirty sites were permanently marked and sampled in nine regional ecosystems. Most of the vegetation burnt comprised eucalypt woodlands, *Melaleuca* open forest, heathlands and peat swamps, and these all showed a remarkable resilience to the fire with no obvious loss of species and substantial vegetative regeneration occurring on many burnt individuals. There was very limited burning in the *Syncarpia hillii, Lophostemon confertus* open forests or rainforests. The fire varied in intensity, and proportion of tree deaths ranged widely but high intensity fire caused widespread death of trees. In some burnt areas, recovery was characterised by mass germination of *Acacia penninervis* var. *longiracemosa*, *A. flavescens* and *Dodonaea viscosa* subsp. *burmanniana* into dense layers.

Across the 19 sites with data from previous sampling in 1995–2005, the post-fire vegetation appears to contain a similar species composition to that recorded from 1995 onwards. These conclusions are largely consistent with those of Meiklejohn *et al.* (2021) and showed that robust recovery is occurring across all of the nine regional ecosystems sampled, apart from the coastal *Casuarina equisetifolia* subsp. *incana* woodlands and *Banksia integrifolia* woodlands on the frontal beach ridges and high dunes. In this latter vegetation, the death of most adult trees and subsequent lack of regeneration of the dominant tree species, despite adequate follow-up rainfall in these areas, is a major concern.

Recommendations

The 2019–20 fires were likely a harbinger of the types of fire seasons that will become more common due to climate change (Chapter 2). Planning for managing plant species diversity into the future will require strategies and actions taken before, during and following severe fire seasons. Planning should involve all relevant authorities, community members, and Indigenous groups, and be tenure-blind. Several of these recommendations require significant and ongoing investment in resources. The costs involved must be traded off against the savings associated with enhanced conservation of plant diversity that is the foundation

Fig. 9.6. Burnt *Casuarina equisetifolia* subsp. *incana* low woodland illustrating standing dead trees on coastal dune near Happy Valley, K'gari. (Photo: V. John Neldner)

of ecosystems across Australia. The alternative is an *ad hoc* reliance on event-driven funding to support emergency recovery action.

Pre-fire actions

- Manage existing and emerging threats such as fragmentation, disease, invasive weeds, grazing, altered hydrology and climate change to increase resilience and reduce risk from fire events (Monks *et al.* 2019; Gallagher *et al.* in press).
- Increase representation of species and populations in *ex situ* conservation. For species with highly restricted distributions or threatened status, the strategic creation, expansion and monitoring of translocated insurance populations or *ex situ* germplasm collections can contribute to long-term persistence and support future recovery actions.
- Enhance databases of species occurrence to allow for more robust assessments of overlaps with future fires and to help protect important populations.
- Improve understanding of species responses to fire. More than 8000 species-level observations of fire response (e.g. whether they can resprout after fire, have fire-cued recruitment, seed bank longevity, and the length of time after fire until plants first reproduce) have been made in Australia (Falster *et al.* 2021). However, this only represents approximately one-third of Australian plant species. A systematic national program for documenting, collating, or robustly predicting the fire response of plant species is urgently needed to ensure emergency response planning accurately captures impacts.
- Develop and implement fuel management plans to protect highly fire-sensitive populations. Fuel management (e.g. planned fire or mechanical slashing) is routinely undertaken in Australia with the aim of reducing the severity of wildfires and to increase opportunities for fire suppression. For some sensitive plant species, expansion of strategic fuel management to enhance protection of significant plant populations, with integration into broader fuel management plans, may enhance population persistence with future fires. However, the conservation risks of conducting or not conducting strategic fuel management should be subject to careful assessment. Indigenous approaches to burning may be locally helpful (McKemey *et al.* 2021) but also need to be assessed for how these align with the appropriate fire regime for a species, and the risk of adding more fire in a system where climate change is driving increased fire frequency.
- Improve forecasting of the risk of fire. Better prediction of which regions are at risk of fire may improve the capacity to organise and respond to fires. Recent advances have been made in modelling and monitoring spatio-temporal dynamics of fuel conditions, aided by satellite imagery and aerial LiDAR (Price and Gordon 2016; Yebra *et al.* 2013). There are considerable opportunities to improve monitoring and forecasting of fuel conditions by adopting these new technologies into formalised assessments of bushfire risk. Forecasting of other disturbances, in particular drought impacts on vegetation, will also improve risk assessments of plant vulnerabilities to fire.

During fire

Responding rapidly to an emerging fire threat is critical for preventing or reducing negative impacts on some vulnerable plant populations. The capacity to respond both rapidly and effectively during fires will be improved by strategic planning in the years and months preceding a severe fire season.

- Provide information for decision making. Mapping of species occurrences and threats is crucially important for rapid assessment of species vulnerability during fire events.

This mapping also needs to identify priorities for action (i.e. supported by information on species traits, conservation status, and time since last fire), identify what fire-fighting actions are permissible where, for example 'go' or 'no-go zones' for constructing fire breaks, and be readily accessible in a format that is easy to interpret during fire emergencies. Critically, decisions involving priority plants need to be integrated into emergency response frameworks.

- Enhance coordination. Often different agencies, or groups within agencies, are responsible for managing threatened species and for managing fire emergencies. Species may also occur across multiple state and territory jurisdictions that have different statutory responsibilities for management. Clear communication and coordination between and within agencies are vital to ensure that the people deciding where to focus fire suppression efforts have accurate information on the location of high priority species.
- Ensure availability of adequate resourcing and capacity. Responding during a fire event requires adequate resourcing and capacity (Legge *et al.* 2022). Demonstrated effectiveness of strategic and highly targeted fire suppression efforts for flora conservation (e.g. Hannam 2020) indicates that continued investment in capacity and bolstering operational experience in biodiversity-focused fire suppression may yield significant conservation benefits.

Post-fire actions

Post-fire responses are largely characterised by managing interacting biotic threats, such as weeds, disease or herbivory, and, less often, abiotic threats such as erosion. Threats such as climate change and short fire intervals require strategic long-term planning to allow effective management.

- Enhance coordination of strategic post-fire recovery actions. Effective management of interacting threats will be improved by coordination among government agencies, across jurisdictions, and with the community. Community groups are often well placed to undertake post-fire interventions, such as fencing and weeding (e.g. Kohout *et al.* 2020). Monitoring and testing the effectiveness of these actions is critical.
- Implement active restoration of areas where species are at high risk of local extinction. The 2019–20 fires had major impacts on the ash-type forests, where the dominant species are 'obligate seeders' (Clarke *et al.* 2015), and as such at risk of local extinction if fire recurs before flowering and seed-set occurs. Within Victoria, ~30% of the 83 000 ha of ash-type forests that burnt were immature (Nolan *et al.* 2021). In response, the state government initiated the single largest reseeding program to aerially sow seed in these areas (Nolan *et al.* 2021). The success of such initiatives will be dependent on the availability of sufficient seed collected before fire, the resources to conduct such operations, and exclusion of fire until the planted trees have had sufficient time to become reproductively mature.

References

Andrew SC, Mokany K, Falster DS, Wenk E, Wright IJ, *et al.* (2021) Functional diversity of the Australian flora: Strong links to species richness and climate. *Journal of Vegetation Science* **32**, 13018. doi:10.1111/jvs.13018

Auld TD, Mackenzie BDE, Le Breton TD, Keith DA, Ooi MKJ, *et al.* (2020) *A Preliminary Assessment of the Impact of the 2019/2020 Fires on* NSW *Plants of National Significance*. NSW Department of Planning, Industry and Environment, Sydney.

Auld TD, Keith DA, Mackenzie BDE, Tozer MG, Ooi MKJ, *et al.* (unpublished) A framework for prioritising the impacts of wildfires on plants and ecosystems.

Barrett S, Yates CJ (2015) Risks to a mountain summit ecosystem with endemic biota in southwestern Australia. *Austral Ecology* **40**, 423–432. doi:10.1111/aec.12199

Barrett S, Hartley R, Yates CJ (2005) 'Stirling Range dryandra (*Dryandra montana*) recovery plan'. Department of Environment and Conservation Western Australia, Perth, <http://www.environment.gov.au/resource/stirling-range-dryandra-dryandra-montana>.

Bell SAJ, Copeland LM (2004) *Commersonia rosea* (Malvaceae s.l.: Lasiopetaleae): a new, rare fire ephemeral species from the upper Hunter Valley, New South Wales. *Telopea* **10**, 581–587.

Bell S, Holzinger B (2015) Wildfire reveals new populations of the endangered *Commersonia rosea* and *Monotaxis macrophylla* in northern Wollemi National Park, NSW. *Australasian Plant Conservation* **23**, 2–4.

Bell SAJ, Woolley T (2021) 'Post-fire surveys for the Endangered *Androcalva rosea* in the Upper Hunter Valley, New South Wales'. Unpublished report to NSW Office of Environment and Heritage.

Clarke PJ, Lawes MJ, Murphy BP, Russell-Smith J, Nano CE, *et al.* (2015) A synthesis of postfire recovery traits of woody plants in Australian ecosystems. *Science of the Total Environment* **534**, 31–42. doi:10.1016/j.scitotenv.2015.04.002

Commander LE, Merritt DJ, Rokich DP, Flematti GR, Dixon KW (2008) Seed germination of *Solanum* spp. (Solanaceae) for use in rehabilitation and commercial industries. *Australian Journal of Botany* **56**, 333–341. doi:10.1071/BT07209

Eco Logical Australia (2021) '*Pterostylis metcalfei* survey – Ebor District, 2021'. Prepared for NSW Department of Planning, Industry and Environment.

Falster D, Gallagher R, Wenk EH, Wright IJ, Indiarto D, *et al.* (2021) AusTraits, a curated plant trait database for the Australian flora. *Scientific Data* **8**, 254. doi:10.1038/s41597-021-01006-6

Gallagher RV (2020) Final national prioritisation of Australian plants affected by the 2019–2020 bushfire season. Commonwealth Department of Agriculture, Water and Environment, <https://www.dcceew.gov.au/sites/default/files/env/pages/289205b6-83c5-480c-9a7d-3fdf3cde2f68/files/final-national-prioritisation-australian-plants-affected-2019-2020-bushfire-season.pdf>.

Gallagher RV, Allen S, Mackenzie BD, Yates CJ, Gosper CR, *et al.* (2021) High fire frequency and the impact of the 2019–2020 megafires on Australian plant diversity. *Diversity & Distributions* **27**, 1166–1179. doi:10.1111/ddi.13265

Gallagher RV, Allen SP, Mackenzie BDE, Keith DA, Nolan RH, *et al.* (in press) An integrated approach to assessing abiotic and biotic threats to post-fire plant species recovery: lessons from the 2019–20 Australian fire season. *Global Ecology and Biogeography.* doi:10.1111/geb.13478.

Ghebrehiwot HM, Kulkarni MG, Light ME, Kirkman KP, Van Staden J (2011) Germination activity of smoke residues in soils following a fire. *South African Journal of Botany* **77**, 718–724. doi:10.1016/j.sajb.2011.03.006

Gosper CR, Miller BP, Gallagher RV, Kinloch J, van Dongen R, *et al.* (2022) Mapping risk to plant populations from short fire intervals via relationships between maturation period and environmental productivity. *Plant Ecology.* doi:10.1007/s11258-022-01229-6

Hannam P (2020) Incredible, secret firefighting mission saves famous 'dinosaur' trees'. *The Sydney Morning Herald*, 15 January, <https://www.smh.com.au/environment/conservation/incredible-secret-firefighting mission-saves-famous-dinosaur-trees-20200115-p53rom.html>.

Kohout M, Coupar P, Elliott M (2020) Battling an 'aggressive pioneer' after fire: *Phytolacca octandra* (Inkweed). *Australasian Plant Conservation* **29**, 7–8.

Legge S, Woinarski J, Scheele B, Garnett ST, Lintermans M, *et al.* (2022) Rapid assessment of the biodiversity impacts of a megafire to guide urgent management intervention and recovery. *Diversity & Distributions* **28**, 571–591. doi:10.1111/ddi.13428

Mackenzie BDE, Auld TD (in press) *Wollemia nobilis*. In *The IUCN Red List of Threatened Species 2022*, <https://www.iucnredlist.org>.

Mackenzie BDE, Clarke SW, Zimmer HC, Liew ECY, Phelan MT, *et al.* (2021) Ecology and conservation of a living fossil: Australia's Wollemi Pine (*Wollemia nobilis*). In *Imperiled: The Encyclopedia of Conservation*. Vol. 1. (Eds D DiPaolo and J Villella) pp. 884–894. Elsevier, Oxford, UK. doi:10.1016/B978-0-12-821139-7.00188-4

McKemey M, Patterson ML, Hunter J, Ridges M, Ens E, *et al.* (2021) Indigenous cultural burning had less impact than wildfire on the threatened Backwater grevillea (*Grevillea scortechinii* subsp. *sarmentosa*) while effectively decreasing fuel loads. *International Journal of Wildland Fire* **30**, 745–756. doi:10.1071/WF20135

Meiklejohn AM, Melzer RI, Hines HB, Laidlaw MJ, Toyne M, *et al.* (2021) 'Post-fire assessment report – natural values: 2020 Duling bushfire, K'gari (Fraser Island), Great Sandy National Park, South East Queensland Bioregion'. Department of Environment and Science, Queensland Government, Brisbane.

Metcalfe P (2017) 'Management of *Diplodium metcalfei* 2017'. Unpublished report prepared for the NSW Office of Environment and Heritage.

Monks L, Barrett S, Beecham B, Byrne M, Chant A, *et al.* (2019) Recovery of threatened plant species and their habitats in the biodiversity hotspot of the Southwest Australian Floristic Region. *Plant Diversity* **41**, 59–74. doi:10.1016/j.pld.2018.09.006

Neldner VJ, Ngugi MR (2021) *Vegetation Recovery after the 2019 and 2020 Bushfires on K'gari (Fraser Island), Great Sandy National Park, Southeast Queensland Bioregion*. Queensland Herbarium, Department of Environment and Science, Brisbane.

Nolan RH, Collins L, Leigh A, Ooi MK, Curran TJ, *et al.* (2021) Limits to post-fire vegetation recovery under climate change. *Plant, Cell & Environment* **44**, 3471–3489. doi:10.1111/pce.14176

Price OF, Gordon CE (2016) The potential for LiDAR technology to map fire fuel hazard over large areas of Australian forest. *Journal of Environmental Management* **181**, 663–673. doi:10.1016/j.jenvman.2016.08.042

Threatened Species Operations (2020) *Initial Analysis of the Impacts of Bushfires (August – December 2019) on Threatened Species in Southern Queensland*. Department of Environment and Science, Queensland Government, Brisbane.

Yebra M, Dennison PE, Chuvieco E, Riano D, Zylstra P, *et al.* (2013) A global review of remote sensing of live fuel moisture content for fire danger assessment: moving towards operational products. *Remote Sensing of Environment* **136**, 455–468. doi:10.1016/j.rse.2013.05.029

Zimmer HC, Auld TD, Benson JS, Baker PJ (2014) Recruitment bottlenecks in the rare Australian conifer *Wollemia nobilis*. *Biodiversity and Conservation* **23**, 203–215. doi:10.1007/s10531-013-0593-2

10

The impacts of the 2019–20 wildfires on Australian fungi

Tom W. May, Payal Bal, Tianxiao Hao, Angela Muscatello, Elle Bowd, David E. A. Catcheside, Pamela S. Catcheside, Naveed Davoodian and Camille Truong

Summary

Context and challenges

- Fungi are megadiverse but poorly documented in Australia and worldwide.
- Major trophic modes are parasitism, saprotrophism and mutualism (e.g. lichens and mycorrhizas).
- Fungi are functionally important in ecosystems, particularly through mutualistic interactions with other biota and as decomposers.
- Many fungi have wide distributions which means that specific knowledge of life history and threats is usually required to assess conservation status.
- Few fungi are formally listed as threatened under Australian legislation.
- Some fungi are adapted to fire through production of heat-resistant spores or underground resting stages.
- Recent metabarcoding studies on fungal communities have detected positive and negative effects of fire on the presence and abundance of different guilds of fungi.

Main findings

- Based on spatial analysis of distributional records, we found that:
  - Only half the fungi known from Australia are represented in accessible distribution databases.
  - Most species with some accessible distribution data are represented by one or few records.
- In relation to the 2019–20 wildfires, we found that:
  - At least part of the range of 3523 species of fungi was burnt.
  - 177 species had at least 50% of their range burnt in fires of any severity.

  - 188 species had at least 20% of their range burnt by fires of high severity.
  - 59 species had all of their known range burnt by fires of high severity. However, they all had only one or two unique records, so their fire overlap values are of low confidence.
- There will also be numerous undescribed species of fungi potentially affected by the 2019–20 wildfires.
- The species identified as having high overlap with fire are a priority for assessment of traits and threat status.
- Future monitoring will benefit from combining single-species approaches with whole-community sampling using metabarcoding.

Introduction

What are fungi?

Kingdom *Fungi* is one of the major radiations in the tree of life. Fungi can be defined simply as organisms that have hyphae and reproduce by spores, although there are exceptions, such as the single-celled yeasts. Hyphae are branching filaments forming a network called a mycelium.

The most familiar fungi are macrofungi with readily visible sporing bodies, such as mushrooms, puffballs, lichens and coral fungi. However, most fungi are microfungi, not readily visible apart from via their symptoms, such as for leaf-spotting fungi. Spores are microscopic and produced sexually and/or asexually in large numbers, often with capacity to travel long distances. The feeding part of the fungus, the mycelium, may be short-lived on ephemeral substrates, or persist for long periods in soil, wood or other substantial substrates. Spore production can be directly on the mycelium or on specialised sporing bodies (such as mushrooms). Sporing bodies can appear annually, often sporadically (as with mushrooms), or may be long-lived (such as the persistent sporing bodies of many bracket fungi).

Fungi are megadiverse but extremely under-documented. There are ~15 000 species already recorded from Australia, from an estimate of between 50 000 and 250 000 expected species (Chapman 2009). Globally, fungal diversity is estimated at between 2.2 and 3.8 million species (Hawksworth and Lücking 2017). Fungi have significant economic impacts that can be negative through animal, human and plant disease, and positive as sources of biologically active compounds (e.g. antibiotics like penicillin) and through the key roles they play in natural environments, as components of food webs, carbon sinks, and symbioses.

Fungi are heterotrophs, requiring carbon from other sources. Key to understanding the role of fungi in ecosystems is appreciation of the diversity of ways in which fungi obtain and share nutrients. The main trophic roles are (1) decomposition (saprotrophism) of a wide range of organic compounds, including those in woody substrates, such as lignin; (2) parasitism of animals, fungi and plants; and (3) mutualisms – the most important of which are mycorrhizas (associations of plant roots and fungi) and lichens (stable associations between fungi and photosynthetic algae or cyanobacteria).

Due to the variety of trophic roles, most species of animals and plants interact with fungi in one way or other. Examples of key ecosystem processes that involve fungi are (1) biotic soil crusts (of which lichens are a major component), stabilising soil and affecting seed germination, and (2) mycorrhizas of canopy-dominant trees and shrubs, including *Allocasuarina*, *Eucalyptus* and *Nothofagus*, which together cover > 77% of native forest area in Australia (SOFR 2018). An example of the complexity of interactions is the diversity of sequestrate (truffle-like) fungi – many forming mycorrhizal partnerships with forest trees and other plants – the underground sporing bodies of which are consumed by mammals, many of which are threatened.

Conservation status

Sixty species of fungi are formally listed as threatened or rare under state legislation for New South Wales (nine), Tasmania (28, all lichens), Victoria (eight) and Western Australia (15, all as 'Priority taxa'). Sixteen species of Australian fungi are in the IUCN Red List of Threatened Species, assessed as Critically Endangered, Endangered, Vulnerable or Near Threatened. No fungi are listed as threatened under the Commonwealth *Environment Protection and Biodiversity Conservation Act 1999*, although several fungi are listed as threatening processes!

The low number of formally threat-listed fungi reflects the paucity of assessments of conservation status rather than lack of threatened species. There is a lack of experts with resources to make the nominations, so nominations are in part community driven and have mostly been *ad hoc* rather than arising from systematic assessment. Few of the fungi identified as threatened in Australia are under active survey or management.

Among other biota, short-range endemics are often among the first species recognised as threatened, due to low extent of occurrence and/or area of occupancy. However, there are few proven short-range endemics among Australian fungi (May 2002). Therefore, for fungi to be listed as threatened generally requires specific knowledge of declines (usually not possible given lack of baseline monitoring), population estimates (technically difficult) or specific threats.

Threats identified for fungi include destruction of habitat and climate change. For obligate symbionts, threats to partner species have been used to identify threats to the fungus, such as for *Cyttaria septentrionalis*, a parasite restricted to Antarctic beech (*Nothofagus moorei*) (itself globally assessed as Vulnerable).

Fungi and fire

Fire affects soil fungal communities directly (fire-induced mortality) or indirectly through plant–soil–microbe interactions (Wardle *et al.* 2004). High-severity wildfire can heat soils to > 500°C, resulting in the combustion of soil organic matter and leaf litter, volatilisation of soil nutrients and increases in soil pH (Certini 2005; Chapter 5). Changes in soil properties, including pH, manganese, magnesium, phosphorus and nitrate, represent some of the potential mechanisms through which fire may alter microbial communities indirectly (Bowd 2020).

McMullan-Fisher *et al.* (2011) reviewed fire and fungi in Australia, concluding that fire effects are variable in space and in relation to taxonomic and trophic groups of fungi. In the first few years after fire there is a flush of fire-adapted fungi – regenerating from sclerotia, the storage bodies formed by genera such as *Laccocephalum* (e.g. native bread, stonemaker), or from heat-resistant spores. There are also fungi that appear to prefer long-unburnt vegetation – for example, the critically endangered *Hypocreopsis amplectens*, found in long-unburnt stands of heathy woodland, and *Roccellinastrum flavescens*, a lichen found exclusively on leaves of pencil pine (*Arthrotaxis cupressoides*) (McMullan-Fisher *et al.* 2011).

The fire regime, in terms of fire intensity and frequency, affects the dynamics of fungal communities (McMullan-Fisher *et al.* 2011; Rincón *et al.* 2014; Egidi *et al.* 2016; Bowd *et al.* 2021). For instance, crown-fire-induced tree mortality can have more severe impacts on ectomycorrhizal and litter-associated fungi than ground-fire-induced losses in soil organic matter (Pérez-Izquierdo *et al.* 2021). The post-fire flush of fungi can be a confounding factor in studies of fire effects, depending on the time since fire across study sites. McMullan-Fisher *et al.* (2011) noted there were few studies that tested the effect of different fire regimes; the studies often lacked replication and they did not yield consistent results.

Ecosystem recovery following fire has dependencies on the resilience of soil fungal communities and maintenance of their interactions with host plants and animal dispersers (Dove and Hart 2017). After fire, fungi have important roles in soil stabilisation and those that produce sporing bodies early in the post-fire succession are often mycorrhizal and support seedling establishment. Sequestrate fungi provide vital food resources for mammals post-fire, as evidenced by frequent diggings. Such diggings help reduce fire frequency and intensity by breaking up hydrophobic layers at the soil surface and improving soil aeration and water penetration (Palmer *et al.* 2020). Some sequestrate fungi appear to produce sporing bodies in greater abundance after fire, but the degree to which fire stimulates sporing body production is unclear, and may vary across ecosystems and fungal lineages.

The persistence of fungal communities post-fire is contingent on species-specific ecological and physiological tolerances and the presence of heat-resistant mycelia and spore banks that allow them to withstand the direct impacts of wildfire (Kurth *et al.* 2013; Glassman *et al.* 2016). Post-fire environments may select for specific taxa including some that thrive due to a lack of competition and the creation of new niches, such as decomposing fire-killed mycelium or fallen trees (Warcup 1990; Rincón *et al.* 2014). In contrast, less fire-tolerant species, including some ectomycorrhizal (ECM) species that specialise in nitrogen uptake from organic sources, may decline after fire for up to a decade (Treseder *et al.* 2004; Pérez-Izquierdo *et al.* 2021). Aside from fire-induced mortality, fire can negatively affect ECM fungal diversity by reducing the diversity and abundance of ECM host trees, such as fire-sensitive *Nothofagus* species, as well as fungal consumers, particularly mammals, that are vital for spore dispersal (Dundas *et al.* 2018).

Recent research on the effect of fire on soil fungal communities

Recently, molecular-based approaches using DNA metabarcoding of environmental samples (e.g. soil, roots, mammal scats) have opened the door for diversity assessment of microbial communities affected by fire. Studies in Australia have focused on wet forests, investigating both logging and fire. In Victoria, Bowd *et al.* (2021) found fire and logging had negative effects on symbiotic fungi, mediated by changes in soil properties and plant communities. In contrast, they reported an increase in saprotrophic fungi with multiple fires. Similarly in Tasmania, Ammitzboll *et al.* (2021) reported that logging and high-severity burning reduced fungal biomass including for some symbionts, but had a positive influence on saprotrophs. They suggested that a range of low to moderate burns contribute to habitat diversity and the persistence of diverse soil microbial communities.

Purpose of this chapter

In this chapter we assess impacts of the 2019–20 wildfires on Australian fungi. We make recommendations arising from the present analysis as well as general recommendations about enhancements to data infrastructure and further research to improve the knowledge base upon which informed management decisions can be made.

Using the pipeline of Hao *et al.* (2021), we created a cleaned and de-duplicated dataset (accessible at https://doi.org/10.5281/zenodo.6324422) on data from the following repositories (downloaded 16 September 2021): (1) Atlas of Living Australia (ALA), (2) Global Biodiversity Information Facility, (3) Fungimap, (4) MycoPortal and (5) iNaturalist. The pipeline discarded records not in native vegetation. The > 1 million records based on environmental DNA from the ALA (a quarter of which are identified to species level) were excluded due to questionable identification accuracy. The final dataset contained records identified to species for Kingdom Fungi (i.e. excluding fungi-like members of other kingdoms).

For the fire overlap analysis, we replicated the methods of Marsh *et al.* (2021; Chapter 11) as applied to invertebrates except that in preparing data: (1) the coordinate uncertainty cut off was > 5 km, although records lacking coordinate uncertainty were retained, (2) specimen and observation data were included (the latter predominantly from Fungimap and iNaturalist, and considered adequate in relation to identifications), and (3) we did not apply a date cut-off, as a significant proportion of data was collected < 1990 or had no date information (33%).

The fire overlap analysis was carried out for the Preliminary Analysis Area (PAA) (see Chapter 1). Overlap with fire was calculated using the GEEBAM fire layer (Department of Agriculture Water and the Environment 2020) at 250 m resolution, which includes classes of severity: (1) no data, (2) unburnt, (3) low and moderate, (4) high, and (5) very high. The fire severity layer was masked so that the overlap analysis only considered fire impacts on native vegetation (see Marsh *et al.* (2021) for details). Total records were converted to unique records across a 2 × 2 km grid. For species with one or two unique records, overlap was in relation to the total number of records (which may be greater than two due to multiple records within the one grid cell). For species with three or more unique records, a range polygon was created using the α hull method with $\alpha = 2$ in the ConR package in R (Dauby *et al.* 2017), and overlap with the fire layer was calculated using these polygons. The overlap analyses were carried out for burnt areas (classes: low and moderate, high, or very high) and severely burnt areas (classes: high or very high). Marsh *et al.* (2021) compared analyses treating GEEBAM class 2 (unburnt) as burnt or unburnt, to account for possibility of undetected burning of the understorey. We treated GEEBAM 2 as unburnt. Lists of species with high overlap with fire (i.e. at least 50% overlap with fire or ≥ 20% overlap with severe fire, corresponding to Tables 10.1 and 10.2) are available for point and polygon data at https://doi.org/10.5281/zenodo.6324422.

We acknowledge a range of caveats about the data and analyses, including (1) misapplication of names of Northern Hemisphere taxa, (2) some exotic species may not have been excluded, (3) highly patchy sampling taxonomically and geographically, and (4) the National Species List for fungi is incomplete, meaning data are sometimes attached to different names for the same species or recently described species are omitted.

Findings

Characteristics of fungi repository data

The initial cleaned dataset shows that there is a significant lack of readily accessible distribution data for many species of fungi (Fig. 10.1). There are in the order of 15 000 named species of fungi known from Australia but just over half (8251) have any data in the sampled repositories. A few species have been widely recorded, such as the Fungimap target species vermilion grisette (*Amanita xanthocephala*) with almost 4000 records

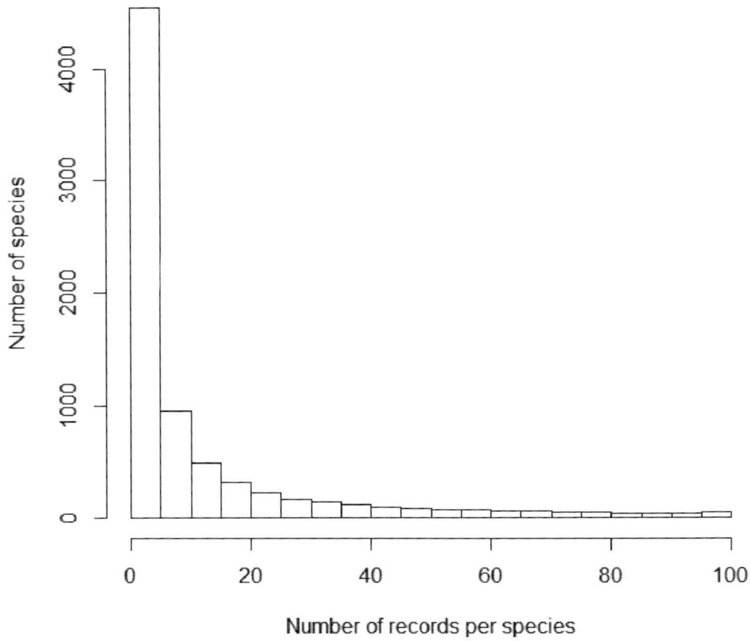

Fig. 10.1. Frequency distribution of the number of records per species for the 7584 species of Australian fungi with at least one record in the sampled repositories. Not shown on this histogram are 667 species that had more than 100 records.

(Fig. 10.2a). However, a quarter of species are represented by just one digitised specimen or observation, and around two-thirds have 10 or fewer records. Although the average number of records is 42, the modal number of records is one. The lack of records is due to a combination of (1) many species are genuinely known only from the type collection and few other collections, and (2) for other species that have more specimens, these are often not fully digitised, particularly the specimens held in non-Australian institutions.

Fire overlap

Some fungi distributions are displayed against the fire extent in Fig. 10.2, including distributions typical of many species, either spanning eastern and western Australia (Fig. 10.2a) or extending from Tasmania to southern Queensland on the east coast (Fig. 10.2b). In relation to overlap with fire, additional distribution information for species with apparently small distributions may increase or decrease the fire overlap, depending on where the distribution extends with additional data.

Of the 7445 species with at least one data point remaining after data cleaning, 6634 had at least one record within the PAA. Of these, 3523 species had some part of their range overlap with the fire extent (2205 species had an overlap of ≥ 1%) and 177 species had at least 50% of their distribution burnt (Table 10.1).

Overlap with fire extent is not the only indication of effect. Fire severity is also an important consideration for impact and recovery. Some 188 species had at least 20% of their distributions burnt by severe fire (Table 10.2). Of these, 59 species had all of their known range burnt. However, all had only one or two unique records, so their fire overlap values are of low confidence.

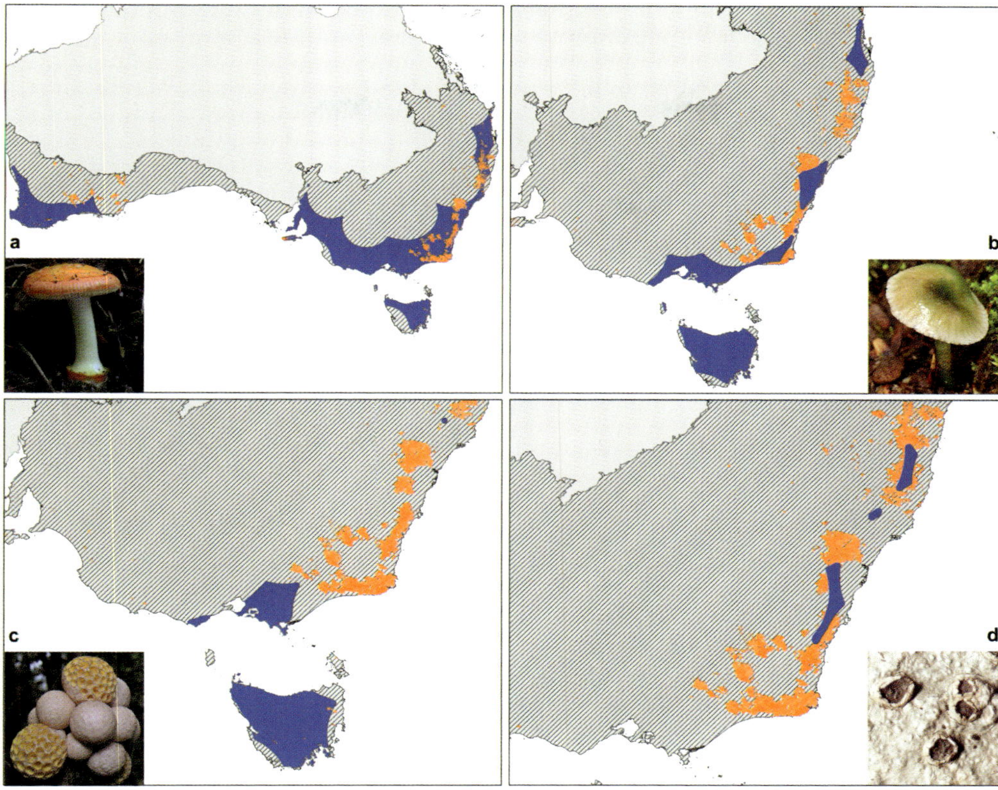

Fig. 10.2. Map of Australia with ranges, as α hull polygons (in blue), for (a) *Amanita xanthocephala*, (b) *Gliophorus graminicolor*, (c) *Cyttaria gunnii* and (d) *Gintarasia megalophthalma*; demonstrating distributions within the Preliminary Analysis Area (cross-hatched). Extent of the 2019–20 wildfires represented in orange (overlaid in (a) and (c) and underlaid in (b) and (d). (Photos of fungi: (a) Tom May, (b, c) Paul George, (d) Bill Malcolm)

Table 10.1. Tallies of the number of fungal species within the Preliminary Analysis Area with extensive (≥ 50%) overlap with the 2019–20 wildfires.

	Number of species		
% distribution overlapping with fire	**Species with 3 or more unique records (α hull range)**	**Species with 1 or 2 unique records (point locations)**	**Total**
100	–	112	112
90–99.9	–	–	–
80–89.9	2	–	2
70–79.9	1	–	1
60–69.9	6	5	11
50–59.9	16	35	51
Total with at least 50% overlap with fire	**25**	**152**	**177**

Table 10.2. Tallies of the number of fungal species within the Preliminary Analysis Area for which ≥ 20% of their distribution overlapped with the 2019–20 wildfires that were severe (class high or very high).

% distribution overlapping with severe fire	Number of species		
	Species with 3 or more unique records (α hull range)	Species with 1 or 2 unique records (point locations)	Total
100	–	59	59
90–99.9	–	–	–
80–89.9	–	–	–
70–79.9	–	–	–
60–69.9	–	5	5
50–59.9	2	20	22
40–49.9	10	–	10
30–39.9	20	2	22
20–29.9	70	–	70
Total with at least 20% overlap with severe fire	**102**	**86**	**188**

When considering morphogroups (groupings of convenience that are similar in overall morphology), many of the species tallied for Tables 10.1 and 10.2 are lichens, with sequestrate fungi and mushrooms the next most prevalent. Among the 86 species that had 50% or greater overlap with severe fire intensity, about half (46) are lichens, 14 are mushrooms, 10 are sequestrate fungi (e.g. *Cortinarius tubercularis*), and the remainder are a mix of microfungi and macrofungi in other morphogroups (such as coral fungi and puffballs).

Among species with ≥ 20% overlap with severe fire, a few species appear to be rare and/or have specific habitat requirements, suggesting potential for high impact of the 2019–20 wildfires. For example, among lichens, *Sticta hypopsiloides* (two points, 100% overlap with severe fire) is a rare species occurring in montane rainforests in Queensland and *Cresponea ancistrosporelloides* (one point, 100% overlap) is known only from the original collection on rock in sclerophyll forest in Western Australia. Lichens known from more than two sites and with high fire overlap include *Anzia minor* (13 points, 42.3% overlap) and *Hypotrachyna subpustulifera* (three points, 52.1% overlap), both restricted to sandstone sites in New South Wales, and *Gintarasia megalophthalma* (44 points, 25.7% overlap), a species occurring in medium altitude wet forests and rainforests of eastern Australia (Fig. 10.2d).

For many species the distribution as currently known is likely to be incomplete. This is especially so for sequestrate fungi because they form underground sporing bodies and thus require specialised survey techniques. However, three sequestrate species with ≥ 20% overlap with severe fire are distinctive morphologically and considered uncommon by experts: *Andebbia pachythrix* (30.9% overlap), *Cortinarius argyrionus* (28.1% overlap) and *Leucogaster meridionalis* (27.0% overlap). Among epigeal fungi, some species showing ≥ 20% overlap with severe fire are known from further collections (not digitised) that indicate a much wider distribution or that there are large areas of potentially suitable habitat between known points (e.g. *Sphaerosoma trispora* and *Galerina lurida*).

None of the 75 species of Australian fungi listed on the IUCN Red List of Threatened Species or under state conservation legislation appear among the species with overlap with fire tabulated in Tables 10.1 and 10.2. However, 21 of the 75 species (including *Entoloma ravinense*, see the case study in Box 10.1) are not present in the cleaned dataset due to data issues including high uncertainty about coordinates (often deliberate, due to status as threatened).

Box 10.1. *Entoloma ravinense*

Entoloma ravinense (Fig. 10.3) is a distinctive, white, fan-shaped, gilled fungus up to 63 mm across with a short stem. The cap is rather woolly in young specimens and the gills become pink as spores mature. Such fan-shaped, almost stemless Entolomas are considered to be very rare and *E. ravinense* is no exception. In spite of intensive surveys throughout much of South Australia it has been found at only two sites on Kangaroo Island, ~30 km distant from each other at the western end of the island. As well as this narrow geographical range, *E. ravinense* has specialised environmental requirements. It is saprotrophic and seemingly specific to the bark of the sugar gum (*Eucalyptus cladocalyx*), as it has not been found associated with any other eucalypt species. Sporing bodies are found only on the undersides of thick, rotten, shed bark. Annual surveys of Kangaroo Island first found *E. ravinense* 3 years after the intense wildfires of

Fig. 10.3. Sporing body of *Entoloma ravinense* growing on bark of *Eucalyptus cladocalyx*. (Photo: David Catcheside)

December 2007. It is possible that thick bark shed after intense fires provides the required substrate. Although this rare fungus may require an appropriate fire regime, fire events that are too frequent and/or severe may be detrimental or even catastrophic, threatening its survival. The particularly severe and extensive wildfires of 2019–20 burnt through the only two known locations where the fungus occurs, killing or damaging the majority of trees. Abundant freshly shed sugar gum bark was seen at both locations in 2020 and 2021, but there was little evidence of fungal activity degrading the bark and no sporing bodies of *E. ravinense* were found. Based on observations after the 2007 fire, *E. ravinense* still might produce sporing bodies once 3 years have elapsed from the recent fires. However, it is possible that the spore and mycelial bank in its known range has been destroyed. In which case, this little fungus, found in 2010 and described in 2016, may be extinct. Continued monitoring of known and potential sites, including stands of *Eucalyptus cladocalyx* on the mainland, is a priority as is obtaining a pure culture for *ex situ* conservation, should the fungus re-appear.

Constraints on interpretation

Without specific knowledge of life history, especially responses to fire and modes of dispersal and colonisation, it is difficult to determine how the 2019–20 wildfires affected the conservation status of fungi, even for species with high overlap with severe fire. Apart from the case study on *Entoloma ravinense* (Box 10.1), our analysis was not able to include detailed assessment of threats in relation to traits. Therefore, it is vital to gain further information on how fungi survive after fire. For some species occurring deep in the soil or having heat-resistant spores, fire may have minimal deleterious effects. For other species that occur only in association with aerial parts of plants or with invertebrates, fire may cause local extinction. This uncertainty points to the importance of investigating how fungi recolonise after fire, and the relative roles of mycelial persistence and spore dispersal, along with the distances over which dispersal is effective.

An important caveat of the fire overlap analysis is that distribution information is both meagre and patchy for many species of fungi, and it is likely that some species with high overlap will be found to be more widely distributed once more data are available, especially those known from few and widely separated locations. Conversely, some species with low overlap may be under-reported in areas of high fire severity. Species adapted to specific fire-sensitive microhabitats (i.e. logs, tree bark) may also be highly vulnerable to fire even though they are widely distributed.

Nevertheless, the list of species that show high fire overlap is a useful first-pass analysis to enable prioritisation of surveys, threat assessments and taxonomic studies. Further fungi identified as of potential conservation concern in relation to the 2019–2020 wildfire boundary in eastern Victoria are listed by Dell *et al.* (2020).

Alternative strategies for determining fire impact

Two other approaches for assessing fire impacts that complement analysis of sporing body records for individual species are to (1) focus on particular ecological communities, and (2) extend metabarcoding to detect fungal community patterns (as implemented by Bowd *et al.* (2021) and Ammitzboll *et al.* (2021)) across a broader range of substrates, ecosystems and geography, including establishment of permanent monitoring plots.

In relation to particular ecological communities, those affected by fire and that contain plants that are important symbiotic partners of fungi are worth investigating. For example, some ECM fungi appear to strongly associate with *Nothofagus*, a main canopy species of Cool Temperate Rainforest (CTR). CTR patches are often larger in Tasmania while stands are often small and scattered elsewhere, but wherever it occurs CTR is often in a matrix of *Eucalyptus* forest. Fire is critical in controlling the boundary between surrounding *Eucalyptus* forest and CTR. In Tasmania, Lunn *et al.* (2018) found that fire affected rain-forest species such as *Nothofagus* more strongly than eucalypts. Increased fire intensity and frequency will favour *Eucalyptus* over *Nothofagus* – this could have significant effects on any fungi restricted to *Nothofagus*. An example of the extent of loss of cover of CTR following fire comes from the O'Shannassy Catchment area to the east of Melbourne, where a 2009 fire resulted in loss of 96% of 889 ha of CTR in areas burnt at moderate to high severity (Tolsma *et al.* 2019). Further studies are needed to identify *Nothofagus*-specific fungi and their role for postfire forest regeneration.

Conclusions

The fire overlap analysis identified some 3523 species of fungi affected by the 2019–20 fires; with several hundred species significantly affected, either by having at least half their range burnt or by having at least 20% of their range burnt by fires of high severity. However, existing distribution data are patchy for many species. In particular, species whose whole range was affected by fires of high severity were all known from very few records. Furthermore, numerous species await formal description. Therefore, to follow up, a strategic approach that combines investigation of individual prioritised species (such as arising from the present analysis) with whole of community monitoring, using metabar-coding to create baseline data against which disturbance effects can be measured, is recommended.

Land and natural resource managers rely on taxon experts to guide response and recovery, particularly after unprecedented events such as the 2019–20 wildfires. For less conspicuous taxa such as fungi, in addition to clarifying taxon delimitation, it is imperative to increase knowledge of their life history. Knowledge of functional significance, distribution, tolerance, dependency and response to episodic events is an essential ingredient for effective management and species preservation. In addition, we know that the resilience of fungal communities is important to ecosystem recovery, through their role in soil stabilisation, seedling establishment and as food resources for animals. Yet we have a limited understanding of how these relationships are affected by large wildfire events, let alone how these relationships may change in a future where fire is more common. In our changing and increasingly volatile landscape, it is essential that research and active management go hand-in-hand. We must be adaptive and responsive so that as we learn more about fungi, as well as their critical role in ecosystem function and recovery from disturbance, we can apply learnings for positive conservation outcomes.

Recommendations

- Carry out national threat assessments of Australian fungi to allow prioritisation of species for direct management and monitoring action.
    - First step – assess at least 50 fungi species that appear to be severely affected by the 2019–20 wildfires.

- Second step – focus on fungal symbionts of other threatened taxa where loss of host will have high impact on the fungus, in particular host-specific (1) ECM fungi of fire sensitive plants, such as *Nothofagus*, (2) leaf-inhabiting fungi of threatened plants, especially fungi that undertake entire life-cycle in/on the leaf, and (3) invertebrate parasites, such as those found within the gut.
- Compile trait database as essential underpinning for assessments. Traits should include trophic mode, substrate, host preference and response to fire.
- Keep the National Species List up-to-date to facilitate access to data.
- Increase the coverage of digitisation of fungi specimens to provide better visualisation of known distributions.
- Implement baseline monitoring of fungi stratified across ecosystems, using a whole-of-community metabarcoding approach.
  - There is a need to sample not only soil, but also woody debris and living plants (especially roots) and invertebrates.
  - It is vital to link molecular operational units (mOTUs) to known taxa (and precise guild); this requires (1) bulking up reference sequence library for described species, in combination with (2) targeted collecting of 'fire sensitive' mOTUs that do not match known taxa, for formal description.
- Utilise and extend existing studies to identify species significantly affected by fire.
- Prioritise the Cool Temperate Rainforest fungal community for survey and documentation because it is particularly susceptible to fire-effects, via reduction of host range.
- Establish *ex situ* living collection (biobank), sampled across taxonomy and trophic groups, focusing on threatened species and keystone species.
- Better understand how fungi recolonise after fire, investigating the spore bank and survival as mycelium.
- Develop guidelines for ecological management and regeneration in fire-prone ecosystems, including the conservation of unburnt areas or biological legacies, such as large, old trees remaining post-fire, which act as 'refugia' and potential inoculum sources for fungal communities. Management practices should avoid soil transfer (due to pathogens) and retain organic litter as a vital substrate for many fungi and invertebrates.

Acknowledgements

Thanks to Teresa Lebel (State Herbarium of South Australia) for providing information about ecology of sequestrate fungi and to Sara Maroske, Melissa Danks (Western Australian Museum) and John Woinarski for helpful feedback on the manuscript.

References

Ammitzboll H, Jordan GJ, Baker SC, Freeman J, Bissett A (2021) Diversity and abundance of soil microbial communities decline, and community compositions change with severity of postlogging fire. *Molecular Ecology* **30**, 2434–2448. doi:10.1111/mec.15900

Bowd EJ (2020) Above and below-ground disturbance responses in Mountain Ash forests. PhD thesis. The Australian National University, Canberra.

Bowd EJ, Banks SC, Bissett A, May TW, Lindenmayer DB (2021) Direct and indirect disturbance impacts in forests. *Ecology Letters* **24**, 1225–1236. doi:10.1111/ele.13741

Certini G (2005) Effects of fire on properties of forest soils: A review. *Oecologia* **143**, 1–10. doi:10.1007/s00442-004-1788-8

Chapman AD (2009) *Numbers of Living Species in Australia and The World.* Australian Biological Resources Study, Canberra.
Dauby G, Stévart T, Droissart V, Cosiaux A, Deblauwe V, *et al.* (2017) ConR: an R package to assist large-scale multispecies preliminary conservation assessments using distribution data. *Ecology and Evolution* **7**, 11292–11303. doi:10.1002/ece3.3704
Dell M, Worley M, Casanova MT, McMullan-Fisher S, Louwhoff S, *et al.* (2020) 'An assessment of conservation priorities and actions for bryophytes, algae and fungi in response to Victoria's 2019–2020 bushfires'. Independent consultant report, Melbourne.
Department of Agriculture Water and the Environment (2020) *Australian Google Earth Engine Burnt Area Map: A Rapid, National Approach to Fire Severity Mapping.* Department of Agriculture Water and the Environment, Canberra.
Dove NC, Hart SC (2017) Fire reduces fungal species richness and *in situ* mycorrhizal colonization: a meta-analysis. *Fire Ecology* **13**, 37–65. doi:10.4996/fireecology.130237746
Dundas SJ, Hopkins AJM, Ruthrof KX, Tay NE, Burgess TI, *et al.* (2018) Digging mammals contribute to rhizosphere fungal community composition and seedling growth. *Biodiversity and Conservation* **27**, 3071–3086. doi:10.1007/s10531-018-1575-1
Egidi E, McMullan-Fisher S, Morgan J, May T, Zeeman B, *et al.* (2016) Fire regime, not time-since-fire, affect soil fungal community diversity and composition in temperate grasslands. *FEMS Microbiology Letters* **363**, fnw196. doi:10.1093/femsle/fnw196
Glassman SI, Levine CR, Dirocco AM, Battles JJ, Bruns TD (2016) Ectomycorrhizal fungal spore bank recovery after a severe forest fire: Some like it hot. *The ISME Journal* **10**, 1228–1239. doi:10.1038/ismej.2015.182
Hao T, Elith J, Guillera-Arroita G, Lahoz-Monfort JJ, May TW (2021) Enhancing repository fungal data for biogeographic analyses. *Fungal Ecology* **53**, 101097. doi:10.1016/j.funeco.2021.101097
Hawksworth DL, Lücking R (2017) Fungal diversity revisited: 2.2 to 3.8 million species. *Microbiology Spectrum* **5**(4). doi:10.1128/microbiolspec.FUNK-0052-2016
Kurth VJ, Fransioli N, Fulé PZ, Hart SC, Gehring CA (2013) Stand-replacing wildfires alter the community structure of wood-inhabiting fungi in southwestern ponderosa pine forests of the USA. *Fungal Ecology* **6**, 192–204. doi:10.1016/j.funeco.2013.01.006
Lunn TJ, Gerwin M, Buettel JC, Brook BW (2018) Impact of intense disturbance on the structure and composition of wet-eucalypt forests: a case study from the Tasmanian 2016 wildfires. *PLoS One* **13**, e0200905. doi:10.1371/journal.pone.0200905
Marsh J, Bal P, Fraser H, Umbers K, Greenville A, *et al.* (2021) 'Assessment of the impacts of the 2019–20 wildfires of southern and eastern Australia on invertebrate species'. NESP Threatened Species Recovery Hub Project 8.3.1. Final Report, Brisbane.
May TW (2002) Where are the short range endemics among Western Australian macrofungi? *Australian Systematic Botany* **15**, 501–511. doi:10.1071/SB01041
McMullan-Fisher SJM, May TW, Robinson RM, Bell TL, Lebel T, *et al.* (2011) Fungi and fire in Australian ecosystems: a review of current knowledge, management implications and future directions. *Australian Journal of Botany* **59**, 70–90. doi:10.1071/BT10059
Palmer BJ, Valentine LE, Page M, Hobbs RJ (2020) Translocations of digging mammals and their potential for ecosystem restoration: a review of goals and monitoring programmes. *Mammal Review* **50**, 382–398. doi:10.1111/mam.12208
Pérez-Izquierdo L, Clemmensen KE, Strengbom J, Granath G, Wardle DA, *et al.* (2021) Crown-fire severity is more important than ground-fire severity in determining soil fungal community development in the boreal forest. *Journal of Ecology* **109**, 504–518. doi:10.1111/1365-2745.13529
Rincón A, Santamaría BP, Ocaña L, Verdú M (2014) Structure and phylogenetic diversity of post-fire ectomycorrhizal communities of maritime pine. *Mycorrhiza* **24**, 131–141. doi:10.1007/s00572-013-0520-0

SOFR (2018) 'Australia's State of the Forests Report.' Australian Bureau of Agricultural and Resource Economics and Sciences. Canberra, <https://www.agriculture.gov.au/abares/forestsaustralia/sofr/sofr-2018>

Tolsma A, Hale R, Sutter G, Kohout M (2019) 'Post-fire dynamics of Cool Temperate Rainforest in the O'Shannassy Catchment'. Arthur Rylah Institute for Environmental Research Technical Report Series No. 298. Department of Environment, Land, Water and Planning, Heidelberg.

Treseder KK, Mack MC, Cross A (2004) Relationships among fires, fungi and soil dynamics in Alaskan Boreal forests. *Ecological Applications* **14**, 1826–1838. doi:10.1890/03-5133

Warcup JH (1990) Occurrence of ectomycorrhizal and saprophytic discomycetes after a wild fire in a eucalypt forest. *Mycological Research* **94**, 1065–1069. doi:10.1016/S0953-7562(09)81334-8

Wardle DA, Bardgett RD, Klironomos JN, Setälä H, van der Putten WH, *et al.* (2004) Ecological linkages between aboveground and belowground biota. *Science* **304**, 1629–1633. doi:10.1126/science.1094875

11

Impacts of the 2019–20 wildfires on Australian invertebrates

Jess R. Marsh, Payal Bal, Hannah Fraser, Aaron Greenville, Tanya Latty, Melinda L. Moir, Libby Rumpff, Kate Umbers and John C. Z. Woinarski

Summary

Context and challenges

- There are far more species of invertebrates than any other component of Australia's biodiversity, and many have highly localised ranges: hence, it is likely that much of the biodiversity loss in the 2019–20 wildfires was experienced by invertebrate species.
- However, profound knowledge gaps render it especially challenging to assess the impacts of the 2019–20 wildfires on invertebrate species, and to direct and prioritise recovery responses.
- Invertebrate species also have low public profiles and low priority in conservation management efforts, so their losses in these wildfires have been overshadowed by those of more iconic vertebrate species.

Main findings

- Based on spatial analysis of distributional records, we found that for the 2019–20 wildfires:
  - At least part of the range of 13 581 invertebrate species was burnt.
  - Three hundred and eighty two species had their entire range burnt. However, all of these were species with only one or two records each, so their fire overlap values are of low confidence, because their actual ranges are so poorly resolved.
  - The numbers of fire-affected invertebrate species are far higher than comparable tallies for vertebrate species.
- From analysis of fire overlap and species' likely susceptibility to fire, we conclude that at least 60 invertebrate species are likely to have become eligible for listing as threatened because of the 2019–20 wildfires; for context, this is a

similar tally to the total number of terrestrial invertebrate species currently listed nationally as threatened (63 species).

- Although extinction is difficult to prove, it is highly likely that the *Banksia montana* mealybug (*Pseudococcus markharveyi*) went extinct because of the 2019–20 fires. It is plausible that the number of fire-related extinctions is much higher, but there have been few post-fire searches that would allow extinction to be corroborated or refuted.
- Many species from ancient lineages are likely to have been particularly impacted by the 2019–20 wildfires.
- The most important recovery management action for fire-affected invertebrate species is to attempt to reduce the likelihood of future comparable wildfires, to mitigate impacts of such fires if they do occur, and to create a patchwork mosaic of fire ages, including areas of longer unburnt habitat.

Introduction

There are ~320 000 invertebrate species in Australia (Chapman 2009) – ~40 times the number of vertebrate species and ~12 times the number of plant species. Many Australian invertebrate species have highly restricted distributions ('short range endemics') (Harvey *et al.* 2011), rendering such species particularly at risk from acute threats. Given their diversity and restricted nature, it is probable that the majority of the biodiversity impact of the extensive 2019–20 Australian wildfires was experienced by invertebrate species (Hyman *et al.* 2020). Yet the public profile of, and responses to, biodiversity losses in the wildfires focused predominantly on vertebrate species, particularly iconic animals such as the koala (*Phascolarctos cinereus*) (e.g. Mo *et al.* 2021; Parrott *et al.* 2021). This taxonomic mismatch of impact and response is consistent with the disproportionately small representation of invertebrates on threatened species lists, and in conservation policy and management in Australia (Braby 2018; Taylor *et al.* 2018) and globally (Régnier *et al.* 2015).

In this chapter, we report on the impacts of the 2019–20 wildfires on Australian invertebrate species. The material presented here is based in part on more detailed reporting in Marsh *et al.* (2021) and Marsh *et al.* (in press). Note that Chapter 6 describes in more detail the impacts of the wildfires on aquatic invertebrates, especially spiny crayfish (*Euastacus* species).

There are long-standing impediments to the conservation of Australian invertebrates, largely relating to major shortcomings in knowledge (Taylor *et al.* 2018), and these challenges also constrain assessments of the impacts of the 2019–20 wildfires (see also Chapter 32). The majority of invertebrate species in Australia are not yet described (Austin *et al.* 2004; Chapman 2009), so assessments of fire impacts on described species substantially underestimate the actual number of fire-affected invertebrate species. For many of the described invertebrate species, there are very few records. For example, our detailed analysis (Marsh *et al.* 2021) collated, across all readily available distributional databases, a total of 342 534 records of 45 529 invertebrate species; but of these species, 31% had only one record and 14% had only two records in these collated databases and, for such species, estimates of fire impact have low confidence.

For most invertebrate species, there are no monitoring programs (Legge *et al.* 2018): this means that it is almost impossible to assess population losses due to these wildfires. There are also disproportionately few Australian invertebrate species listed as threatened

(Walsh *et al.* 2013), and few subject to conservation management programs. As a consequence, imperilled invertebrate species were generally not prioritised for protection in operational actions during the 2019–20 wildfires, and there is little knowledge of management actions that may support post-fire recovery of fire-affected invertebrates.

Relative to the amount of knowledge derived from research on vertebrates and plants, there have been few studies of the responses of Australian invertebrate species to fires (Saunders *et al.* 2021). For example, a review of the responses to fire of Australian biodiversity noted that:

> *the patterns of fire responses exhibited by invertebrates are ... even more difficult to detect than those ... for vertebrates and plants ... invertebrates deserve far more attention than they have received to date in fire ecology ... given the less-than-ideal state of the data relating to fire-response patterns exhibited by invertebrates, it is not surprising that our understanding ... is poor. (Whelan et al. 2002, pp. 98, 111)*

Furthermore, most of this limited knowledge of the impacts of fire on invertebrates relates to mild fires rather than fires of the severity of the 2019–20 wildfires (Friend 1994). Of these few studies, many also reported on the fire responses of higher taxonomic groupings of invertebrates (families or orders) rather than individual species, thereby potentially obscuring the impacts on that subset of species that are fire sensitive (Majer *et al.* 1997).

In part because such studies typically report the persistence of some invertebrate groups after (mild) fires, there is now something of an established paradigm that invertebrates are relatively impervious to fires. For example, Bradstock (2008, p. 812) reported:

> *Most invertebrate groups are little affected by fires, principally because individuals in varied life cycle phases are able to avoid lethal heat via shelter in soils, tree stems and rocks. The abundance and density of differing invertebrate groups typically display rapid resilience after fires of varied type and intensity.*

Although this statement may be valid for some groups, it is unlikely to represent invertebrates that are restricted to fire sensitive habitats, have small ranges or population sizes, occur in microhabitats that provide little protection from fire, or have life history characteristics (e.g. low reproductive output, limited dispersal capability) that constrain recruitment. For species with such characteristics, much of the population may be killed during fire and recovery may be challenging. In this chapter, we focus on such species, identifying those whose conservation status and outlook was most detrimentally affected by the 2019–20 wildfires, and hence are most in need of targeted management that supports recovery.

Our assessment focuses mainly on the consequences of the 2019–20 wildfires on individual invertebrate species, but such a focus should also be more broadly contextualised. One such context relates to fire regimes, and the consequences more broadly of repeated fires. In many cases, fire impacts may be magnified (and recovery thwarted) by short intervals between successive fires, or a history of fires that each burn part of a habitat but collectively burn most or all of it (Moir 2021). Any presumption that invertebrates may be resilient to fires is unlikely to apply if regimes change to more frequent and severe fires.

Furthermore, the impacts of the 2019–20 wildfires should not be considered in isolation, because they may compound or be compounded by other threats that collectively may imperil a species far more than the influence of any factor alone. For example, the Bogong moth (*Agrotis infusa*) was historically super-abundant (with billions of individuals), but has declined dramatically in recent years, due to land-use changes, drought and other factors (Green *et al.* 2021). The 2019–20 wildfires added incrementally to this pattern of decline (Green *et al.* 2021).

In a broader context, the impact of the fires on invertebrates will have far-reaching consequences beyond the invertebrates themselves: the fires will affect the pivotal ecological functions that invertebrates provide (Waldbauer 2004). The 2019–20 wildfires caused marked reductions in the population size, biomass and diversity of many invertebrates over extensive areas, and such losses are likely to cause long-lasting ecological degradation and complex repercussions for ecosystems. Furthermore, some invertebrate species have cultural significance: for example, the Bogong moth is not only a keystone species (Braby *et al.* 2021), but has also long been important to Indigenous Australians for ceremony and sustenance (Stephenson *et al.* 2020).

To assess the impacts of the 2019–20 wildfires on Australian invertebrates, we collated species' distributional data from many sources (mostly the Atlas of Living Australia) and created range polygons for all species with at least three records. We then overlaid – for every species – the individual point records and polygons with maps that categorised fire severity (Department of Agriculture Water and the Environment 2020) to calculate 'fire overlap' – the proportions both of its records and of its modelled distribution that was burnt (Marsh *et al.* 2021). Our analysis was restricted to those regions of eastern and southern Australia that experienced exceptional fire events in 2019–20 (the 'Preliminary Analysis Area' (PAA): see Chapter 1).

Quantifying fire overlap is relatively straightforward, but assessment of impact is more challenging. To complement our assessments of fire overlap, we developed and applied a trait database for invertebrate species that comprised information on life history characteristics, threats and management needs. We then used the trait database to identify species likely to have experienced the most loss in the 2019–20 fires. Marsh *et al.* (2021) present these methods and analyses in detail. Ideally, this approach for estimating fire impact should be tested by post-fire surveys to corroborate the predicted extent of population loss; however, there has been very limited post-fire sampling for most invertebrate groups (but see Box 11.1 for one example).

Box 11.1. Kangaroo Island assassin spider

The Kangaroo Island assassin spider (*Zephyrarchaea austini*) (Fig. 11.1) is a small spider (< 3 mm long), whose known range is restricted to western Kangaroo Island. At the time of the 2019–20 fires, it was only known from a single conservation reserve (Western River Wilderness Protection Area). The species lives in a microhabitat (elevated leaf litter suspended among low-lying vegetation) that is highly flammable, even in low-intensity fire, and offers little shelter from fire. The entirety of Western River Wilderness Protection Area burnt in severe wildfires in 2019, with the exception of a small section that had been subject to a fuel reduction burn in 2015. Surveys of this latter area found that, despite 6 years passing post-burn, the elevated leaf litter layer had not yet accumulated, ensuring that the habitat was not suitable for the assassin spider. This example highlights the lag time required before burnt habitat will be suitable for such species and also the challenges in balancing fire management actions and conservation, especially for species susceptible to even low-intensity fires. Happily, after many fruitless months of intensive post-fire searches of unburnt habitat in and near the known range, a single individual was located in September 2021 (J. Marsh, *pers. obs.*), providing some hope for the future of the species (see also Chapter 34).

Fig. 11.1. There is no known photograph of a live Kangaroo Island assassin spider. Here, a related species, the eastern massif assassin spider (*Zephyrarchaea robinsi*). (Photo: M. Harvey, © Western Australian Museum)

Findings

We collated distributional data (342 534 records) for 45 528 Australian invertebrate species, of which 32 163 species occurred in the PAA, with a total of 238 633 records. We found that the 2019–20 wildfires burnt at least part of the range of 13 581 of these species. For 787 species, at least 50% of their range was burnt in fires of any severity and/or at least 30% of their range was burnt by fires of high severity. For 382 species, *all* of the known range was burnt, although we note that this tally comprised only species with one or two occurrence records each, so fire overlap values are of low confidence. These results corroborate and extend to national scale an earlier finding, for a subset of taxonomic groups of invertebrates at more regional scale (New South Wales), that all of the known range of many invertebrate species was burnt in the 2019–20 wildfires (Hyman *et al.* 2020).

These tallies are substantial underestimates of the actual number of invertebrate species affected, as our analysis was mostly restricted to described species, which comprise only about one-third of the estimated number of Australian invertebrates (Chapman 2009), and because we could find no acceptable records (e.g. of sufficient locational precision) for many described species known to occur in the PAA.

The numbers of invertebrate species impacted by fire are far higher than the comparable tallies for vertebrate species (Table 11.1), with invertebrate species contributing 98% of the animal species with at least 90% distributional overlap with the 2019–20 wildfires. For example, we found that 2186 invertebrate species had higher proportions of their distributions burnt than the koala (17%), an iconic vertebrate species accorded pre-eminent public profile as a casualty of the 2019–20 fires, and a major focus of the conservation response (Parrott *et al.* 2021). Although the values are not readily accessible, we doubt that the total recovery investment spread across these 2000+ invertebrate species more fire-affected than the koala would be within an order of magnitude of the effort directed towards the koala (Marsh *et al.* in press).

Fire overlap is not synonymous with fire impact. Some invertebrate species occurring in burnt areas may nonetheless have experienced little immediate mortality because the

Table 11.1. The number of species of invertebrate and vertebrate taxa with extensive overlap with the 2019–20 wildfires.

Tallies in brackets for invertebrates exclude those species with only one or two records in our analysis. Data sourced from Marsh *et al.* (2021).

% overlap with fire	No. of species					
	Invertebrates	Fish	Frogs	Reptiles	Birds	Mammals
100%	382 (0)	4	0	0	0	0
90–99.9%	6 (6)	1	1	1	0	1
80–89.9%	25 (25)	0	1	1	0	1
70–79.9%	26 (26)	0	3	1	0	1
60–69.9%	43 (43)	0	3	1	17	1
50–59.9%	305 (103)	4	8	1	4	5
Total 50+	**787 (203)**	**9**	**16**	**5**	**21**	**9**

entire population or particular life stages are sheltered from fire – for example, by living deep enough underground. However, fire may affect even such species if the post-fire environment offers fewer resources, higher rates of predation or more extreme microclimates.

For the invertebrate species with at least 50% overlap with fire or at least 30% overlap with severe fire (defined in the fire mapping as canopy scorched or consumed), we compiled a database of life history, ecological and other traits related to fire susceptibility. In our trait database we also included 19 expert-nominated species for which there were sufficient private distributional data and a further nine threatened species with 30–50% overlap with all fires. Combining trait information with fire overlap data, we concluded that at least 60 of these invertebrate species were likely to have suffered at least 30% reduction of their population due to the 2019–20 fires, and hence were likely to be now eligible for, and have sufficient information to justify, listing as threatened (see Supplementary Table S11 online at doi:10.6084/m9.figshare.20132213). If such species were indeed to be listed nationally as threatened, it would result in an almost 100% increase in the number of terrestrial invertebrate species currently listed nationally as threatened (63 species).

The actual number of invertebrate species that are now imperilled because of the 2019–20 wildfires is likely to be far more than 60, but information shortcomings constrain listing assessments for many species. For example, some species were represented in our analysis by only one or two records, while for others information on traits was unavailable. For such species, particularly where the fire impact was suspected to be severe, filling information deficiencies should be prioritised, such as through post-fire surveys to better assess the extent of population loss and to safeguard any detected surviving populations.

Invertebrates across most taxonomic groups were affected by the 2019–20 wildfires. Impacts were especially pronounced for groups with (1) many short-range endemic species, (2) many species occurring in habitats that are fire sensitive and (3) many species occurring in micro-habitats that offer little protection from fire. For some groups, such as millipedes, land snails, flightless ground beetles and a few spider families, many of their species fit in all three of these categories. Note also that many of these groups are of great antiquity (Rix and Harvey 2012), and the increasing severity and incidence of wildfire are likely to cause ongoing loss and retreat of these old lineages, as areas long offering them protection from fire become increasingly less secure.

The 2019–20 wildfires are likely to have caused the extinctions of some invertebrate species. However, extinction is difficult to prove, especially so for species that are poorly known (e.g. those for which locational information was meagre before the fire) and hard to detect. Nonetheless, there is compelling evidence that the 2019–20 wildfires caused the extinction of at least one species, the *Banksia montana* mealybug (*Pseudococcus markharveyi*) (Moir 2021) (see Box 11.2).

Box 11.2. *Banksia montana* mealybug

The *Banksia montana* mealybug is, or was, a tiny (1–3 mm) true bug of underwhelming charisma (Fig. 11.2). It had some characteristics that should have given it some conservation security. From not long after its discovery (in 2007) and description (in 2013), it has been listed as Critically Endangered at global, national and state (Western Australia) levels, nominally affording it some conservation profile, protection and action. Its ecology was reasonably well known and its distribution precisely known due to targeted surveys. The mealybug's entire range was within a high-profile conservation reserve (Stirling Range National Park) with a well-established fire management plan. Much intensive conservation management was directed towards its host plant species (including translocations, application of phosphite to mitigate dieback and caging to reduce pressure from mammalian herbivores). Fire was a recognised threat to the species and, as a consequence, its known locations were delineated in fire plans and highlighted for protection during fire operations. Some conservation efforts had been specifically implemented for the mealybug to try to

Fig. 11.2. *Banksia montana* mealybug (square and inset) within foliage of its host plant. (Photos: © Melinda Moir)

reduce its extinction risk, such as through attempted translocations, but unfortunately these resulted in multiple failures, in part due to predators and parasitoids (Threatened Species Scientific Committee 2018).

However, the mealybug also had characteristics that rendered it especially susceptible. It was entirely dependent on a single host plant species, *Banksia montana*, that also happened to be extremely rare, highly localised and fire sensitive; it occurred only on old (at least 9 and up to 30 years) individual plants. It also had very limited dispersal ability, so could not readily recolonise burnt areas from unburnt patches. Indeed, it could not disperse unless host plants had interlocking foliage, with the bugs walking over 'bridges' of crossed branches. This adaptation was probably suitable when the mountain ridges were covered in a dense tall heath featuring abundant *Banksia montana*, but the history of fires meant that such conditions no longer exist. These imperilling characteristics ended up outweighing the conservation efforts. By 2012, it was known to occur in only two small sites, 7 km apart. In 2018, a control burn, designed to reduce the likelihood of wildfire, burnt one of these sites, destroying the population of mealybugs and rendering the host plants there unsuitable for many years. Following that fire, the entire mealybug population occurred only on six plants present at one site. From 26 December 2019, a series of severe fires that burnt 40 000 ha of the national park killed all of these plants and, as a consequence, the mealybug is now likely to be extinct (Moir 2021).

This case shows the devastating impact of a single fire event, but also illustrates the cumulative impacts of successive fires (iteratively depleting populations) and the importance of fire regimes generally. It also provides the hard lesson that conservation objectives for fire management – in this case a recommended interval between successive fires of at least 18 years, and exclusion of fire from the sole known population – may be challenging or impossible to achieve, even in landscapes largely devoted to conservation.

The likely extinctions, and the extent of population losses for so many other invertebrate species, indicate that the 2019–20 wildfires had a catastrophic impact on the Australian invertebrate fauna. For some of the extant but severely fire-affected species, recovery may take many years and may never be realised if comparable wildfires recur in the near future. For other fire-affected species recovery may be possible, but it will require targeted and effective management investment at a scale that vastly exceeds the limited effort that has typically been devoted to invertebrate conservation in Australia.

We applied information from our trait database to identify the main threats to, and thus indicate the likely management needs of, the fire-affected invertebrate species (Marsh *et al.* 2021). For most species, the most significant ongoing threat is the likely increased frequency of severe wildfire, and the action most needed to support recovery and to limit future losses is targeted fire management. In many cases this involves excluding fire from sites holding surviving populations and attempting to prevent recurrence of comparable severe fires in areas burnt in the 2019–20 wildfires.

Given the major knowledge gaps that currently constrain conservation of Australian invertebrates, monitoring programs should be implemented for fire-affected invertebrate species. Such monitoring can help document the pace and extent of recovery, assess the effectiveness of management actions (see Box 11.3), and provide a more robust baseline from which to assess the impacts of future fire events.

Box 11.3. Green carpenter bee

The green carpenter bee (*Xylocopa aeratus*) (Fig. 11.3) provides a contrast to the two other case study species in that historically it was widespread, from northern New South Wales to South Australia. However, due to the compounding impacts of habitat loss and fragmentation, and inappropriate fire regimes, many subpopulations have been extirpated and it is now considered regionally extinct in Victoria. In South Australia, this large and spectacular native bee survives solely on Kangaroo Island, in old-growth (> 35 years old, but typically 50–90 years old) heathlands. This habitat association is largely because it is reliant on the soft wood of large dead trunks of *Banksia* spp. and dried flower stalks of grass-trees (*Xanthorrhoea* spp.) for nesting substrates. For the Kangaroo Island subpopulation, extensive fires in 2007 severely reduced old-growth habitat, and the fires of 2019–20 burnt ~95% of the remaining habitat. The Kangaroo Island bees now survive precariously only in some small unburnt patches. Every such severe fire locally annihilates bees and destroys old-growth nesting substrate, which takes at least 35 years to recover, and such subpopulation losses cumulatively increase the overall risk of extinction. Researchers have trialled, with some success, artificial breeding sites to help maintain and recover bee subpopulations, but the most critical management requirement is to seek to ensure that habitat remains unburnt for many decades – a formidable challenge (Hogendoorn *et al.* 2021).

Fig. 11.3. Green carpenter bee (*Xylocopa aeratus*) from Kangaroo Island, showing female (above) and male (below). (Photo: Dr Richard V. Glatz)

Our analysis focused on the impacts of the 2019–20 wildfires on individual invertebrate species; however, assessment of the impacts and consequent management needs of thousands of invertebrate species poses formidable challenges, and the history of conservation efforts for Australian invertebrates does not provide much reassurance that such challenge will be embraced. Furthermore, a conservation response focused solely on the needs of individual species recognised to be fire-affected does little for the conservation of the most poorly known species (including those that have not yet been described) or for ecosystem functioning. A complementary approach that may be more strategic is to identify sites in the landscape that are of exceptional significance for the conservation of invertebrates more broadly, and to factor the protection of such sites into management planning and operations. Given the propensity of invertebrates in some taxonomic groups to be characterised by very small ranges (Harvey 2002), this planning and protective management is especially relevant for sites that contain many co-occurring, short-range endemic species (centres of endemism). Exceptionally many species may be safeguarded, or lost, in these key locations: they are 'irreplaceable sites' (*sensu* Pressey *et al.* 1994) for conservation. To date, few such key sites have been identified for Australian invertebrates (Moir *et al.* 2009; Barrett and Yates 2015). A more comprehensive systematic attempt to identify and circumscribe such sites, recognise these as key assets, and then include them in fire planning and operations should help to increase protection for many imperilled invertebrate species.

Conclusion

The 2019–20 wildfires had a devastating impact on many Australian invertebrate species. Some of these losses are unrecoverable, and reductions in population size for many invertebrates are likely to cause long-lasting ecological detriment. Recovery for many species will require targeted investment, at a scale far more substantial than the desultory efforts typically allocated to invertebrate conservation in Australia to date. The most important ongoing management action is to reduce the likelihood of future comparable wildfire. However, as illustrated in the case studies, the achievement of such an objective will be challenging. The fire management objective is also unlikely to be realised unless the root cause of such wildfires – climate change – is controlled effectively. Without such control, catastrophic wildfires will recur, and the Australian invertebrate fauna, including some lineages of great antiquity, will continue to be diminished.

Recommendations

- At least 60 invertebrate species for which there is sufficient evidence of severe impacts due to the 2019–20 wildfires should be assessed for national listing as threatened.
- Surveys should be undertaken for these species, and other more poorly known fire-affected species, to allow for robust assessments of population loss, and to identify important populations remaining after fire.
- Targeted management to support the recovery of fire-affected invertebrates must be identified in recovery plans, conservation advices and regional fire management plans, and implemented, with the scale of investment substantially increased from the meagre levels characterising previous conservation efforts for Australian invertebrates.
- The main priority for such management should be to reduce the likelihood of future comparable wildfires, and to maintain areas of long-unburnt habitat.
- Monitoring programs for fire-affected invertebrate species should be implemented to chart recovery, assess and refine management effectiveness, and provide a more robust baseline from which impacts of future fires can be measured.
- Important sites for the conservation of invertebrates (e.g. centres of endemism) should be identified and their protection mandated in policy and legislation (e.g as critical habitat). They should be recognised as priorities for protection in fire management plans and operations.
- Major knowledge gaps that constrained assessments of the impacts of the 2019–20 wildfires on the Australian invertebrate fauna should be filled. There is a need for better consolidation of distributional records, taxonomic resolution of poorly known groups, more targeted ecological studies (particularly of responses to fire, and to define preferred fire regimes), assessment of other threats, and development and maintenance of a database of species' traits related to fire susceptibility.
- Increased effort should be directed towards management of other key threats, to increase their resilience under a future marked by increased likelihood of wildfire.

- Raising the profile of invertebrate species at risk is required to elicit more public support and resourcing for their conservation. Such action should aim to encompass the suite of imperilled invertebrate groups rather than reinforcing bias towards those groups of most public appeal.
- The disproportionately low representation of invertebrates in threatened species lists and in conservation investment urgently needs to be rectified.
- More effective and urgent commitments to reducing greenhouse gas emissions are needed, for otherwise future fire regimes are likely to exert even more significant impacts on Australian invertebrates, with diminishing prospects of recovery.

Acknowledgements

The data collation and analysis of this chapter was supported by the Australian Government's National Environmental Science Program. We thank many invertebrate experts for their contributions to this work, and the project guidance of Jason Ferris and Fiona Woods (Department of Agriculture, Water and the Environment).

References

Austin AD, Yeates DK, Cassis G, Fletcher MJ, La Salle J, *et al.* (2004) Insects 'Down Under'– diversity, endemism and evolution of the Australian insect fauna: examples from select orders. *Australian Journal of Entomology* **43**, 216–234. doi:10.1111/j.1326-6756.2004.00448.x

Barrett S, Yates CJ (2015) Risks to a mountain summit ecosystem with endemic biota in southwestern Australia. *Austral Ecology* **40**, 423–432. doi:10.1111/aec.12199

Braby MF (2018) Threatened species conservation of invertebrates in Australia: an overview. *Austral Entomology* **57**, 173–181. doi:10.1111/aen.12324

Braby MF, Yeates DK, Taylor GS (2021) Population declines and the conservation of insects and other terrestrial invertebrates in Australia. *Austral Entomology* **60**, 3–8. doi:10.1111/aen.12519

Bradstock RA (2008) Effects of large fires on biodiversity in south-eastern Australia: disaster or template for diversity? *International Journal of Wildland Fire* **17**, 809–822. doi:10.1071/WF07153

Chapman AD (2009) *Numbers of Living Species in Australia and the World.* Australian Biological Resources Study, Canberra.

Department of Agriculture Water and the Environment (2020) *Australian Google Earth Engine Burnt Area Map: A Rapid, National Approach to Fire Severity Mapping.* Department of Agriculture Water and the Environment, Canberra.

Friend GR (1994) Fire ecology of invertebrates – implications for nature conservation, fire management and future research. In *Fire and Biodiversity: the Effects and Effectiveness of Fire Management.* pp. 155–162. Department of the Environment Sport and Territories Biodiversity Unit, Canberra.

Green K, Caley P, Baker M, Dreyer D, Wallace J, *et al.* (2021) Australian Bogong moths *Agrotis infusa* (Lepidoptera: Noctuidae), 1951–2020: decline and crash. *Austral Entomology* **60**, 66–81. doi:10.1111/aen.12517

Harvey MS (2002) Short-range endemism amongst the Australian fauna: some examples from non-marine environments. *Invertebrate Systematics* **16**, 555–570. doi:10.1071/IS02009

Harvey MS, Rix MG, Framenau VW, Hamilton ZR, Johnson MS, *et al.* (2011) Protecting the innocent: studying short-range endemic taxa enhances conservation outcomes. *Invertebrate Systematics* **25**, 1–10. doi:10.1071/IS11011

Hogendoorn K, Glatz RV, Leijs R (2021) Conservation management of the green carpenter bee *Xylocopa aerata* (Hymenoptera: Apidae) through provision of artificial nesting substrate. *Austral Entomology* **60**, 82–88. doi:10.1111/aen.12510

Hyman I, Ahyong ST, Köhler F, McEvey SF, Milledge GA, *et al.* (2020) Impacts of the 2019–2020 bushfires on New South Wales biodiversity: a rapid assessment of distribution data for selected invertebrate taxa. *Technical Reports of the Australian Museum Online* **32**, 1–17. doi:10.3853/j.1835-4211.32.2020.1768

Legge S, Scheele BC, Woinarski JCZ, Garnett ST, Keith DA, *et al.* (2018) Summary: monitoring extent and adequacy for threatened species. In *Monitoring Threatened Species and Ecological Communities.* (Eds S Legge, DB Lindenmayer, NM Robinson, BC Scheele, DM Southwell and BA Wintle) pp. 127–133. CSIRO Publishing, Clayton.

Majer J, Recher HF, Wellington AB, Woinarski JCZ, Yen A (1997) Invertebrates of eucalypt formations. In *Eucalypt Ecology: Individuals to Ecosystems.* (Eds JE Williams and JCZ Woinarski) pp. 278–302. Cambridge University Press, Cambridge.

Marsh J, Bal P, Fraser H, Umbers K, Greenville A, *et al.* (2021) *Assessment of the Impacts of the 2019–20 Wildfires of Southern and Eastern Australia on Invertebrate Species.* Threatened Species Recovery Hub of the National Environmental Science Program, Brisbane.

Marsh JR, Bal P, Fraser H, Umbers K, Latty T, *et al.* (in press) Accounting for the neglected: invertebrate species and the 2019-20 Australian megafires. *Global Ecology and Biogeography.*

Mo M, Roache M, Reid T, Oliver DL, Broome L, *et al.* (2021) Corporate support for threatened species recovery efforts: three case studies from the 2019–20 Australian bushfire season. *Australian Zoologist* **41**, 186–193. doi:10.7882/AZ.2020.031

Moir ML (2021) Coextinction of *Pseudococcus markharveyi* (Hemiptera: Pseudococcidae): a case study in the modern insect extinction crisis. *Austral Entomology* **60**, 89–97. doi:10.1111/aen.12506

Moir ML, Brennan KEC, Harvey MS (2009) Diversity, endemism and species turnover of millipedes within the south-western Australian global biodiversity hotspot. *Journal of Biogeography* **36**, 1958–1971. doi:10.1111/j.1365-2699.2009.02137.x

Parrott ML, Wicker LV, Lamont A, Banks C, Lang M, *et al.* (2021) Emergency response to Australia's Black Summer 2019–2020: the role of a Zoo-based conservation organisation in wildlife triage, rescue, and resilience for the future. *Animals (Basel)* **11**, 1515. doi:10.3390/ani11061515

Pressey RL, Johnson IR, Wilson PD (1994) Shades of irreplaceability: towards a measure of the contribution of sites to a reservation goal. *Biodiversity and Conservation* **3**, 242–262. doi:10.1007/BF00055941

Régnier C, Achaz G, Lambert A, Cowie RH, Bouchet P, *et al.* (2015) Mass extinction in poorly known taxa. *Proceedings of the National Academy of Sciences of the United States of America* **112**, 7761–7766. doi:10.1073/pnas.1502350112

Rix MG, Harvey MS (2012) Phylogeny and historical biogeography of ancient assassin spiders (Araneae: Archaeidae) in the Australian mesic zone: evidence for Miocene speciation within Tertiary refugia. *Molecular Phylogenetics and Evolution* **62**, 375–396. doi:10.1016/j.ympev.2011.10.009

Saunders ME, Barton PS, Bickerstaff JRM, Frost L, Latty T, *et al.* (2021) Limited understanding of bushfire impacts on Australian invertebrates. *Insect Conservation and Diversity* **14**, 285–293. doi:10.1111/icad.12493

Stephenson B, David B, Fresløv J, Arnold LJ, Delannoy J-J, *et al.* (2020) 2000 year-old Bogong moth (*Agrotis infusa*) Aboriginal food remains, Australia. *Scientific Reports* **10**, 1–10. doi:10.1038/s41598-020-79307-w

Taylor GS, Braby MF, Moir ML, Harvey MS, Sands DPA, *et al.* (2018) Strategic national approach for improving the conservation management of insects and allied invertebrates in Australia. *Austral Entomology* **57**, 124–149. doi:10.1111/aen.12343

Threatened Species Scientific Committee (2018) *Conservation Advice* Pseudococcus markharveyi *Banksia Montana Mealybug.* Department of Environment and Energy, Canberra.

Waldbauer G (2004) *What Good are Bugs? Insects in the Web of Life.* Harvard University Press, Cambridge, MA.

Walsh JC, Watson JEM, Bottrill MC, Joseph LN, Possingham HP (2013) Trends and biases in the listing and recovery planning for threatened species: an Australian case study. *Oryx* **47**, 134–143. doi:10.1017/S003060531100161X

Whelan RJ, Rodgerson L, Dickman CR, Sutherland EF (2002) Critical life cycles of plants and animals: developing a process-based understanding of population changes in fire-prone landscapes. In *Flammable Australia: the Fire Regimes and Biodiversity of a Continent.* (Eds RA Bradstock, JE Williams and MA Gill) pp. 94–124. Cambridge University Press, Cambridge, UK.

12

Estimating the number of wild animals affected by Australia's 2019–20 wildfires

Lily M. van Eeden and Christopher R. Dickman

Summary

Context and challenges

- The 2019–20 wildfires in Australia were unprecedented in their extent and severity, affecting individuals and populations of many fauna species.
- We sought to assess these impacts by estimating the numbers of native vertebrate animals likely to have been present within the fire impact area.
- For several major taxon groups, datasets on pre-fire population densities were limited so we were necessarily conservative in the estimates that we made for these groups.
- Similarly, limited available data on the ability of different taxon groups to survive fire meant that we could not estimate mortality rates.
- Robust, consistent and targeted long-term monitoring is required to inform assessments of the impacts of fire on individual animals and populations, to identify key resources needed by fire-susceptible species and to assess the effectiveness of management interventions.

Main findings

- Our estimate of the numbers of reptiles, birds, mammals, and frogs affected by the 2019–20 wildfires rose from over 1 billion individuals in January 2020 while the fires still raged to 2.8 billion individuals after the fires were extinguished.
- The magnitude of the estimates stimulated huge interest from national and global media about the impacts of these fires upon biodiversity. Media focus on numbers of animals killed, rather than affected, by the fires made nuanced messaging difficult, but perhaps helped to catalyse management action, liberation of funds and subsequent inquiries into the impacts of the fires on wildlife.

Introduction

Fire is a common feature of Australian landscapes, with many plant and animal species evolving adaptations that allow them to survive, benefit from and even depend on fire. However, fire extent, frequency and intensity have increased in recent decades due to the changing global climate (Bowman *et al.* 2020). As such, Australian ecosystems are experiencing fires that may have long-term consequences for flora and fauna, with detrimental impacts to conservation of biodiversity (Lindenmayer and Taylor 2020). The wildfires that occurred in 2019–20 were unprecedented in their extent and severity (Bowman *et al.* 2020; Wintle *et al.* 2020; Collins *et al.* 2021; Chapter 2), including burning vegetation types such as rainforest that are not normally exposed to fire. It was quickly recognised that these fires were having major impacts on native fauna and flora, both on the survival of individual animals and on the resilience of populations and communities of conservation concern.

As the fires burnt across increasingly large areas of temperate forest and woodland, news of the impact of the fires on Australia's wildlife focused on a subset of charismatic fauna in both print and digital media. Most graphically, harrowing images of burnt koalas (*Phascolarctos cinereus*) and kangaroos captured attention in Australia and then globally (Fig. 12.1), as did the rescue efforts being made for these animals. This prompted us to think about the impacts of the fires on individual animals, recognising that there were vastly more individuals of many more species behind these icons that were also being affected. We wondered whether it was possible to estimate numbers of *all* vertebrates affected by the fires (not just the icons) to more fully appreciate the potential toll and to explore whether numbers of individuals in the path of the fires could translate to impacts at the population and species levels, or even to the ecological communities to which they belonged.

One of us (CD) made preliminary estimates in January 2020 that hundreds of millions of vertebrate animals had been impacted by the fires, with the numbers rising rapidly to

Fig. 12.1. Emotive scenes showing the direct impacts of fire on the welfare of charismatic fauna, such as koalas, generated international media interest in the Australian wildfires and their impacts on wild animals. (Photo: The Footage Company/Nine Network Australia)

more than 1 billion animals as the fires continued to rage (Box 12.1). The fires burnt for many weeks following this estimate, finally being extinguished in eastern Australia in March 2020. Although the notion that over a billion animals could have been in the path of the fires was shocking (Dickman and McDonald 2020), this figure was based on population density estimates that had been collected previously to assess the impacts of land clearing on wildlife in New South Wales (Johnson *et al.* 2007) and was clearly rough.

Box 12.1. Your number's up: media interest in the effects of the fires on wildlife

Anyone walking along an Australian forest trail on a hot summer day would be likely to see small lizards scuttling among the leaf litter, chance upon a sun-baking snake, and enjoy the sounds and sights of many birds before flushing a kangaroo or spotting a sleeping koala. Yet intense media focus on the marsupials gave the impression that they were the primary casualties of the 2019–20 megafires; other wildlife was ignored. Just after Christmas 2019 one of us (CD) brooded about this imbalance in reporting and used some readily available information on animal densities, contained in a 2007 report for WWF-Australia (Johnson *et al.* 2007), to estimate that some 480 million mammals, birds and reptiles would have been in the path of fires that had raged in New South Wales to that time. With some trepidation, and several caveats, CD provided the estimate in an interview with *The Times* on the 27 December 2019. The number appeared in *The Times* on the same day and swiftly went viral, appearing in stories about the impacts of the fires in Australia and globally in the days before the New Year. When the University of Sydney switchboard opened on 2 January 2020, the operators were inundated with calls from concerned people, including wildlife carers who had been rescuing animals from fire fronts, and especially from journalists from radio, television, podcasters, social media and other news outlets requesting updates and interviews. On 3 January 2020, CD issued a media release to outline the background rationale for his estimate of 480 million animals, updating this 5 days later to acknowledge the greatly increased area burnt by the fires since the original estimate to more than 800 million animals affected in New South Wales and more than a billion animals if burnt areas in Victoria were also included. Despite the continuing passage of the fires, no further updates were made: the estimate of 'over a billion' fire-affected animals was more than enough to sustain media attention. On 3 January 2020 there were 242 media articles quoting CD in 41 countries; on 9 January this peaked at 907 media articles, and by the end of January 2020 the numerical toll of Australia's wildlife had been noted in at least 6495 media articles in 115 countries (Fig. 12.2). Over 100 interviews were given in that month. Although media attention diminished when no further updates on numbers were provided, there were further peaks in interest when our preliminary updated numbers became available in July 2020 and later in the year when our final report was published (van Eeden *et al.* 2020). By year's end the estimates of animal numbers affected by the 2019–20 fires had been reported in at least 7539 media articles, in 36 languages, and reached an estimated audience of 25.6 billion people – more than three times the global population (V. Reiner, Media Office, University of Sydney, *pers. comm.*).

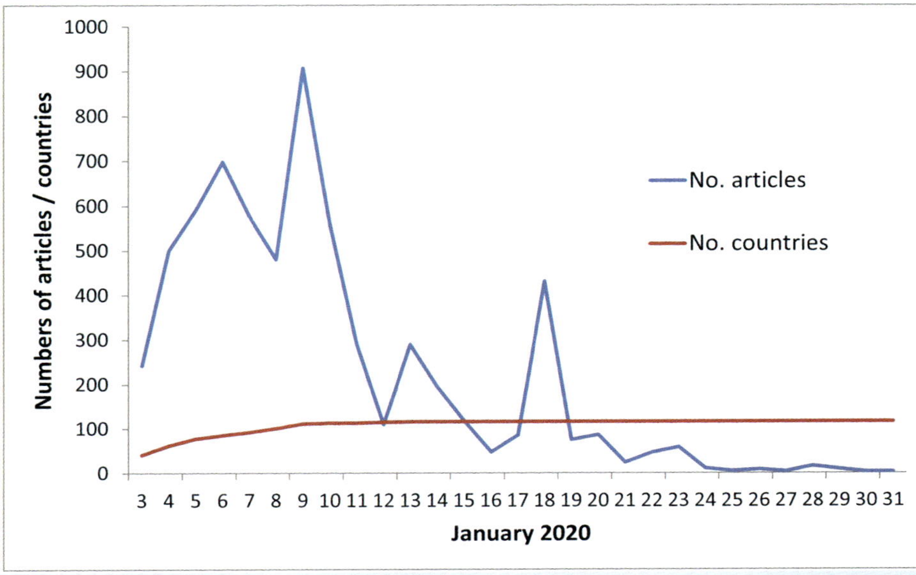

Fig. 12.2. Daily numbers of articles reporting estimates of animal numbers affected by Australia's fires, and cumulative number of countries in which the reports appeared, in January 2020. (Source: V. Reiner, Media Office, University of Sydney)

Communicating nuance to the media is sometimes difficult. Our observation of the media interest was that it was hard to maintain the clear message that our estimates were not a death toll but rather represented an estimate of the number of animals within the fire impact area that might have been affected in a variety of ways (e.g. direct mortality, displacement, increased competition for resources). The media's (and public's) black-and-white interpretation of our estimate as a death toll probably contributed to the incredible extent of the media interest.

Nonetheless, this high visibility may have benefitted the post-fire conservation cause. After the release of our initial numerical estimates and later report, recovery funds were set up by different levels of government and further funds were donated to many non-government organisations; there was a senate inquiry and Royal Commission into National Natural Disaster Arrangements (with a focus on bushfires), and further inquiries in New South Wales and Victoria.

Following the release of the preliminary estimate, we were commissioned by WWF-Australia to extend our analyses to cover the final impact area of the fires and include a more comprehensive suite of faunal groups (van Eeden *et al.* 2020). We defined the impact area to include the National Indicative Aggregated Fire Extent Dataset (NIAFED, Feb 2020 release: DAWE 2020) and an additional 120 000 ha of rainforest burnt in northern Australia. This excluded areas where fires would not be considered unusual (e.g. in savanna or grassland ecosystems in northern and central Australia). We initially explored whether we could estimate mortality rates based on existing faunal mortality data post-fire and mapping of fire severity. However, we soon found that data on mortality rates in and immediately after fire were limited and, where available, dominated by mammal species (Jolly *et al.* 2022). Similarly, at the time we conducted our analysis, fire severity mapping was available only for New South Wales. Such mapping has since been made available for the entire extent of the 2019–20 wildfires (DAWE 2020), but still has not been ground-truthed. As such, our scope was limited to estimating the number of individuals likely to have been present within the fire-impact area (in our analysis, 11.46 million ha), and precluded estimates of actual mortality. Nonetheless, although we could not ascribe rates of mortality, we note that many observers have recorded at least some instances of many animals being killed in previous Australian wildfires (e.g. Hood 1941; Fox 1978).

The methods by which we generated our density estimates differed between the four major groups that we focused on: amphibians, reptiles, birds and mammals. All were based largely on known estimates of population densities of native species within fire-affected bioregions. These methods are summarised briefly below, and greater detail can be accessed in van Eeden *et al.* (2020):

- For frogs, Michael Mahony mapped the distributions of 67 species to stream and non-stream (e.g. wetland) habitats. Population densities for these species were then estimated based on a literature review and expert knowledge, with these estimates then used to calculate the number of individual frogs within the fire impact area. This estimate was confined to Victoria and New South Wales (which encompassed most of fire-affected regions) due to lack of available data on frog densities for other fire-impacted areas.
- A modelling approach was taken for reptiles, led by Dale Nimmo. The model predicted squamate (i.e. snake and lizard) reptile densities as a function of broad-scale environmental variables and species body size, and was informed by a global database of reptile population densities (Santini *et al.* 2018).
- BirdLife Australia generated estimates for birds using population density estimates from almost 104 000 standardised bird count surveys from their Birdata database. These density estimates were stratified by vegetation type and bioregion to calculate estimates of the numbers of individuals by bioregion and within the overall fire impact area.
- For mammals, we began with an existing database on species population densities in forest and woodland habitats that was used to assess the effects of land clearing in New South Wales (Johnson *et al.* 2007). We then added to this database with new estimates sourced from a literature review and contact with taxon experts. Population density estimates for mammals were categorised into 11 species groups and then averaged per group for each of the fire-affected bioregions to generate estimates of population numbers per hectare within the fire impact area.

We aimed to include as many faunal groups as possible; however, there were many species and groups for which too few data were available to inform our estimates. For example, we were not able to make estimates for invertebrates because population density estimates for invertebrates are lacking in Australia. As invertebrates represent by far the majority of animal species (Chapman 2009; Chapter 11), this is a major omission in our estimates. Indeed, Chris Reid estimated in January 2020 that up to 240 trillion arthropods were likely to have been impacted in 8 million ha of burnt habitat (C. Reid, Australian Museum, *pers. comm.*). There is little known about how invertebrates are impacted by wildfire and taxonomic biases in the data that are available, hindering our ability to understand the impacts of major fire events on invertebrates and the predicted increases in frequency and severity of fires under future climate scenarios (Chapter 11; Saunders *et al.* 2021).

Similarly, we did not make estimates for several other vertebrate groups, including fish, flying-foxes, platypus, and all non-squamate reptiles (e.g. turtles) due to lack of available density estimates.

Key findings

In total, we estimated that around 2.8 billion individuals among the included taxa were likely to have been present within the 2019–20 wildfire impact area. This estimate comprised 52 million frogs, 2.5 billion reptiles, 180 million birds and 143 million mammals (Table 12.1). The methods used and nature of the density data available meant that confidence intervals could not be calculated for all groups and that, even for reptiles and birds, the confidence intervals are large.

As noted above, these numbers are not estimates of the death toll, but rather of how many individuals were likely to have been present within the fire impact area. Differences in species mobility, ability to shelter or escape from fire, landscape characteristics (e.g. availability of shelter), and fire intensity and severity would all shape mortality rates (Nimmo *et al.* 2019, 2021). Animals present within the fire impact areas would have been affected in a variety of ways, including direct mortality, indirect mortality due to post-fire succession processes (e.g. increased predation due to lack of available shelter), heat stress and smoke inhalation, displacement to neighbouring unburnt areas, and increased competition for resources. Lack of available data on population densities pre-fire and different species' ability to survive different fire severities meant that we could not estimate rates of direct mortality, indirect mortality, or non-lethal impacts. Nonetheless, using assessments in Garnett and Baker (2021), Garnett *et al.* (Chapter 15) estimated that 1.53 million individual threatened or Near Threatened birds died in the fires; much greater numbers of individuals of common (non-listed) bird species would also have been killed.

There are several limitations to these estimates and their interpretation. For all taxa, there were few data available on population densities, and what was available was often decades old and thus may over- or under-represent population sizes at the time of the fires. There were no available density estimates for some taxa in some bioregions where they are known to occur, and thus conservative estimates were made which are likely to under-represent species prevalence and abundance. In contrast, density estimates are more likely to have been collected in areas where taxa were targeted because they were known to be more abundant than elsewhere within a bioregion. The model for reptiles relied largely on density estimates from overseas (Santini *et al.* 2018) and estimates for some frog species were based on expert knowledge due to lack of available published data. For birds and

Table 12.1. Number of individuals estimated to have been present within the 2019–20 wildfire impact areas (from van Eeden *et al.* 2020). CI = confidence interval.

Taxon group	Estimate (millions)	Lower CI (millions)	Upper CI (millions)
Frogs	52	N/A	N/A
Reptiles	2457	779	9622
Birds	180	151	207
Mammals	143	N/A	N/A
Total	2832		

mammals, we assumed that each taxonomic group present in a bioregion occurred throughout at the same average density. While necessary, this simplification does not account for variation in densities that would occur both spatially and temporally (e.g. seasonally), nor for population densities that would have been lower in the predominantly dry conditions that had affected ecosystems in eastern Australia in the 3 years leading up to the wildfires.

Overall, we consider that the estimates for each taxon group are likely to be conservative. Each bioregion contains a range of land uses, but the fires occurred primarily in areas that probably provided good habitat for wildlife. Furthermore, we confined our estimates to burnt areas, but impacts are likely to have extended into unburnt areas – for example, through soil run-off into waterways, heat stress and smoke inhalation, and influxes of animals fleeing from neighbouring burnt areas.

Given this range of limitations, largely centred on a lack of available data, we consider that making estimates about future fire impacts would be significantly improved by implementing a systematic monitoring program of population sizes and densities for a diverse range of species. This would also provide a framework to facilitate and capture data from incidental monitoring of species populations before and after fire to assess mortality rates under different fire severities. The current paucity of data has been identified by a range of authors as limiting conservation knowledge and recovery estimates and has prompted calls to expand the network of national monitoring sites (Legge *et al.* 2018; Lindenmayer *et al.* 2014). A greater and more reliably sustained flow of biodiversity data could be collated by a centralised national information system and then used to guide management actions, support decision makers, and gain and maintain public confidence. Such a system could be set up as a biodiversity analogue of Australia's federal Bureau of Meteorology. Indeed, a 'Bureau of Biodiversity' has been advocated by the Australian Academy of Science (Academy of Science 2020) and the idea elaborated by Dickman (2021).

Since our report (van Eeden *et al.* 2020) was released, several studies have documented reductions in the abundance of some species from before to after fire, or reduced abundance in burnt areas relative to matched unburnt areas (e.g. Box 12.2); although not conclusive, these observations are strongly indicative of substantial mortality during the fire or in its aftermath. There have been several assessments targeting arboreal marsupials such as the greater glider (*Petauroides volans*) and yellow-bellied glider (*Petaurus australis*) (see also Chapter 16). In one survey in the Blue Mountains in New South Wales, for example, these species were reported to have disappeared in some areas where fires reached the tree canopy (Smith and Smith 2021). Surveys in Victoria detected greater gliders at 57% (DELWP 2021), 29% (Rutter and Blake 2020) and 8% (Burns and Atkins 2021) of sites where they were previously known to occur pre-fire. Yellow-bellied gliders have been detected at around 50% (DELWP 2021), 47% (Burns and Atkins 2021) and 42% (Rutter and Blake 2020) of sites where they were previously known to occur in Victoria, again pre-fire.

Box 12.2. Assessing the extent of wildlife population loss in Victoria

South-eastern Australia experienced some of the most severe fires of the 2019–20 fire season. In Victoria, more than 1.5 million ha burnt across the eastern side of the state. The Victorian Government invested in monitoring and recovery efforts, seeking to assess the impacts of the fires on species and communities most at risk and to maximise recovery efforts where they were most needed. The government estimated that 244 species of flora and fauna had more than 50% of their modelled habitat burnt, and species and locations were then prioritised for monitoring based on the degree of fire impact, conservation status, and logistics such as seasonal detectability and site access (DELWP 2021). Where monitoring was possible, the existence of contemporary pre-fire data for species was vital for effectively assessing the impacts of fire on species populations.

Post-fire population monitoring found reductions in species persistence for most species assessed (DELWP 2021). These included:

- *Mammals*: In the Eastern Alpine National Park, broad-toothed rats (*Mastacomys fuscus*) persisted at 67% of unburnt sites but only at 25% of burnt sites where they had occurred in the past decade. Among 104 sites assessed, 41 of which had been burnt, spotted-tailed quolls (*Dasyurus maculatus*) were detected at only two sites in the Alpine National Park, both of which were unburnt. Greater gliders persisted at 57% of sites and yellow-bellied gliders persisted at 48–53% of sites within and adjacent to burnt areas in highland forests of East Gippsland where they were known to have occurred previously. Gliders were more frequently detected in unburnt forest or forest that experienced low or medium severity fire than high-severity fire.
- *Birds*: Within highland forests of East Gippsland, there were generally low detection rates for sooty owls (*Tyto tenebricosa*) at surveyed burnt sites (detected at three of 37 sites),

Fig. 12.3. Monitoring before and after the 2019–20 wildfires allowed researchers to document a reduction of local glossy black-cockatoo populations in Victoria to less than a fifth of pre-fire levels (DELWP 2021). (Photo: Peter Menkhorst, DELWP)

powerful owls (*Ninox strenua*) (two of 37 sites), and masked owls (*Tyto novaehollandiae*) (no confirmed detections). Pre-fire surveys of randomly selected habitat identified feeding sign for glossy black-cockatoos (*Calyptorhynchus lathami*) (Fig. 12.3) at 35.7% of sites. This reduced to 6.4% detection in burnt areas after the fires, less than a fifth of pre-fire detection rates. Cockatoo frequency did not change at unburnt sites, suggesting that birds did not disperse from burnt to neighbouring unburnt areas.

- *Frogs*: Spotted tree frogs (*Litoria spenceri*) were detected at three of seven sites where they were previously known to occur and Booroolong frogs (*Litoria booroolongensis*) were detected at both surveyed sites. Martin's toadlet (*Uperoleia martini*) was detected at 36% of sites, green and golden bell frogs (*Litoria aurea*) were detected at 63% of sites, and Keferstein's tree frog (*Litoria dentata*) was not detected at the three surveyed wetlands where each species was known to occur. Blue Mountains tree frogs (*Litoria citropa*) and leaf-green tree frogs (*Litoria nudidigita*) were detected at 57% and 73%, respectively, of sites at which they were previously known to occur, including severely burnt sites.
- *Fish, crayfish and mussels*: All targeted aquatic species in moderate to high intensity burn areas persisted but were affected, with some galaxiids declining in abundance by up to 70%.

An alternative approach to estimating faunal losses in the 2019–20 wildfires was provided by Legge *et al.* (in press). These authors estimated the proportional population loss in the wildfires for many vertebrate species, based on spatial overlap with fires of varying severity and expert elicitation of proportional population loss in fires of varying severity. For the few vertebrate species with plausible total population estimates, this proportional loss can be used to estimate the numbers of individuals killed in the 2019–20 wildfires. One of the few such examples is for the koala. Legge *et al.* (in press) reported that 16.9% of the koala's distribution was burnt in the 2019–20 wildfires (with 6.9% burnt in fires of high severity); they then estimated that, as a consequence, the pre-fire total population of koalas was reduced by 7.15% (with lower and upper bounds of 4.3% and 9.4%). Given that Adams-Hosking *et al.* (2016) estimated the total koala population in 2012 to be 330 000 individuals, it follows that the estimated number of koalas killed in the fires is 23 595 (14 169 to 31 020). Although it is highly likely that an ongoing declining trend for koalas (Woinarski and Burbidge 2020) meant that the population was smaller immediately before the fires than the estimated population in 2012, an estimate of at least 10 000 koalas killed in the fires is plausible. Given that several hundred mammal species occur in the fire-affected regions (Chapter 16), many of which would have far larger population sizes than the koala, and many of these had a higher proportion of their range burnt, it is likely that at least hundreds of thousands, and probably millions, of mammals were killed directly by these fires.

Nonetheless, for these species and most others, there are still not enough available data to make generalisations about immediate or post-fire mortality that might support or conflict with our estimates. Overseas, attempts have been made to assess direct mortality due to fire by systematically counting carcasses immediately after fire, with the magnitude of these tallies demonstrating that wildlife mortality can be very high; for example, extrapolation of transect sampling of corpses after fire estimated that > 15 million vertebrates were killed in a single recent fire event in the Pantanal, South America (Tomas *et al.* 2021).

However, this approach would provide reliable estimates of mortality only for larger and more conspicuous vertebrate species. Otherwise, the existence of pre-fire monitoring data complemented by post-fire assessments is the only way to accurately assess the magnitude of population loss due to fire on resident fauna.

Conclusion

The megafires that ravaged Australian landscapes in 2019–20 focused national and international attention on the causes of the fires and, particularly, their impacts on people and wildlife. Initial concern for wildlife was directed mostly at large charismatic animals such as koalas and kangaroos, but after our estimates that over a billion vertebrates could have been in the path of the fires, much attention focused on the magnitude of the numbers across a broad suite of species. Owing to the paucity of pre-fire surveys the actual impact of the fires on animal mortality could not be quantified directly, and our estimates therefore were based on extrapolations of average animal densities in the habitats that burnt. Restrictions imposed by the COVID-19 pandemic slowed the roll-out of post-fire surveys, but results from current work suggest that the occupancy and abundance of many vertebrate species is much less in burnt than in unburnt forest, and less than before the fires at sites where pre-fire information was available, with this reduction especially pronounced in areas burnt at high severity (Box 12.2). The reduced numbers probably result from animal deaths in the fires and from mortality afterwards in the open and resource-depleted post-fire environment, although the relative contribution of each source of mortality is not known. A key lesson to emerge from our 'critter counts' and attendant media is that wildlife monitoring in much of Australia is sadly lacking, and that a greatly expanded program of ongoing general and targeted surveys must be established if we are to understand and effectively manage the effects of fires and other disturbances, and measure the extent of recovery, in future. In addition to the specific recommendations listed below, we suggest that a dedicated Bureau of Biodiversity be set up to receive, curate and interpret survey and monitoring data so that our management of wildlife can be informed and proactive.

Recommendations

The following recommendations derive mainly from Dickman (2021) and van Eeden *et al.* (2020):

- Undertake long-term species monitoring, including invertebrates, using nationally standardised measures (which may differ for different taxa) in all bioregions, prioritising areas and species or populations at greatest risk of future wildfires, and recovering from past fires.
- Develop strategies to reduce the likelihood of fires of high severity in these areas along with plans to protect these species and communities if such fires occur. This might include improving habitat connectivity to ensure access to fire refuges for mobile species or retrieving threatened species for conservation in captivity until post-fire induced threats are abated (e.g. Ayres *et al.* 2012, DELWP 2021, Lamont *et al.* 2021).
- Identify key resources necessary for the survival of prioritised species and communities (e.g. food or shelter) and compounding threats to these species and communities during post-fire succession (e.g. increased predation risk). We should then be prepared to manage these threats or provide necessary resources to mitigate fire risks and post-fire mortality.

- Experimentally evaluate the effectiveness of different post-fire management methods to ensure effective management of prioritised species and communities in the event of fire.
- Adopt population viability analysis (PVA) for selected species as a useful tool for modelling the effects of disturbance events (in this case, wildfire) where basic demographic traits are known (Glen and Dickman 2013). At present, we make assumptions about the impacts of fires on species based on their habits and the extent of their range that has been burnt, but PVA would allow stronger quantification. PVA appears to be as yet underutilised for assessments of fire impacts.

Acknowledgements

Dale Nimmo, Michael Mahony, Gilad Bino, Glenn Ehmke, Joris Driessen, Kerryn Herman and James O'Connor conducted analyses or prepared components of the WWF-Australia report, John Woinarski kindly contributed some additional text on the impacts of the wildfires on koalas, Vivienne Reiner collated data on media interest, and Jemma Cripps and Tim O'Brien shared data on post-fire monitoring in Victoria. We are grateful to WWF-Australia for funding this project.

References

Academy of Science (2020) Environment reform must include a 'Biodiversity BOM'. Australian Academy of Science, <https://www.science.org.au/files/userfiles/support/position-statements/environment-reform-must-include-biodiversity-bom.pdf>.

Adams-Hosking C, McBride MF, Baxter G, Burgman M, De Villiers D, *et al.* (2016) Use of expert knowledge to elicit population trends for the koala (*Phascolarctos cinereus*). *Diversity & Distributions* **22**, 249–262. doi:10.1111/ddi.12400

Ayres R, Nicol M, Raadik T (2012) *Establishing New Populations for Fire-affected Barred Galaxias (*Galaxias fuscus*): Site Selection, Trial Translocation and Population Genetics. Black Saturday Victoria 2009 – Natural Values Fire Recovery Program*. Arthur Rylah Institute for Environmental Research, Department of Sustainability and Environment, Heidelberg,. <https://www.ari.vic.gov.au/__data/assets/pdf_file/0023/34952/VBRRA-P14b-web.pdf>.

Bowman D, Williamson G, Yebra M, Lizundia-Loiola J, Pettinari ML, *et al.* (2020) Wildfire: Australia needs a national monitoring agency. *Nature* **584**, 188–191. doi:10.1038/d41586-020-02306-4

Burns PA, Atkins ZS (2021) 'Gliding from the ashes: Post-fire surveys for the greater glider and yellow-bellied glider in Far East Gippsland'. Unpublished report for Aussie Ark Australian Endangered Species Recovery Fund.

Chapman AD (2009) 'Numbers of living species in Australia and the world. Report for the Australian Biological Resources Study, 2nd edition'. Department of the Environment, Water, Heritage, and the Arts. Australian Government, Canberra, <https://www.environment.gov.au/system/files/pages/2ee3f4a1-f130-465b-9c7a-79373680a067/files/nlsaw-2nd-complete.pdf>.

Collins L, Bradstock RA, Clarke H, Clarke MF, Nolan RH, *et al.* (2021) The 2019/2020 mega-fires exposed Australian ecosystems to an unprecedented extent of high-severity fire. *Environmental Research Letters* **16**, 044029. doi:10.1088/1748-9326/abeb9e

DAWE (2020) *Australian Google Earth Engine Burnt Area Map: A Rapid, National Approach to Fire Severity Mapping.* Department of Agriculture, Water and the Environment (DAWE), Canberra.

DELWP (2021) *Biodiversity Bushfire Response Supplementary Report: Bushfire Impacts on Species in Victoria*. Department of Environment, Land, Water and Planning (DELWP), Melbourne.

Dickman CR (2021) Ecological consequences of Australia's 'Black Summer' bushfires: managing for recovery. *Integrated Environmental Assessment and Management* **17**, 1162–1167. doi:10.1002/ieam.4496

Dickman CR, McDonald T (2020) Some personal reflections on the present and future of Australia's fauna in an increasingly fire-prone continent. *Ecological Management & Restoration* **21**, 86–96. doi:10.1111/emr.12403

Fox A (1978) The '72 fire of Nadgee Nature Reserve. *Parks & Wildlife* **2**, 5–24.

Garnett ST, Baker GB (Eds) (2021) *The Action Plan for Australian Birds 2020*. CSIRO Publishing, Melbourne.

Glen AS, Dickman CR (2013) Population viability analysis shows spotted-tailed quolls may be vulnerable to competition. *Australian Mammalogy* **35**, 180–183. doi:10.1071/AM12045

Hood JB (1941) Birds and bushfires. *South Australian Ornithologist* **15**, 125–127.

Johnson C, Cogger H, Dickman C, Ford H (2007) 'Impacts of landclearing: the impacts of approved clearing of native vegetation on Australian wildlife in New South Wales'. Report prepared for WWF-Australia, Sydney.

Jolly CJ, Dickman CR, Doherty TS, van Eeden LM, Geary WL, *et al.* (2022) Animal mortality during fire: a global systematic review. *Global Change Biology* **28**, 2053–2065. doi:10.1111/gcb.16044

Lamont A, Parrott M, Wicker L, Banks C, Lang M, *et al.* (2021) Building wildlife resilience in disasters. *Australian Journal of Emergency Management* **36**, 12–14.

Legge S, Lindenmayer DB, Robinson NM, Scheele BC, Southwell DM, *et al.* (Eds) (2018) *Monitoring Threatened Species and Ecological Communities*. CSIRO Publishing, Melbourne.

Legge S, Rumpff L, Woinarski JCZ, Whiterod NS, Ward M, *et al.* (in press) The conservation impacts of ecological disturbance: time-bound estimates of population loss and recovery for fauna affected by the 2019–20 Australian megafires. *Global Ecology and Biogeography*, doi:10.1111/geb.13473.

Lindenmayer DB, Burns E, Thurgate N, Lowe A (Eds) (2014) *Biodiversity and Environmental Change: Monitoring, Challenges and Direction*. CSIRO Publishing, Melbourne.

Lindenmayer DB, Taylor C (2020) New spatial analyses of Australian wildlife highlight the need for new fire, resource, and conservation policies. *Proceedings of the National Academy of Sciences of the United States of America* **117**, 12481–12485. doi:10.1073/pnas.2002269117

Nimmo DG, Avitabile S, Banks SC, Bliege Bird R, Callister K, *et al.* (2019) Animal movements in fire-prone landscapes. *Biological Reviews of the Cambridge Philosophical Society* **94**, 981–998. doi:10.1111/brv.12486

Nimmo DG, Carthey AJR, Jolly CJ, Blumstein DT (2021) Welcome to the Pyrocene: animal survival in the age of megafire. *Global Change Biology* **27**, 5684–5693. doi:10.1111/gcb.15834

Rutter N, Blake B (2020) 'Post-fire assessment of Greater Gliders in East Gippsland: interim report'. WWF-Australia, Sydney.

Santini L, Isaac NJB, Ficetola GF (2018) TetraDENSITY: a database of population density estimates in terrestrial vertebrates. *Global Ecology and Biogeography* **27**, 787–791. doi:10.1111/geb.12756

Saunders ME, Barton PS, Bickerstaff JRM, Frost L, Latty T, *et al.* (2021) Limited understanding of bushfire impacts on Australian invertebrates. *Insect Conservation and Diversity* **14**, 285–293. doi:10.1111/icad.12493

Smith P, Smith J (2021) 'Post-fire survey of arboreal mammals at Wombeyan and Jenolan – interim report. Report prepared For Glideways'. The Great Eastern Ranges and WWF-Australia, Sydney.

Tomas WF, Berlinck CN, Chiaravalloti RM, Faggioni GP, Strussmann C, *et al.* (2021) Counting the dead: 17 million vertebrates directly killed by the 2020's wildfires in the Pantanal Wetland, Brazil. *Scientific Reports* **11**, 23547. doi:10.1038/s41598-021-02844-5 van Eeden LM, Nimmo D, Mahony M, Herman K, Ehmke G, *et al.* (2020) 'Impacts of the unprecedented 2019–2020 bushfires on Australian animals'. Report prepared for WWF-Australia, Sydney.

Wintle BA, Legge S, Woinarski JCZ (2020) After the megafires: what next for Australian wildlife? *Trends in Ecology & Evolution* **35**, 753–757. doi:10.1016/j.tree.2020.06.009

Woinarski J, Burbidge AA (2020) *Phascolarctos cinereus* (amended version of 2016 assessment). The IUCN Red List of Threatened Species 2020: e.T16892A166496779, <https://www.iucnredlist.org/species/16892/166496779>.

13

The impacts of the 2019–20 wildfires on Australian frogs

Michael J. Mahony, Harry B. Hines, Frank Lemckert, David Newell, J. Dale Roberts, Jodi J. L. Rowley, Ben C. Scheele and Matt West

Summary

Context and challenges

- Frogs depend on moist microclimates or burrowing to survive fire.
- The impact of wildfires on frog populations depends on fire severity, and on traits that allow survival and recovery after fire.
- To explore likely indicators of risk to fire, we categorised frog species as occurring in fire-adapted or fire-sensitive vegetation communities; by use of refuge type (burrow, tree hollow, under debris, dense stream side vegetation, permanent wetlands); and by ecological group (breeding in terrestrial sites, ephemeral ponds, permanent ponds and streams and bogs/soaks). We combined this information with estimates of the proportion of species' distributions that overlapped with the 2019–20 wildfires (Legge *et al.* 2022, in press), to consider which species have been most heavily affected. As well as the species considered by Legge *et al.* (in press), we added several species from the south of Western Australia.

Main findings

- Species occurring in fire-adapted vegetation communities use refuges that are more secure against fire, including burrows, tree hollows or under debris. Species in fire-sensitive vegetation communities use refuges (dense shrubs, leaf litter) that are likely to be consumed by fire.
- Range-restricted species that occur in fire-sensitive vegetation have likely suffered the highest impact from the 2019–20 wildfires (e.g. *Assa darlingtoni*, *Litoria kroombitensis*, *L. spenceri*, *Philoria pughi* and *Pseudophryne pengilleyi*).
- Several threatened species with wide but fragmented distributions in fire-adapted vegetation, with high wildfire overlap (*Heleioporus australiacus flavopunctatus*, *L. watsoni* and *L. littlejohni*), were also likely impacted. Early evidence from field monitoring confirms this.

- For most species with broad distributions but not considered threatened, the impacts of wildfire on abundance are largely unknown. We note the value of widespread monitoring using citizen science, e.g. FrogID, in providing information on impacts of major environmental disturbances.
- The short-term effects of wildfire on tadpoles and their aquatic habitats are largely unknown.
- Management and research attention needs to focus on species in fire-sensitive vegetation, and range-restricted species in fire-adapted vegetation communities.

Introduction

The 2019–20 wildfires burnt large areas in eastern and southern Australia (Collins *et al.* 2021). While fires are a common feature in many Australian ecosystems, particularly in eucalypt forests, the extent and severity of the 2019–20 fires was unprecedented (Collins *et al.* 2021). Areas not usually subject to fire also burnt, including significant areas of rainforest; for example, half of the Gondwana Rainforests of Australia World Heritage Area (GRAWHA) burnt (NSW Department of Planning, Industry and Environment 2020; Chapter 4).

Among the affected biota are over 100 species of amphibians, all anurans (frogs) (Legge *et al.* 2022). We consider the impacts of the 2019–20 fires on this group: we (1) summarise what is known generally about the impacts of fire on frogs in Australia, (2) make inferences about fire impacts on species persistence, (3) given these inferences, estimate the impact of the 2019–20 wildfires on frogs, and (4) provide recommendations for research and management of fire-affected species.

Previous studies on the impacts of fire on frogs

Eighteen field studies have investigated the effect of fire on Australian frogs, 11 on low to moderate and seven on severe fires. Most included pre- and post-fire monitoring of abundance (Recher *et al.* 1987; Bamford 1992; Wardell-Johnson *et al.* 1995; Driscoll and Roberts 1997, 1998; Roberts *et al.* 1999; Lemckert 2000; Bamford and Roberts 2003; Lemckert *et al.* 2004; Penman *et al.* 2006; Daly and Craven 2007; Edwards and Roberts 2011; Gillespie and West 2012; Westgate *et al.* 2012, 2018; Lowe *et al.* 2013; Potvin *et al.* 2017; Roberts 2018; Heard *et al.* 2021). Additional studies have used surrogate models to measure heat penetration (Penman *et al.* 2006), fundamental niche models and predicted climate change and fire occurrence (Penman *et al.* 2005b), population genetics (Potvin *et al.* 2017), citizen science involvement (Rowley *et al.* 2020), and two expert elicitations to assess the impact of fire on abundance (Legge *et al.* 2022, in press). Collectively, these studies show that low-severity fires produce minimal or short-term declines in abundance, with several traits (reproductive mode, burrowing) associated with differing sensitivity among species. For example, short-term declines in abundance occurred in *Geocrinia lutea* and *G. vitellina* after their breeding bogs were burnt, with evidence that adults and eggs were killed from heat exposure and desiccation (Driscoll and Roberts 1997). Populations of a congener *G. alba* and another sedentary bog-breeding species (*Spicospina flammocaerulea*) were not affected or abundance increased after fires, at least in the short term (Wardell-Johnson *et al.* 1995; Driscoll and Roberts 1997; Bamford and Roberts 2003; Edwards and Roberts 2011), possibly due to their reliance on secure ground water sources that protect them from radiant heat, and that supported reproduction post-fire.

Five studies of high-severity fires showed strong impacts on abundance. An intense wildfire on the Dorrigo Plateau in mid-eastern New South Wales caused the loss of a small isolated population, and a > 80% decline in a larger isolated population of the litter-dwelling *Assa darlingtoni*, a frog not generally threatened by wildfire since it occurs in temperate rainforests that typically do not burn (Lemckert 2000). In the fire-adapted vegetation community of upland heath, an intense wildfire in the upper Shoalhaven River catchment resulted in declines of adults and tadpoles of *L. littlejohni* (= *L. watsoni*) (Daly and Craven 2007). This is the only study to report impacts on tadpole abundance as well as adults, but it is unknown whether the tadpoles were present in pools and then died from direct fire impacts, or whether few tadpoles were produced because adults sheltering in the riparian zone were killed by fire before breeding. Surveys were conducted in fire-adapted vegetation after the Victorian Black Saturday fires (2009) for three frogs with long-term pre-fire monitoring data. Using standard visual encounter surveys, the abundances of two pond breeding tree frogs (*Litoria ewingi* and *L. paraewingi*) were not significantly different pre- and post-fire. However, using molecular genetic methods, it was found that there was a marked reduction in effective population size and genetic diversity (Potvin *et al.* 2017). In *L. spenceri*, a stream-breeding species, juvenile recruitment was lower immediately post-fire, but adult survivorship increased compared with before-fire estimates (Gillespie and West 2012).

Following the extensive fires of 2019–20, a rapid assessment was undertaken to prioritise species for management intervention (Legge *et al.* 2022). Assessment focussed on nationally and globally threatened species (*Environment Protection and Biodiversity Conservation Act 1999* (EPBC Act) and IUCN Red List) plus additional species that experts considered to be at risk. A risk score was calculated for each species from the pre-fire imperilment and extent of fire overlap, and a set of traits was used to estimate variability in mortality. Risk score and trait variability were combined to identify species with highest priority for urgent management intervention. Twenty-four threatened species had > 10% of their distribution fire-affected and a further 11 unlisted species had > 30% fire overlap. A subsequent analysis used expert elicitation to estimate population loss and recovery for 66 frog taxa (with > 10% fire overlap) (Legge *et al.* in press). This assessment predicted that six frog taxa would experience > 30% decline 10 years/three generations after the fire. Coupled with other parameters of extinction risk, seven taxa were considered eligible for up-listing of threatened status, and eight for threatened species listing.

Inferring fire impacts on frogs

The distribution and abundance of frogs is determined by the availability of water in the environment (Hillman *et al.* 2008). Frogs do not drink, absorbing water through their skin and losing it by evaporation. Frogs must find moisture in the environment to maintain water balance (e.g. ponds, streams, wet soil, damp leaf litter, condensation). In humid environments, water loss from the skin and lungs is limited, but in dry conditions it is high and frogs use a range of adaptations to avoid desiccation, typically moving to protective microhabitats (Tracy *et al.* 2011). The greatest threat to their survival is dry air and a dry environment; both become prevalent during drought and wildfire events. In wildfires, radiant heat kills exposed individuals and destroys protective habitats, and the aftermath is a dry landscape that threatens survival. Frogs are small, have low mobility, constantly rely on moisture, and their life cycles are tied to terrestrial and aquatic habitats that are impacted by fire in different ways. Yet they occur in fire-adapted vegetation communities, so they must have specific adaptive traits that enable them to survive fire.

Fire has long been a component of eucalypt and heath ecosystems, both fire-adapted vegetation communities; but rainforest, without a history of fire, is fire sensitive (Hardesty *et al.* 2005). Most Australian frogs occur in fire-adapted vegetation, and have behavioural adaptations to avoid temperature and moisture stress, that may also protect them from wildfire. For example, arboreal tree frogs shelter in hollows or under bark; those that stay closer to ground may shelter in dense vegetation, in reed beds or under debris (Lowe *et al.* 2013; Mo 2015). Most ground frogs shelter by burrowing, moving under debris, into deep soil cracks, or into reed beds (Bamford and Roberts 2003). In fire-sensitive communities, where fire is normally excluded because of the high moisture content in soil, leaf litter and plants, frogs are presumably not adapted to wildfire, but their response is unknown because wildfires have not previously occurred in these habitats (Collins *et al.* 2021).

The 2019–20 wildfires

Thirty-eight frog species occur in, with 11 species restricted to, relictual temperate and subtropical rainforest of the GRAWHA. This area represents the second highest level of phylogenetic endemism for Australian frogs (Rosauer *et al.* 2009). About 53% of this area burnt in the 2019–20 wildfires (NSW Department of Planning 2020).

We aimed to understand the reliance that frogs have on specific habitat features, mostly refuges, in order to help identify actions and research needs for fire-impacted frogs. We considered those frog species whose distributions overlapped with the 2019–20 fires. We used our collective experience to categorise their dependence on fire-sensitive and fire-adapted vegetation communities. Fire-sensitive communities are 'vegetation communities that have no history of fire' (Hardesty *et al.* 2005), including rainforest and permanent coastal wetlands. Wet sclerophyll vegetation is fire-adapted but with a very long return interval. We considered traits that could provide protection from fire and enable population recovery, including refuge site and breeding mode (ecological group). Full details on the taxa covered, their vegetation community category and refuge sites are available in Supplementary Table S13 online at doi:10.6084/m9.figshare.20132186. We follow Legge *et al.* (in press) for estimates of fire overlap, and also by considering fire impacts on species and on subspecies when they were differentially impacted by fire, and using 'taxa' when referring to a mix of species and subspecies, and 'species' when referring only to that taxonomic rank. Our analysis is also informed by several unpublished monitoring projects on frog occupancy undertaken since the 2019–20 wildfires.

Findings

Occurrence in fire-sensitive and fire-adapted vegetation communities

Most frog species in our analysis occur in fire-adapted vegetation communities (92 of 103 species), with only 12 restricted to fire-sensitive communities (Table 13.1). Six species of frogs from fire-sensitive vegetation communities were listed as threatened at the time of the 2019–20 wildfires (*Mixophyes fleayi*, *Philoria frosti*, *P. kundagungan*, *P. richmondensis*, *P. sphagnicolus*, *Taudactylus pleione*). A further two (*Assa wollumbin*, *Philoria knowlesi*) have been recognised since then, with application of the IUCN Red List criteria (IUCN 2019) indicating they should be listed as threatened (Mahony *et al.* 2020, 2021). Monitoring of *P. richmondensis*, *P. kundagungan* and *P. knowlesi* (pre- and post-fire) has revealed lower abundances at burnt sites (Heard *et al.* 2021). Changes in habitat structure due to loss of

Table 13.1. Dependence on fire-adapted vegetation communities (percentage classes) against the proportion of their distributions that overlapped with the 2019–20 fires, for 103 species of frogs.

Within wildfire overlap categories the table is presented with taxa arranged in families, Limnodynastidae, Myobatachidae and Pelodryadidae. The table is presented in a risk assessment format such that the species with the highest level of fire overlap (> 60% of distribution), which also depend on fire-adapted vegetation communities, are highlighted in red, and those with moderate overlap (> 40%) in orange. Threatened species (EPBC Act and IUCN Red List) are highlighted in bold.

Wildfire overlap	Dependence on fire-adapted vegetation communities				
	≤ 20	**21 ≤ 40%**	**41 ≤ 60%**	**61 ≤ 80%**	**81 ≤ 100%**
0–20%	*Philoria frosti, P. loveridgei, P. richmondensis*[b]				*Heleioporus albopuncatatus, H. psammophilus, Limnodynastes convexiusculus, Lim. dumerili dumerilii, Lim. d variagatus, Lim. peronii, Lim. salmini, Lim. tasmaniensis, Neobatrachus albipes, Platyplectron ornatus.*
	Assa darlingtoni, ***A. wollumbin, Mixophyes fleayi, Taudactylus pleione***	***Mixophyes iteratus***			*Crinia georgiana, C deserticola, C. parinsignifera, C. signifiera, C. tinnula, Myobatrachus gouldii, Pseudophryne australia, P. biboroni*[c]*, P. major, P. semimarmorata,* **Uperoleia mahonyi,** *U. martini, U. rugosa*
			Litoria brevipalmata, ***Litoria kroombitensis***[a]	*L. wilcoxi*	*Cyclorana albogutatta, C. brevipes, Litoria moorei/cyclorhyncha*[d]**, *L. castanea, L. raniformis*,** *L. fallax, L. caerulea, L. ewingi, L. verreauxi, L. inermis, L. latopalmata, L. nasuta, L. peronii, L. rothii, L. belatus, L. rubella*
21–40%	*Lechriodus fletcheri, Philoria kundagungan*	*Litoria pearsoniana*		*A. brevis*	*Lim. dumerilii insularis*
				Geocrinia victoriana, Paracrinia haswelli, Pseudophryne coriacea, P. corroboree	*Crinia pseudinsignifera, C. subinsignifera, Metacrinia nichollsi*[b]*, Pseudophryne raveni, Uperoleia fusca, U. laevigata, U. tyleri*
			Litoria aurea[b]*, L. barringtonensis, L. chloris, L. nudidigita, L. revelata*	*Litoria cooloolensis, L. booroolongensis, L. lesueuri,*	***Litoria olongburensis****, L. gracilenta,* ***L. piperata****, L. jervisiensis, L. freycineti, L. tyleri, L. quiritatus*

continued

Table 13.1. Continued

Wildfire overlap	Dependence on fire-adapted vegetation communities				
	≤ 20	21 ≤ 40%	41 ≤ 60%	61 ≤ 80%	81 ≤ 100%
41–60%	***Philoria knowlesi***				*Heleioporus a. australiacus*
		Mixophyes balbus, M. fasciolatus		***P. pengilleyi***	
			Litoria phyllochroa		*Litoria spenceri*
61–80%	*Philoria sphagicolus*			*L. citropa*	*Limnodynastes d. fryi, L. d. grayi*
		Litoria daviesae	*Litoria subglandulosa*		***Litoria littlejohni***, *L. dentata, L. belatus, L. quiritatus*
81–100%	*Philoria pughi*				***Heleioporus a. flavopunctatus***
					Litoria watsoni

[a] Unaffected by 2019–20 wildfire but affected by 2018 wildfire.
[b] Genetically distinct population in the Stirling Ranges (Edwards *et al.* 2008)
[c] Species occurs in isolated population on Kangaroo Island, South Australia; > 50% of the island burnt.
[d] Hybrid populations

canopy and streamside vegetation, and feral pig invasion, are likely to cause further drying of critical breeding habitats.

Of 92 species found in fire-adapted habitats, 80 are non-threatened. Wildfires may have reduced their population size (Table 13.2). Sixty species reliant on dense riparian vegetation or leaf litter for protection were most likely affected (Table 13.3). Unplanned monitoring of these species has occurred through FrogID (Rowley *et al.* 2020), and numerous species persist (Box 13.1); however, the extent of impact is not known. The highest impacts were probably experienced by the 12 threatened taxa that occupy fire-adapted communities, notably the ground frogs *Pseudophryne corroboree*, *P. pengilleyi* and *Mixophyes balbus*, and tree frogs *Litoria daviesae*, *L. littlejohni*, *L. piperata*, *L. spenceri* and *L. watsoni* (Table 13.1).

Box 13.1. The value of citizen science in understanding the impact of fire on frogs

In the wake of the 2019–20 wildfires, there was an urgent need to understand the impacts on frogs, and subsequent conservation priorities. The enormous logistical challenges involved in rapidly collecting post-fire data across a large geographic area were further compounded by COVID-19 travel restrictions. However, the national citizen science project, FrogID (Rowley *et al.* 2019), gathered data on calling frog records across the fire zone, providing the first published data on the persistence of frogs through the fires (Fig. 13.1; Rowley *et al.* 2020). This study demonstrated

Fig. 13.1. (A) Use of smartphone with the FrogID application to record frog calls; (B) *Mixophyes balbus* (arrow) at entrance of refuge hollow in base of partly burnt rainforest tree, Gibraltar Range National Park, NSW; (C) burnt peat swamp, breeding habitat of *H. a. australiacus*, Newnes Plateau, NSW; and (D) burnt upland rainforest, stream habitat of *P. kundagungan*, Main Range National Park, Qld. (Photos: (A, B) J. Rowley, (C) M. Mahony, (D) H. Hines)

widespread short-term persistence of frog species, with 45 frog species detected within 125 days post-fire. All 33 frog species with more than five records from the months of December–March pre-fire were detected post-fire. As the FrogID database grows, the longer-term impacts of the fires can be assessed, and the database now holds over 10 000 records of over 100 frog species from areas burnt in the 2019–20 wildfires. While most of these records are of common, widespread species of low conservation concern (i.e. 17% *C. signifera*), they also include threatened species and species of conservation concern post-fire (e.g. *A. wollumbin*, *P. pughi*). Alongside traditional scientific monitoring, citizen science projects such as FrogID offer a powerful new tool to understand the impact of fires on Australia's biodiversity, while also engaging the public in biodiversity monitoring.

Table 13.2. Habitat refuges used by frogs during wildfire.

Only species within the fire-overlap zone are included. Arranged by family (Limnodynastidae, Myobatrachidae and Pelodryadidae in separate rows) and genera with number of species in brackets.

Protective refuge	Number of species	Families	Genera and species
Burrows	21	All ground frogs, Myobatrachidae and Limnodynastidae. Some tree frogs, Pelodryadidae	*Heleioporus* (5), *Limnodynastes* (4 sp., 6 taxa), *Myobatrachus* (1), *Neobatrachus* (1), *Philoria* (1), *Platyplectron* (1)
			Mixophyes (4), *Myobatrachus* (1), *Taudactylus* (1)
			Cyclorana (2)
Hollows	11	All tree frogs, Pelodryadidae	*Litoria* (11)
Under debris on ground, under or between rocks	47	Approximately equal number of ground and tree frogs	*Adelotus* (1), *Limnodynastes* (4)
			Crinia (5), *Metacrinia* (1), *Pseudophryne* (8), *Uperoleia* (5)
			Litoria lesueuri sp. group (3)*, *Litoria* (20)
Leaf litter	12	All ground frogs	*Philoria* (5), *Lechriodus* (1)
			Assa (2), *Geocrinia* (1), *Paracrinia* (1), *Pseudophryne* (1), *Uperoleia* (1)
Dense streamside vegetation	7	All tree frogs, Pelodryadidae	*Litoria* (7)
Within wetland	6	One ground frog Myobatrachidae and 5 tree frogs, Pelodryadidae	*Crinia* (1)
			Litoria (5)
Total number of species	104		

* Species groups are from Tyler and Davies (1978)

Table 13.3. Numbers of species in ecological groups (after Murray *et al.* 2011) and occurrence in fire-adapted or fire-sensitive vegetation communities.

Species and taxa with tadpoles that are not free-living or have direct development are in bold, all others have free-living tadpoles.

Ecological group	Total number of species or taxa	Highly dependent on fire-sensitive vegetation community. Number of species or taxa	Highly dependent on fire-adapted vegetation community. Number of species or taxa
Terrestrial breeders	22	***Philoria* (7), *Assa* (2), *Geocrinia* (1), *Metacrinia* (1), *Myobatrachus* (1),** *Pseudophryne* (9), *Taudactylus* (1)	
Ephemeral water bodies	34	*Lechriodus* (1)	*Heleioporus* (4), *Limnodynastes* (3), *Neobatrachus* (1), *Platyplectron* (1), *Crinia* (5), *Uperoleia* (5), *Cyclorana* (2), *L. brevipalmata* (1), *L. caerulae* sp. group (3), *L. rubella* sp. group (4), *L. latopalmata* sp. group (4)
Permanent water bodies (ponds and wetlands)	28		*Adelotus* (1), *Limnodynastes dumerilli* (5), *Lim. peronii* (1), *Lim. terrareginae* (1), *Crinia tinnula* (1), *C. parinsignifera* (1), *Paracrinia* (1), *Uperoleia tyleri* (1), *Spicospina* (1), *Litoria adelaidensis* (1), *Litoria aurea* sp. group (4), *L. fallax* sp. group (3), *L. ewingi* sp. group (4), *L. peronii* sp. group (3)
Streams	17	*Mixophyes* (4)	*L. phyllochroa* sp. group (10), *L. lesueuri* sp. group (3)

Importance of refuges

More than 78 frog species shelter in refuges that likely provide thermal and moisture protection from wildfire (Table 13.2). Forty-seven taxa shelter under terrestrial debris: decaying logs and rocks with crevices for a frog to squeeze under where the soil is moist. Twelve forest-dwelling frogs burrow in soil or squeeze into soil cracks (Penman *et al.* 2005a, 2006; Westgate *et al.* 2012). Only 10 species, all tree frogs, rely on arboreal hollows or bark for shelter (Lemckert 1999; Taylor *et al.* 2003; Mo 2015). Field observations show that species found in fire-adapted vegetation and that burrow or move into rock crevices are least affected by wildfire; for example, several taxa appeared on the surface to breed after heavy rainfall that extinguished the fires.

Several taxa shelter in dense riparian vegetation and under leaf litter, and this may make them vulnerable to severe wildfire (Table 13.2). Seven of these occur in fire-adapted vegetation communities (e.g. *L. ewingii* species group) and post-fire monitoring indicates significant impacts on some species (e.g. *L. littlejohni* and *L. watsoni*) (Stock, Wallace and Klop-Toker, unpublished data). Thirteen species that shelter under leaf litter or in the topsoil are also vulnerable to severe wildfire, and several (*Assa* (2), *Philoria* (6), and *Lech. fletcheri*) occur in fire-sensitive vegetation communities (Tables 13.1, 13.2). Where the 2019–20 wildfires penetrated fire-sensitive vegetation communities, post-fire monitoring has indicated a reduction in abundance of *A. darlingtoni*, *P. pughi* and *P. sphagnicolus* (M. Mahony, *pers. obs.*) and *P. knowlesi*, and *P. kundagungan* (Heard *et al.* 2021).

Ecological group

The ecological group of frogs indicates their reliance on landscape topographical features, and thus their vulnerability to severe fires. Forty-five species use permanent streams, ponds, and wetlands for the tadpole stage of the life cycle (Table 13.3); these frogs have a constant source of moisture with cooler, stable temperatures (Driscoll *et al.* 2010; Collins *et al.* 2012; Westgate *et al.* 2018). These habitats are common in fire-adapted vegetation. The fire-caused mortality rate of frogs using these habitats is unknown. Thirty species predominantly inhabit the riparian zones of permanent streams (Table 13.3), and an intense fire in this zone may kill many individuals, especially species that take refuge in leaf litter and understorey vegetation. In some species adults move well away from the riparian zone seasonally to forage, disperse and shelter (Lemckert and Brassil 2000, 2003; Lemckert *et al.* 2006; Newell *et al.* 2013; Quick *et al.* 2015), and if this coincides with intense wildfires, they may be directly exposed, or, protected (e.g. by large boulders). Up to 28 species of frogs breed in permanent ponds and wetlands, mostly in fire-adapted vegetation communities (Table 13.3). While these habitats provide moisture, many frogs that use them shelter in dense surrounding vegetation (tree frogs), in nearby terrestrial refuges (ground frogs), and a small number of species use emergent reed beds. These shelter sites are particularly susceptible to intense fire. Permanent ponds and wetlands may provide protection for tadpoles, but an intense fire in wetland vegetation can lead to the death of many frogs. A wildfire in 2018 burnt through sedges and reeds of permanent wetlands in an estuarine complex on Kooragang Island (NSW) and many individual frogs were observed dead, without any signs of physical burning, apparently killed by heat exposure (M. Mahony, *pers. obs.*).

Thirty-four taxa breed in ephemeral situations (Table 13.3) and these would appear particularly vulnerable to severe wildfire. Most have wide ranges, can be locally abundant, and occupy fire-adapted vegetation, but there is little information on wildfire impact. Several species are arboreal (*L. rubella* sp. group. *L. gracilenta*, *L. brevipalmata*), and crown fires would impact adults. Others are mostly terrestrial (*Crinia* (5), *Limnodynastes* (3), *Uperoleia* (5)) and shelter under debris or in cracks where they may be protected. Post-fire monitoring found *C. signifera* breeding in severely burnt forests once rains occurred (J. Rowley, *pers. obs.*), presumably having been well protected in refuges during the fire.

Twenty-two taxa breed in terrestrial sites (Knowles *et al.* 2004; Willacy *et al.* 2015; Heard *et al.* 2021; Bolitho *et al.* 2021), mostly in fire-sensitive vegetation. Several species of *Geocrinia* in south-western Australia breed in soaks or bogs. One species, *G. vitellina*, which relies on groundwater, recovered well after intense wildfire burnt a large proportion of its habitat in 1997 (Roberts *et al.* 1999). Experimental studies indicate *G. lutea* can cope with low intensity fire but also persists in areas that have been subject to multiple high intensity fires (Bamford and Roberts 2003).

Vegetation preference, refuge types, and ecological group

Of the 12 frog taxa (out of 103 assessed; Table 13.1) that depend on fire-sensitive rainforest and wetland vegetation communities, most do not have traits likely to protect them during wildfire (Tables 13.2, 13.3). *Taudactylus pleione* is the only species known to shelter in rock crevices. Most shelter in leaf litter (*Philoria* (7), *Assa* (2), *Lech. fletcheri* and *M. fleayi*) and would be killed if their habitats burnt. In several of these taxa, larval stages occur underground, depending on ground water to maintain soaks, and are unlikely to be impacted by wildfire. Heard *et al.* (2021) showed a decline in the occupancy and number of calling males of *P. kundagungan* and *P. knowlesi* in areas burnt during 2019–20. Eight of the 12 taxa living in fire-sensitive vegetation are listed as threatened, but fortunately none experienced a high overlap with wildfires (Table 13.1). Wildfires in rainforests pose a great threat, not only because the frog fauna lack adaptations to avoid fire, but because they also have small ranges with derived reproductive modes, and severe wildfires could eliminate small, isolated populations in one event, as recorded by Lemckert (2000).

Threatened species and wildfire

Frogs of conservation concern, listed mostly for threats other than wildfire (Gillespie *et al.* 2020), occurred in fire-sensitive and fire-adapted vegetation communities (Table 13.1). Of most concern among the threatened taxa are 25 that occur in fire-adapted vegetation; several had considerable overlap with the wildfires (Table 13.1), including frogs from montane vegetation communities (*P. corroboree*, *P. pengilleyi*), eucalypt forests (*L. booroolongensis*, *L. daviesae*, *L, raniformis*, *L. littlejohni*, *L. spenceri*, *L. watsoni*, *H. a. australiacus*, *H. a. flavopunctatus*, *M. balbus* and *U. martini*) and wallum wetlands (*L. olongburensis*).

Conclusions

Range-restricted species that use fire-sensitive vegetation and shelter in vegetation or leaf litter are likely to have suffered the highest impact from the 2019–20 wildfires (e.g. *Assa* spp., *L. kroombitensis*, *L. spenceri*, *P. pughi*, *Pseud. pengilleyi*). Such species are vulnerable to single wildfire events that can severely reduce abundance and destroy a large percentage of habitat. For threatened species such as *P. pengilleyi*, *T. pleione*, and *L. kroombitensis*, catastrophic events substantially increase the risk of extinction. Several threatened taxa with wide but often fragmented distributions, occurring in fire-adapted communities, had large wildfire overlaps and may also have suffered large losses (*H. a. flavopunctatus*, *L. watsoni*, *L. littlejohni*). Post-fire monitoring has shown persistence of some populations and losses of others, presumably related to differences in local fire severity or drought impacts.

Knowing how frogs will respond to the predicted increase in wildfire occurrence and severity is essential for future management. Previous reviews of fire management for maintaining biodiversity values have focused on fire frequency rather than intensity or severity (Whelan *et al.* 2002; Driscoll *et al.* 2010). Understanding whether frog refuge sites affect their survival in wildfire will indicate which frogs are most vulnerable to fire. This is not to downplay the importance of suitable habitat attributes after the wildfire for supporting population recovery, but we focused here on surviving the fire as the first issue to be understood. Maintaining, protecting, connecting and possibly enhancing habitat refuges could enhance future protection.

Recommendations

To understand and mitigate the impact of future wildfire we recommend:

- For species with restricted or fragmented ranges: map critical habitat on fire management digital layers to inform planned burning and wildfire response. Species of greatest concern: *L. kroombitensis*, *L. littlejohni*, *L. spenceri*, *L. watsoni*, *Assa* spp., *P. corroboree*, *P. pengilleyi*, *Philoria* spp., *T. pleione*.
- For species with large distributions, occurring in fire-adapted vegetation communities and for whom large areas of habitat were destroyed by wildfires: frogs likely survived in

areas that burnt at low–moderate fire severity, and populations are expected to recover over time. This assumption should be verified with monitoring, while habitat is managed to maintain connectivity.

- For all frogs: establish long-term monitoring of representative species using demographic and occupancy approaches (e.g. Bolitho *et al.* 2021), in order to improve our understanding of fire sensitivity. This could be achieved using 'habitat based' groupings of frogs to reduce complexity.
- Undertake rapid assessment of population fitness (fecundity, connectivity and effective population size) using population genetics, to inform conservation actions. While cost-effective, the challenge for this approach is that rapid genetic assessment requires targeted collection and storage of pre-and post-fire samples.
- Support non-invasive citizen science projects (e.g. FrogID), that allow the rapid collection of data on frog species in response to threatening processes across wide spatial scales.
- Undertake research to identify refuges used by frogs and determine how to protect or enhance them.

References

Bamford MJ (1992) The impact of fire and increasing time after fire upon *Heleioporus eyrei*, *Limnodynastes dorsalis* and *Myobatrachus gouldii* (Anura: Leptodactylidae) in *Banksia* woodland near Perth, Western Australia. *Wildlife Research* **19**, 169–178. doi:10.1071/WR9920169

Bamford MJ, Roberts JD (2003) The impact of fire on frogs and reptiles in south-western Australia. In *Fire in Ecosystems of South-West Western Australia: Impacts and Management*. (Eds I Abbott and N Burrows). pp. 349–361. Backhuys Publishers, Leiden.

Bolitho LJ, Rowley JJ, Hines HB, Newell D (2021) Occupancy modelling reveals a highly restricted and fragmented distribution in a threatened montane frog (*Philoria kundagungan*) in subtropical Australian rainforests. *Australian Journal of Zoology* **67**, 231–240. doi:10.1071/ZO20037

Collins L, Bradstock RA, Tasker EM, Whelan RJ (2012) Impact of fire regimes, logging and topography on hollows in fallen logs in eucalypt forest of south eastern Australia. *Biological Conservation* **149**, 23–31. doi:10.1016/j.biocon.2012.01.065

Collins L, Bradstock RA, Clarke H, Clarke MF, Nolan RH, *et al.* (2021) The 2019/2020 mega-fires exposed Australian ecosystems to an unprecedented extent of high-severity fire. *Environmental Research Letters* **16**, 044029. doi:10.1088/1748-9326/abeb9e

Daly G, Craven P (2007) Monitoring populations of Heath Frog *Litoria littlejohni* in the Shoalhaven region on the south coast of New South Wales. *Australian Zoologist* **34**, 165–172. doi:10.7882/AZ.2007.014

Driscoll DA, Roberts JD (1997) Impact of fuel-reduction burning on the frog *Geocrinia lutea* in southwest Western Australia. *Australian Journal of Ecology* **22**, 334–339. doi:10.1111/j.1442-9993.1997.tb00679.x

Driscoll DA, Roberts JD (1998) Corrigenda to: impact of fuel reduction burning on the frog *Geocrinia lutea* in south-west Western Australia. *Australian Journal of Ecology* **23**, 598. doi:10.1111/j.1442-9993.1998.tb00772.x

Driscoll DA, Lindenmayer DB, Bennett AF, Bode M, Bradstock RA, *et al.* (2010) Fire management for biodiversity conservation: key research questions and our capacity to answer them. *Biological Conservation* **143**, 1928–1939. doi:10.1016/j.biocon.2010.05.026

Edwards DL, Roberts JD (2011) Genetic diversity and biogeographic history inform future conservation management strategies for the rare sunset frog (*Spicospina flammocaerulea*). *Australian Journal of Zoology* **59**, 63–72. doi:10.1071/ZO11005

Edwards DL, Roberts JD, Keogh JS (2008) Climatic fluctuations shape the phylogeography of a mesic direct-developing frog from the south-western Australian biodiversity hotspot. *Journal of Biogeography* **35**, 1803–1815. doi:10.1111/j.1365-2699.2008.01927.x

Gillespie GR, West M (2012) *Evaluation of Impacts of Bushfire on the Spotted Tree Frog* Litoria spenceri *in the Taponga River Catchment, Northeast Victoria: Black Saturday.* Department of Sustainability and Environment, East Melbourne.

Gillespie GR, Roberts JD, Hunter D, Hoskin CJ, Alford RA, *et al.* (2020) Status and priority conservation actions for Australian frog species. *Biological Conservation* **247**, 108543. doi:10.1016/j.biocon.2020.108543

Hardesty J, Myers R, Fulks W (2005) Fire, ecosystems, and people: a preliminary assessment of fire as a global conservation issue. *The George Wright Forum* **22**, 78–87.

Heard G, Bolitho L, Newell F, Hines H, McCall H, *et al.* (2021) 'Post-fire impact assessment for priority frogs: northern Philoria'. NESP Threatened Species Recovery Hub Project 8.1.3 report, Brisbane.

Hillman SS, Withers PC, Drewes RD, Hillyard SD (2008) *Ecological and Environmental Physiology of Amphibians.* Oxford University Press, Oxford.

IUCN (2019) 'Guidelines for using the IUCN Red List categories and criteria'. Version 14. Prepared by the Standards and Petitions Subcommittee,. <http://www.iucnredlist.org/documents/RedListGuidelines.pdf>.

Knowles R, Mahony M, Armstrong J, Donnellan S (2004) Systematics of sphagnum frogs of the genus *Philoria* (Anura: Myobatrachidae) in eastern Australia, with the description of two new species. *Records of the Australian Museum* **56**, 57–74. doi:10.3853/j.0067-1975.56.2004.1391

Legge S, Woinarski JCZ, Scheele BC, Garnett ST, Lintermans M, *et al.* (2022) Rapid assessment of the biodiversity impacts of the 2019–2020 Australian megafires to guide urgent management intervention and recovery and lessons for other regions. *Diversity & Distributions* **28**, 571–591. doi:10.1111/ddi.13428

Legge S, Rumpff L, Woinarski JCZ, Whiterod NS, Ward M, *et al.* (in press) The conservation impacts of ecological disturbance: time-bound estimates of population loss and recovery for fauna affected by the 2019–20 Australian megafires. *Global Ecology and Biogeography.* doi:10.1111/geb.13473

Lemckert F (1999) Impacts of selective logging on frogs in a forested area of northern New South Wales. *Biological Conservation* **89**, 321–328. doi:10.1016/S0006-3207(98)00117-7

Lemckert F (2000) Observations on the effects of fire on the hip-pocket frog, *Assa darlingtoni. Herpetofauna* **30**, 32–33.

Lemckert F, Brassil T (2000) Movements and habitat use of the endangered giant barred river frog (*Mixophyes iteratus*) and the implications for its conservation in timber production forests. *Biological Conservation* **96**, 177–184. doi:10.1016/S0006-3207(00)00066-5

Lemckert F, Brassil T (2003) Movements and habitat use by the giant burrowing frog, *Heleioporus australiacus. Amphibia-Reptilia* **24**, 207–211. doi:10.1163/156853803322390453

Lemckert FL, Brassil T, Haywood A (2004) Effects of a low intensity fire on populations of pond breeding anurans in mid-northern New South Wales, Australia. *Applied Herpetology* **1**, 183–196.

Lemckert F, Mahony M, Brassil T, Slatyer C (2006) The biology of the threatened Green-thighed frog *Litoria brevipalmata* (Anura: Hylidae) in the central and mid-north coastal areas of New South Wales. *Australian Zoologist* **33**, 337–344. doi:10.7882/AZ.2006.007

Lowe K, Castley JG, Hero J-M (2013) Acid frogs can stand the heat: amphibian resilience to wildfire in coastal wetlands of eastern Australia. *International Journal of Wildland Fire* **22**, 947–958. doi:10.1071/WF12128

Mahony M, Moses B, Mahony SV, Lemckert F, Donnellan S (2020) A new species of frog in the *Litoria ewingii* species group (Anura: Pelodryadidae) from south-eastern Australia. *Zootaxa* **4858**, 201–230. doi:10.11646/zootaxa.4858.2.3

Mahony MJ, Hines HB, Mahony SV, Moses B, Catalano SR, *et al.* (2021) A new hip-pocket frog from mid–eastern Australia (Anura: Myobatrachidae: *Assa*). *Zootaxa* **5057**, 451–486. doi:10.11646/zootaxa.5057.4.1

Mo M (2015) Frogs using cossid moth galleries: an opportunity in Flooded Gum plantations in the New South Wales North Coast bioregion. *Australian Zoologist* **37**, 501–507. doi:10.7882/AZ.2015.010

Murray KA, Rosauer D, McCallum H, Skerratt LF (2011) Integrating species traits with extrinsic threats: closing the gap between predicting and preventing species declines. *Proceedings. Biological Sciences* **278**, 1515–1523. doi:10.1098/rspb.2010.1872

Newell DA, Goldingay RL, Brooks LO (2013) Population recovery following decline in an endangered stream-breeding frog (*Mixophyes fleayi*) from subtropical Australia. *PLoS ONE* **8**, e58559. doi:10.1371/journal.pone.0058559

NSW Department of Planning, Industry and Environment (2020) *Wildfire and Conservation Bushfire Recovery: Immediate Response*. Environment, Energy and Science: Department of Planning, Industry and Environment, NSW State Government, Sydney.

Penman T, Lemckert F, Slade C, Mahony M (2005a) Non-breeding habitat requirements of the giant burrowing frog, *Heleioporus australiacus* (Anura: Myobatrachidae) in south-eastern Australia. *Australian Zoologist* **33**, 251–257. doi:10.7882/AZ.2005.022

Penman T, Mahony M, Towerton AL, Lemckert F (2005b) Bioclimatic analysis of disjunct populations of the giant burrowing frog, *Heleioporus australiacus*. *Journal of Biogeography* **32**, 397–405. doi:10.1111/j.1365-2699.2005.01205.x

Penman T, Lemckert F, Mahony M (2006) A preliminary investigation into the potential impacts of fire on a forest dependent burrowing frog species. *Pacific Conservation Biology* **12**, 78–83. doi:10.1071/PC060078

Penman TD, Keith DA, Elith J, Mahony MJ, Tingley R, *et al.* (2015) Interactive effects of climate change and fire on metapopulation viability of a forest-dependent frog in south-eastern Australia. *Biological Conservation* **190**, 142–153. doi:10.1016/j.biocon.2015.05.020

Potvin DA, Parris KM, Smith KL, Date D, Keely CC, *et al.* (2017) Genetic erosion and escalating extinction risk in frogs with increasing wildfire frequency. *Journal of Applied Ecology* **54**, 945–954. doi:10.1111/1365-2664.12809

Recher HF, Shields J, Kavanagh R, Webb G (1987) Retaining remnant mature forest for nature conservation at Eden, New South Wales. In *Nature Conservation: The Role of Remnants of Native Vegetation*. (Eds DA Saunders, GW Arnold, AA Burbidge and AJM Hopkins) pp. 177–194. Surrey Beatty & Sons, Chipping Norton.

Quick G, Goldingay RL, Parkyn J, Newell DA (2015) Population stability in the endangered Fleay's barred frog (*Mixophyes fleayi*) and a program for long-term monitoring. *Australian Journal of Zoology* **63**, 214–219. doi:10.1071/ZO14106

Roberts J (2018) Conservation of frogs in South-western Australia. In *Status of Conservation and Decline of Amphibians: Eastern Hemisphere*. (Eds H Heatwole and JJL Rowley) pp. 73–75. CSIRO Publishing, Melbourne.

Roberts JD, Conroy S, Williams K (1999) Conservation status of frogs in Western Australia. In *Declines and Disappearances of Australian Frogs*. (Ed. A Campbell) pp. 177–184. Environment Australia, Canberra.

Rosauer D, Laffan SW, Crisp MD, Donnellan SC, Cook LG (2009) Phylogenetic endemism: a new approach for identifying geographical concentrations of evolutionary history. *Molecular Ecology* **18**, 4061–4072. doi:10.1111/j.1365-294X.2009.04311.x

Rowley JJ, Callaghan CT, Cutajar T, Portway C, Potter K, *et al.* (2019) FrogID: citizen scientists provide validated biodiversity data on frogs of Australia. *Herpetological Conservation and Biology* **14**, 155–170.

Rowley JJ, Callaghan CT, Cornwell WK (2020) Widespread short-term persistence of frog species after the 2019–2020 bushfires in eastern Australia revealed by citizen science. *Conservation Science and Practice* **2**, e287. doi:10.1111/csp2.287

Taylor R, Woinarski J, Chatto R (2003) Hollow use by vertebrates in the Top End of the Northern Territory. *Australian Zoologist* **32**, 462–476. doi:10.7882/AZ.2002.024

Tracy CR, Laurence N, Christian K (2011) Condensation onto the skin as a means for water gain by tree frogs in tropical Australia. *American Naturalist* **178**, 553–558. doi:10.1086/661908

Tyler MJ, Davies M (1978) Species-groups within the Australopapuan hulid frog genus *Litoria* Tschudi. *Australian Journal of Zoology Supplementary Series* **26**, 1–47. doi:10.1071/AJZS063

Wardell-Johnson G, Roberts J, Driscoll D, Williams K (1995) *Orange-Bellied and White-Bellied Frogs Recovery Plan*. Department of Conservation and Land Management, Perth.

Westgate MJ, Driscoll DA, Lindenmayer DB (2012) Can the intermediate disturbance hypothesis and information on species traits predict anuran responses to fire? *Oikos* **121**, 1516–1524. doi:10.1111/j.1600-0706.2011.19863.x

Westgate MJ, MacGregor C, Scheele B, Driscoll DA, Lindenmayer BD (2018) Effects of time since fire on frog occurrence are altered by isolation, vegetation and fire frequency gradients. *Diversity & Distributions* **24**, 82–91. doi:10.1111/ddi.12659

Whelan RJ, Rodgerson L, Dickman CR, Sutherland EF (2002) Critical life cycles of plants and animals: developing a process-based understanding of population changes in fire-prone landscapes. In *Flammable Australia: the Fire Regimes and Biodiversity of a Continent*. (Eds RA Bradstock, JE Williams and AM Gill) pp. 94–124. Cambridge University Press, Cambridge, UK.

Willacy RJ, Mahony M, Newell DA (2015) If a frog calls in the forest: bioacoustic monitoring reveals the breeding phenology of the endangered Richmond Range mountain frog (*Philoria richmondensis*). *Austral Ecology* **40**, 625–633. doi:10.1111/aec.12228

14

The impacts of the 2019–20 wildfires on Australia's lizards and snakes

Dale Nimmo, Chris Jolly, Renee Catullo, Matthew Greenlees, Jane Melville, Damian Michael, Eridani Mulder, Darren Southwell and Jonathan Webb

Summary

Context and challenges

- Australia harbours a disproportionate share of global squamate diversity (lizards and snakes). The areas impacted by the 2019–20 wildfires comprise diverse squamate communities that include many threatened and narrowly distributed species.
- The impact of fire on most Australian squamates is not well understood. Reptile populations are monitored less than any other vertebrate group, and many species have unresolved taxonomy.

Main findings

- Very few studies have considered the effects of fire on squamate mortality or how attributes of fire regimes affect squamates in areas impacted by the 2019–20 wildfires.
- The habitat of 445 squamate species from 11 families was within the footprint of the 2019–20 wildfires, representing nearly 40% of Australia's described species. This included 29 species listed as threatened nationally or globally.
- One species, Kate's leaf-tailed gecko (*Saltuarius kateae*), had its entire known (highly restricted) range burnt, but post-fire sampling indicated that the species persists and is breeding in the firegrounds.
- Species in most need of monitoring and conservation following the 2019–20 wildfires are narrowly distributed species with moderate-to-high fire overlap and traits that make them vulnerable to fire and post-fire conditions.
- Assessments of within-species diversity across 14 priority species identified 19 evolutionarily significant units and more than 18 management units.
- Assessment of the on-ground effects of the 2019–20 wildfires is hampered by a lack of long-term monitoring and investment. There is an urgent need to break a cycle of chronic under-funding of reptile monitoring and conservation in Australia, particularly given the increasing prevalence of large, high severity wildfires.

Introduction

Although Australia accounts for only 5% of the world's landmass, it supports ~10% (~1130) of the ~11 690 reptile species on Earth (Uetz *et al.* 2021), most of which are terrestrial squamates (lizards and snakes). From remote rock atolls in frigid seas off the coast of Tasmania to the canopies of wet tropical rainforests through Far North Queensland, and from alpine meadows to arid deserts, lizards and snakes inhabit the entire Australian continent, and 96% are found nowhere else (Chapple *et al.* 2019). Australia has had three documented squamate extinctions since European invasion (Woinarski *et al.* 2019), and 6–11 more have a high likelihood of extinction in the next 20 years (Geyle *et al.* 2021). Currently, 7.1% of Australian squamates are threatened with extinction (IUCN 2021). Lizards and snakes (collectively squamates) comprise most of Australian reptile species and are the focus of this chapter. We recognise that the 2019–20 wildfires also affected some of Australia's freshwater turtle species, but these are not considered here (but see Chapter 6 for information on impacts of these fires on aquatic systems, and Chapter 27 for information on one fire-affected turtle species).

While many Australian ecosystems have a deep evolutionary history with fire, most populations are not adapted to fire *per se*, but to *fire regimes* (frequency, interval, season, intensity, severity, and spatial aspects of fire; Gill 1975). Fire regimes vary across Australia from rainforests that experience fires once every 100 years, to tropical savanna ecosystems that experience fires every 1–5 years (Murphy *et al.* 2013). Globally, fire regimes have shifted over recent centuries, and altered fire regimes are regarded as an extinction threat to thousands of species (Kelly *et al.* 2020). Of the 60 Australian terrestrial squamates considered most at risk of extinction, approximately a third have altered fire regimes listed as a threat (Geyle *et al.* 2021). Altered fire regimes are regarded as a threat to 79 squamate species, and many of these occur within the regions affected by the 2019–20 wildfires (Tingley *et al.* 2019).

Key findings

Australian squamates and fire: what is the state of knowledge?

We undertook a global review of studies of the immediate impacts of fire: those that track the fate of squamate reptiles through the path of a fire and assess what proportion of animals live and die (see Jolly *et al.* 2022 for details). We found six such studies, all from North America (n = 4) or Australia (n = 2). These studies tracked an average of 14 individuals (range = 1–61) through fire, reporting mortality rates of 0–50% (Durbian 2006; Hellgren *et al.* 2010), with an average mortality rate of 15%. Both studies conducted in Australia are noteworthy. One study of the pygmy bluetongue lizard (*Tiliqua adelaidensis*) tracked eight individuals through an intense grassfire (Fenner and Bull 2007). The lizards sheltered in spider burrows during the fire, and all survived. Another tracked frill-necked lizards (*Chlamydosaurus kingii*) through two fires in northern Australia (Griffiths and Christian 1996). The first was an early dry season fire, which are usually of low intensity. All 17 of the lizards that were radio-tracked survived the fire by remaining in the tree canopy. The second was an intense, late dry season fire. This time, animals that remained in treetops died, whereas animals that escaped to termite mounds survived. Overall, a quarter (6) of the 24 frilled lizards died during the late dry season fire, with another succumbing to fire-caused injuries sometime later. Animal survival during fire is the outcome of a complex interplay between animal behaviour, the environment (e.g. the provision of refuges) and fire attributes (e.g. intensity, rate of spread) (Jolly *et al.* 2022).

We identified 69 papers from across Australia that examined the direct or indirect impacts of fire on squamates (Fig. 14.1). Most of the research (> 90%) compared the occurrence or composition of squamates at sites that differed in their most recent fire history (e.g. burnt *v.* unburnt sites or sites that differ in their time since fire). These studies have demonstrated that fire can impact squamates for decades (Dixon *et al.* 2018), primarily through its impact on habitat (i.e. vegetation structure), although, in some instances, recovery is rapid. Far less is known regarding how other elements of the fire regime affect squamate populations or communities. Of the 69 published studies, only three (4%) considered the impacts of fire severity, four considered fire intensity (6%), 11 considered fire frequency (16%), three considered fire interval (4%), seven considered fire season (10%), and nine considered spatial aspects of fire (13%) (Fig. 14.1). While there is a good spread of studies across Australia that have examined how the most recent fire affects squamate populations and communities – including numerous studies within the analysis area (Fig. 14.1) – studies of other aspects of the fire regime are few and geographically clustered. This lack of published research severely hampers our capacity to understand the likely ramifications of the 2019–20 wildfires on squamate species and communities.

The 2019–20 wildfires: which species were most affected?

The simplest means of assessing which species were potentially affected by the 2019–20 wildfires is to assess 'fire overlap', which is the proportion of a species' geographical range within the fire footprint. It is important to differentiate this notion of 'affected' species from the on-ground impacts of the fires. Species may have high fire overlap, but how that plays out on the ground – the actual impacts of the fires on individuals and populations – could range from minimal to substantial. With this in mind, the 2019–20 wildfires affected 445 squamate species, constituting 39% of Australia's 1128 described reptile species and 42% of Australia's 1051 terrestrial squamate species (Uetz *et al.* 2021). The fires overlapped the distributions of squamate species from 11 families (Fig. 14.2): Scincidae (n = 199; 43%), Elapidae (n = 67; 61% of terrestrial elapids), Diplodactylidae (n = 45;

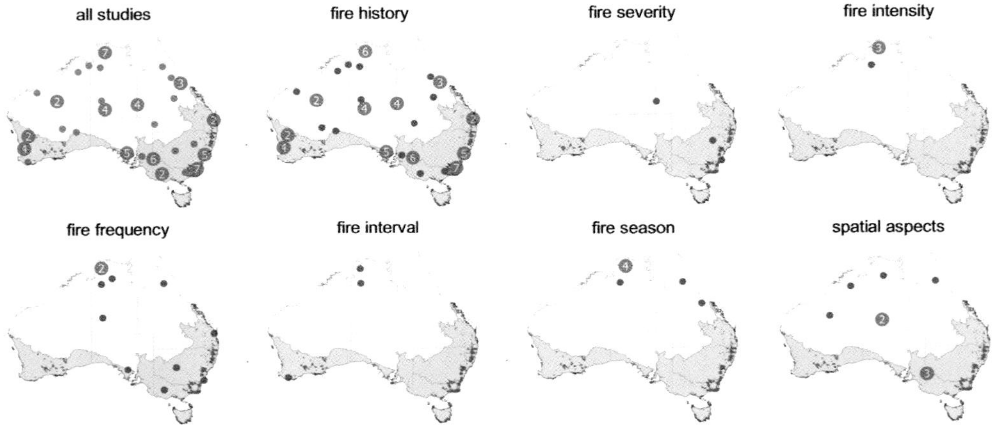

Fig. 14.1. Studies investigating the responses of Australian terrestrial squamate reptile populations or communities to fire. Light grey = analysis area. Dark grey = 2019–20 wildfires. Circles represent the location of studies and numbers within circles refer to the number of studies (when > 1). 'Fire history' refers to studies that consider only the most recent fire (e.g. time since fire/burnt versus unburnt/before versus after fire), and no other elements of the fire regime.

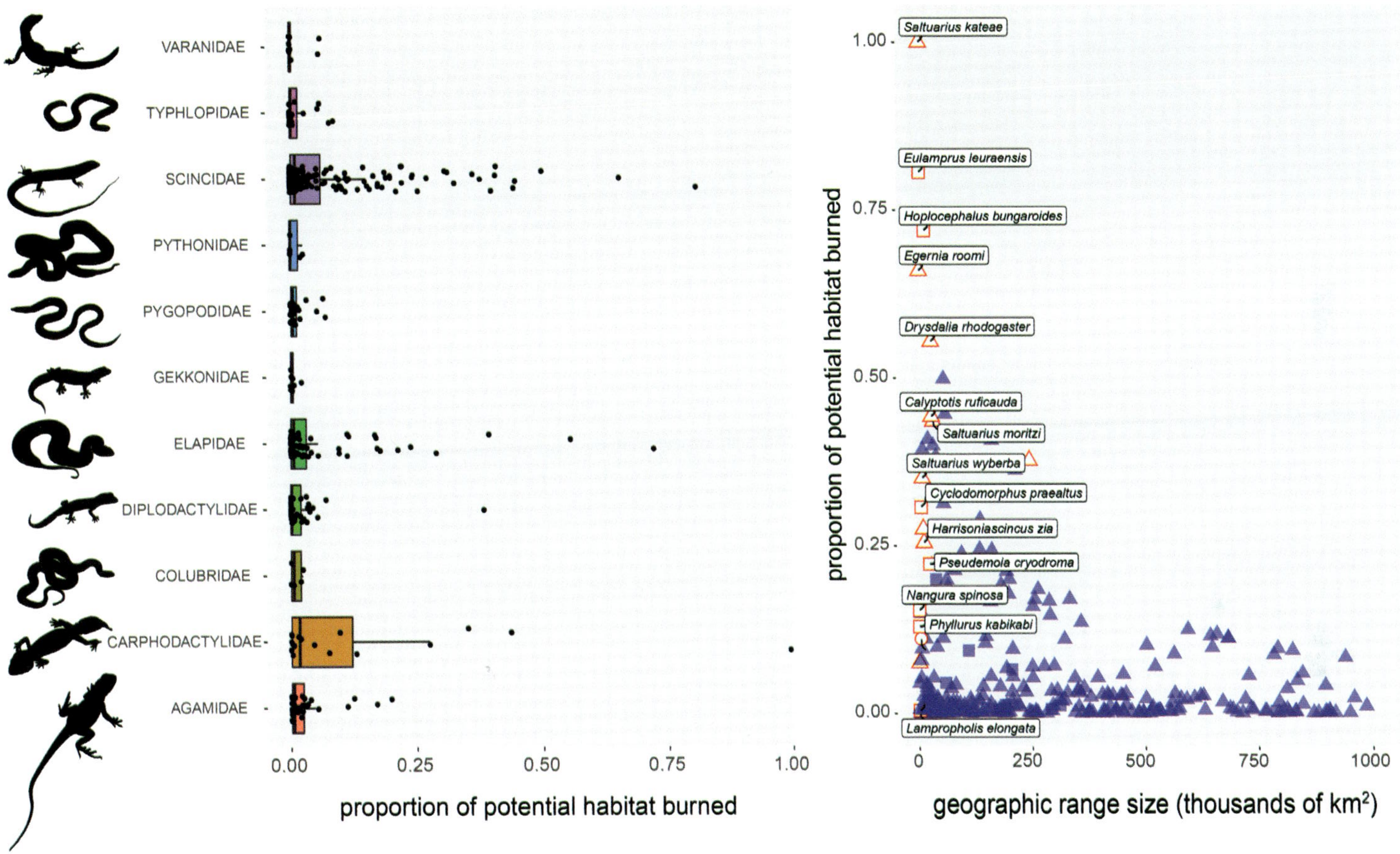

Fig. 14.2. Summary of overlap between Australian terrestrial squamate species' geographical ranges and the 2019–20 wildfires. (Left) Species' overlap by taxonomic family, and (right) comparison of species' overlap with species' geographical range size (for species with geographical ranges < 1 million km^2). Priority species are hollow with a red outline, non-priority species are dark blue. Squares = species threatened under either the EPBC Act or IUCN, triangles = unlisted species, circles = data deficient species (according to IUCN). Examples of priority species are labelled. Overlap estimates are derived from IUCN range maps and species distribution models when available.

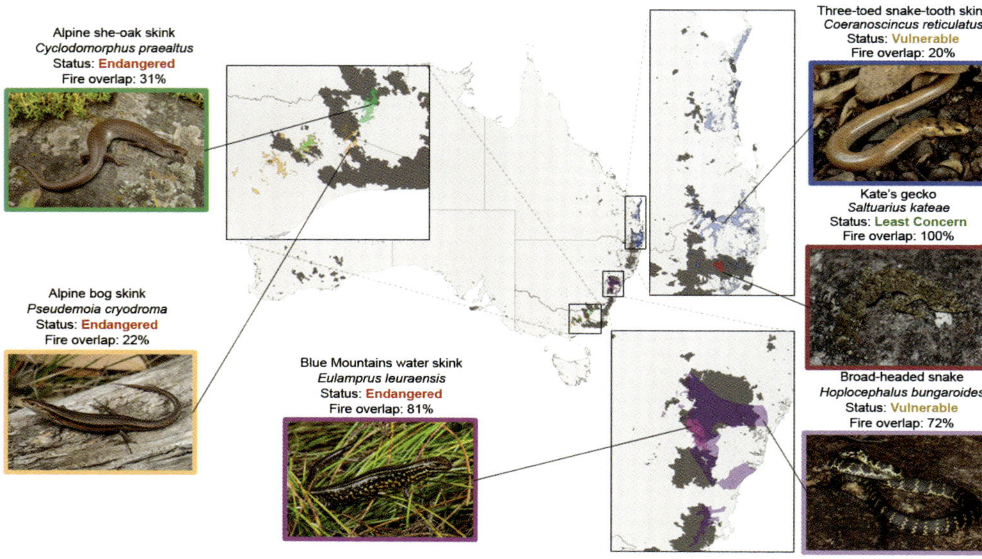

Fig. 14.3. Overlap between the Australian 2019–20 wildfires and a subset of the federal government's priority reptile species. Dark grey = the 2019–20 wildfires (from NIAFED), and different colours show the geographical range of different species. Conservation status is derived from the EPBC Act or the IUCN Red List. (Photos: Brendan Schembri (*C. reticulatus* and *E. leuraensis*), Jules Farquhar (*C. praealtus* and *P. cryodroma*), Chris Jolly (*H. bungaroides*), Matt Greenlees (*S. kateae*))

44%), Agamidae (*n* = 38; 35%), Pygopodidae (*n* = 28; 62%), Typhlopidae (*n* = 20; 41%), Carphodactylidae (*n* = 18; 56%), Varanidae (*n* = 11; 37%), Gekkonidae (*n* = 8; 11%), Pythonidae (*n* = 6; 40%), and Colubridae (*n* = 5; 100%). This included 29 species listed as threatened under either the *Environment Protection and Biodiversity Conservation Act 1999* (EPBC Act) or the International Union for Conservation of Nature (IUCN) Red List. However, these measures do need to be interpreted with caution. Of the 445 species with some fire overlap, the majority (*n* = 243 species) had < 1% of their range affected (Fig. 14.2). Given the coarse nature of both species range maps (mostly IUCN range polygons) and the national fire maps (the National Indicative Aggregated Fire Extent Datasets, DAWE 2020a), these latter species may equate to 'rounding errors' that bear little relationship to the on-ground situation. More relevant are those estimates of overlap that exceed 10%. Here, we find 44 species with 10–30% of their range affected, 15 species with 30–50% of their range affected, three species with 50–80% of their habitat affected, and two species with > 80% of their range affected, including one (Kate's leaf-tailed gecko) whose entire known geographical range overlapped with the fire footprint (Fig. 14.3, Box 14.1, Fig. 14.4).

Box 14.1. Kate's leaf-tailed gecko

Situated in the Northern Rivers region of New South Wales, between the Richmond Range and the Clarence River, lies a unique escarpment known as the Kangaroo Creek Sandstones – home to Kate's leaf-tailed gecko. Kate's leaf-tailed gecko is a relatively large gecko, ~20 cm from nose to tail. Like other members of this genus, its 'flat' body allows it to shelter in the narrow cracks and crevices in the sandstone outcrops with which it is exclusively associated (Fig. 14.4). Until 2020, there were fewer than 20 official records of the species from a tiny area around Mount Marsh and Mount Neville. The entire area from which this species was known burnt during the 2019–20 wildfires. This made Kate's leaf-tailed gecko the only known vertebrate species in Australia whose entire range was affected by the fires. Federal and state government conservation agencies and the local community were all greatly concerned about the species. Following the fires, the Border Ranges Richmond Valley Landcare Group won funding to conduct surveys for Kate's leaf-tailed gecko to determine the impact of the fires. Before the fires, there was only a limited understanding of where the gecko occurred, and no information on trends in its population or threats to the species. Since late 2020, surveys conducted in the area have determined that the geckos fared remarkably well despite the fires. In larger expanses of the outcrop, geckos were recorded in numbers that were equivalent to the maximum number detected in an area when the species was first discovered and described. Juvenile geckos were detected too, suggesting that the fires had not prevented recruitment. Surveys have detected them at sites where they had not previously been recorded, expanding the known distribution of the species. Despite their small range, the geckos' strong association with the rocky habitat seems to have afforded them protection. While in smaller outcrops there may have been some direct impact of the fires, deep crevices in rocky escarpment and outcrops appear to have acted as vital refuges for the geckos. Continued monitoring of this species will be required to determine whether the longer-term, indirect effects of the fires affect population persistence.

Fig. 14.4. Burnt habitat of Kate's leaf-tailed gecko. (Photo: Matt Greenlees)

In February 2020, the Australian Government released a rapid assessment of the impacts of the 2019–20 wildfires, which included a list of priority reptile species in need of urgent reconnaissance surveys and management interventions (Legge *et al.* 2020). This rapid assessment considered the proportion of a species' range that overlapped with fire (divided into overlap categories) and species' national conservation status (i.e. EPBC Act or IUCN status). Species in the highest fire overlap categories (i.e. 50–80% or > 80%) that were listed as threatened (under either the EPBC Act or IUCN) were ranked most in need of attention, while non-threatened species with high overlap were included if their traits or vulnerabilities suggested acute, negative responses to fire and/or the immediate post-fire landscape (Legge *et al.* 2020). Finally, species were added if they had very narrow distributions within or in very close proximity to the firegrounds, regardless of their traits (e.g. Kate's leaf-tailed gecko); state agencies made a specific case for their inclusion based on expert knowledge; or if taxonomic considerations suggested possible effects on significant ecological units. The priority species were typically narrowly distributed with relatively high fire overlap (Figs 14.2, 14.3).

The rapid assessment used maps of nationally threatened species held by the Department of Agriculture, Water and Environment (DAWE), and otherwise relied on IUCN range maps. A subsequent project (Legge *et al.* in press) refined these maps for several species using species distribution models (SDMs), and resulted in some marked changes in the estimated fire overlap. For instance, overlap for the Blue Mountains water skink (*Eulamprus leuraensis*) – a species of great concern given its narrow distribution, threatened status (Endangered under IUCN) and declining population trajectory (Chapple *et al.* 2019) – increased from two-thirds using IUCN maps to 81% using SDMs. The changes in overlap in some instances moved species from one overlap category into another. For instance, the mustard-bellied snake (*Drysdalia rhodogaster*) moved from the 30–50% overlap category when using IUCN maps to the 50–80% category when using SDMs, while an increase in fire overlap for the Blue Mountains water skink moved it from the second highest (50–80%) to the highest (> 80%) category. These changes, while relatively rare, would have influenced how species were ranked during the rapid prioritisation (Legge *et al.* 2020), highlighting the importance of high-quality distribution data to allow overlap to be accurately measured.

Fire severity was not considered during the initial rapid assessment because national fire severity maps were not available, but DAWE later developed a nation-wide fire severity map (DAWE 2020b). Assessing the extent of overlap between species' ranges and fire severity classes added nuance to the emerging picture of how species were likely affected by the wildfires. For instance, although the entire range of Kate's leaf-tailed gecko burnt, more than half burnt at low severity. Similarly, the narrowly distributed Kaputar rock skink (*Egernia roomi*) had substantial fire overlap (66%), but 95% of areas within the fire boundary burnt at low severity. These species may not be as severely affected as initially feared.

Assessing species and within species diversity

Assessment of the impact of fires on individual species can be challenging when overarching questions about taxonomy are unresolved. The 2019–20 wildfires impacted wet forests that are naturally fragmented in eastern Australia, and the distributions of many of the reptile species endemic to these forests are also naturally fragmented (Rosauer *et al.* 2015). Naturally fragmented habitats can lead to many cryptic species that are not yet recognised by science nor conservation. Recognising and naming these species is challenging – cryptic species often require genetic data to identify species status, and vouchers are needed for species description. Below the species level, identifying and protecting within-species genetic diversity is important for managing threatened species. Highly fragmented species will have populations whose genetic distinctiveness should be preserved to ensure the longevity of the species, as genetic diversity contributes to prevention of inbreeding depression, is important in developing captive populations, and is essential to enabling species to adapt to change (reviewed in Pierson *et al.* 2016).

Genetic data were unavailable to assess the taxonomic status and within-species diversity for many of the priority squamate species. A review of the 14 priority squamates for which landscape genetic data did exist (Catullo *et al.* 2021; Catullo and Moritz 2020) identified three that included undescribed or candidate species. When unrecognised species are included in assessments of fire impact, it can completely shift estimated fire overlap. Splitting species inherently reduces range size, meaning the estimated impact can be greatly increased for one species and reduced, even to zero, for other species. An example of shifting impacts is seen in the *Eulamprus kosciuskoi/leuraensis* species complex, which is distributed in three disjunct locations (Fig. 14.5). Genetic assessment of this complex (Pepper *et al.* 2018) identified that the two geographically discrete populations of *E. kosciuskoi* were not monophyletic and hence were likely distinct species. All three lineages within the species complex were distributed almost entirely within burnt areas and failure to recognise these as distinct would have resulted in an underestimation of fire impacts on both species currently within *E. kosciuskoi*. Assessments of within-species diversity across the 14 species identified 19 evolutionarily significant units and more than 18 management units, indicating that reptile species within the fire-affected regions were highly fragmented with distinct

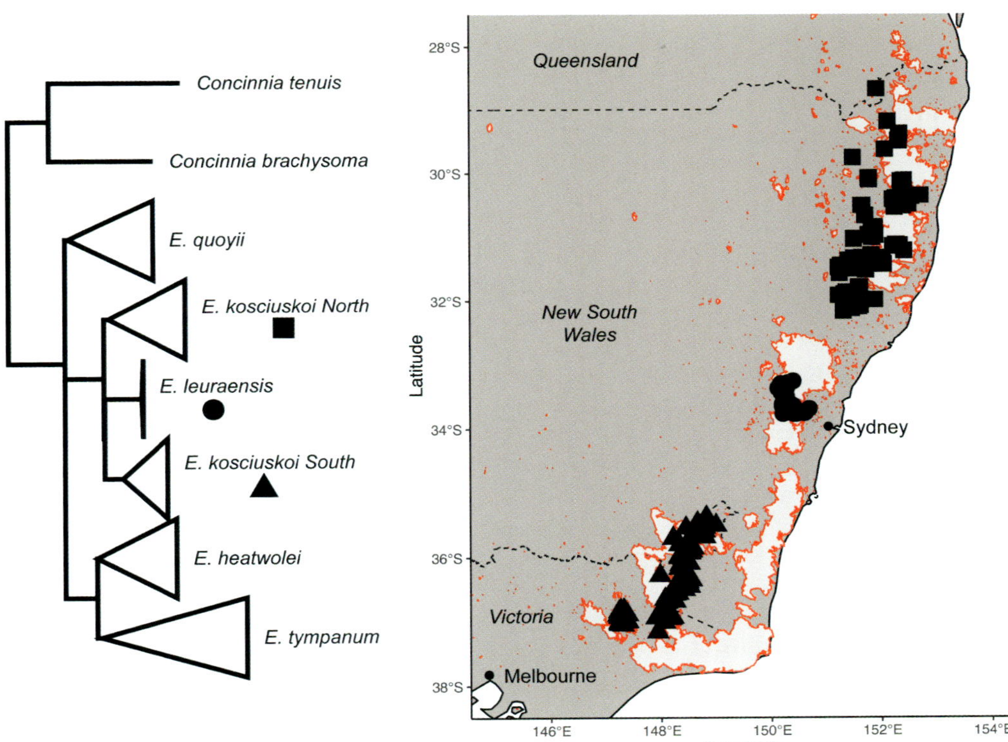

Fig. 14.5. Phylogenetic tree of *Eulamprus* taxa (modified from Pepper *et al.* 2018), which identified species-level divergence across each of the three geographically distinct clades in the *E. kosciuskoi/leuraensis* complex. The distribution of clades (shapes) is shown relative to the 2019–20 wildfires (pale areas outlined in red), which shows that each clade is independently fire impacted and requires assessment and management of impacts.

populations that require careful and independent overlap assessment and management planning.

Beyond fire overlap: engaging experts

Quantifying overlap between the 2019–20 wildfires and species ranges, once the taxonomic units are defined, is relatively straightforward, but the obvious question is: how have the fires impacted populations on the ground? Answering this question is complicated by the paucity of information on the natural history of squamate reptiles, a lack of research on the responses of lizards and snakes to fire (outlined above), and – with some notable exceptions (Box 14.2, Fig. 14.6) – long-term monitoring of squamate populations within the fire-affected regions. Monitoring programs are lacking even for Australia's most imperilled reptiles: nearly 40% of Australia's threatened reptiles lack any form of population monitoring (Woinarski 2018). Threatened reptile populations are monitored the least of all terrestrial vertebrate groups (Scheele *et al.* 2019).

Box 14.2. Broad-headed snake

Broad-headed snakes (*Hoplocephalus bungaroides*) are small (< 90 cm snout–vent length), live-bearing, venomous elapid snakes, which are listed as Vulnerable under the EPBC Act. They have a relatively small geographical distribution – within a ~250 km radius of Sydney – and require intact sandstone rock outcrops to persist (Fig. 14.6). During cooler months (May–October), broad-headed snakes occupy the edge of exposed sandstone plateau where they shelter under small, exposed rocks. However,

Fig. 14.6. Broad-headed snake in its sandstone habitat. (Photo: Chris Jolly)

during the warmer months (November–April), they leave the outcrops for the shelter of old-growth eucalypt forests where they shelter in tree hollows (Webb and Shine 1998). Adult snakes are most vulnerable to wildfires during this period (Webb *et al.* 2021). Approximately 72% of their geographical range burnt during the 2019–20 wildfires (Figs 14.2, 14.3). A long-term (30 years) study of a population in Morton National Park provides an insight into the effects of the 2019–20 wildfires on this species. The region was burnt by high-intensity fires in November and December 2002 (the Touga Fire) and in January and February 2020 (the Morton Fire). Estimates of abundance showed that the study population declined by 34% and 26% after the Touga and Morton fires, respectively (Webb *et al.* 2021). The population recovered 1 year after the Touga Fire, but no population recovery was apparent 1 year following the Morton Fire (Webb *et al.* 2021).

The broad-headed snake offers an illuminating case study on the importance of context when assessing the impacts of wildfire on threatened species. The major threats to broad-headed snakes are the removal of natural rocks for the landscaping industry, and removal of snakes for the illegal pet trade. Rock removal results in a loss of thermally suitable shelter sites for broad-headed snakes and one of their major prey species, the velvet gecko (*Amalosia lesueurii*) (Shine *et al.* 1998). In the Morton National Park population, adult survival rates were 20% lower in years when humans had overturned rocks while searching for snakes, suggesting that snake collectors had removed snakes from the wild for private hobbyist collections. The snake's slow life history (maturity at 6 years), small litter size and infrequent reproduction (reproducing every second or third year) means that the removal of adult females can drive small populations extinct (Jolly *et al.* 2021). The population has declined by ~60% over 30 years, from 127 in 1992 to only 50 snakes in 2020. If one extrapolates to the entire study plateau, the total population currently sits at just 300 snakes, half the number that was reported in 2002 (Webb *et al.* 2002). This result highlights how declines wrought by wildfires need to be considered among other threatening processes. In the case of the broad-headed snake, poaching and bush rock removal pose far greater threats to the persistence of the Morton National Park population than do single wildfires. However, while stable populations of broad-headed snakes may recover from severe fire events, small declining populations may not, and increases in the frequency and intensity of wildfires could hasten local extinctions.

Given the scarcity of prior research and baseline monitoring for most species – and the time needed to assess impacts and recovery – alternative approaches are needed to estimate which species are most acutely affected by the fires. One means of temporarily filling this gap is expert elicitation. To assess the impacts of the 2019–20 fires on reptiles, a structured approach, known as the IDEA (Investigate, Discuss, Estimate, Aggregate) protocol (Hemming *et al.* 2018), was used to elicit knowledge. Experts identified the rainforest cool-skink (*Harrisoniascincus zia*), alpine she-oak skink (*Cyclodomorphus praealtus*), and long sun skink (*Lampropholis elongata*) as being most adversely affected by high severity fires. These species occupy quite distinct habitats – from alpine grasslands to rainforest. What they share is a tendency to shelter in or under flammable retreat sites, such as grass tussocks, leaf litter and logs. By contrast, some species were predicted to be relatively resistant to the immediate effects of fire, such as some leaf-tailed geckos (*Saltuarius* spp.) and the Kaputar rock skink, which make greater use of non-flammable shelters such as rock crevices.

Estimates of fire impacts were used to extrapolate the range-wide impacts of the 2019–20 wildfires on each species, while also considering each species' pre-fire population trajectory. The species predicted to suffer the largest population declines immediately after the wildfire included the Blue Mountains water skink (~25% decline) and the alpine she-oak skink (~20%), due to a combination of vulnerability to, and substantial overlap with, high severity fire. Despite experts predicting that the narrowly distributed Kate's leaf-tailed gecko would suffer less fire-induced mortality than many other species, it nonetheless ranked among the most impacted species due to its entire range being within the boundary of the wildfires.

Most species were predicted to undergo further declines between 1 week and 1 year after the wildfires, probably reflecting the harsh conditions typical of post-fire landscapes, such as reduced resources and increased predation risk. Projecting forward to 10 years or three generations post-fire, the Blue Mountains water skink and the alpine she-oak skink were still projected to have declined most substantially, with both predicted to decline by ~30%. In these taxa, the overall population size is predicted to be ~10% less than it would have been had the fires not occurred. Based on this analysis, Legge *et al.* (in press) suggested up to nine reptile species should have their conservation status assessed for potential listing as threatened, including the unlisted Kate's leaf-tailed gecko, the Kaputar skink and the long sun skink.

Future assessment and management implications

Future rapid assessments to large wildfires will also be heavily reliant on maps, but there is ample room for improvement in the kinds of maps that are used. First, the comparisons of IUCN range maps with more refined SDMs highlight the need for a national database of up-to-date species distribution maps, generated using high-quality, contemporary data, and vetted by experts with on-ground knowledge. Second, Australia urgently needs a national monitoring agency responsible for generating and cataloguing fire events across the country in a consistent manner (Bowman *et al.* 2020). Reliable, nation-wide maps of long-term fire histories would have allowed rapid assessments to incorporate not only the amount of a species' range affected by the 2019–20 wildfires but also the underlying fire history. Similarly, high-quality maps of other threats could aid in identifying important refuges while guiding management actions.

Woinarski (2018) described a vicious cycle that reinforces the inadequacy of Australia's reptile monitoring and conservation. A lack of monitoring data for reptiles results in fewer species with population data necessary to support their listing under the EPBC Act (e.g. rate of population decline). A lack of listing leads to a lack of investment in reptile monitoring and conservation, which perpetuates the cycle (Woinarski 2018). This cycle intersects with the rapid assessment described here in several ways. First, given that species' conservation status was a major component of the species ranking (Legge *et al.* 2022), imperilled species that are not formally listed may have been overlooked. Second, a lack of survey and monitoring effort means that the distributional limits of many species remain uncertain, complicating the calculation of overlap measures. Third, a lack of investment in monitoring means we lack baseline data on the status of populations leading up to and following the 2019–20 wildfires. This is further compounded by a lack of studies on squamate responses to fire, particularly concerning aspects of the fire regime beyond the most recent fire event (outlined above). Hence, future investment in monitoring Australia's squamate fauna is fundamental to their conservation, particularly given the increasing frequency of large-scale disturbance events. There is also a need to understand the effectiveness of post-fire interventions aimed at limiting decline or hastening recovery.

In many instances, it may be most effective to address co-occurring threats (Box 14.2). With these threats removed, species may be more resilient to large wildfires.

Conclusion

This chapter has considered the impacts of the unprecedented 2019–20 wildfires on Australia's unique squamate fauna. While the wildfires burnt the habitats of hundreds of squamate species, and over 60 species have ranges that overlapped with the fires by > 10%, our systematic review revealed how little research has been undertaken that can help us predict the outcomes for squamate populations on the ground. This knowledge gap was filled using structured expert elicitation, but empirical data are needed; this will involve a step change in investment in squamate monitoring, research and conservation. Management interventions that will aid in the recovery of squamates will involve addressing key threats to squamates, including invasive predators, feral herbivores, unsustainable development, land clearing, and, notably, inappropriate fire regimes.

We have identified several obstacles that need to be addressed to improve upon further rapid assessments in the wake of future megafires (see 'Recommendations').

Recommendations

- Research that documents the immediate and indirect effects of fire on Australian squamate reptiles, including multiple aspects of the fire regime, is urgently needed. This will require funding for new projects and better use of existing data.
- Broad-scale, population-level genetic sampling to clarify taxonomic uncertainty and within-species genetic diversity in Australian reptiles, to facilitate more accurate assessments of fire impacts and improve success of recovery efforts, such as translocations, is needed.
- Substantial investment in statistically robust and powerful monitoring for lizard and snake populations is needed to document trajectories and to assess how species are affected by fire events and regimes, as well as other potential threats.
- Systematic curation of vetted geographical records of species observations is required to refine understanding of the distributions of Australian reptile species, including the development of species distribution models where possible.
- Drawing upon enhanced monitoring, species assessments and conservation advice must be current for Australian reptile species, to help break the cycle of underfunding reptile conservation.
- Further investment in the conservation of Australia's threatened reptile species, including funding for conservation interventions and threat abatement, is needed.

Acknowledgments

We would like to thank Sarah Legge and John Woinarski. We also thank the Department of Agriculture, Water and Environment and especially the Geospatial and Information Analytics (GAIA) Branch, which undertook the initial calculations of fire overlap. We thank Libby Rumpff and Hayley Geyle for playing key roles in the expert elicitation work, and the many experts that contributed to the rapid assessment and expert elicitation. We thank Jules Farquhar and Brendan Schembri for permission to use their photos.

References

Bowman D, Williamson G, Yebra M, Lizundia-Loiola J, Pettinari ML, *et al.* (2020) Wildfires: Australia needs a national monitoring agency. *Nature* **584**, 188–191. doi:10.1038/d41586-020-02306-4

Catullo RA, Moritz CC (2020) Project 8.3.3: Genetic assessment of bushfire-impacted vertebrate species. National Environmental Science Program Threatened Species Recovery Hub, Brisbane, <https://www.nespthreatenedspecies.edu.au/projects/genetic-assessment-of-priority-taxa-and-management-priorities>.

Catullo RA, Schembri R, Tedeschi LG, Eldridge MDB, Joseph L, *et al.* (2021) Benchmarking taxonomic and genetic diversity after the fact: Lessons learned from the catastrophic 2019–2020 Australian bushfires. *Frontiers in Ecology and Evolution* **9**, 645820. doi:10.3389/fevo.2021.645820

Chapple D, Tingley R, Mitchell N, Macdonald S, Keogh JS, *et al.* (2019) *The Action Plan for Australian Lizards and Snakes 2017.* CSIRO Publishing, Melbourne.

DAWE (2020a) National indicative aggregated fire extent dataset. Department of Agriculture, Water and the Environment, Canberra, <http://www.environment.gov.au/fed/catalog/search/resource/details.page?uuid=%7B9ACDCB09-0364-4FE8-9459-2A56C792C743%7D>

DAWE (2020b) Australian Google Earth Engine burnt area map. A rapid, national approach to fire severity mapping. Department of Agriculture, Water and the Environment, Commonwealth of Australia, Canberra, <https://www.environment.gov.au/system/files/pages/a8d10ce5-6a49-4fc2-b94d-575d6d11c547/files/ageebam.pdf>

Dixon KM, Cary GJ, Worboys GL, Gibbons P (2018) The disproportionate importance of long-unburned forests and woodlands for reptiles. *Ecology and Evolution* **8**, 10952–10963. doi:10.1002/ece3.4561

Durbian FE (2006) Effects of mowing and summer burning on the massasauga (*Sistrurus catenatus*). *American Midland Naturalist* **155**, 329–334. doi:10.1674/0003-0031(2006)155[329:EOMASB]2.0.CO;2

Fenner AL, Bull CM (2007) Short-term impact of grassland fire on the endangered pygmy blue-tongue lizard. *Journal of Zoology* **272**, 444–450. doi:10.1111/j.1469-7998.2007.00287.x

Geyle HM, Tingley R, Amey AP, Cogger H, Couper PJ, *et al.* (2021) Reptiles on the brink: identifying the Australian terrestrial snake and lizard species most at risk of extinction. *Pacific Conservation Biology* **27**, 3–12. doi:10.1071/PC20033

Gill AM (1975) Fire and the Australian flora: a review. *Australian Forestry* **38**, 4–25. doi:10.1080/00049158.1975.10675618

Griffiths AD, Christian KA (1996) The effects of fire on the frillneck lizard (*Chlamydosaurus kingii*) in northern Australia. *Australian Journal of Ecology* **21**, 386–398. doi:10.1111/j.1442-9993.1996.tb00625.x

Hellgren EC, Burrow AL, Kazmaier RT, Ruthven DC (2010) The effects of winter burning and grazing on resources and survival of Texas horned lizards in a thornscrub ecosystem. *The Journal of Wildlife Management* **74**, 300–309. doi:10.2193/2009-090

Hemming V, Burgman MA, Hanea AM, McBride MF, Wintle BC (2018) A practical guide to structured expert elicitation using the IDEA protocol. *Methods in Ecology and Evolution* **9**, 169–180. doi:10.1111/2041-210X.12857

IUCN (2021) The IUCN Red List of Threatened Species. Version 2021–2, <https://www.iucnredlist.org>.

Jolly CJ, von Takach B, Webb JK (2021) Slow life history leaves endangered snake vulnerable to illegal collecting. *Scientific Reports* **11**, 5380. doi:10.1038/s41598-021-84745-1

Jolly CJ, Dickman CR, Doherty TS, van Eeden LM, Geary WL, *et al.* (2022) Animal mortality during fire. *Global Change Biology* **28**, 2053–2065. doi:10.1111/gcb.16044

Kelly LT, Giljohann KM, Duane A, Aquilué N, Archibald S, *et al.* (2020) Fire and biodiversity in the Anthropocene. *Science* **370**, eabb0355. doi:10.1126/science.abb0355

Legge S, Woinarski JCZ, Garnett ST, Nimmo DG, Scheele BC, *et al.* (2020) 'Rapid analysis of impacts of the 2019–20 fires on animal species, and prioritisation of species for management response. Report prepared for the Wildlife and Threatened Species Bushfire Recovery Expert Panel, 9 February'. Department of Agriculture, Water and the Environment, Canberra.

Legge S, Woinarski JCZ, Scheele BC, Garnett ST, Lintermans M, *et al.* (2022) Rapid assessment of the biodiversity impacts of the 2019–2020 Australian megafires to guide urgent management intervention and recovery and lessons for other regions. *Diversity & Distributions* **28**, 571–591. doi:10.1111/ddi.13428

Legge S, Rumpff L, Woinarski JCZ, Whiterod NS, Ward M, *et al.* (in press) The conservation impacts of ecological disturbance: Time-bound estimates of population loss and recovery for fauna affected by the 2019–2020 Australian megafires. *Global Ecology and Biogeography* doi:10.1111/geb.13473

Murphy BP, Bradstock RA, Boer MM, Carter J, Cary GJ, *et al.* (2013) Fire regimes of Australia: a pyrogeographic model system. *Journal of Biogeography* **40**, 1048–1058. doi:10.1111/jbi.12065

Pepper M, Sumner J, Brennan IG, Hodges K, Lemmon AR, *et al.* (2018) Speciation in the mountains and dispersal by rivers: molecular phylogeny of *Eulamprus* water skinks and the biogeography of Eastern Australia. *Journal of Biogeography* **45**, 2040–2052. doi:10.1111/jbi.13385

Pierson JC, Coates DJ, Oostermeijer JGB, Beissinger SR, Bragg JG, *et al.* (2016) Genetic factors in threatened species recovery plans on three continents. *Frontiers in Ecology and the Environment* **14**, 433–440. doi:10.1002/fee.1323

Rosauer DF, Catullo RA, VanDerWal J, Moussalli A, Moritz C (2015) Lineage range estimation method reveals fine-scale endemism linked to Pleistocene stability in Australian rainforest herpetofauna. *PLoS One* **10**, e0126274. doi:10.1371/journal.pone.0126274

Scheele BC, Legge S, Blanchard W, Garnett S, Geyle H, *et al.* (2019) Continental-scale assessment reveals inadequate monitoring for threatened vertebrates in a megadiverse country. *Biological Conservation* **235**, 273–278. doi:10.1016/j.biocon.2019.04.023

Shine R, Webb JK, Fitzgerald M, Sumner J (1998) The impact of bush-rock removal on an endangered snake species, *Hoplocephalus bungaroides* (Serpentes: Elapidae). *Wildlife Research* **25**, 285–295. doi:10.1071/WR97022

Tingley R, Macdonald SL, Mitchell NJ, Woinarski JC, Meiri S, *et al.* (2019) Geographic and taxonomic patterns of extinction risk in Australian squamates. *Biological Conservation* **238**, 108203. doi:10.1016/j.biocon.2019.108203

Uetz P, Freed P, Aguilar R, Hošek J (2021) The reptile database, <http://www.reptile-database.org>.

Webb JK, Shine R (1998) Ecological characteristics of a threatened snake species, *Hoplocephalus bungaroides* (Serpentes, Elapidae). *Animal Conservation* **1**, 185–193. doi:10.1111/j.1469-1795.1998.tb00028.x

Webb JK, Brook BW, Shine R (2002) Collectors endanger Australia's most threatened snake, the broad-headed snake *Hoplocephalus bungaroides*. *Oryx* **36**, 170. doi:10.1017/S0030605302000248

Webb JK, Jolly CJ, Hinds M, Adams C, Cuartas-Villa S, *et al.* (2021) Effects of the Australian 2019–2020 megafires on a population of endangered broad-headed snakes *Hoplocephalus bungaroides*. *Austral Ecology* aec.13118. doi:10.1111/aec.13118

Woinarski JCZ (2018) The extent and adequacy of monitoring for Australian threatened reptile species. In *Monitoring Threatened Species and Ecological Communities* (Eds SM Legge, DB Lindenmayer, NM Robinson, BC Scheele, DM Southwell and B Wintle) pp. 69–84. CSIRO Publishing, Melbourne.

Woinarski JCZ, Braby MF, Burbidge AA, Coates D, Garnett ST, *et al.* (2019) Reading the black book: the number, timing, distribution and causes of listed extinctions in Australia. *Biological Conservation* **239**, 108261. doi:10.1016/j.biocon.2019.108261

15

The impacts of the 2019–20 wildfires on Australian birds

Stephen T. Garnett, Michelle J. Ensbey, Joshua Lee, Rebecca L. Boulton, Jody A. Gates, Karleah K. Berris, Damon L. Oliver, George Malolakis, David McCreery, David Bain and Sarah M. Legge

Summary

Conservation context and challenges

- There is little information on the fate of individual birds in and following fire, but those of more mobile species may have fled, with reports of forest species in suburbs and other unusual habitats in the days and months after the fire.
- Many also died, with burnt birds as large as eagles washed up on beaches. Those that survived lost food and shelter. There is also likely to have been a net loss of tree hollows used for nesting by many species; replacement will take a long time.
- Studies to date suggest that recolonisation is rapid as habitat recovers its structure, but recovery may take years and possibly decades for species associated with rainforest and other wetter habitats.
- Active management is expected to help a few species to recover, particularly through nest box erection.

Main findings

- The ranges of 180 million individual birds from more than 350 terrestrial bird taxa intersected with the 2019–20 megafires.
- The 2019–20 wildfires burnt > 25% of the range of 51 bird taxa, much of it severely; the most extreme overlap (68% of range burnt) was for the Kangaroo Island southern emu-wren (*Stipiturus malachurus halmaturinus*).
- Of the most fire-affected birds, eight were meeting the IUCN Red List criteria as threatened or Near Threatened before the fire. Another 19 were assessed as meeting the criteria directly because of the fires.
- The IUCN Red Index for all Australian birds declined by over 9% as a direct result of the fires.

- Bird communities in burnt rainforests were at lower densities after the fires and were recovering more slowly than those in adjacent burnt dry sclerophyll communities.
- Populations of the glossy black-cockatoo (*Calyptorhynchus lathami*) did better than expected in the year after fire, though may face future food shortages.

Introduction

The 2019–20 wildfires were exceptional in their extent and severity (Chapter 2). Moreover, they represent the most extreme manifestation, to date, of changes in fire severity and size that have been occurring for several decades. Ferocious fires were anticipated by research published before the fires occurred (Di Virgilio *et al.* 2019) and may be representative of a new fire regime being imposed by human-induced climate change (van Oldenborgh *et al.* 2021). Understanding the responses of birds to these exceptional fires will therefore help focus attention on those species, habitats and processes that will need greatest attention from conservation managers in coming decades, both before and after fires (Box 15.1).

For the general public, there was evidence that many birds were killed by the fire – they were found on beaches where the tidal wrack was studded with the coloured corpses of land birds that died fleeing over the sea (Fig. 15.1; Butt and McManus 2020). While some eastern bristlebirds (*Dasyornis brachypterus*) were taken into captivity from the last population known in Victoria as the fire approached (Selwood *et al.* 2022), people were mostly helpless bystanders as the fires overwhelmed attempts to slow their advance. Unlike with some mammal species (Chapter 27), few injured birds were rescued for rehabilitation after the fires and there were only very limited local efforts to provision birds with food or water (Sharrod 2020).

Box 15.1. Birds in post-fire landscapes

Bird occupation of post-fire landscapes is contingent on the functional traits of birds, the recovery capacity of affected ecosystems and the time since fire. In the first 4 months after the 2019–20 wildfires, a continent-wide investigation into bird occurrence using citizen science data concluded that 22 species had decreased, 30 species increased, and 24 species occurred at similar frequencies, although this may have been confounded by changes in detectability and dispersion following the fires (Lee *et al.* in press). Birds with smaller body sizes, smaller range sizes and specialised diets were less common after the fires. In ecosystems with little previous exposure to fire such as rainforests, there may be a much slower recovery. Surveys in Gondwanan rainforests that burnt in the 2019–20 wildfires showed that bird communities were less diverse in burnt rainforests than unburnt rainforests a year after fire (Lee *et al.* 2022). Frugivores, folivores and insectivores were less abundant in burnt rainforests, raising concerns for fire-affected rainforests that rely on seed dispersal via avian frugivores for recovery. Birds with traits and life histories that are ill-suited to post-fire recolonisation and species that are dependent on rarely burnt ecosystems may require more conservation attention in the face of a growing threat of extreme fire events.

Fig. 15.1. Burnt birds, here a rainbow lorikeet (*Trichoglossus haematodus*), washed up on Mallacoota beaches after the 2019–20 megafires. (Photo: Justin McManus/*The Age*)

Fire is not infrequent in the habitat of many of the bird species affected by the recent fires (Clarke 2020). Even rainforest, usually a fire refuge, sometimes burns, judging from the presence of emergent eucalypts that can only have recruited after fires in the distant past (Bowman *et al.* 2013). For heathland, and the birds that depend on it, fires can prevent a transition to more woody habitats (Loyn 1997). However, despite a capacity among Australian birds to 'cope' with fire, the 2019–20 fires were exceptional, and we need to understand their impacts (Clarke 2020) in order to predict whether the inevitable fires of the near future will lead to long-term declines or if bird populations are sufficiently resilient to persist even if the fire regime changes.

In this chapter, we summarise the impacts of the 2019–20 fires on birds. To do so we build on information from earlier studies of the responses of birds to fire in southern Australia, present results from expert elicitation and preliminary findings since the fires, and summarise the impact of the recent fires on extinction risk. We conclude with recommendations on the management likely to benefit birds in the face of the increasingly hostile fire environment anticipated as the climate heats up (Herold *et al.* 2018) and droughts intensify (Evans *et al.* 2017).

Key findings

Overviews of the impacts of the 2019–20 wildfires on Australian birds were derived from three sources. First van Eeden *et al.* (2020) (see also Chapter 12) estimated that ~180 million birds were present in burnt areas at the time of the fires, the density estimates being derived from over 100 000 standardised surveys extracted from BirdLife Australia's Birdata database and stratified by vegetation type and bioregion. The proportion of these birds that died is unknown.

Initial intersections of distribution maps of bird taxa with maps of the areas burnt in the 2019–20 wildfires identified 378 bird taxa as having been affected (Ward *et al.* 2020). Subsequently, Legge *et al.* (in press) refined and extended this analysis, identifying 51 taxa with > 25% of their range burnt by the fires, including all 16 taxa endemic to Kangaroo Island. The taxon with the highest proportion burnt was the Kangaroo Island southern emu-wren, of which 68% of the modelled distribution was burnt.

The third measure of the fires' impact was on the IUCN Red List status (IUCN 2012) of Australian birds. In August 2020, 6 months after the fires, the status of all 1285 Australian bird taxa was assessed using the IUCN Red List criteria (Garnett and Baker 2021). Of 51 bird taxa with ≥ 25% of their ranges burnt, eight had met the criteria as threatened or Near Threatened in 2011 (Garnett *et al.* 2011). The fires resulted in all but two of these meeting the criteria for listing at a higher threat level in 2020 (Kangaroo Island glossy black-cockatoo (*Calyptorhynchus lathami halmaturinus*) and western ground parrot (*Pezoporus wallicus flaviventris*) were already highly threatened). An additional 19 taxa were deemed to meet the IUCN Red List criteria as threatened in 2020 as a result of the fire. Twelve birds with 5–24% of their habitat burnt also met the IUCN Red List criteria as threatened or Near Threatened in 2020 (Garnett and Baker 2021), but for none of them did the fires contribute significantly to their extinction risk.

The Red List status can also be used to calculate changes to the Red List Index (RLI; Butchart *et al.* 2007), a measure of the conservation status and its change across a group of species. The RLI (which ranges from 1 if all taxa are Least Concern to 0 if all taxa are Extinct) for all Australian birds was 0.894 at the end of 2020. Had the fires in 2019–20 not occurred the RLI would have been 0.902, meaning the fires caused a 7.2% decline in the national Red List Index based on an analysis of the 973 taxa living in terrestrial habitats (so excluding seabirds, shorebirds and mangrove taxa). For the 51 taxa with ≥ 25% of their ranges burnt, the RLI fell by over 21.2% from 2001–2020, from 0.957 to 0.753 (Fig. 15.2). In

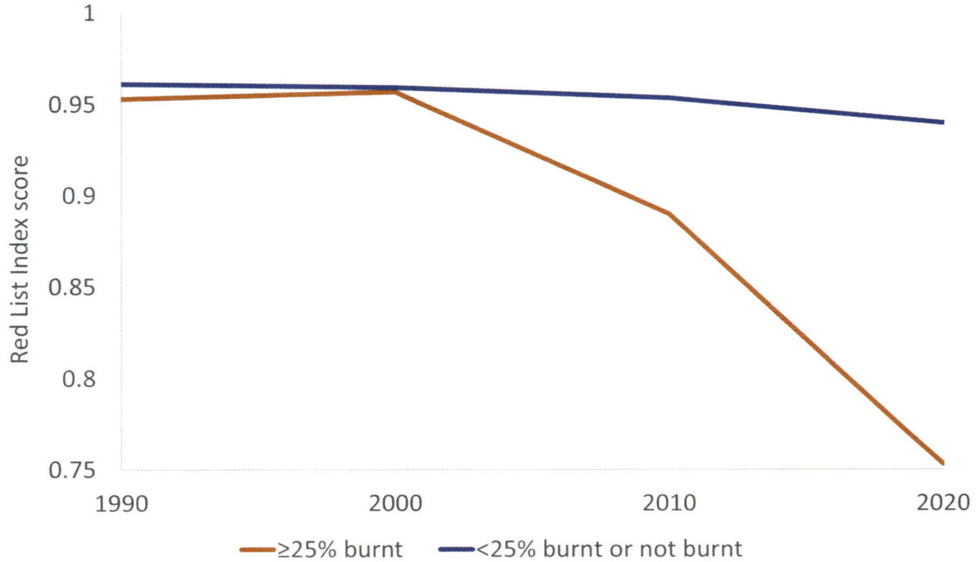

Fig. 15.2. Red List Index for Australian bird taxa from terrestrial habitat for which ≥ 25% of the habitat burnt in 2019–20 (51 taxa) and all other taxa (931 taxa).

contrast, the RLI for the remaining 922 Australian bird taxa living in terrestrial habitats fell by just 1.2% to 0.940 over the same period.

For the 26 taxa that moved to a higher extinction risk category because of the fires (Garnett and Baker 2021), the numbers killed by fire can be estimated. Based on estimates of population size of each taxon before the fires derived from range and density measures, and of mortality during them based on fire extent and severity, an estimated 1.53 million individuals of these taxa died, ~45% of their combined populations before the fires occurred (Table 15.1). Expert elicitation was also used to estimate the mortality immediately after the fire for 12 of the same taxa (Legge *et al.* in press). Assuming the same starting populations, the experts estimated 160 600 died, compared with 164 600 when calculated from Garnett and Baker (2021) for the same 12 taxa. The experts also considered that, across a wider group of 24 birds of concern, declines would continue from immediately after fire (18.5%) to 1 year after fire (24.1%), with only a slight recovery to 17.6% below pre-fire population a decade (or three generations) after the fire (Legge *et al.* in press).

Table 15.1. Estimated mortality of birds for which the IUCN Red List status was affected by the 2019–20 fires (status and data from Garnett and Baker (2021) unless indicated).

Taxon	Status post-fire	No. estimated to have died	% decline of estimated pre-fire population
Kangaroo Island glossy black-cockatoo (*Calyptorhynchus lathami halmaturinus*)*	Endangered	25	9.1
South-eastern glossy black-cockatoo (*Calyptorhynchus l. lathami*)	Vulnerable	400	5.1
Gang-gang cockatoo (*Callocephalon fimbriatum*)	Vulnerable	3000	10.7
Kangaroo Island crimson rosella (*Platycercus elegans melanopterus*)	Vulnerable	104 000	34.6
Western ground parrot (*Pezoporus wallicus flaviventris*)	Critically Endangered	40	28.6
Eastern ground parrot (*Pezoporus w. wallicus*)	Near Threatened	6700	15.2
Southern rufous scrub-bird (*Atrichornis rufescens ferrieri*)	Endangered	1500	28.3
Northern rufous scrub-bird (*Atrichornis r. rufescens*)	Endangered	500	15.9
Kangaroo Island superb fairy-wren (*Malurus cyaneus ashbyi*)	Endangered	269 000	53.1
Kangaroo Island southern emu-wren (*Stipiturus malachurus halmaturinus*)	Endangered	73 000	59.3
Kangaroo Island New Holland honeyeater (*Phylidonyris novaehollandiae campbelli*)	Vulnerable	334 000	46.0
Kangaroo Island brown-headed honeyeater (*Melithreptus brevirostris magnirostris*)	Endangered	47 000	48.2
Kangaroo Island white-eared honeyeater (*Nesoptilotis leucotis thomasi*)	Endangered	27 000	54.0

Taxon	Status post-fire	No. estimated to have died	% decline of estimated pre-fire population
Kangaroo Island little wattlebird (*Anthochaera chrysoptera halmaturina*)	Vulnerable	9500	47.5
Kangaroo Island red wattlebird (*Anthochaera carunculata clelandi*	Vulnerable	48 000	38.8
Kangaroo Island purple-gaped honeyeater (*Lichenostomus cratitius cratitius*)	Vulnerable	62 000	35.9
Lowland pilotbird (*Pycnoptilus floccosus sandlandi*)	Vulnerable	77 000	23.2
Upland pilotbird (*Pycnoptilus f. floccosus*)	Vulnerable	11 000	25.2
Kangaroo Island shy heathwren (*Calamanthus cautus halmaturina*)	Vulnerable	2000	47.9
Southern yellow-throated scrubwren (*Sericornis citreogularis citreogularis*)	Near Threatened	160 000	20.8
Kangaroo Island spotted scrubwren (*Sericornis frontalis ashbyi*)	Endangered	187 000	51.9
Kangaroo Island striated thornbill (*Acanthiza lineata whitei*)	Vulnerable	120 000	47.0
Kangaroo Island brown thornbill (*Acanthiza pusilla zietzi*)	Endangered	124 000	65.1
Kangaroo Island western whipbird (*Psophodes nigrogularis lashmari*)	Endangered	13 000	51.6
Kangaroo Island grey currawong (*Strepera versicolor halmaturina*)	Vulnerable	27 000	41.3
Western Bassian thrush (*Zoothera lunulata halmaturina*)	Endangered	4100	58.7
Total		**1 530 000**	**45.1**

* Estimated loss from Southern Rivers flock (Berris *et al.* 2020)

Finally, we assessed the current threat from fire, relative to other threats, to Australia's birds. Using the approach of Garnett *et al.* (2019) on the data assembled in Garnett and Baker (2021), an increase in the frequency, scale or intensity of fire constitutes a threat to 88 of all 221 threatened and Near Threatened Australian bird taxa. This fire threat makes up 15.6% of the total threat burden to Australian birds, sitting between the closely associated 'Increased frequency or length of droughts' (16.3%) and 'Rising temperatures and heat waves' (12.3%) as the most important threats to Australian avifauna.

For some taxa, such as the regent honeyeater (*Anthochaera phrygia*), losses may have been underestimated because their distribution is so poorly known (Crates *et al.* 2022). For others, better than expected survival surprised observers (see the case studies for Kangaroo Island glossy black-cockatoo and western whipbird (*Psophodes nigrogularis lashmari*) (Box 15.2), eastern ground parrot (*Pezoporus wallicus wallicus*) and eastern bristlebird (Box 15.3)). There are two lessons apparent from such studies. One is that monitoring needs to occur for at least several years after fires to place the immediate response post-fire in context. For example, the glossy black-cockatoos on Kangaroo

Island survived fire far better than expected, breeding successfully even in badly burnt areas in the first few months post-fire. However, a more recent survey (November 2021) suggests that the inexperienced young birds may not have survived well and the population may have declined. The second lesson is that each species responds to fire differently. Thus, the bristlebirds, whipbirds and ground parrots that survived the fire in unburnt refuges were able to make use of regenerating vegetation within 12 months post-fire, whereas southern emu-wren (*Stipiturus malachurus halmaturinus*), white-eared honeyeater (*Nesoptilotis leucotis thomasi*) and Bassian thrush (*Zoothera lunulata halmaturina*) still had populations well below pre-fire estimates after the same period. In all areas, drought-breaking rains soon after the fires may have had a major influence on the rate of recovery.

Box 15.2. Response of Kangaroo Island birds to catastrophic fire

Woodland bird community

The majority of bird taxa severely affected by the 2019–20 wildfire were endemic taxa from Kangaroo Island. On 3 January 2020, almost the entire area of the island's largest remaining wooded habitat, in Flinders Chase, was incinerated with a blaze of unprecedented ferocity. Most birds in the fire ground probably died. However, surveys immediately after the fire found survivors in tiny remnant patches that had somehow escaped being burnt, including highly fire-sensitive species like the western whipbird and southern emu-wren (Boulton *et al.* 2020). Six surveys since have all confirmed the importance of these unburnt patches. Eighteen months after the fire, whipbirds were defending territories over 2 km from the nearest unburnt vegetation and shy heathwrens (*Calamanthus cautus halmaturina*) had become widespread. However, the recovering vegetation did not yet have enough complexity for emu-wrens, which persisted only in unburnt vegetation, and some species that prefer taller, moister vegetation were also still very scarce (Boulton and Gates 2021). The lessons are that recovery can be surprisingly fast, but that there are major differences among species depending on how quickly habitat structure recovers.

Glossy black-cockatoos

The fires that incinerated most of western Kangaroo Island in January 2020, threatened to wipe out decades of conservation effort for glossy black-cockatoo (Fig. 15.3), including protecting nests from possums, providing artificial nest boxes and plantings of the species' sole food tree, the drooping she-oak (*Allocasuarina verticillata*). This effort had helped to bring the population from 150 birds in 1995 to more than 350 in 2016 (Berris *et al.* 2018).

Spatial analysis found that 54% of the drooping she-oak feeding habitat had been burnt. Nevertheless, 59% of nests the following winter breeding season were successful, higher than the long-term average, although fewer nests were recorded overall in areas that lost significant amounts of feeding habitat. A thorough census in September 2020 found over 450 individuals, many of them young of the year. Just one of the island's flocks was smaller than before the fires, by ~25 birds. However, the 2021 census recorded only 377 birds, and in some areas almost all unburnt drooping she-oak trees had been heavily foraged upon by the birds indicating that food may now be limiting. The burnt drooping she-oak woodlands could take 15 years or longer to reach

Fig. 15.3. The Kangaroo Island glossy black-cockatoo. (Photo: Karleah Berris, Kangaroo Island Landscape Board)

maturity and resume production of seed. Therefore, although direct mortality from the fires was lower than feared, indirect mortality after the fires may be occurring over several years if post-fire food shortages have reduced survival or constrained breeding (Berris and Welz 2021).

Box 15.3. Post-fire recovery of eastern bristlebirds and ground parrots at Nadgee Nature Reserve

After almost 40 years without fire, more than 90% of the heaths of Nadgee Nature Reserve in coastal southern New South Wales burnt at extreme intensity on 28 February 2020. Ten weeks after the fire, the ground cover and shrub layer were recovering rapidly and two bird species that have been the focus of management, the eastern bristlebird and ground parrot, were detected (Fig. 15.4; Oliver and Malolakis 2020). By September 2020 at least 36 eastern bristlebirds were found in small semi-burnt patches of heath and heathy woodland and drainage lines along all four heath transect sites monitored regularly, but only a few ground parrots were detected, at a single site, compared to the usual 30–50 parrots detected across four survey points. Six months later, in March 2021, the grasses and sedges on which ground parrots feed were

Fig. 15.4. Coastal heath, Nadgee Nature Reserve, 12 months after the fire had recovered to pre-fire levels. (Photo: Damon Oliver/DPIE)

flowering and seeding freely. In this survey, 25 parrots were detected at the four sites, an abundance close to the long-term average from 2004–2019 (Oliver and Malolakis 2020).

Broad context for impact and recovery

Much is already known about the response of Australian forest bird communities to fire (Woinarski and Recher 1997), with fire frequency having the most important influence on responses (Bradstock 2008). Several studies show that a regime of control burns produces little change in avifauna composition (Wooller and Calver 1988; Smith and Smith 2017; Kuchinke *et al.* 2020; Wills *et al.* 2020). However, competitive and other relationships among species can change after fire (Rainsford *et al.* 2021) and increasingly rare, long-unburnt landscapes appear to be favoured by many species that are also disadvantaged by anthropogenic processes like habitat fragmentation (Davis *et al.* 2016; Wills *et al.* 2020). Also, even single fires can be detrimental to bird communities for extended periods. Although birds can fly, heat-related mortality is still high during severe fires (Pescott 1983; Geary *et al.* 2020; McKechnie *et al.* 2021), especially for less mobile species. The loss of nests, hollows and other fire susceptible habitat can be detrimental to the recovery or recolonisation of bird species after large fires, a process hindered further by logging or habitat fragmentation (Geary *et al.* 2020; Lindenmayer and Sato 2018; Lindenmayer *et al.* 2021; Chapter 19). In mountain ash (*Eucalyptus regnans*) forest the abundance of only one species, the flame robin (*Petroica phoenicea*), was found to have recovered 10 years after the 2009 fires, with significant declines of bird species associated with tree cavities in combination with logging (Lindenmayer and Sato 2018; Lindenmayer *et al.* 2021). Other resources provided by large old trees, such as hanging bark, large-scale flowering in the canopy and undergrowth, were also affected. Diet and habitat specialisation, size and mobility all affect the ability of birds to survive in the post-fire environment, as does loss of resources in general (Loyn 1997; Lindenmayer *et al.* 2008; Franklin *et al.* 2021). Birds with a plant-based diet, such as seed and nectar, suffer the greatest initial declines (Loyn 1997) while distance from unburnt refuges can influence rate of recolonisation (Lindenmayer *et al.* 2008; Watson *et al.* 2012; Murphy *et al.* 2021). Such refuges can be critical for persistence of some species in a landscape, generally harbouring higher abundances of birds even several years after a fire and probably being the source of most recolonisation of surrounding burnt habitat (Robinson *et al.* 2014; Murphy *et al.* 2021). Such refuges are likely to come under increasing pressure as the climate changes.

There are few studies of the responses of marked birds to fire. One study reported significant population changes for 13 of 20 studied species from before to 3 years after a high-severity fire in forest near Canberra (Baker *et al.* 1997). Even relatively mild fires may kill birds either directly or after the fire as a result of food shortages. For example, only 22% of birds banded before a mild fire in a dry sclerophyll forest in south-western Australia were re-trapped over the 3 years after fire (Wooller and Calver 1988). However, none of these studies determined how individuals survive fires. Instead, accounts tend to be anecdotal, like the record of up to 20 superb lyrebirds (*Menura novaehollandiae*) sheltering from the 2019 fire beside a farm dam (Fig. 15.5; Millington 2020).

Greater understanding of individual responses to fire will improve management responses, which are currently highly limited. Expert elicitation attempting to determine the relative benefits likely to accrue after the fires from current management suggests that, for most bird taxa, little can currently be achieved with existing knowledge (Legge *et al.*, unpublished data; Fig. 15.6). Much of the management is not directed specifically at reducing the risk from fires. For example, provision of nest boxes for Kangaroo Island glossy black-cockatoos is a response to a general lack of hollows on the island and also benefits the birds after fires (Berris *et al.* 2020). Similarly, decades of research and management have enabled re-establishment of eastern bristlebirds in many heaths in New South Wales from which the species had been extirpated. This reduces risk from wildfire by reducing the probability that all populations will be burnt simultaneously. However, establishing numerous subpopulations is also a sound strategy for reducing many forms of risk, not just fire, and has enabled down-listing of the taxon despite the fire risk (Bain *et al.* 2021).

Fig. 15.5. Superb lyrebirds, usually solitary, gather around a dam as fire rages in the background in a valley near Wollombi, New South Wales. (Photo: P. J. Wallis)

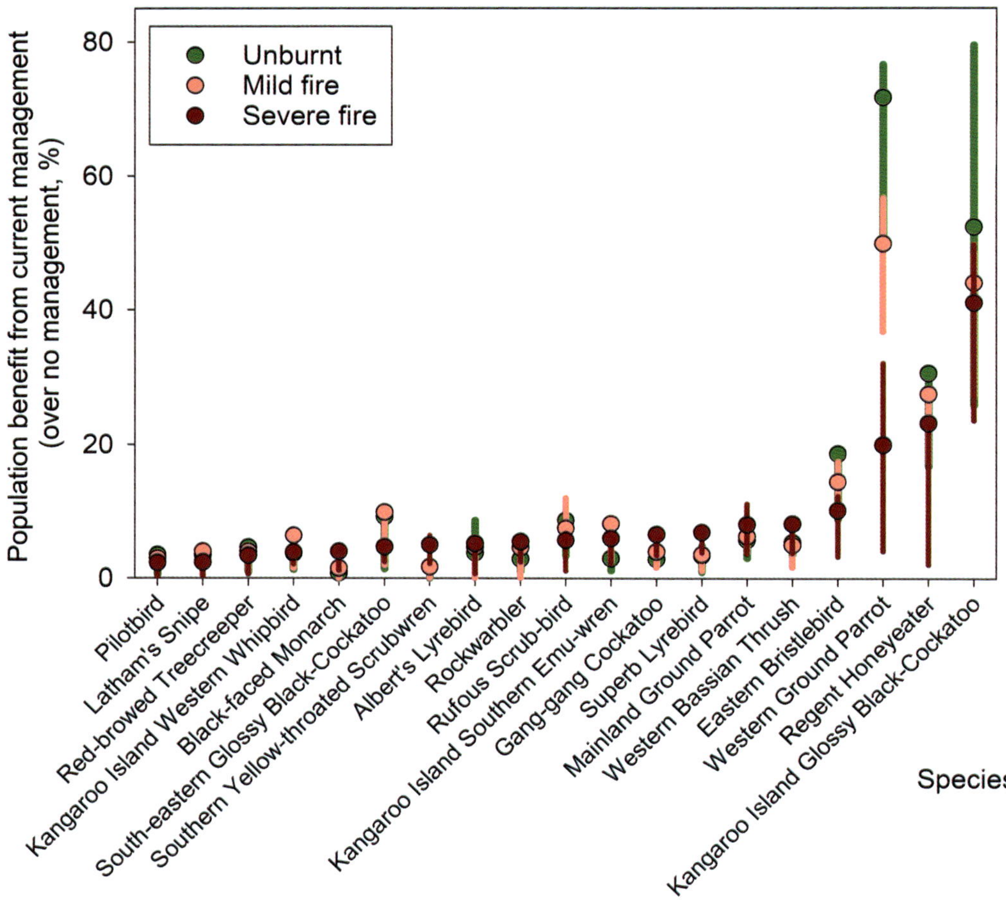

Fig. 15.6. Expert-elicited estimates of the benefits of current management, after mild and severe fire, and if there are no fires, for a selection of Australian bird taxa affected by the 2019–20 fires. The population benefit is expressed as the percentage increase in the population with current management in place, compared to a scenario without that management (Legge *et al.*, unpublished data).

Conclusion

The intense fires of 2019–20 undoubtedly killed many birds. Immediately after the fire greatest concern was for species from low, dense habitats, like eastern bristlebirds or western whipbirds, but most have returned to burnt areas as ground cover has regenerated. The tiny population of Kangaroo Island glossy black-cockatoos also attracted attention. It initially fared better than expected, but 21 months post-fire its population has declined and there is evidence of food shortages in burnt areas, so the potential for ongoing declines over the next few years remains a concern. Of greater concern are the birds of rainforests and other damp, dense habitats as these burn very infrequently, so can take much longer to recover the vegetative structure required by habitat specialists (although see Loyn 1997). Where hollows limit nesting opportunities, post-fire provision of nest boxes may accelerate recovery, although this is expensive, localised and may be species or location specific. Otherwise, management needs to concentrate on improving landscape fire management and fire suppression, including responding more rapidly and effectively to future fires so they are neither as intense nor widespread. After extensive fire, protection of long-unburnt habitat becomes imperative as these patches become increasingly scarce in the landscape. Such places should be protected not just from wildfire but also from prescribed burning.

Recommendations

Fundamental to protection of Australian birds from extreme drought and vast uncontrollable fires will be mitigation of climate change. Without global cooperation to limit greenhouse gas emissions, the task of fire and biodiversity managers will become increasingly difficult.

In the meantime, research and management can help moderate the impacts of extensive fires on birds in these ways:

- Increase the amount and quality of monitoring to assess how birds respond to major disturbances and other threats, such as fire and changes to fire regime.
- Study how birds survive fires of different intensities using new technologies for tracking survival and mobility so landscape elements that promote survival can be conserved or provided.
- Continue existing management that reduces extinction risk, including risk from fire – the larger the population before a fire, and the more places in a landscape that are occupied, the greater the probability of persistence. Strategies can include intensive management (e.g. nest boxes), assisted migration and habitat restoration.
- Prepare contingency plans for responses to fire, identifying particularly those actions that need to be implemented quickly even if the impacts are not immediate (e.g. planting or protecting casuarinas that may not provide seed in under a decade).
- Identify long-unburnt parts of the landscape and develop and implement plans to protect them, with priority especially for the protection of large contiguous areas of long-unburnt vegetation.
- Ensure the value of important threatened bird habitat, particularly where fragmented, is recognised alongside the values for people and property.
- Plan for climate change (e.g. Molloy *et al.* 2020). Do not assume that historical frequencies of drought and fire will continue. Use modelling to estimate what is likely to occur within a feasible planning frame and adapt fire management strategies accordingly.

Acknowledgements

We are grateful to the many people who have contributed to surveys of the birds after the fires and to efforts to restore habitats and their bird populations.

References

Bain D, Clarke RH, Oliver D, Bramwell MD, MacGregor C, *et al.* (2021) Southern eastern bristlebird *Dasyornis brachypterus brachypterus*. In *The Action Plan for Australian Birds 2020*. (Eds ST Garnett and GB Baker) pp. 584–587. CSIRO Publishing, Melbourne.

Baker G, Dettmann B, Wilson S (1997) Fire and its impact on avian population dynamics. *Pacific Conservation Biology* **3**, 206–212. doi:10.1071/PC970206

Berris KK, Welz T (2021) 'Kangaroo Island Glossy Black-Cockatoo census report'. Kangaroo Island Landscape Board. KILB Publication Number KILB-2021–11–R02. Kangaroo Island Landscape Board, Kingscote.

Berris K, Barth M, Mooney T, Paton D, Kinloch M, *et al.* (2018) From the brink of extinction: successful recovery of the glossy black-cockatoo on Kangaroo Island. In *Recovering Australian Threatened Species: A Book of Hope.* (Eds S Garnett, P Latch, D Lindenmayer and J Woinarski) pp. 75–84, CSIRO Publishing, Melbourne.

Berris K, Barth M, Welz T, Mooney PA (2020) 'Kangaroo Island Glossy Black-Cockatoo 2020 census report'. Report to Landscape South Australia Kangaroo Island, Kingscote and the Australian Wildlife and Nature Recovery Fund, Sydney.

Boulton RL, Gates J (2021) 'Kangaroo Island rapid bird assessments. Draft August 2021'. Department of Environment and Water, Adelaide.

Boulton RL, Hunt TJ, Ireland LJ, Thomas JL (2020) 'Kangaroo Island rapid bird assessments, February 2020'. Birdlife Australia, Melbourne.

Bowman DMJS, Murphy BP, Boer MM, Bradstock RA, Cary GJ, *et al.* (2013) Forest fire management, climate change, and the risk of catastrophic carbon losses. *Frontiers in Ecology and the Environment* **11**, 66–67. doi:10.1890/13.WB.005

Bradstock RA (2008) Effects of large fires on biodiversity in south-eastern Australia: disaster or template for diversity? *International Journal of Wildland Fire* **17**, 809–822. doi:10.1071/WF07153

Butchart SHM, Akçakaya HR, Chanson J, Baillie JEM, Collen B, *et al.* (2007) Improvements to the red list index. *PLoS One* **2**, e140. doi:10.1371/journal.pone.0000140

Butt C, McManus J (2020) 'It's a sorry sight': Dead birds wash up on Mallacoota's beaches. *The Age*, 7 January, <https://www.theage.com.au/national/victoria/it-s-a-sorry-sight-dead-birds-wash-up-on-mallacoota-s-beaches-20200107-p53plm.html>.

Clarke MF (2020) Our birds will be ok, they've evolved to cope with fire ... haven't they? *Emu-Austral Ornithology* **120**, 184–186. doi:10.1080/01584197.2020.1750964

Crates RL, Rayner D, Stojanovic BC, Scheele A, Roff J, *et al.* (2022) Poor-quality monitoring data underestimate the impact of Australia's megafires on a critically endangered songbird. *Diversity & Distributions* **28**, 506–514. doi:10.1111/ddi.13385

Davis RA, Doherty TS, van Etten EJ, Radford JQ, Holmes F, *et al.* (2016) Conserving long unburnt vegetation is important for bird species, guilds and diversity. *Biodiversity and Conservation* **25**, 2709–2722. doi:10.1007/s10531-016-1196-5

Di Virgilio G, Evans JP, Blake SA, Armstrong M, Dowdy AJ, *et al.* (2019) Climate change increases the potential for extreme wildfires. *Geophysical Research Letters* **46**, 8517–8526. doi:10.1029/2019GL083699

Evans JP, Argueso D, Olson R, Di Luca A (2017) Bias-corrected regional climate projections of extreme rainfall in south-east Australia. *Theoretical and Applied Climatology* **130**, 1085–1098. doi:10.1007/s00704-016-1949-9

Franklin MJ, Major RE, Bedward M, Bradstock RA (2021) Relative avian mobility linked to use of fire-affected resources in forested landscapes. *Forest Ecology and Management* **497**, 119484. doi:10.1016/j.foreco.2021.119484

Garnett ST, Baker GB (Eds) (2021) *The Action Plan for Australian Birds 2020.* CSIRO Publishing, Melbourne.

Garnett S, Szabo J, Dutson G (2011) *Action Plan for Australian Birds 2010.* CSIRO Publishing, Melbourne.

Garnett ST, Butchart SHM, Baker GB, Bayraktarov E, Buchanan K, *et al.* (2019) Metrics of progress in the understanding and management of threats, and their application to Australian birds. *Conservation Biology* **33**, 456–468. doi:10.1111/cobi.13220

Geary WL, Doherty TS, Nimmo DG, Tulloch AI, Ritchie EG (2020) Predator responses to fire: a global systematic review and meta-analysis. *Journal of Animal Ecology* **89**, 955–971. doi:10.1111/1365-2656.13153

Herold N, Ekström M, Kala J, Goldie J, Evans JP (2018) Australian climate extremes in the 21st century according to a regional climate model ensemble: implications for health and agriculture. *Weather and Climate Extremes* **20**, 54–68. doi:10.1016/j.wace.2018.01.001

IUCN (International Union for Conservation of Nature) (2012) *IUCN Red List Categories and Criteria*. IUCN, Gland, Switzerland.

Kuchinke D, Di Stefano J, Sitters H, Loyn R, Gell P, *et al.* (2020) Prescribed burn severity has minimal effect on common bird species in a fire-prone forest ecosystem. *Forest Ecology and Management* **475**, 118437. doi:10.1016/j.foreco.2020.118437

Lee JS, Callaghan CT, Cornwell WK (in press) Using citizen science to measure recolonisation of birds after the Australian 2019–2020 mega-fires. *Austral Ecology*. doi:10.1111/aec.13105

Lee JS, Cornwell W, Kingsford R (2022) Rainforest bird communities threatened by extreme fire. *Global Ecology and Conservation* **33**, e01985. doi:10.1016/j.gecco.2021.e01985

Legge SM, Rumpff L, Woinarski JCZ, Whiterod NS, Ward M, *et al.* (in press) The conservation impacts of ecological disturbance: time-bound estimates of population loss and recovery for fauna affected by the 2019–2020 Australian megafires. *Global Ecology and Biogeography* doi:10.1111/geb.13473.

Lindenmayer DB, Sato C (2018) Hidden collapse is driven by fire and logging in a socioecological forest ecosystem. *Proceedings of the National Academy of Sciences of the United States of America* **115**, 5181–5186. doi:10.1073/pnas.1721738115

Lindenmayer DB, Wood JT, Cunningham RB, Macgregor C, Crane M, *et al.* (2008) Testing hypotheses associated with bird responses to wildfire. *Ecological Applications* **18**, 1967–1983. doi:10.1890/07-1943.1

Lindenmayer D, Blair D, McBurney L, Banks S, Bowd E (2021) Ten years on – a decade of intensive biodiversity research after the 2009 Black Saturday wildfires in Victoria's Mountain Ash forest. *Australian Zoologist* **41**, 220–230. doi:10.7882/AZ.2020.041

Loyn RH (1997) Effects of an extensive wildfire on birds in far eastern Victoria. *Pacific Conservation Biology* **3**, 221–234. doi:10.1071/PC970221

McKechnie AE, Rushworth IA, Myburgh F, Cunningham SJ (2021) Mortality among birds and bats during an extreme heat event in eastern South Africa. *Austral Ecology* **46**, 687–691. doi:10.1111/aec.13025

Millington B (2020) 'Solitary' lyrebirds band together to save themselves in 'incredible' show of unity under bushfire threat. *ABC News* (online), 30 January, <https://www.abc.net.au/news/2020-01-30/lyrebirds-band-together-to-avoid-approaching-bushfire/11910666>.

Molloy SW, Burbidge AH, Comer S, Davis RA (2020) Using climate change models to inform the recovery of the western ground parrot *Pezoporus flaviventris*. *Oryx* **54**, 52–61. doi:10.1017/S0030605318000923

Murphy MJ, Jones HA, Koen T (2021) A seven-year study of the response of woodland birds to a large-scale wildfire and the role of proximity to unburnt habitat. *Austral Ecology* **46**, 1138–1155. doi:10.1111/aec.13049

Oliver D, Malolakis G (2020) 'Rapid post-fire assessment of upland and coastal heaths and surrounding forest vegetation at Nadgee NR and implications for threatened heathland avifauna'. Report to Department of Planning, Industry and Environment, Sydney.

Pescott T (1983) Beach-washed birds after the Ash Wednesday fire. *Geelong Naturalist* **20**, 17–19.

Rainsford FW, Kelly LT, Leonard SWJ, Bennett AF (2021) Post-fire habitat relationships for birds differ among ecosystems. *Biological Conservation* **260**, 109218. doi:10.1016/j.biocon.2021.109218

Robinson NM, Leonard SWJ, Bennett AF, Clarke MF (2014) Refuges for birds in fire-prone landscapes: The influence of fire severity and fire history on the distribution of forest birds. *Forest Ecology and Management* **318**, 110–121. doi:10.1016/j.foreco.2014.01.008

Selwood KE, Antos M, Bramwell M, Lee A, Lynch M, *et al.* (2022) Emergency conservation interventions during times of crisis: a case study for a threatened bird species in the Australian Black Summer bushfires. *Conservation Science and Practice* **4**, e606. doi:10.1111/csp2.606

Sharrod J (2020) Water stations help Australia's wildlife recover from bushfires. The International Fund for Animal Welfare, <https://www.ifaw.org/au/journal/water-for-wildlife-australia>.

Smith P, Smith J (2017) Influence of fire regime and other habitat factors on a eucalypt forest bird community in south-eastern Australia in the 1980s. *Australian Journal of Zoology* **64**, 312–326. doi:10.1071/ZO16053

van Eeden L, Nimmo D, Mahoney M, Herman K, Ehmke G, *et al.* (2020) *Australia's 2019–2020 Bushfires: The Wildlife Toll*. World Wide Fund for Nature, Melbourne.

van Oldenborgh GJ, Krikken F, Lewis S, Leach NJ, Lehner F, *et al.* (2021) Attribution of the Australian bushfire risk to anthropogenic climate change. *Natural Hazards and Earth System Sciences* **21**, 941–960. doi:10.5194/nhess-21-941-2021

Ward M, Tulloch AIT, Radford JQ, Williams BA, Reside AE, *et al.* (2020) Impact of 2019–2020 mega-fires on Australian fauna habitat. *Nature Ecology & Evolution* **4**, 1321–1326. doi:10.1038/s41559-020-1251-1

Watson SJ, Taylor RS, Nimmo DG, Kelly LT, Clarke MF, *et al.* (2012) The influence of unburnt patches and distance from refuges on post-fire bird communities. *Animal Conservation* **15**, 499–507. doi:10.1111/j.1469-1795.2012.00542.x

Wills AJ, Liddelow G, Tunsell V (2020) Wildfire and fire mosaic effects on bird species richness and community composition in south-western Australia. *Fire Ecology* **16**, 5. doi:10.1186/s42408-019-0065-5

Woinarski JCZ, Recher HF (1997) Impact and response: a review of the effects of fire on the Australian avifauna. *Pacific Conservation Biology* **3**, 183–205. doi:10.1071/PC970183

Wooller RD, Calver MC (1988) Changes in an assemblage of small birds in the understorey of dry sclerophyll forest in southwestern Australia after fire. *Wildlife Research* **15**, 331–338. doi:10.1071/WR9880331

16

Impacts of the 2019–20 wildfires on native mammals

John C. Z. Woinarski, Jemma Cripps, Louise Durkin, Brad Law, Sarah M. Legge, Phoebe V. Macak, Jenny Nelson and Libby Rumpff

Summary

Context and challenges

- The fate of some mammal species in the 2019–20 wildfires was a high-profile – emblematic – concern for the community, with such attention catalysing concerted responses by many organisations.
- The impacts of the wildfires were superimposed on a prevalent pattern of decline for many Australian mammals.
- Post-fire recovery of many mammal species will be challenging, given the individual and compounding impacts of other threats and a changing fire regime likely to be characterised by more frequent comparable fires.
- On-ground research following the 2019–20 wildfires has provided some evidence about population-level impacts at some sites.

Main findings

- The 2019–20 wildfires burnt at least 50% of the distributions of five mammal species and 30–50% of the distributions of a further 18 mammal species.
- Including subspecies, the most extreme fire overlap (95% of range burnt) was for the Kangaroo Island dunnart (*Sminthopsis griseoventer aitkeni*).
- Post-fire sampling has confirmed marked population losses in burnt areas for a range of mammals, especially where fires were of high severity.
- Population losses due to the 2019–20 wildfires are likely to have been of sufficient magnitude to render five mammal taxa newly eligible for listing as threatened.
- Many government agencies and other groups instituted management interventions in the immediate aftermath of the fires. These efforts are likely to have achieved some local benefits.
- Most of the fire-affected mammal species are unlikely to recover to their pre-fire population size over the next decade; and ongoing decline of many species is likely should comparable fires recur in the future.

Introduction

Mammals attracted much of the public and media concern about the impacts of the 2019–20 wildfires. Particularly prominent were rescue efforts for koalas (*Phascolarctos cinereus*) and kangaroos injured in the fires (Fig. 16.1; see also Chapter 26), and food supplementation for brush-tailed rock-wallabies (*Petrogale penicillatus*) in some burnt areas (see Chapter 24). Reflecting this profile and concern, among its other investments in conservation of fire-affected species, the Australian Government allocated $14 million to post-fire recovery of one mammal species, alone, the koala (Chapter 22). However, even the plight of mammals hitherto largely unknown to the public – most notably the Kangaroo Island dunnart – was heavily profiled. In this chapter, we describe the impacts on mammals occurring in the fire-affected regions. We also briefly describe some management actions that were implemented, and ongoing management required to support recovery and reduce the likelihood of such loss in any comparable future fire. Note that this chapter considers responses of native mammals only; information on some introduced mammals is given in Chapter 17, in the context of compounding threats.

The 2019–20 Australian wildfires were without modern precedent (Chapter 2), and their impacts on mammals are likely to have been comparably exceptional. However, when considering impact, it is informative to use a broader frame of reference than that year's fires alone. Many mammal species in the regions burnt by these fires were declining before 2019–20 (e.g. Woinarski *et al.* 2015; Wagner *et al.* 2020). Furthermore, the 2019–20 wildfires formed part of a changing regime, of wildfires occurring more frequently, in some cases causing transitions in vegetation beyond the previously prevailing successional cycle (Lindenmayer and Taylor 2020b). The return of mammal fauna to their state before the

Fig. 16.1. Eastern grey kangaroo (*Macropus giganteus*) in a landscape extensively burnt by the 2019–20 wildfires, Kangaroo Island. (Photo: Nicolas Rakotopare/Threatened Species Recovery Hub)

2019–20 wildfires may therefore no longer be possible as some habitats and resources are unlikely to recover before the next fires occur (Camac *et al.* 2017).

Many of the short-term impacts on mammals of the 2019–20 wildfires have not yet been evaluated directly, and longer-term impacts may take many years to be realised. Hence, in this chapter we try to contextualise the impacts of the 2019–20 wildfires with the knowledge base accruing from many previous studies in southern and eastern Australia of responses of mammals to fire (e.g. Bradstock 2008). Such previous studies have been conducted in most of the habitats that were burnt in 2019–20, including montane ash forests (Lindenmayer *et al.* 2021), alpine and subalpine environments (Green and Sanecki 2006), heathlands (Fox 1982), and coastal and foothill forests (Lunney 1987). Nonetheless, some of the information derived from these previous studies may provide only limited inference on the impacts of, and recovery potential following, the 2019–20 wildfires given the exceptional scale of the 2019–20 wildfires, and the extent to which they burnt habitats (notably rainforests) that have rarely been burnt previously (Collins *et al.* 2021).

Findings

Impacts of the 2019–20 wildfires

Three main approaches have been used to evaluate the impacts of the 2019–20 wildfires on mammals. The first approach, described in Chapter 12, used spatial modelling of mammal densities to estimate the number of mammals (albeit excluding bats) that occurred pre-fire in areas burnt in the 2019–20 wildfires. This tally was 143 million native mammals, mostly comprising rodents (50 million), possums (39 million) and small dasyurids (37 million). Undoubtedly, many of these individual mammals were killed by fire, but that proportion is unresolved.

The second approach involved spatial analysis to superimpose mapping of the areas burnt in the 2019–20 wildfires on the known distributions of mammal species (Legge *et al.* 2022). This analysis found that five mammal species – long-footed potoroo (*Potorous longipes*), Hastings River mouse (*Pseudomys oralis*), parma wallaby (*Notomacropus parma*), brush-tailed rock-wallaby and brown antechinus (*Antechinus stuartii*) – had at least 50% of their distributions burnt in the 2019–20 wildfires and a further 18 species had 30–50% burnt (Table 16.1). Seventeen mammal species or subspecies that were recognised as threatened before the fire had at least 10% of their distributions burnt, with six of these having at least 50% burnt. The maximum fire overlap was for the threatened Kangaroo Island dunnart, with 95% of its distribution burnt, including 90% at high severity. Notably, the 2019–20 fires affected a very broad suite of mammals, including species narrowly endemic to rainforests (e.g. black-tailed antechinus (*Antechinus arktos*)), the aquatic platypus (*Ornithorhynchus anatinus*) (Bino *et al.* 2021), and highly dispersive species such as grey-headed flying-fox (*Pteropus poliocephalus*) (Baranowski *et al.* 2021).

The proportion of a species' population killed in fires will generally be less than the proportion of the distribution burnt, because some individuals in burnt areas survive fire. Legge *et al.* (in press) used expert elicitation to estimate proportional population losses for each species at sites exposed to mild fires and to severe fires. Experts estimated this proportion immediately after fire, at 1-year post-fire and at 10 years (or three generations, whichever longer) post-fire. This assessment concluded that the 2019–20 wildfires caused an immediate population reduction of 65% for Kangaroo Island dunnart, 39% for long-footed potoroo, 34% for Hastings River mouse, and 20–30% for six other taxa: Kangaroo

Table 16.1. Summary of distributional overlap with fire (for high severity fires and total fire), and estimated percentage population loss for the mammal taxa most affected by the 2019–20 fires (modified from Legge *et al.* (in press)).

All taxa with at least 50% overlap with fire are included, along with threatened (and Near Threatened) taxa with at least 10% overlap with fire. Conservation status is as before the 2019–20 wildfires: CR Critically Endangered, EN Endangered, VU Vulnerable, NT Near Threatened, LC Least Concern. Note that % population decline is expressed relative to a standardised population size of 100 immediately pre-fire (i.e. -64.8 indicates the total population declined by an estimated 64.8%); decline was not estimated for two taxa (nd). Non-threatened species with 30–49.9% overlap with fire (hence not included in the body of this Table) comprise sugar glider (*Petaurus breviceps*) (47.8% overlap), eastern broad-nosed bat (*Scotorepens orion*) (46.8%), eastern forest bat (*Vespadelus pumilus*) (43.5%), long-nosed bandicoot (*Perameles nasuta*) (43.3%), red-necked pademelon (*Thylogale thetis*) (42.7%), greater broad-nosed bat (*Scoteanax rueppellii*) (41.6%), short-eared possum (*Trichosurus caninus*) (40.2%), mainland dusky antechinus (39.4%), white-footed dunnart (*Sminthopsis leucopus*) (37.5%), swamp rat (36.0%), agile antechinus (*Antechinus agilis*) (35.0%), eastern pygmy possum (*Cercartetus nanus*) (34.2%), eastern false pipistrelle (*Falsistrellus tasmaniensis*) (33.8%), feathertail glider (*Acrobates pygmaeus*) (33.8%) and bush rat (*Rattus fuscipes*) (30.8%).

		% overlap with fire		Estimated % population decline			Pre-fire conservation status	
Common name	**Scientific name**	**severe**	**total**	**1-week**	**1-yr**	**10-yr**	**EPBC**	**IUCN**
Kangaroo Island dunnart	*Sminthopsis fuliginosus aitkeni*	90.2	95.2	–64.8	–64.9	–46.4	EN	CR
Long-footed potoroo	*Potorous longipes*	38.1	81.4	–38.9	–46.3	–36.7	EN	VU
Hastings River mouse	*Pseudomys oralis*	22.1	73.7	–34.0	–36.0	–30.2	EN	VU
Kangaroo Island echidna	*Tachyglossus aculeatus multiaculeatus*	53.2	63.2	–27.7	–29.4	–23.8	EN	LC (as species)
Parma wallaby	*Notomacropus parma*	17.5	59.4	–20.4	–24.2	–15.9		NT
Long-nosed potoroo (NSW, Vic)	*Potorous tridactylus trisulcatus*	28.8	56.6	–26.0	–35.8	–33.4	VU (as *P. t. tridactylus*)	NT (as species)
Brush-tailed rock-wallaby	*Petrogale penicillata*	20.0	50.6	–10.4	–16.2	–22.7	VU	VU
Brown antechinus	*Antechinus stuartii*	20.8	50.3	–23.0	–26.3	–12.2		LC
New Holland mouse	*Pseudomys novaehollandiae*	15.5	45.6	–13.8	–23.9	–26.1	VU	VU
Southern brown bandicoot (SE Aust)	*Isoodon obesulus obesulus*	21.7	42.7	nd	nd	nd	EN	LC (as species)
Yellow-bellied glider (SE Aust)	*Petaurus australis*	16.9	41.4	–17.4	–21.0	–25.0		NT (as species)
Long-nosed potoroo (Qld, NSW)	*Potorous tridactylus tridactylus*	11.3	41.3	–14.4	–24.0	–24.7	VU (as *P. t. tridactylus*)	NT (as species)

Common name	Scientific name	% overlap with fire		Estimated % population decline			Pre-fire conservation status	
		severe	total	1-week	1-yr	10-yr	EPBC	IUCN
Spotted-tail quoll (SE mainland)	*Dasyurus maculatus maculatus* (SE mainland)	16.9	40.4	–12.7	–18.0	–17.9	EN	NT (as species)
Greater glider	*Petauroides volans*	14.6	35.6	–17.7	–21.7	–31.5	VU	VU
Platypus	*Ornithorhynchus anatinus*	11.4	27.4	–2.9	–8.4	–13.7		NT
Broad-toothed rat (mainland)	*Mastacomys fuscus mordicus*	14.3	27.2	–14.8	–17.9	–22.2	VU	NT (as species)
Smoky mouse	*Pseudomys fumeus*	12.5	27.1	–9.3	–17.0	–17.8	EN	VU
Large-eared pied bat	*Chalinolobus dwyeri*	9.9	26.6	nd	nd	nd	VU	NT
Koala	*Phascolarctos cinereus*	6.9	16.9	–7.1	–9.9	–25.8	VU (popn in Qld, NSW, ACT)	VU
Mountain pygmy-possum	*Burramys parvus*	13.3	15.7	–5.2	–9.2	–13.5	EN	CR
Grey-headed flying-fox	*Pteropus poliocephalus*	6.1	15.2	–3.3	–10.2	–15.2	VU	VU

Island echidna (*Tachyglossus aculeatus multiaculeatus*), parma wallaby, the south-eastern mainland subspecies of long-nosed potoroo (*Potorous tridactylus trisulcatus*), brown antechinus, mainland dusky antechinus (*Antechinus mimetes*) and swamp rat (*Rattus lutreolus*) (Table 16.1). The elicitation further predicted that most of the fire-affected species will show little recovery over the decade after fire, even under the unlikely presumption of no comparable wildfires in this period. In part, this predicted lack of recovery is because fire impacts were superimposed on a prevailing pattern of decline associated with other threats.

Based on these estimates of population losses, Legge *et al.* (in press) considered that five mammal taxa (mainland dusky antechinus, parma wallaby, platypus, long-nosed bandicoot and the south-eastern Australian subspecies of yellow-bellied glider (*Petaurus australis*)) are likely to be newly eligible for listing as threatened under the Australian *Environment Protection and Biodiversity Conservation Act 1999* (EPBC Act) and a further eight already listed taxa (mainland subspecies of broad-toothed rat (*Mastacomys fuscus mordicus*), greater glider (*Petauroides volans*), koala, long-footed potoroo, the two mainland subspecies of long-nosed potoroo (*Potorous tridactylus tridactylus* and *P. t. trisulcatus*), New Holland mouse (*Pseudomys novaehollandiae*) and Kangaroo Island dunnart) are likely to be eligible for up-listing (Box 16.1). To our knowledge, over recent centuries, no other single event has caused such an acute and sudden deterioration in the conservation status of so many mammal species. For the most fire-affected mammals, the process for listing as threatened or up-listing under the EPBC Act is underway, providing formal recognition of their extent of loss and some additional protection.

Box 16.1. Greater glider and yellow-bellied glider

Australia's two largest gliders, the greater glider (Fig. 16.2) and yellow-bellied glider, occur extensively across the forests of eastern Australia. Both had much of their range burnt in the 2019–20 wildfires (36% for greater glider and 41% for yellow-bellied glider). Both are dependent on large tree hollows for denning, with trees typically forming such hollows only when > 100 years old (Lindenmayer *et al.* 2013). These two glider species are undergoing broad-scale decline, mostly as a result of vegetation clearance, timber-harvesting, climate change and fire (McLean *et al.* 2018; Wagner *et al.* 2020; Lefoe *et al.* 2022). Both have low reproductive rates, so will recover very slowly from major disturbance. Greater gliders also have limited dispersal capability (Tyndale-Biscoe and Smith 1969), so may take many years to recolonise areas from which a population has been lost (Smith and Smith 2020). Greater gliders are obligate folivores, so may face major food shortages in burnt landscapes: early observers reported them 'crawling weakly over charred ground' weeks after the Victorian 1939 wildfire, and those that survived the fire 'died subsequently of starvation' (Fleay 1947). In contrast, yellow-bellied gliders have broader diets and occupy larger home ranges (Goldingay and Kavanagh 1991) so may persist somewhat better in largely burnt landscapes.

To investigate the impact of the 2019–20 wildfires, spotlight surveys were undertaken in high elevation forests in East Gippsland, Victoria, 1 year after the fires at 30 sites with pre-fire (mostly since 2010) records of greater gliders. At each site, a 500 m transect was surveyed using double-observer spotlighting methods (Cripps *et al.* 2021). An owl call playback survey was also undertaken 100 m away from each transect to target forest owls. Yellow-bellied gliders are known to respond to owl calls and the playback of owl calls probably increased detectability for this species. Greater

Fig. 16.2. The 2019–20 wildfires had a severe impact on the greater glider population. (Photo: Andrew Geschke)

gliders and yellow-bellied gliders were detected at 57% and 53% of sites respectively, and were detected more frequently in unburnt forest or forest burnt at low or medium severity than in forest burnt at high severity (Table 16.2).

Other post-fire surveys in burnt lower-elevation forests in East Gippsland found very low detections of greater gliders at previously occupied sites. Rutter and Blake (2020) detected greater gliders at only 29% of 24 previously occupied sites, and Burns and Atkins (2021) detected greater gliders at only one of 12 (8%) previously occupied sites. Yellow-bellied gliders appear to be more able to persist within the fire footprint, and were detected at 42% (10 of 24) (Rutter and Blake 2020) and 47% (14 of 30) of sites surveyed (Burns and Atkins 2021).

Comparable declines and losses due to the 2019–20 wildfires have also been reported for greater gliders in the Blue Mountains of New South Wales. At 16 sites known to be occupied by greater gliders before the fires, no gliders were recorded in sampling at 10 and 17 months post-fire at the two severely burnt sites, and they were considered extirpated from areas burnt with high severity; there was a mean decline of 43% in six moderately burnt sites; but also a mean decline of 34% at eight unburnt sites, attributed to extreme heat and drought (Smith and Smith 2021).

Table 16.2. Detections of the greater glider and the yellow-bellied glider during surveys in Victoria, summarised by fire severity.

All sampled sites had pre-fire greater glider records.

Species	Unburnt sites surveyed	Unburnt sites detected	Low–medium severity sites surveyed	Low–medium severity sites detected	High severity sites surveyed	High severity sites detected
Greater glider	5	4 (80%)	18	11 (61%)	7	2 (29%)
Yellow-bellied glider	5	3 (60%)	18	11 (61%)	7	2 (29%)

These spatial overlap assessments may be conservative, as fire-related impacts on mammals may also have extended beyond the burnt area. In an intriguing but disconcerting study, Peters *et al.* (2021) noted that nine smoky mice (*Pseudomys fumeus*) in a captive colony held outdoors 20 km from the nearest fire died 3 to 30 days after the approach of a 2019–20 wildfire, with pathological and clinical evidence demonstrating that these individuals died due to smoke inhalation. Currently, there is no information on the extent to which the susceptibility to smoke of the unfortunately named smoky mouse is representative of other mammal species, or of the extent to which the impact of smoke attenuates over distance or duration. However, the death of these mice does indicate that at least some mammal populations occurring in unburnt patches within or on the periphery of the fire scar may not persist, potentially subverting the previously assumed refugial value of such patches.

The third approach to estimating impacts of the 2019–20 wildfires has involved on-ground surveys to directly compare the abundance of some mammal species in matched burnt and unburnt areas, or at sites sampled before and after the 2019–20 wildfires, or a combination of these (a Before-After-Control-Impact (BACI) design). A constraint on the Before-After sampling approach is the limited amount of monitoring established for most Australian mammals. This direct approach has the advantage of collection of robust data rather than inference or expert opinion but, because of its resource demands, has been undertaken for only some species, in some parts of their range. Many of the results have not yet been analysed or published, and for obvious reasons this approach has provided data to date only on the short-term responses to the 2019–20 wildfires. Examples of this approach are presented in Boxes 16.1 to 16.3; all demonstrate that populations declined in burnt areas, with decline most marked in areas where fire was of high severity.

Box 16.2. Golden-tipped bat

The impacts of wildfire on insectivorous bats are varied (Law *et al.* 2018). One species, the golden-tipped bat (*Phoniscus papuensis*) (Fig. 16.3) has particular ecological specialisations that render it especially susceptible to fire. This bat roosts in suspended nests made by yellow-throated scrubwren (*Sericornis citreogularis*) or brown gerygone (*Gerygone mouki*), typically located in the forest understorey: these are highly likely to be destroyed with any fire. It also forages mostly on spiders taken from webs strung among shrubs, and this resource is also likely to be severely depleted by fire. The habitats mostly favoured by golden-tipped bats – rainforest and wet eucalypt forest gullies – rarely burn, but this was not the case in the 2019–20 wildfires, at least in part because drought preceding the fires probably increased the flammability of such patches. Extensive trapping surveys were conducted across 35 burnt and 26 unburnt rainforest sites 1 year after the 2019–20 fires (Law *et al.* in press). Site occupancy was found to be > 90% at sites surrounded by extensive rainforest and gully systems or unburnt forest, but just 20% in comparable sites where 100% of the surrounding area was burnt (regardless of severity). Radio-tracking also revealed bats did not roost where rainforest was burnt by high severity fire. Maternal bats did not avoid roosting in rainforest burnt by low severity fire in northern New South Wales, though roost selection avoided rainforest burnt by such fires in southern New South Wales,

Fig. 16.3. Golden-tipped bat. (Photo: Brad Law)

suggesting some variability in responses to fire. These results illustrate the vulnerability of rainforest-dependent species to extreme fire events, even though such areas are normally protected from fire.

Box 16.3. Koala

There have been several studies aiming to assess the impacts of the 2019–20 wildfires on koala populations. In north-eastern New South Wales, 123 sites sampled (using pellet counts) for koalas before the 2019–20 fires were re-sampled 2–7 months after fire. Post-fire site occupancy of koalas declined markedly (median reduction of 71%) relative to pre-fire occupancy. Fire severity affected the extent of koala loss, with survival fivefold higher at sites burnt in fires of low to medium severity than at sites exposed to severe fire (Phillips *et al.* 2020). In contrast, a different sampling approach (using acoustic monitoring) of koalas in hinterland forests of north-eastern New South Wales reported that fire had no effect on site occupancy rate, but a dramatic (at least short-term) effect on density, with 100% reduction in density where fire was severe within a 1 km radius of the sample site, 50% reduction where fire severity was moderate and no change where it was low severity (Law *et al.* 2022). Some recovery was evident one year after fire in areas where high severity fire dominated. In southern

New South Wales, Cristescu *et al.* (in press) used drones with thermal imagery ~30 days post-fire to compare koala density at two sets of paired burnt and unburnt sites, reporting 24% and 26% lower densities in the two burnt sets of sites relative to their unburnt controls. That study also sampled two sets of paired burnt and unburnt sites ~100 days post-fire in south-eastern Queensland, based on searches for scats by detection dogs. That research found that koala density was 71% lower in burnt areas relative to unburnt control areas in one set, but 317% higher in burnt sites relative to unburnt controls at the other set, noting that this result may be due in part to poorly-matched controls. This study noted that koalas rapidly used fire-affected trees, feeding (within 3 months of fire) on epicormic growth, and could maintain home ranges within fire-affected areas.

The 2019–20 wildfires reduced populations of many mammal species. There may be important ramifications of such losses, as many of these affected species (such as flying-foxes, bandicoots and potoroos) perform important ecological roles. Reduced populations of these species may compromise the recovery of ecosystem health and functionality in burnt landscapes.

Broad context for impact and recovery

With caveats due to the exceptional size of the 2019–20 wildfires, the evidence base built from many decades of research on the responses of mammals to previous wildfires (e.g. Arthur *et al.* 2012) can be used to infer the longer-term consequences of the 2019–20 fires. Unsurprisingly, this research body has demonstrated that there are marked differences among mammal species in their susceptibility to population loss in fires, and in the time taken to recover (Whelan *et al.* 2002). This variation is associated in part with differences among species in ecological and life history traits. Species that are less capable of fleeing fire, and that have no protective shelter (e.g. burrows), are more likely to suffer higher rates of mortality during fire. Species with more pronounced dietary or habitat specialisation, or that have more dependence on resources that are prone to depletion by fire, will tend to have lower rates of survival in the aftermath of fire. Species with low reproductive rates and more limited dispersal ability are likely to take longer to recover. Impacts vary also with characteristics of the fire and with the spatial and temporal patterning of fires (the fire regime). Impacts also vary according to the evolutionary history of the environment, with typically more tolerance to fire, and more likelihood of recovery, in many eucalypt forests than in rainforests (Box 16.2). Fire impacts are also likely to be more pronounced and long-lasting for some mammals in montane ash forests, where the dominant eucalypt trees are killed by fire and tree regrowth is from seed, than in other eucalypt forests and woodlands, where trees are more likely to survive fire, producing post-fire epicormic foliage.

Many mammals are likely to be killed in the fire itself. Nonetheless, survival rates may be high for some species even in severely burnt areas (Banks *et al.* 2011b) and, for some mammal species, recovery post-fire is largely a result of recruitment from the residue of those animals that survived in burnt areas (Banks *et al.* 2011a). In landscapes that are mostly burnt, unburnt patches may provide refuges and source areas for recolonisation into burnt areas (Berry *et al.* 2015); however, the edges of burnt and unburnt vegetation may be targeted by feral cats (Hradsky 2020). Native mammals that survive in their burnt home ranges may also be more exposed to higher predation rates after fire (Arthur *et al.* 2012) because the fire has removed vegetation cover and because higher-risk foraging may be necessary in an environment with fewer food resources. Fire may also reduce the abundance of den sites such as tree hollows, a critical resource for many mammal species and one that may take more than a century to restore (Lindenmayer *et al.* 2013). Very extensive fires, such as the 2019–20 wildfires, also tend to homogenise the landscape, reducing its ecological variability and hence its capability to support diverse assemblages of mammals, and the variety of resources it can offer to individual species. This latter factor may be especially important for species such as Leadbeater's possum (*Gymnobelideus leadbeateri*) which requires a mix of resources that may be most readily attained in habitat mosaics of variable time since fire (Nitschke *et al.* 2020).

Furthermore, although many mammal species may suffer major losses in individual fires, it is the patterning of successive fires that may most critically affect viability of mammal populations. Population modelling has demonstrated that an increasing frequency and extent of fires – especially where accompanied by other threats – is likely to drive precipitous population declines for many mammal species, including koala (Lunney *et al.* 2007) and Leadbeater's possum (Todd *et al.* 2016). Because of changing disturbance regimes (fire and timber-harvesting), older-aged forests are a diminishing component of the landscape (Chapter 19). These habitats may be critical conservation strongholds for hollow-dependent mammals, and hence should be recognised as assets for protection in fire planning and operations (Lindenmayer and Taylor 2020a).

Management responses

Some management actions taken before and after the 2019–20 wildfires may have reduced the impacts of these fires on mammals and helped support their post-fire recovery. In some regions, sustained control programs for introduced predators led (pre-fire) to population increases for some threatened mammals such as long-footed potoroo (e.g. Robley *et al.* 2014). These programs probably contributed to an increased likelihood that sufficient individuals survived fire to maintain population viability. Furthermore, the establishment of such predator control programs also allowed for their rapid reimplementation following the 2019–20 fires, to help reduce compounding impacts. Survey and monitoring programs established for some threatened mammals before the 2019–20 fires provided a baseline from which to assess the extent of loss in the fires and to provide a target for post-fire recovery. In some cases, research undertaken before the 2019–20 fires resulted in the development of effective survey protocols for species with low detectability (e.g. Hohnen *et al.* 2019), thereby allowing for post-fire surveys to most efficiently locate and protect populations that survived the fires. Although there proved to be no need in the 2019–20 fires, there are now well-established and successful captive-breeding programs for some threatened mammals, such as mountain pygmy-possum (*Burramys parvus*). Such insurance populations provide a potential source for re-establishment of populations that may be lost in any future comparable fire (e.g. Chapter 27).

Management actions were taken soon after the 2019–20 wildfire to aid surviving mammal individuals, to reduce other threats and to support recovery. Rapid responses were made by many groups to find, treat and provide shelter for mammals injured in the fires (see Chapter 26). Previous post-fire research had provided evidence of reasonable survivorship for treated koalas released back to the wild (Lunney *et al.* 2004), so such animal welfare and rehabilitation efforts can provide at least a local benefit to recovery. Targeted and broad-scale control programs, mostly poison baiting, for introduced predators were implemented at many sites following the 2019–20 wildfires (see Chapter 17). A population of the Kangaroo Island dunnart that was discovered in a partly burnt landscape was protected post-fire by the rapid construction of predator-proof fencing (see Chapters 24, 25). At some sites with surviving dunnart populations, at which fire had removed vegetation cover, artificial shelters were also provided to reduce post-fire predation risks (Fig. 16.4).

There were also some high-profile programs undertaken to supply food for some mammal populations persisting in largely burnt landscapes, including aerial food drops for rock-wallabies (Chapter 24) and provision of nutrient-rich biscuits for mountain pygmy-possums (Chapter 27). In many largely burnt landscapes, in which the abundance of hollow-bearing trees had been markedly reduced by fire, conservation organisations and community groups provided nest boxes or artificial hollows for hollow-dependent mammals and birds. Conversely, some management actions taken after fire are likely to have had inimical impacts on mammal survival and recovery; most notably, salvage logging will have compounded the losses of hollows caused through fire (see Chapter 19).

Even with the large body of research that has been undertaken on the impacts of fire on Australian mammals, there remain some important knowledge gaps that should be addressed. These include (1) the relative cost-effectiveness of management actions taken after fire to help recover mammals; (2) whether critical sites for the conservation of mammals can be identified, circumscribed and included as assets for protection during fire control operations; (3) the extent to which smoke causes mortality in mammals beyond the fire perimeter; (4) what mechanisms can reduce the likelihood of future catastrophic fire, and the extent to which these mechanisms themselves cause detriment to mammals;

Fig. 16.4. Artificial shelters were constructed in some burnt landscapes to reduce risks of predation for native mammals after fire, in this case for Kangaroo Island dunnart. (Photo: Nicolas Rakotopare/Threatened Species Recovery Hub)

and (5) the capability for post-fire recovery of mammals, especially those restricted to habitats that are ill-adapted to fire (notably rainforests).

Conclusions

The 2019–20 wildfires caused severe population losses of many mammal species. These losses have acutely compounded the impacts of other threats that were already driving decline in much of the distinctive Australian mammal fauna. Comparable catastrophic fires are likely to recur more frequently in the future, due to global climate change. Many of the mammal species most affected by the 2019–20 wildfires are unlikely to recover to their pre-fire population size, especially if recovery is interrupted by comparable fires within 10–20 years. Conservation efforts taken in the period since the fires have provided some local and regional benefits, but the extent of such benefit is not yet well resolved.

Recommendations

- Increase the likelihood of recovery (and build resilience) for fire-affected mammal species through the more pervasive and effective implementation of programs for the control of threats that are more tractable to manage (especially habitat loss, timber-harvesting, and introduced predators).
- Expand monitoring programs to more effectively chart recovery and evaluate the effectiveness of recovery actions taken after the 2019–20 wildfires.
- Undertake research on the impact of fire (and fire regimes) on mammal species for which fire responses are poorly known, to help inform assessments of fire impacts in the future.
- Undertake research on the impacts of smoke on mammal species, to allow for a more robust estimate of mortality beyond the fire perimeter.
- Within largely burnt regions, instigate active management programs that aim to protect unburnt (especially long-unburnt) patches from further fire.
- Identify other key sites, such as natural fire refugia, for the conservation of threatened mammals and seek to recognise these in fire management plans as assets for protection during fire operations.
- Implement legislation, policy and practice that more effectively and rapidly reduces greenhouse gas emissions – else, there will be an escalating rate of destructive fires that will be beyond the capacity of many mammal species to survive or recover from.

References

Arthur AD, Catling PC, Reid A (2012) Relative influence of habitat structure, species interactions and rainfall on the post-fire population dynamics of ground-dwelling vertebrates. *Austral Ecology* **37**, 958–970. doi:10.1111/j.1442-9993.2011.02355.x

Banks SC, Dujardin M, McBurney L, Blair D, Barker M, *et al.* (2011a) Starting points for small mammal population recovery after wildfire: recolonisation or residual populations? *Oikos* **120**, 26–37. doi:10.1111/j.1600-0706.2010.18765.x

Banks SC, Knight EJ, McBurney L, Blair D, Lindenmayer DB (2011b) The effects of wildfire on mortality and resources for an arboreal marsupial: resilience to fire events but susceptibility to fire regime change. *PLoS One* **6**, e22952. doi:10.1371/journal.pone.0022952

Baranowski K, Faust CL, Eby P, Bharti N (2021) Quantifying the impacts of Australian bushfires on native forests and gray-headed flying foxes. *Global Ecology and Conservation* **27**, e01566. doi:10.1016/j.gecco.2021.e01566

Berry LE, Driscoll DA, Banks SC, Lindenmayer DB (2015) The use of topographic fire refuges by the greater glider (*Petauroides volans*) and the mountain brushtail possum (*Trichosurus cunninghami*) following a landscape-scale fire. *Australian Mammalogy* **37**, 39–45. doi:10.1071/AM14027

Bino G, Hawke T, Kingsford RT (2021) Synergistic effects of a severe drought and fire on platypuses. *The Science of the Total Environment* **777**, 146137. doi:10.1016/j.scitotenv.2021.146137

Bradstock RA (2008) Effects of large fires on biodiversity in south-eastern Australia: disaster or template for diversity? *International Journal of Wildland Fire* **17**, 809–822. doi:10.1071/WF07153

Burns PA, Atkins ZS (2021) *Gliding from the Ashes: Post-fire Surveys for the Greater Glider and Yellow-bellied Glider in Far East Gippsland*.' Aussie Ark Australian Endangered Species Recovery Fund.

Camac JS, Williams RJ, Wahren CH, Hoffmann AA, Vesk PA (2017) Climatic warming strengthens a positive feedback between alpine shrubs and fire. *Global Change Biology* **23**, 3249–3258. doi:10.1111/gcb.13614

Collins L, Bradstock RA, Clarke H, Clarke MF, Nolan RH, *et al.* (2021) The 2019/2020 mega-fires exposed Australian ecosystems to an unprecedented extent of high-severity fire. *Environmental Research Letters* **16**, 044029. doi:10.1088/1748-9326/abeb9e

Cripps JK, Nelson JL, Scroggie MP, Durkin LK, Ramsey DSL, *et al.* (2021) Double-observer distance sampling improves the accuracy of density estimates for a threatened arboreal mammal. *Wildlife Research* **48**, 756–768. doi:10.1071/WR19136

Cristescu RH, Gardiner R, Terraube J, McDonald K, Powell D, *et al.* (in press) Difficulties of assessing the impacts of the 2019–2020 bushfires on koalas. *Austral Ecology* doi:10.1111/aec.13120.

Fleay D (1947) *Gliders of the Gum Trees*. Bread and Cheese Club, Melbourne.

Fox BJ (1982) Fire and mammalian secondary succession in an Australian coastal heath. *Ecology* **63**, 1332–1341. doi:10.2307/1938861

Goldingay RL, Kavanagh RP (1991) The Yellow-bellied Glider: a review of its ecology, and management considerations. In *Conservation of Australia's forest fauna*. (Ed. D Lunney) pp. 365–375. Royal Zoological Society of New South Wales, Sydney.

Green K, Sanecki G (2006) Immediate and short-term responses of bird and mammal assemblages to a subalpine wildfire in the Snowy Mountains, Australia. *Austral Ecology* **31**, 673–681. doi:10.1111/j.1442-9993.2006.01629.x

Hohnen R, Murphy B, Gates J, Legge S, Dickman C, *et al.* (2019) Detecting and protecting the threatened Kangaroo Island dunnart (*Sminthopsis fuliginosus aitkeni*). *Conservation Science and Practice* **1**, e4. doi:10.1111/csp2.4

Hradsky BA (2020) Conserving Australia's threatened native mammals in predator-invaded, fire-prone landscapes. *Wildlife Research* **47**, 1–15. doi:10.1071/WR19027

Law B, Doty A, Chidel M, Brassil T (2018) Bat activity before and after a severe wildfire in Pilliga forests: Resilience influenced by fire extent and landscape mobility? *Austral Ecology* **43**, 706–718. doi:10.1111/aec.12617

Law BS, Gonsalves L, Burgar J, Brassil T, Kerr I, *et al.* (2022) Fire severity and its local extent are key to assessing impacts of Australian mega-fires on koala (*Phascolarctos cinereus*) density. *Global Ecology and Biogeography* **31**, 714–726. doi:10.1111/geb.13458

Law BS, Madani G, Lloyd A, Gonsalves L, Hall L, *et al.* (in press) Habitat occupancy of a rainforest-dependent bat after Australia's 2019–20 mega-fires. *Animal Conservation*.

Lefoe M, Rendall AR, McKinnon F, Whisson DA (2022) Logging and wildfire limit the distribution of a vulnerable arboreal mammal. *Forest Ecology and Management* **503**, 119773. doi:10.1016/j.foreco.2021.119773

Legge S, Woinarski JCZ, Scheele BC, Garnett ST, Lintermans M, *et al.* (2022) Rapid assessment of the biodiversity impacts of the 2019–2020 Australian megafires to guide urgent management intervention and recovery and lessons for other regions. *Diversity & Distributions* **28**, 571–591. doi:10.1111/ddi.13428

Legge SM, Rumpff L, Woinarski JCZ, Whiterod NS, Ward M, *et al.* (in press) The conservation impacts of ecological disturbance: time-bound estimates of population loss and recovery for fauna affected by the 2019–2020 Australian megafires. *Global Ecology and Biogeography* doi:10.1111/geb.13473.

Lindenmayer D, Taylor C (2020a) Extensive recent wildfires demand more stringent protection of critical old growth forest. *Pacific Conservation Biology* **26**, 384–394. doi:10.1071/PC20037

Lindenmayer DB, Taylor C (2020b) New spatial analyses of Australian wildfires highlight the need for new fire, resource, and conservation policies. *Proceedings of the National Academy of Sciences of the United States of America* **117**, 12481–12485. doi:10.1073/pnas.2002269117

Lindenmayer DB, Blanchard W, McBurney L, Blair D, Banks SC, *et al.* (2013) Fire severity and landscape context effects on arboreal marsupials. *Biological Conservation* **167**, 137–148. doi:10.1016/j.biocon.2013.07.028

Lindenmayer D, Blair D, McBurney L, Banks S, Bowd E (2021) Ten years on – a decade of intensive biodiversity research after the 2009 Black Saturday wildfires in Victoria's Mountain Ash forest. *Australian Zoologist* **41**, 220–230. doi:10.7882/AZ.2020.041

Lunney D (1987) Effects of logging, fire and drought on possums and gliders in the coastal forests near Bega, NSW. *Australian Wildlife Research* **14**, 263–274. doi:10.1071/WR9870263

Lunney D, Gresser SM, Mahon PS, Matthews A (2004) Post-fire survival and reproduction of rehabilitated and unburnt koalas. *Biological Conservation* **120**, 567–575. doi:10.1016/j.biocon.2004.03.029

Lunney D, Gresser S, O'Neill LE, Matthews A, Rhodes J (2007) The impact of fire and dogs on koalas at Port Stephens, New South Wales, using population viability analysis. *Pacific Conservation Biology* **13**, 189–201. doi:10.1071/PC070189

McLean CM, Kavanagh RP, Penman T, Bradstock R (2018) The threatened status of the hollow dependent arboreal marsupial, the greater glider (*Petauroides volans*), can be explained by impacts from wildfire and selective logging. *Forest Ecology and Management* **415–416**, 19–25. doi:10.1016/j.foreco.2018.01.048

Nitschke CR, Trouvé R, Lumsden LF, Fedrigo M, Bennett LT, *et al.* (2020) Spatial and temporal dynamics of habitat availability and stability for a critically endangered arboreal marsupial: implications for conservation planning in a fire-prone landscape. *Landscape Ecology* **35**, 1553–1570. doi:10.1007/s10980-020-01036-2

Peters A, Hume S, Raidal S, Crawley L, Gowland D (2021) Mortality associated with bushfire smoke inhalation in a captive population of the Smoky Mouse (*Pseudomys fumeus*), a threatened Australian rodent. *Journal of Wildlife Diseases* **57**, 199–204. doi:10.7589/JWD-D-20-00026

Phillips S, Wallis K, Lane A (2020) *Quantifying the Impacts of Bushfire on Populations of Wild Koalas (*Phascolarctos cinereus*): Insights from the 2019/20 Fire Season.* WWF-Australia, Sydney.

Robley A, Gormley A, Forsyth D, Triggs B (2014) Long-term and large-scale control of the introduced red fox increases native mammal occupancy in Australian forests. *Biological Conservation* **180**, 262–269. doi:10.1016/j.biocon.2014.10.017

Rutter N, Blake B (2020) 'Post-fire assessment of Greater Gliders in East Gippsland: Interim Report'. WWF-Australia, Sydney.

Smith P, Smith J (2020) Future of the greater glider (*Petauroides volans*) in the Blue Mountains, New South Wales. *Proceedings of the Linnean Society of New South Wales* **142**, 55–66.

Smith P, Smith J (2021) 'Post-fire survey of arboreal mammals at Wombeyan and Jenolan – interim report. Report Prepared for Glideways'. The Great Eastern Ranges and WWF-Australia, Sydney.

Todd CR, Lindenmayer DB, Stamation K, Acevedo-Cattaneo S, Smith S, *et al.* (2016) Assessing reserve effectiveness: application to a threatened species in a dynamic fire prone forest landscape. *Ecological Modelling* **338**, 90–100. doi:10.1016/j.ecolmodel.2016.07.021

Tyndale-Biscoe CH, Smith RFC (1969) Studies on the marsupial glider, *Schoinobates volans* (Kerr): III. Response to habitat destruction. *Journal of Animal Ecology* **38**, 651–659. doi:10.2307/3041

Wagner B, Baker PJ, Stewart SB, Lumsden LF, Nelson JL, *et al.* (2020) Climate change drives habitat contraction of a nocturnal arboreal marsupial at its physiological limits. *Ecosphere* **11**, e03262. doi:10.1002/ecs2.3262

Whelan RJ, Rodgerson L, Dickman CR, Sutherland EF (2002) Critical life cycles of plants and animals: developing a process-based understanding of population changes in fire-prone landscapes. In *Flammable Australia: the Fire Regimes and Biodiversity of a Continent*. (Eds RA Bradstock, JE Williams and MA Gill) pp. 94–124. Cambridge University Press, Cambridge, UK.

Woinarski JCZ, Burbidge AA, Harrison PL (2015) The ongoing unravelling of a continental fauna: decline and extinction of Australian mammals since European settlement. *Proceedings of the National Academy of Sciences of the United States of America* **112**, 4531–4540. doi:10.1073/pnas.1417301112

17

How introduced animals compound the effects of fire on native plants and animals

Sarah M. Legge, David H. Duncan, David M. Forsyth, Kate Giljohann, Katja Hogendoorn, Rosemary Hohnen, Bronwyn Hradsky and Mark Lintermans

Summary

- Introduced fauna threaten Australia's native species via herbivory, predation, and competition. After fire, these effects can be amplified.
- Post-fire impacts from introduced animals can cause mortality in the months following fire to exceed mortality caused during the fire. The impacts of introduced animals can persist years after the fire event – for example, if fires allow expansion of introduced species into new areas.
- Introduced herbivores influence the greatest range of native species post-fire, affecting plants (consumption, trampling), terrestrial animals (competition, removal of cover), and aquatic animals (trampling riparian vegetation, fouling streams).
- Some introduced species are a particular threat for specific taxa. For example, introduced terrestrial predators threaten many mammals, and introduced fish threaten native fish.
- Control programs for introduced species began quickly after the 2019–20 wildfires, and at increased scales. The open post-fire canopy improved visibility over extensive areas, briefly allowing aerial shooting and baiting to occur in places where such techniques are not usually feasible. The wildfires also reduced populations of introduced species, providing opportunities to achieve suppression.
- Extensive, severe fires provide opportunities to expand introduced species control, such as preventing the reinvasion of introduced fish into streams where they have been extirpated by fire-related sediment pulses; or by delaying apiary activities in forests recovering from fire, so that native species recolonise tree hollows before feral bees.

- Predicting the outcomes of interactions between introduced species and wildfire is crucial for shaping effective and efficient management in future fires, but such predictions will remain challenging until long-term monitoring programs are established that track how populations of introduced and native species respond to wildfires, and if/how management of introduced species supports recovery.

Introduction

Images of blackened tree stems stretching to the horizon and reports of charred animal bodies littering the ground following fires are confronting. Nonetheless many Australian ecosystems have mechanisms to survive fire events: plants may resprout or regenerate from canopy or soil-stored seed (Chapter 9), and many animal species from fire-prone ecosystems can survive the fire event itself either by moving away or sheltering from the flames, heat and smoke. However, after fires, regenerating or recruiting plants, and surviving animals, need to survive extended periods in a post-fire environment with changed habitat structure, resource availability, competition and predation risks. The ability of plant and animal species to survive the post-fire period can strongly depend on their sensitivity to other threats, including introduced animals.

Introduced animal species have had, and continue to have, profound effects on many Australian plants and animals. Introduced herbivores, terrestrial predators, aquatic predators (fish) and invertebrates (such as honey bees) can all exacerbate pressure on populations of native species after fire. Below we present the contextual evidence for compounding effects from key groups of introduced animals, specific observations relating to the 2019–20 wildfires, management responses that were instigated, and priorities for future research and management. At the end of the chapter we summarise the shared themes across these groups of introduced animals.

Introduced herbivores

Introduced herbivores, such as house mice (*Mus musculus*), pigs (*Sus scrofa*), deer (*Cervus* spp.) and feral horses (*Equus caballus*), can have strong direct effects on native species through grazing and browsing, or indirect impacts like trampling, wallowing, compaction and competition for resources. Hooved species (ungulates) are functionally distinct from native herbivores and are typically considered to pose the greatest threat in forested environments, but locally 'overabundant' native herbivores can also become a problem. In some degraded woody ecosystems, herbivores may pose a critical threat even at low to moderate densities, by preventing ecosystem-structuring plants from regenerating (Bennett *et al.* 2020).

Herbivore impacts can be amplified when combined with other key landscape drivers, such as fire and drought. Herbivores may shelter in unburnt areas and then exploit sensitive recovering growth in surrounding burnt areas (Fig. 17.1) (Robinson *et al.* 2013). Larger herbivores also target recently burnt areas for the higher quality forage (Fuhlendorf *et al.* 2009). Post-fire herbivory can reduce vegetation regeneration, recruitment, survival and cover; influence plant community composition (e.g. when more palatable species are targeted); and increase competition for resources with co-occurring native fauna (Tuft *et al.* 2012). Pre- and post-fire, droughts stress ecological communities and may exacerbate herbivore impacts (Giljohann *et al.* 2017). Fires may reduce the abundance of some

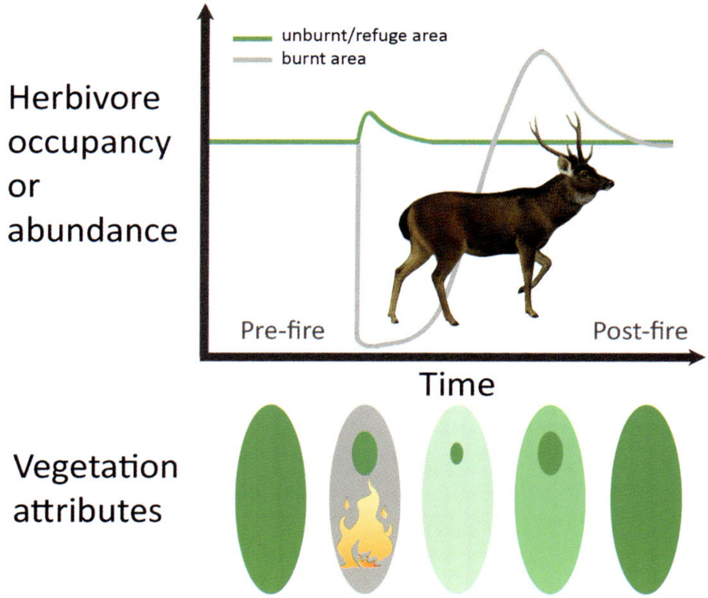

Fig. 17.1. Conceptual model of herbivore response to extensive fire. Herbivore occupancy/abundance is reduced in burnt areas (grey line) but is unchanged (or slightly increased) in unburnt refuges (green line). The post-fire flush of fast-growing, palatable understorey species in burnt areas attracts herbivores. The high food quality and quantity increase herbivore survival and reproductive rates (immigrants and survivors) in burnt areas, maximising population growth there. The herbivore population eventually stabilises, then declines as food quality and quantity change due to forest succession and the pressure of herbivory on understorey plants. The changing vegetation attributes post-fire are indicated by darkening green shading.

introduced herbivores, but only temporarily. For example, Forsyth *et al.* (2012) reported that introduced sambar deer (*Cervus unicolor*) abundance was substantially reduced immediately after the 2009 Victorian wildfires, but that nearly all burnt habitat was reoccupied 24 months post-fire. Similarly, introduced rodents recolonise burnt areas and increase rapidly after fire, outcompeting native rodents for limited food resources and slowing plant recovery (Friend 1993).

Major disturbance events, such as extensive wildfires, may radically alter the risk from herbivores. For example, of 92 ecosystems highly impacted by the 2019–20 wildfires, 11 were considered highly exposed (> 50% distribution) to introduced herbivores post-fire (Keith *et al.* 2022). These included rainforests, wet eucalypt forests, peatlands and subalpine woodlands. European rabbits (*Oryctolagus cuniculus*) and deer were the most pervasive introduced herbivore taxa across affected ecosystems, although feral horses and feral pigs were locally important. In a separate assessment, 97 plant species were considered highly vulnerable to introduced herbivore impacts after the 2019–20 wildfires, and 44 additional plant species were considered at medium risk (Gallagher *et al.* 2021). Excluding large introduced herbivores from regenerating and unburnt vegetation to maintain habitat condition and reduce competition was considered an urgent management action for 62 of 119 priority animal species following the 2019–20 wildfires, and possibly required for an additional 14 priority animal species (Legge *et al.* 2022).

Introduced herbivore control after the 2019–20 wildfires

Following the 2019–20 wildfires, introduced herbivore control was a management focus across many fire-affected areas. For example, the Commonwealth Government invested $67 million across seven fire-affected regions; in six of those regions, local stakeholders considered that 20–30% of their region's funding should be directed to introduced animal control, with most or all of that for introduced herbivores (https://www.dcceew.gov.au/environment/biodiversity/bushfire-recovery/funding-support/regional-delivery-program).

State governments also invested heavily to reduce introduced herbivores impacts. Large-scale aerial shooting programs for deer, pigs, goats (*Capra hircus*) and feral cattle (*Bos taurus*) began immediately after the fires (DELWP 2020; DPIE 2020). Helicopter-based shooting of ungulates can be highly effective in non-forested habitats but is less effective in forests because animals are less detectable. However, severe fires removed the canopy over large areas, presenting a brief opportunity for landscape-scale aerial control in areas where this was not previously possible. State agencies sought not just to reduce immediate post-fire impacts, but also to reduce the populations of ungulates for longer-term benefit (DPIE 2021). For example, on Kangaroo Island the 2019–20 wildfires killed many feral pigs, and the South Australian Government is leveraging this opportunity to try to eradicate pigs from the island by 2023 (Korcz 2020).

Future research and management

The value of reducing herbivore populations post-fire is widely assumed, but has not been quantified. We know little about how population density, movement dynamics, or occupancy of introduced herbivores change after extensive fire, or how these attributes relate to herbivore impacts post-fire (but see Forsyth *et al.* 2012). These knowledge gaps limit optimal allocation of resources to remediation actions across the landscape after extensive fires. There are also few studies that explore how fire–herbivore interactions are conditioned by other threats (e.g. timber harvesting, disease, predation), variation in fire characteristics (e.g. extent, severity), and changing climate (including rainfall or temperature anomalies). Severe fires often follow years of below-average rainfall (Chapter 2) and, as novel conditions resulting from Anthropogenic climate change continue, studies that evaluate the interaction between herbivores and fire under a range of conditions are needed.

Field surveys can reveal the immediate impacts of fire and post-fire herbivore activity; however, only long-term monitoring can uncover whether these immediate responses will have ongoing detrimental impacts on native species or communities. However, given the stochastic nature of extensive and severe fire, and the complexity of interacting factors, modelling and simulation are also critical for investigating the knowledge gaps outlined here.

Introduced terrestrial predators

Two introduced predators (red fox (*Vulpes vulpes*), feral cat (*Felis catus*)) have had substantial impacts on Australian native species, especially terrestrial small to medium-sized mammals, ground-dwelling birds, and other species that live on or near the ground (Stobo-Wilson *et al.* 2022). Fire may amplify the predation risk for these animals, for several reasons (Fig. 17.2):

- Understorey vegetation, leaf litter, coarse woody debris and hollow trees provide important shelter for native species; their consumption by fire may increase the hunting success of predators.

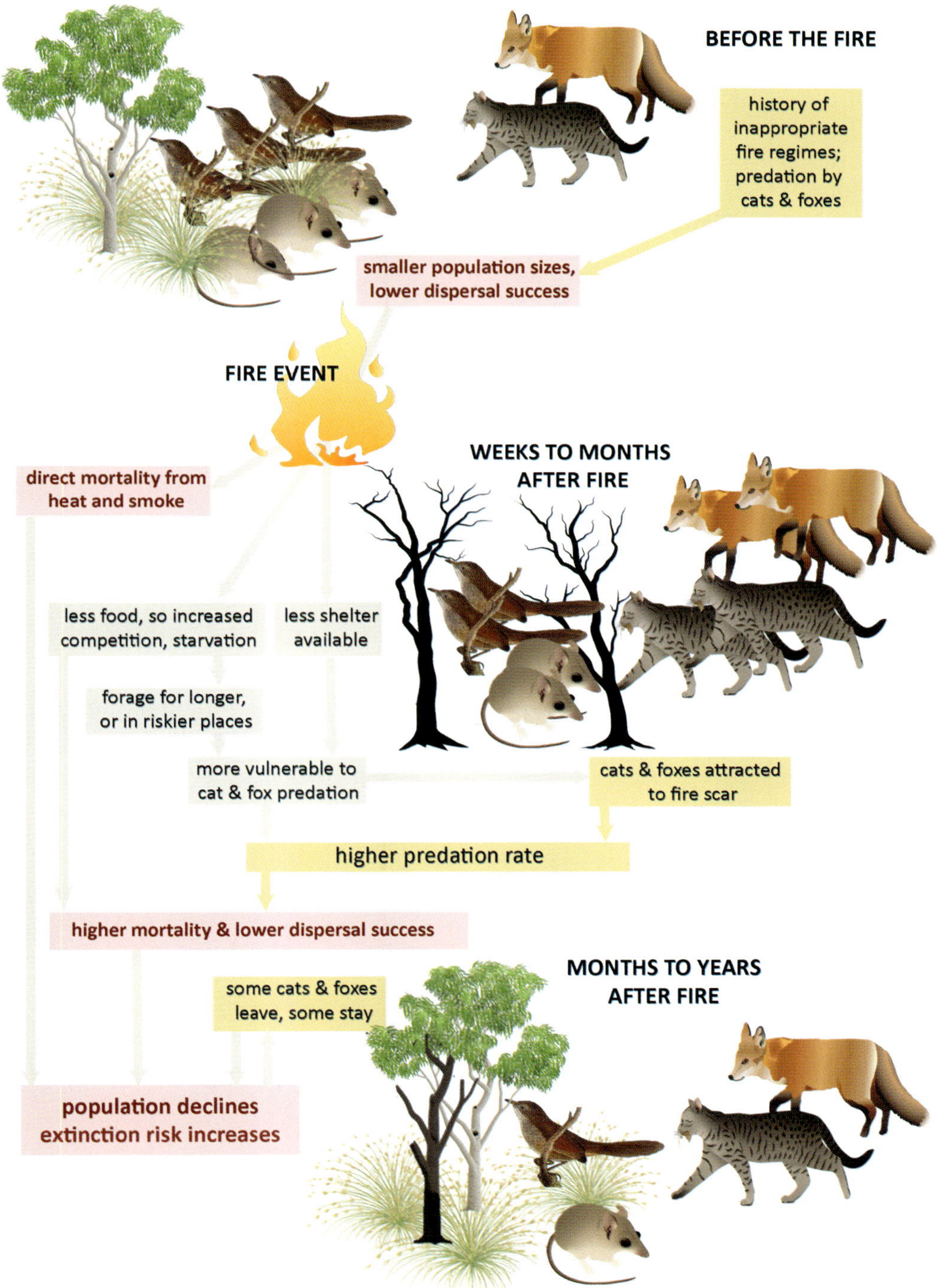

Fig. 17.2. A summary of the mechanisms by which introduced terrestrial predators affect native species pre- and post-fires.

- If fire reduces food availability for native species (or increases competition), individuals may forage for longer or in riskier locations, again increasing the hunting success of predators.
- Opportunities for greater hunting success may, in turn, attract predators to burnt areas, further increasing predation pressure on native species.
- Reductions in native animal population size or population connectivity post-fire (from any cause), reduces the resilience of surviving populations to ongoing predation pressure.
- Populations of native species may already be small and fragmented pre-fire due to prevailing threats including predation from introduced predators, drought, and inappropriate fire regimes. Small, isolated populations are intrinsically more vulnerable to additional losses, caused directly or indirectly by fire, and less likely to recover through reproduction or immigration.

Foxes and cats use fire scars intensively in the weeks post-fire. In south-eastern Australia, foxes with home ranges that overlapped prescribed burns intensified their use of these areas immediately post-fire (Hradsky *et al.* 2017). Similarly, in northern Australian savannas, cats selectively used burnt areas in the first two months post-fire, with individuals travelling up to 13 km from their home range to access some fire scars (McGregor *et al.* 2016). Unburnt patches within large fire scars may be especially attractive if prey densities along the burnt-unburnt edges are high (Parkins *et al.* 2018). The attractiveness of fire scars to foxes and cats appears short-lived – in the order of weeks to a couple of years – perhaps reflecting declines in prey availability from population suppression or declines in hunting success from habitat recovery (Hradsky 2020).

Diet studies have shown that foxes increase their relative consumption of mammals post-fire (Green and Sanecki 2006), concentrating on medium-sized native mammals such as bandicoots, possums and echidnas (Robley *et al.* 2016; Hradsky *et al.* 2017). The effects of fire on cat diet have not been studied, but in north-western Australia, the small mammal mortality rate increased 20-fold after severe fire due to predation, including by cats (Leahy *et al.* 2016), and cats hunted more efficiently in burnt areas (McGregor *et al.* 2015); so the mammal component of cat diet may have increased post-fire.

Since introduced predators are attracted to burnt areas and hunt more efficiently there, it is unsurprising that several studies have shown that fire increases the vulnerability of small to medium-sized native mammals, reptiles, and ground-dwelling birds to foxes, cats, and also to native predators (e.g. Leahy *et al.* 2016; Moore *et al.* 2018; Comer *et al.* 2020). These studies occurred after much smaller fires than the 2019–20 wildfires. Very large, severe fires may also cause substantial mortality to cats and foxes, potentially lowering predation risk to native animals that survive the fire. However, if prey populations are reduced, and are more accessible to predators, the net predation risk after large severe fire may still increase. For example, on Kangaroo Island, cat activity decreased by ~60% after severe fire and had not recovered a year later (Box 17.1). Cats on Kangaroo Island were attracted to small unburnt habitat patches that harboured refuge populations of native mammals (Chapter 24), suggesting that cat impacts in these areas could exceed pre-fire impacts. In another example, long-footed potoroo (*Potorous longipes*) populations in Victoria were much less likely to occur in areas burnt severely in the 2019–20 wildfires compared to unburnt locations. They were also substantially more likely to occur at sites with lower predicted fox densities (DELWP 2021).

Predation is probably the main pathway through which foxes and cats affect native prey species post-fire, but other effects are possible. Native predators including goannas

(*Varanus* spp.) and spotted-tail quolls (*Dasyurus maculatus*) may experience increased competition for prey, if all predators must compete for a smaller resource base. In another example, foxes and cats may capitalise on introduced rodent and rabbit populations if these recover more quickly than native prey post-fire. This might dampen the population increase of the introduced prey, but could also maintain fox and cat populations at elevated densities, delaying recovery by native species. We know very little about these potential pathways, but some likely interactions have been described elsewhere in Australia (e.g. Glen 2014; Pedler *et al.* 2016).

Box 17.1. Post-fire changes in feral cat abundance on Kangaroo Island

Much research into the response of feral cats to fire has occurred in northern Australia, with fewer studies from southern Australia, where fires are less frequent but typically much more intense. Assessing the impacts of large, intense fires on feral cats is inherently difficult as the location and timing of these fires is difficult to predict, so such studies require a chance overlap between fires and an area where predator monitoring has occurred.

The 2019–20 wildfires were the largest in Kangaroo Island's recorded history, burning 46% of the island, including most of the conservation reserve network, and mostly at very high severity. Opportunistically, we were able to examine the impact of this fire on the island's feral cat population, because of a baseline sample just before the fire (Hohnen *et al.* 2021). Sampling of these same sites was conducted 3–8 months post-fire (five sites in burnt areas; two sites in unburnt areas) (Fig. 17.3), and three of these sites were resampled one year after the fire

Three to eight months post-fire, at sites that were burnt by the severe fire, the number of individual cats declined by 57% and the nightly detection rates declined by 65%, compared to pre-fire levels. At the two burnt sites resampled a year post-fire, cat activity was still less than half that of pre-fire levels, suggesting no or limited recovery.

Although post-fire cat activity was lower in burnt areas, the remaining cats could still inhibit recovery of (much-reduced) native wildlife populations, particularly if they

Fig. 17.3. A feral cat on Kangaroo Island.

targeted small unburnt vegetation patches, where surviving wildlife likely congregated. On Kangaroo Island, and other areas burnt by high-severity fires, feral cat control may be important to support recovery of populations of fire-affected species. Such emergency interventions are likely to be required more frequently as the climate warms and fire frequency, severity and extent increase.

Introduced predator control after the 2019–20 wildfires

Of 76 mammal, bird, reptile and frog species prioritised by the Commonwealth Government for management intervention after the 2019–20 wildfires, introduced predator control was considered urgent for 30 species, and possibly required for an additional 23 species (Legge *et al.* 2022).

Given the threat from introduced predators, government and non-government organisations intensified their control programs post-fire. In South Australia, the Kangaroo Island Landscape Board, in conjunction with the National Parks and Wildlife Service, began controlling cats in state-owned areas on the island, while a local non-profit organisation (Kangaroo Island Land for Wildlife) did so within a network of private landholders (see Chapters 24, 25). Trapping and poison-baiting occurred across the fire scar to reduce cat impacts on fire-affected threatened species, including the Kangaroo Island dunnart (*Sminthopsis griseoventer aitkeni*) (DEW 2020). Over 1100 cats were removed in the 2 years following the fires (M. Heard, P. Jennings, H. Groffen, P. Hodgens, *pers. comm.*). One unburnt site with surviving dunnarts was encircled by a cat-proof fence (Chapter 24), and artificial ground shelters were installed on private properties to provide refuge for native animals from feral cats (Dickman and McDonald 2020).

In New South Wales, the National Parks and Wildlife Service increased the spatial scale of poison-baiting fivefold immediately after the 2019–20 fires, and have maintained baiting operations since then (DPIE 2020, 2021). Preliminary monitoring data show that wild dog and fox activity decreased in some areas, although decoupling declines from seasonal patterns is challenging. Cat activity has not shown a consistent change in response to the baiting (P. Mahon, *pers comm.*). In Cape Arid National Park, on the south coast of Western Australia, the Department of Biodiversity, Conservation and Attractions intensified cat control around the western ground parrot (*Pezoporus wallicus flaviventris*) population, with additional aerial and ground baiting, followed by trapping and culling, in the summer and spring following the fires (S. Comer, *pers. comm.*).

In Victoria, the Department of Environment, Land, Water and Planning used biodiversity models and expert workshops to develop a strategic fire response with prioritised cost-effective actions. As part of a theme of 'Intensified and sustained management of threats', > $3 million was invested to control foxes and feral cats in the first 3 years post-fire. The control programs are collaborative and landscape-scale; they target burnt and adjacent unburnt areas that are likely to give the greatest return on investment and benefit the full range of species across Victoria. Actions include re-establishing Southern Ark (a landscape-scale fox management program in East Gippsland), and infilling and expanding this program by another 100 000 ha; expanding the Barry Mountains fox control program, with a focus on protecting the endangered long-footed potoroo; controlling foxes to protect threatened species such as brush-tailed rock-wallabies (*Petrogale penicillata*) in the Upper Snowy River and Little River regions; baiting and trapping for foxes across public and private land in fire-affected areas; and piloting feral cat control programs within priority areas.

Future research and management

The efficacy of predator control for supporting the persistence of native species after fire is poorly studied (Bendall *et al.* 2021). Thus, monitoring data (of predators and prey) from the predator control programs currently underway will inform responses to future fires. Given what we know about fire and introduced predator interactions, landscape-scale control of introduced predators, coupled with intensive management (including artificial refuges, fencing) at sites with small populations of susceptible species, seems the best combination of post-fire actions. However, although this approach may be feasible for foxes, there is no landscape-scale control approach for cats in eastern Australia, and this is a serious implementation gap. Improving our understanding of how to manage post-fire predation pressure to support native species survival, recruitment, population density and persistence remains a priority.

Introduced invertebrates: the honey bee

Honey bees (*Apis mellifera*) were introduced into Australia in the 1820s. Helped by our mild winters and abundant, nectar-rich trees, honey bees have become feral in much of temperate and subtropical Australia. Feral honey bees can reach high abundance if they have access to fresh water and high densities of old eucalypt trees, whose hollows are perfect for nesting bee colonies; densities can exceed 160 colonies/ha (Wood and Wallis 1998).

Honey bees affect Australian native biota in several ways. They pollinate many native plant species, but they also pollinate weed species, alter plant-pollinator networks and seed-set patterns, and transmit bee and plant pathogens (Mallinger *et al.* 2017). Feral honey bees compete for hollows with birds and mammals (Berris and Barth 2020). In addition, because of their sheer numbers, their ability to forage at lower temperatures than solitary bee species, and because they can rapidly recruit foragers to rich resources through dancing, honey bees can outcompete nectarivorous birds, native bees and other insects for floral resources (Paton 2000; Mallinger *et al.* 2017). Such resource depletion could in turn reduce food availability for insectivorous vertebrates, which generally do not feed on honey bees due to their powerful defensive stings. However, this has not been investigated.

The effect of fire on feral honey bees is little studied. Overseas, opportunities to examine these issues are limited, as feral bee densities are often extremely low due to the effects of *Varroa* mite and associated viruses (Oldroyd 1999). While the effects of fire on bee composition and plant pollinator networks have been investigated in several countries (Ne'eman *et al.* 2000; Banza *et al.* 2021), the studies generally pertain to small burnt areas (60–450 ha), much smaller than the 10 km honey bee foraging and dispersal range. However, some inferences can be made from the limited information available. During intense fires, colonies in trees will burn, heat can melt the wax, or smoke can kill bees directly. However, feral colonies in caves and under cliffs may be unaffected. Feral hives high up in the trees may survive low severity fires that pass under them. After the fire, the burnt landscape will temporarily contain few floral resources. This is likely to have caused some feral honey bee populations to decline after the 2019–20 wildfires.

Feral honey bee recolonisation rates after fire depend on whether fresh water and nesting opportunities (tree hollows) are available (Fig. 17.4), the size of the area burnt, the size of the source population of managed and feral colonies at the edge of the burnt area, and whether

Fig. 17.4. Feral honey bees compete with native species for tree hollows. (Photo: Keith J. Smith/ Alamy)

managed hives are allowed into the recovering bushland. Fire can promote tree hollow creation, causing more hollows to be available for feral honey bees. However, fires, especially repeated fires, can remove hollows as trees with existing hollows burn readily. This would increase the competition for remaining hollows. Hollow recolonisation is likely to be rapid: depending on the conditions, colonies produce 0.9–3.6 swarms/season (Oldroyd *et al.* 1997), and swarms can disperse up to 10 km (Villa 2004). In Australia, re-occupation rate after eradicating all colonies in a 500 ha area was 15 colonies/km^2 per year (Oldroyd *et al.* 1997).

Once floral resources recover, feral honey bees are likely to rapidly recolonise forested areas after burns. Where fire increases hollow availability, bees are likely to return in higher densities than before; where fire reduces hollow availability, competition between honey bees and native hollow-nesters will increase. Thus, recolonisation is likely to affect the recovery of native ecosystems after fire through competition with native species for floral resources and hollows, and through pollination of weeds.

Future management

Eradication of feral honey bees in native vegetation is not possible (Oldroyd 1998). Prohibiting beekeeping within and near burnt forests could help delay recolonisation by feral honey bees.

Aquatic systems: fire and introduced fish

Post-fire rainfall events can devastate fish populations when high surface flows and erosion causes ash and sediment slugs in streams that travel 50–80 km downstream of the burnt area (Lyon and O'Connor 2008; Silva *et al.* 2020; Chapter 6). Sedimentation can cause immediate fish mortality events, and can have longer-term impacts by simplifying substrates and infilling pools. These effects can interact with introduced, invasive fish species either negatively or positively and an example of each is presented below.

Fire impacts can alter instream habitats for years or decades, disadvantaging species with specialist habitat requirements and benefitting generalist species. For example, fish species that require particular substrates as breeding or refuge sites (e.g. cobble for two-spined blackfish (*Gadopsis bispinosus*); Lintermans 2007), or high habitat diversity for a range of ecological needs (such as feeding, courtship and spawning) will be disadvantaged, particularly in aquatic environments impacted by other anthropogenic stressors such as flow regulation (Leprieur *et al.* 2008). Invasive species are often more 'generalist' than native species, with wider temperature, salinity and physiological tolerances; larger body size, dispersal capacity, longevity, higher fecundity and delayed maturation. They can adapt to living in altered habitats, and rapidly populate degraded areas (Liu *et al.* 2017). Consequently, invasive species can potentially take advantage of altered ecological conditions and widely disperse into such environments following disturbances (Dunham *et al.* 2003). Following the 2019–20 wildfires, trout (*Oncorhynchus mykiss, Salmo trutta*) abundance rapidly rebounded within 12 months after initially being severely reduced (~80%) post-fire (M. Lintermans, unpublished data). Similarly, in lowland rivers, highly invasive carp (*Cyprinus carpio*) and gambusia (*Gambusia holbrooki*) are expected to rapidly rebound to the detriment of native species (Koehn 2004).

For 16 fish species prioritised by the Commonwealth Government for management intervention following the 2019–20 wildfires, controlling invasive fish was considered a potential action for eight species. Invasive fish were also considered to threaten up to eight of 22 priority spiny crayfish species, and up to five frog species (Legge *et al.* 2022).

Alternatively, fire impacts on invasive fish can present conservation opportunities. In upland environments, invasive trout prey upon small native fauna such as galaxias and spiny crayfish (McCormack 2012; Jarvis *et al.* 2019). Surveys in 2020 of fire-impacted streams in the NSW Alps recorded 'fishless' streams, where trout had probably been extirpated by ash and sediment slugs. If such extirpations occur upstream of barriers to trout recolonisation, these fishless streams present opportunities to reintroduce native fish, including the many threatened galaxias species (Lintermans *et al.* 2020) that generally cannot coexist with trout (TSS 2006; Jarvis *et al.* 2019). Occasionally, galaxias can naturally recolonise streams after trout are removed, if upstream refugial populations of galaxias persist (Lintermans 2000).

Future research and management

Identifying instream barriers to invasive fish colonisation would inform threatened fish reintroductions (Jones *et al.* 2021) and help prioritise native species reintroductions. New remote-sensing methodologies for detecting barriers (LIDAR and drones) are promising (Allan and Lintermans 2021). If invasive fish populations are extirpated upstream of partial barriers, these barriers could be augmented, but the action would need to be rapid (trout are very mobile and would soon recolonise fire-affected streams once partial barriers are breached (Dunham *et al.* 2003)). Where barriers prevent natural recolonisation (e.g. waterfalls > 2 m height, large dams), then recreational fish stocking and other human activities that bypass barriers should be prevented. Trout stocking programs to boost reduced populations are common following natural 'disasters' such as drought or fire, so dialogue between fish stocking agencies, angler groups and threatened fish managers is needed to capitalise on opportunities for fish conservation.

Conclusions

Fire tends to amplify the impacts of herbivory, predation and competition by introduced animals on native species. Our understanding of the interactions between introduced species and fire is based on studies of fire events much smaller than those of 2019–20. However, increasingly large and severe fires will be likely to worsen the impacts of introduced animals. Across animal groups, introduced herbivores affect the greatest range of native species. Of the 119 animal species prioritised for management intervention after the 2019–20 wildfires by the Commonwealth Government (Legge *et al.* 2022), introduced herbivore control was a key management action for 62–76 species across vertebrates and invertebrates; introduced terrestrial predator control was a key action for 30–53 species (mammals, birds, reptiles, frogs); and introduced fish control was identified as a key action for 3–21 frogs, fish and spiny crayfish species.

Fires, especially extensive, severe fires, can also provide opportunities to control introduced species more effectively, if access is improved (e.g. aerial shooting of ungulates in severely burnt forests); if the introduced species' population size is much reduced (e.g. feral pigs on Kangaroo Island); or if the introduced species becomes concentrated in areas and can be targeted more efficiently (e.g. cats hunting in unburnt fragments). In some situations, fires could provide novel conservation opportunities when populations of introduced species are extirpated, such as when post-fire sediment influxes to streams kill trout populations, making sites available for reintroductions of native species that were previously eliminated by trout.

For most Australian biodiversity, there is little or no long-term monitoring, making it difficult to assess the impacts of unexpected events. High-severity fire is likely to be increasingly frequent in temperate and subtropical forests; we urgently need monitoring programs to understand how populations respond to these fires, how threats such as introduced species affect that response, and how introduced animal management can support recovery.

Recommendations

- Interactions between fire and introduced animal species are a key pathway by which wildfire affects native species; therefore enhanced landscape-scale management of introduced species will help build resilience of native species to wildfire.
- After fire, landscape-scale control of introduced animal species across the burnt areas and adjacent unburnt areas, coupled with intensive management at sites with small populations of susceptible native species, needs to occur rapidly (days to weeks).
- Pre- and post-fire monitoring should be substantially enhanced to understand the impact of fires on the distribution, density and movement patterns of introduced species; their impacts on native species; and the efficacy of management interventions.
- We need a better understanding of how the interactions between fire and introduced species are themselves affected by other threats.
- A landscape-scale control option that is cost-effective for feral cats needs to be developed.
- Beekeeping within and near sites recovering from fire should be prohibited.
- The existing barriers to invasive fish should be identified, and partial/incomplete barriers augmented to create river reaches that could support native fish reintroductions when invasive fish populations are extirpated by fire.
- Recreational fish stocking policy and behaviours should be modified to support rather than compromise native fish reintroductions.

Acknowledgements

We thank Paul Mahon, Sarah Comer, Damien McMaster, M. Heard, P. Jennings, H. Groffen and P. Hodgens for providing unpublished information; and an anonymous reviewer, John Woinarski, Richard Glatz and Remko Leijs for helpful comments. Figures 17.1 and 17.2 made with symbols courtesy from the NESP Resilient Landscapes Hub (nesplandscapes.edu.au) and from the Integration and Application Network (ian.umces.edu/media-library).

References

Allan H, Lintermans M (2021) *Investigating the Utility of Drones to identify Potential Fish Refugia*. Report to Australian Society for Fish Biology Threatened Fishes Committee.

Banza P, Evans DM, Medeiros R, Macgregor CJ, Belo AD (2021) Short-term positive effects of wildfire on diurnal insects and pollen transport in a Mediterranean ecosystem. *Ecological Entomology* **46**, 1353–1363. doi:10.1111/een.13082

Bendall ER, Robinson NM, Lindenmayer DB, Morgain R, Legge S, *et al.* (2021) 'Conserving fauna in fire-prone landscapes: A review of fire-associated management actions that affect fauna conservation and recovery. Project report 8.4.4'. NESP TSR Hub, Brisbane.

Bennett A, Duncan DH, Rumpff L, Vesk PA (2020) Disentangling chronic regeneration failure in endangered woodland ecosystems. *Ecosphere* **11**, e02998. doi:10.1002/ecs2.2998

Berris KK, Barth M (2020) PVC nest boxes are less at risk of occupancy by feral honey bees than timber nest boxes and natural hollows. *Ecological Management & Restoration* **21**, 155–157. doi:10.1111/emr.12414

Comer S, Clausen L, Cowen S, Pinder J, Thomas A, *et al.* (2020) Integrating feral cat (*Felis catus*) control into landscape-scale introduced predator control to improve conservation management for threatened fauna: a case study from the south coast of Western Australia. *Wildlife Research* **47**, 762–778. doi:10.1071/WR19217

DELWP (2020) *Victoria's Bushfire Emergency: Biodiversity Response and Recovery, Version 2. August 2020*. Department of Environment, Land, Water and Planning, Victorian Government, Melbourne

DELWP (2021) 'Managing the Great Dividing Range Long-footed Potoroo population – fact sheet'. Department of Environment, Water and Planning, Heidelberg, Victoria.

DEW (2020) *Removing Feral Cats from Parks on Kangaroo Island*. Environment SA News. South Australian Department for Environment and Water, Adelaide.

Dickman C, McDonald T (2020) Some personal reflections on the present and future of Australia's fauna in an increasingly fire-prone continent. *Ecological Management & Restoration* **21**, 86–96. doi:10.1111/emr.12403

DPIE (2020) *Wildlife and Conservation Bushfire Recovery: Immediate Response January 2020*. Department of Planning, Industry and Environment, NSW Government, Sydney.

DPIE (2021) *Wildlife and Conservation Bushfire Recovery: Medium-term Response Plan*. Department of Planning, Industry and Environment, NSW Government, Sydney.

Dunham JB, Young MK, Gresswell RE, Rieman BE (2003) Effects of fire on fish populations: landscape perspectives on persistence of native fishes and nonnative fish invasions. *Forest Ecology and Management* **178**, 183–196. doi:10.1016/S0378-1127(03)00061-6

Forsyth DM, Gormley AM, Woodford L, Fitzgerald T (2012) Effects of large-scale high-severity fire on occupancy and abundances of an invasive large mammal in south-eastern Australia. *Wildlife Research* **39**, 555–564. doi:10.1071/WR12033

Friend GR (1993) Impact of fire on small vertebrates in mallee woodlands and heathlands of temperate Australia: a review. *Biological Conservation* **65**, 99–114. doi:10.1016/0006-3207(93)90439-8

Fuhlendorf SD, Engle DM, Kerby J, Hamilton R (2009) Pyric herbivory: rewilding landscapes through the recoupling of fire and grazing. *Conservation Biology* **23**, 588–598. doi:10.1111/j.1523-1739.2008.01139.x

Gallagher RV, Allen S, Mackenzie BD, Yates CJ, Gosper CR, *et al.* (2021) High fire frequency and the impact of the 2019–2020 megafires on Australian plant diversity. *Diversity & Distributions* **27**, 1166–1179. doi:10.1111/ddi.13265

Giljohann KM, McCarthy MA, Keith DA, Kelly LT, Tozer MG, *et al.* (2017) Interactions between rainfall, fire and herbivory drive resprouter vital rates in a semi-arid ecosystem. *Journal of Ecology* **105**, 1562–1570. doi:10.1111/1365-2745.12768

Glen AS (2014) Fur, feathers and scales: interactions between mammalian, reptilian and avian predators. In *Carnivores of Australia: Past, Present and Future*. (Eds AS Glen and CR Dickman) pp. 279–299. CSIRO Publishing, Collingwood.

Green K, Sanecki G (2006) Immediate and short-term responses of bird and mammal assemblages to a subalpine wildfire in the Snowy Mountains, Australia. *Austral Ecology* **31**, 673–681. doi:10.1111/j.1442-9993.2006.01629.x

Hohnen R, Woinarski J, Legge S, Dickman C, Murphy B (2021) 'Post-fire changes in feral cat density across burnt and unburnt landscapes, Kangaroo Island. NESP Threatened Species Recovery Hub Project 8.4.1 report'. NESP Threatened Species Recovery Hub, Brisbane.

Hradsky BA (2020) Conserving Australia's threatened native mammals in predator-invaded, fire-prone landscapes. *Wildlife Research* **47**, 1–15. doi:10.1071/WR19027

Hradsky BA, Mildwaters C, Ritchie EG, Christie F, Di Stefano J (2017) Responses of invasive predators and native prey to a prescribed forest fire. *Journal of Mammalogy* **98**, 835–847. doi:10.1093/jmammal/gyx010

Jarvis M, Closs G, Dedual M, Dorsey L, Fulton W, *et al.* (2019) The introduced trout and char of Australia and New Zealand. In *Trout and Char of the World*. (Eds JL Kershner, JE Williams, RE Gresswell and J Lobon-Cervia) pp. 573–604. American Fisheries Society, Bethesda, MD.

Jones PE, Tummers JS, Galib SM, Woodford DJ, Hume JB, *et al.* (2021) The use of barriers to limit the spread of aquatic invasive animal species: a global review. *Frontiers in Ecology and Evolution* **9**, 611631. doi:10.3389/fevo.2021.611631

Keith DA, Allen SP, Gallagher RV, Mackenzie BDE, Auld TD, *et al.* (2022) Fire-related threats and transformational change in Australian ecosystems. *Global Ecology and Biogeography*. doi:10.1111/geb.13500

Koehn JD (2004) Carp (*Cyprinus carpio*) as a powerful invader in Australian waterways. *Freshwater Biology* **49**, 882–894. doi:10.1111/j.1365-2427.2004.01232.x

Korcz M (2020) *Kangaroo Island Feral Pig Eradication Action Plan*. Government of South Australia, Adelaide.

Leahy L, Legge S, Tuft K, McGregor HW, Barmuta LA, *et al.* (2016) Amplified predation after fire suppresses rodent populations in Australia's tropical savannas. *Wildlife Research* **42**, 705–716. doi:10.1071/WR15011

Legge S, Woinarski JCZ, Scheele BC, Garnett ST, Lintermans M, *et al.* (2022) Rapid assessment of the biodiversity impacts of the 2019–20 Australian megafires to guide urgent management intervention and recovery, and lessons for other regions. *Diversity & Distributions* **28**, 571–591. doi:10.1111/ddi.13428

Leprieur F, Beauchard O, Blanchet S, Oberdorff T, Brosse S (2008) Fish invasions in the world's river systems: when natural processes are blurred by human activities. *PLoS Biology* **6**, e28. doi:10.1371/journal.pbio.0060028

Lintermans M (2000) Recolonization by the mountain galaxias *Galaxias olidus* of a montane stream after the eradication of rainbow trout *Oncorhynchus mykiss*. *Marine and Freshwater Research* **51**, 799–804. doi:10.1071/MF00019

Lintermans M (2007) *Fishes of the Murray-Darling Basin: An Introductory Guide*. Murray-Darling Basin Commission, Canberra.

Lintermans M, Geyle HM, Beatty S, Brown C, Ebner BC, *et al.* (2020) Big trouble for little fish: identifying Australian freshwater fishes in imminent risk of extinction. *Pacific Conservation Biology* **26**, 365–377. doi:10.1071/PC19053

Liu C, Comte L, Olden JD (2017) Heads you win, tails you lose: life-history traits predict invasion and extinction risk of the world's freshwater fishes. *Aquatic Conservation* **27**, 773–779. doi:10.1002/aqc.2740

Lyon JP, O'Connor JP (2008) Smoke on the water: can riverine fish populations recover following a catastrophic fire-related sediment slug? *Austral Ecology* **33**, 794–806. doi:10.1111/j.1442-9993.2008.01851.x

Mallinger RE, Gaines-Day HR, Gratton C (2017) Do managed bees have negative effects on wild bees? A systematic review of the literature. *PLoS One* **12**, e0189268. doi:10.1371/journal.pone.0189268

McCormack RB (2012) *A Guide to Australia's Spiny Freshwater Crayfish.* CSIRO Publishing, Melbourne.

McGregor HW, Legge S, Jones ME, Johnson CN (2015) Feral cats are better killers in open habitats, revealed by animal-borne video. *PLoS One* **10**, e0133915. doi:10.1371/journal.pone.0133915

McGregor HW, Legge SM, Jones ME, Johnson CN (2016) Extraterritorial hunting expeditions to intense fire scars by feral cats. *Scientific Reports* **6**, 22559. doi:10.1038/srep22559

Moore D, Kearney MR, Paltridge R, McAlpin S, Stow A (2018) Feeling the pressure at home: predator activity at the burrow entrance of an endangered arid-zone skink. *Austral Ecology* **43**, 102–109. doi:10.1111/aec.12547

Ne'eman G, Dafni A, Potss SG (2000) The effect of fire on flower visitation rate and fruit set in four core-species in east Mediterranean scrubland. *Plant Ecology* **146**, 97–104. doi:10.1023/A:1009815318590

Oldroyd BP (1998) Controlling feral honey bee, *Apis mellifera* L. (Hymenoptera: Apidae), populations in Australia: methodologies and costs. *Australian Journal of Entomology* **37**, 97–100. doi:10.1111/j.1440-6055.1998.tb01553.x

Oldroyd BP (1999) Coevolution while you wait: *Varroa jacobsoni*, a new parasite of western honeybees. *Trends in Ecology & Evolution* **14**, 312–315. doi:10.1016/S0169-5347(99)01613-4

Oldroyd B, Thexton E, Lawler S, Crozier R (1997) Population demography of Australian feral bees (*Apis mellifera*). *Oecologia* **111**, 381–387. doi:10.1007/s004420050249

Parkins K, York A, Di Stefano J (2018) Edge effects in fire-prone landscapes: ecological importance and implications for fauna. *Ecology and Evolution* **8**, 5937–5948. doi:10.1002/ece3.4076

Paton DC (2000) Disruption of bird-plant pollination systems in southern Australia. *Conservation Biology* **14**, 1232–1234. doi:10.1046/j.1523-1739.2000.00015.x

Pedler RD, Brandle R, Read JL, Southgate R, Bird P, *et al.* (2016) Rabbit biocontrol and landscape-scale recovery of threatened desert mammals. *Conservation Biology* **30**, 774–782. doi:10.1111/cobi.12684

Robinson NM, Leonard SW, Ritchie EG, Bassett M, Chia EK, *et al.* (2013) Refuges for fauna in fire-prone landscapes: their ecological function and importance. *Journal of Applied Ecology* **50**, 1321–1329. doi:10.1111/1365-2664.12153

Robley A, Moloney P, Moxham C, Neave G, Friend GR, *et al.* (2016) 'The effects of interaction between planned burning and foxes on the presence of native mammals. Technical Report No. 273'. Arthur Rylah Institute for Environmental Research, 1760472050, Department of Environment, Land, Water and Planning, Heidelberg.

Silva LG, Doyle KE, Duffy D, Humphries P, Horta A, *et al.* (2020) Mortality events resulting from Australia's catastrophic fires threaten aquatic biota. *Global Change Biology* **26**, 5345–5350. doi:10.1111/gcb.15282

Stobo-Wilson A, Murphy B, Legge S, Hernan C-E, Chapple D, *et al.* (2022) Counting the bodies: estimating the numbers and spatial variation of Australian reptiles, birds and mammals killed by two invasive mesopredators. *Diversity & Distributions* **28**, 976–991. doi:10.1111/ddi.13497

TSS (2006) *Recovery Plan: Tasmanian Galaxiidae 2006–2010*. Threatened Species Section, Department of Primary Industries, Water, Hobart.

Tuft KD, Crowther MS, McArthur C (2012) Fire and grazing influence food resources of an endangered rock-wallaby. *Wildlife Research* **39**, 436–445. doi:10.1071/WR11208

Villa JD (2004) Swarming behavior of honey bees (Hymenoptera: Apidae) in southeastern Louisiana. *Annals of the Entomological Society of America* **97**, 111–116. doi:10.1603/0013-8746(2004)097[0111:SBOHBH]2.0.CO;2

Wood MS, Wallis RL (1998) Potential competition for nest sites between feral European honeybees (*Apis mellifera*) and common brushtail possums (*Trichosurus vulpecula*). *Australian Mammalogy* **20**, 377–381. doi:10.1071/AM98377

18

The compounding impacts of disease and weeds after the 2019–20 wildfires on Australian vascular plants and communities

Greg Keighery, Carl R. Gosper, Sarah Barrett, David Coates and Robert O. Makinson

Summary

- Many Australian plant species and ecological communities are threatened by weeds and diseases. The diseases affecting the most species are *Phytophthora* dieback (mostly *Phytophthora cinnamomi*) and myrtle rust (*Austropuccinia psidii*).
- For many plant species and ecological communities, weeds and disease compound the impacts of fire. Following the 2019–20 wildfires, 186 fire-affected plant species were considered to be highly vulnerable to the compounding impacts of disease, and three species to the compounding impacts of weeds.
- The interactions of fire, weeds and disease may be complex. The impacts of *Phytophthora* and myrtle rust can be magnified by fire, partly through greater vulnerability of regenerating plant tissues, and because post-fire environments provide conditions more conducive to *Phytophthora* and myrtle rust. Fires may mobilise the sporeload of myrtle rust, facilitating its broad-scale dispersal.
- At least transiently, recently burnt areas may provide suitable conditions favouring some weed species, through flushes of nutrients, reduced competition with native species and increased light, and spread of weeds may then suppress post-fire recovery of native plants. Weed impacts after fire may be especially pronounced at or near modified environments. In some cases, weeds may increase fire severity; invasive grasses may trigger changes in fire regimes and transform ecosystems. Post-fire germination of weeds may deplete their soil seed bank, providing a time-bound opportunity to then implement effective management.
- There are risks of spreading or facilitating disease and weeds associated with fire control operations (e.g. movements of vehicles from infected to uninfected

areas, and application of retardants), and such risks need to be managed through appropriate planning, operations and standards.

- Management of disease and weeds has many challenges, but adequate resourcing of pre- and post-fire monitoring and management can improve conservation outcomes.

Introduction

The decline of a significant number of Australia's vascular plants, especially threatened species, has been attributed to several key threats and threatening processes. Widespread land clearing and habitat loss, anthropogenic habitat fragmentation, climate change, invasive species including introduced diseases, inappropriate fire regimes and hydrological change are all major factors in the significant range contractions and reduction in population size of many plant species (Burgman *et al.* 2007; Silcock and Fensham 2018). Superimposed on these individual threatening processes and of potentially greater impact are the compounding effects of interactions between threats or threat syndromes (Burgman *et al.* 2007; Silcock and Fensham 2018). For example, climate change, when combined with threats such as changed fire regimes, habitat fragmentation and disease, is predicted to substantially increase the magnitude of adverse impacts on the flora (Yates *et al.* 2010). Furthermore, the exposure and/or impact of many threats is likely to differ depending on landscape age and evolutionary history (Hopper 2009; Gosper *et al.* 2021).

While the impact of major threats on the Australian flora has been broadly considered, here we focus on the interactive or compounding threats of fire, disease and weeds in the context of the 2019–20 wildfires (see Gallagher *et al.* in press). Many species, particularly those in temperate Australia, are threatened by particular combinations of fire interval, intensity and season that interrupt life cycle processes critical to survival, and/or reproduction (Keith 1996; Burgman *et al.* 2007; Shedley *et al.* 2018). Narrowly endemic serotinous obligate-seeder shrubs are especially vulnerable (see Yates *et al.* 2003): of the 418 Australian plant species recognised by Silcock and Fensham (2018) as both threatened and declining, there is a documented impact of inappropriate fire regimes for 98 taxa, with 44 considered at risk of extinction.

Particular mention has recently been made of introduced disease given the impact on numerous threatened plant species throughout Australia, especially within the conservation reserve system where some other threats may be less pervasive (Silcock and Fensham 2018; Monks *et al.* 2019). Two plant diseases are of greatest concern: *Phytophthora* dieback (in particular *P. cinnamomi*), a soil-borne water mould pathogen, and myrtle rust, which is spread by wind, water, and other vectors. Both *Phytophthora* dieback and myrtle rust have significant impacts on diverse native plant communities and many rare and threatened species, and are also characterised by major difficulties in their effective control (Barrett and Yates 2015; Makinson 2018; Makinson *et al.* 2020).

Weed invasion, regarded here as plants establishing self-propagating populations in native vegetation outside of their natural range, is a significant consequence of the extensive land clearing and habitat fragmentation that occurred across large areas of southern Australia. There are > 1000 species of weeds occurring in Australian ecosystems, among a broader group of several thousand introduced plant species naturalised in Australia, and more weeds make the transition from gardens or agriculture, or are accidently introduced, annually (Csurches and Edwards 1998). With such a diversity of problematic weed species, management often involves a combination of approaches directed at control of specific species (e.g. the Weeds of National Significance list) and protection of biodiversity assets from suites of weed species (Gosper *et al.* 2015).

Disturbance in the form of fire is considered a major factor exacerbating the impact of disease (Barrett *et al.* 2008; Barrett and Yates 2015; Fensham *et al.* 2020; Pegg *et al.* 2020) and weed invasion (Graham and Taylor 2018). As such, disease and weeds compound the impacts of the 2019–20 wildfires on the persistence and recovery of populations of the estimated > 9000 plant species and the structure, function and floristics of numerous ecological communities affected (see Chapters 8, 9).

Key findings

Disease–fire interactions

Plant pathogens can cause rapid and transformational changes to ecosystems and increase the risk of extinction of plant species, through increasing mortality and reducing health and reproductive output in susceptible species. In addition to direct impacts on plants, plant pathogens have cascading impacts on fauna through affecting the distribution and abundance of plant hosts, nectar, seed and other foods, and other habitat resources.

A range of plant pathogens is having significant impacts in Australian ecosystems, and the threat of new pathogen arrival or emergence is ever-present (e.g. ABARES 2021). In addition to *Phytophthora* dieback and myrtle rust, various canker fungi are impacting the Proteaceae in south-western Australia and causing stem cankers in *Eucalyptus* species, while *Armillaria* root rot (*A. luteobubulina*) causes plant death in a range of native species in south-eastern and south-western Australia (Shearer *et al.* 1997; Yates *et al.* 2021).

For both *Phytophthora* and myrtle rust, recent research has shown that impacts can be magnified by fire events, partly through greater vulnerability of plant tissues regenerating after fire. For *Phytophthora*, observations of seedling mortality after fires suggest that the post-fire environment provides conditions more conducive to the pathogen due to loss of vegetation cover and leaf litter, increased soil temperatures and changes in hydrology, while the unsuberised roots of seedlings are more vulnerable to disease (Moore 2005). In the case of myrtle rust, post-fire regrowth in resprouting plants provides optimal tissues for the pathogen (Pegg *et al.* 2020). Interactive effects of most plant pathogens and fire in Australian ecosystems are poorly known, although the combined impacts of canker and short fire intervals is an acute threat to *Banksia verticillata* (Yates *et al.* 2021), with the 2019–20 wildfires overlapping in extent with known areas of canker impact on *B. brownii* populations in the Stirling Range National Park (WA).

Spatial analyses of 2019–20 wildfire extent and known *Phytophthora* records, and modelled suitable habitat for myrtle rust, indicate that ~56% of the fire footprint intersects areas of plant pathogen occurrence (Gallagher *et al.* in press). Integrating information on plant species susceptibility to disease, plant species geographical range, fire extent and disease occurrence suggested that 186 plant species were highly vulnerable to the compounding impacts of 2019–20 wildfires and either or both *Phytophthora* and myrtle rust

(Gallagher *et al.* in press; Chapter 9). Similarly, for ecosystems, 10 ecosystem types were highly exposed to the compounding impacts of the 2019–20 wildfires and plant pathogens, having > 50% of their mapped extent burnt and with pathogen occurrence. Most of these ecosystem types were rainforests and wet eucalypt forests in eastern Australia highly exposed to fire and myrtle rust, along with the Eastern Stirling Range Montane Heath and Thicket and the Montane Mallee Thicket of the Stirling Range, highly exposed to fire and *Phytophthora* (Keith *et al.* in press; Chapter 8). These predicted impacts are being validated with field survey (see Boxes 18.1 and 18.2).

Box 18.1. *Phytophthora cinnamomi* and fire interaction: Stirling Range National Park

The threat posed by *Phytophthora* is well documented and it was listed as a Key Threatening Process under the Commonwealth *Environment Protection and Biodiversity Conservation Act 1999* (EPBC Act) in 2000 with a national threat abatement plan for dieback first produced in 2001 and revised in 2018 (Commonwealth of Australia 2018). The severity of *P. cinnamomi* impact is perhaps best illustrated by the work of Shearer *et al.* (2004), which estimated that 2284 (40%) described plant species and up to 50% of threatened plants in the Southwest Australian Floristic Region are susceptible. Mortality of infected individuals in the most susceptible species approaches 100% (Shearer *et al.* 2004; Barrett *et al.* 2008). Species highly susceptible to *Phytophthora* include species-rich and community dominant groups such as *Banksia* and other Proteaceae, many peas (Fabaceae) and heaths (e.g. *Leucopogon* and other Ericaceae).

The 1160 km^2 Stirling Range National Park (SRNP) contains more than 1500 plant taxa including some 80 endemic to the park, contributing to the SRNP being a centre of flora richness and endemism in the Southwest Australian Floristic Region global biodiversity hotspot. Disease caused by *P. cinnamomi* is considered the foremost threatening process for this flora (Barrett *et al.* 2008) with an estimated 36% of species considered to be susceptible (Shearer *et al.* 2004). The spatial extent of the disease in the SRNP has been well documented with more than two-thirds of the park considered to be infested. Other ecological processes interact with *Phytophthora* dieback to increase the risk of extinction of species and collapse of ecological communities (Barrett *et al.* 2008). The interaction between fire and *P. cinnamomi* has only recently been explored, finding increased disease impact in the SRNP after fire, particularly for fire-killed species (Moore *et al.* 2015; Barrett and Yates 2015).

A wildfire in late December 2019 burnt over 36 000 ha of the eastern SRNP. This followed a 4000 ha fire in the western SRNP (December 2019) and a 14 000 ha fire in the eastern SRNP (May 2018). Several *Phytophthora*-susceptible ecological communities were significantly impacted by these fires including the Montane Mallee Thicket of the Stirling Range and the Eastern Stirling Range Montane Heath and Thicket (see also Chapter 8). There was also a significant impact on 26 listed Threatened Flora species, many of which are highly susceptible to the pathogen and at very high risk of extinction (Barrett *et al.* 2008). Monitoring of seedling survival in susceptible species since the 2018–19 fires showed relatively low levels of mortality due to *P. cinnamomi* infection

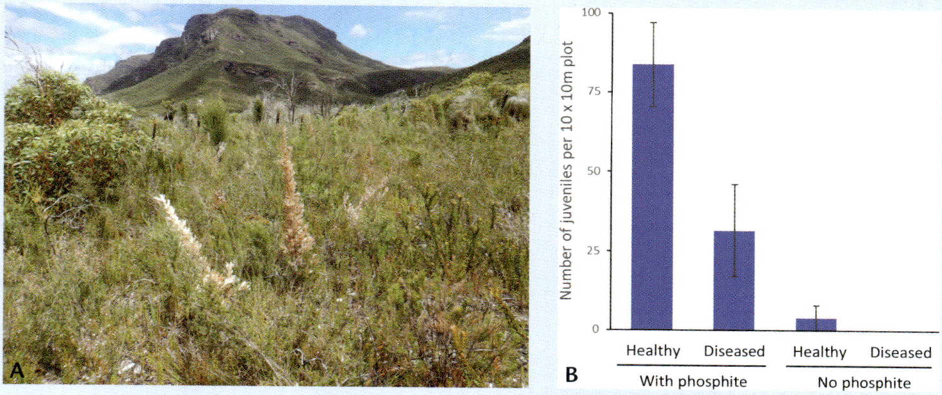

Fig. 18.1. (A) *Banksia anatona* deaths in recently burnt *Phytophthora*-infested habitat, Stirling Range National Park; (B) impact of recurrent phosphite spraying on survival of *B. anatona* recruits (mean ± standard error) in 2022 after the 2018–19 wildfires, in plots that were originally deemed *Phytophthora*-free ('Healthy') and *Phytophthora*-infested ('Diseased') when established in 2014–16 (Barrett and Rathbone 2018). (Photo: Sarah Barrett. Graph: Carl Gosper)

during an initial period of relatively low rainfall, but very high winter–spring rainfall in 2021 has seen the onset of significant mortality of recruits in susceptible species (Fig. 18.1).

The fungicide phosphite has been used to successfully mitigate the impact of *P. cinnamomi* in the SRNP after fires in 1991 and 2000 (Barrett and Rathbone 2018), and this success appears to have continued after phosphite applications following the 2018–19 fires (Fig. 18.1). Ongoing phosphite application, disease hygiene management, and flora and ecological community monitoring are important to mitigate and understand impacts in high priority areas.

Box 18.2. Myrtle rust

The myrtle rust pathogen (*Austropuccinia psidii*), an introduced fungus, affects plants of the Myrtaceae family. A single biotype was detected in Australia in 2010, and is now naturalised along almost the entire eastern seaboard east of the Great Dividing Range and in parts of the Northern Territory. The Australian host range exceeds 350 species, ~15% of which are known or thought likely to be at serious risk. Several hosts in catastrophic decline are listed as Critically Endangered under Queensland, New South Wales or Commonwealth legislation, with others likely to follow (Makinson 2018; Makinson *et al.* 2020). Other biotypes of *A. psidii* represent an additional biosecurity threat; one strain with a particular affinity for eucalypts has caused major economic losses in South American plantations.

Myrtle rust occurs mainly in mesic vegetation types but can extend into drier sclerophyll associations in suitable microhabitats and seasons. The pathogen is biotrophic, attacking new growth (seedlings, seasonal flush, and post-fire resprouts;

Fig. 18.2), and in highly susceptible species can cause death within a few reinfection cycles. The life cycle is rapid (as little as 10 days), enabling rapid increase when conditions permit. The spores are highly mobile on wind, animals and via human vectors.

Fire will destroy infected plants and a proportion of rust spores, but dry conditions and fire updraught will also mobilise sporeload high into the air column, facilitating dispersal over tens or hundreds of kilometres. Recolonisation may easily occur once weather conditions are right, from unburnt patches within a fire footprint or adjacent areas. Regenerative growth, usually on multiple hosts, provides an ideal trophic base. The dominance of Myrtaceae in forest, heath and woodland systems tends to ensure that suitable tissue is available for most of the year, increasing the rust's capacity to re-establish after fire.

Studies of post-fire interactions of myrtle rust disease and its hosts have been limited to subtropical east-coast heathlands and *Melaleuca quinquenervia* floodplain woodlands. Floristic changes and increased weed invasion are known as non-fire-related effects of repeated severe infection in a wet sclerophyll/rainforest transition system (Pegg *et al.* 2017), and can be expected to be amplified after fire on fertile substrates. Fensham and Radford-Smith (2021) noted the potential for pre-fire interactions between myrtle rust and drought in such systems to affect post-fire outcomes.

In a study of *Melaleuca* woodlands, which often fringe or overstand freshwater wetlands, Pegg *et al.* (2018) found regeneration failure accentuated by post-fire

Fig. 18.2. (A) Post-fire sucker resprouts of scrub turpentine (*Rhodamnia rubescens*) trees showing severe infection by myrtle rust (*Austropuccinia psidii*), north-east NSW, 2020–21; (B) myrtle rust infection on reshooting fire-affected needlebark stringybark (bastard tallowwood) (*Eucalyptus planchoniana*), north-east NSW, 2020–21. (Photos: Geoff Pegg)

additive effects of myrtle rust and invertebrate herbivores, particularly mirid bugs (*Eucerocoris suspectus*). Highly variable levels of *M. quinquenervia* seedling resistance to myrtle rust within and between populations are known (Pegg *et al.* 2018), and the interactions of fire, pathogen, insects and local genetics need to be further evaluated.

In a study of coastal heathland after the 2019–20 fires, Pegg *et al.* (2020) found some level of infection by myrtle rust on nearly all regenerating Myrtaceae species present, and on 69% of all myrtaceous regenerants within 3 months, noting that 'the speed at which we observed re-introduction of *A. psidii* on regenerating Myrtaceae from surrounding reservoirs was remarkable'.

Weed–fire interactions

Weeds are a major threat to plant diversity and ecosystem function, having impacts as diverse as competition for resources (e.g. light, water, nutrients) and pollinator and seed disperser fauna, facilitation of feral honeybee dominance, inducing changes in nutrient cycling and the availability of habitat for fauna.

Weeds can have a direct influence on fire regimes and events, through changing the flammability of vegetation via differences in the spatial and temporal distribution of fuel and fuel dryness. Weedy grasses, in particular, can foster reduced fire intervals and greater fire occurrence when established in forests, woodlands and shrublands with naturally shrubby understoreys – a globally significant phenomenon known as the grass–fire cycle (D'Antonio and Vitousek 1992). For many weeds, fires enhance spread and abundance through providing a seed germination stimulus, temporarily eliminating competition with established native vegetation, and initiating a flush of nutrients from ash.

The relationship between fire and weed invasion is complex and highly dependent on context; hence the likelihood that the 2019–20 wildfires facilitated or enhanced problematic weed invasions will be spatially variable. Weed invasions into native vegetation are strongly associated with proximity to, and dispersal pathways for, weed propagules (seed and vegetative material) and disturbance, so tend to be greater on native vegetation edges and closer to towns, agricultural land, roads and mining activity (Gosper *et al.* 2011, 2015). Land uses that lead to elevated soil nutrient levels, such as agriculture and settlements, strongly facilitate invasions (Fisher *et al.* 2006, 2009). Indeed, many Australian ecosystems with highly infertile soils are resistant to weed invasion, even after fire, as long as soil nutrient levels are not elevated (Gosper *et al.* 2011). The commonly used fire retardant, Phos-Chek, can contribute to elevated soil nutrient levels, with consequent implications for weed invasion (Bell *et al.* 2005). Thus, locations where the 2019–20 fires are likely to have had the greatest impact on facilitating weed invasions are along the interface between native vegetation and agriculture, urban areas and roads.

National desktop assessments following the 2019–20 wildfires suggested that the impacts of weeds post-fire across the full ranges of plant species and ecological communities was less than other important threats (with three fire-affected plant species considered highly vulnerable to compounding impacts of weeds), perhaps partly a function of poor representation of weed distributions in spatial datasets (Gallagher *et al.* in press; Keith *et al.* in press).

Nevertheless, significant impacts of weeds on native flora and ecological communities are emerging post-fire, including on orchids (Chapter 9), warm temperate rainforests of Victoria (Chapter 8) and threatened Swan Coastal Plain ecological communities (see Box 18.3).

Although fire can facilitate weed invasion, it also offers an opportunity for weed control. As fires provide a major germination stimulus, they deplete weed soil seed banks, which, if coupled with effective post-fire weed management before new seed input, can contribute to the restoration of ecosystem composition and function (Brown and Brooks 2002). The timing of post-fire weed management to achieve effective weed control varies between weed species based on their maturation time, but for weedy grasses and herbs the first growing season post-fire is crucial.

Box 18.3. Weed invasion post-fire in threatened ecological communities of the Swan Coastal Plain

In December 2019, a wildfire in the Yanchep area north of Perth impacted stands of the EPBC Act-listed tuart (*Eucalyptus gomphocephala*) woodlands and forests of the Swan Coastal Plain threatened ecological community. Tuart is endemic to the Swan Coastal Plain. Studies of the impacts of previous major high-severity bushfires in tuart woodlands provide a basis for understanding post-fire weed issues arising from the 2019 Yanchep fire.

Landscape-scale wildfires have impacted the Yalgorup National Park area south of Perth in 1996, 2006 and 2016, with the 2016 fire being very intense. Quadrat-based resampling was conducted in 2017, in the second spring after the 2016 fire, of plots established in 2007 (Keighery *et al.* 2009). The 2017 sampling recorded significant seedling growth and resprouting of larger tuarts. However, it was noticeable that many of the older tuarts (over 37%) that contained hollows had died in the fire. Weedy species were more noticeable, numerous and more species diverse post-fire. Prior to the fire, the mean richness (per quadrat) of weed species was 7.7. After fires mean weed richness increased to 11, including annual grasses (*Bromus diandrus, Lolium rigidum*), weeds germinating from kangaroo faeces (*Lotus suaveolens, Medicago* spp., *Stellaria media*) and wind-dispersed weeds (*Arctotheca calendula, Urospermum picroides*). Mass flowering and seeding of several previously inconspicuous bulbous weeds occurred (*Freesia* hybrids and *Morea flaccida*).

In the clay pans of the Swan Coastal Plain threatened ecological community, the 2016 wildfire in Yalgorup National Park initiated population collapse of a community dominant, *Leptomeria ellytes*, and led to significant increases in weeds. Pre-fire sampling showed that this community normally has a species richness of 40 native plant species, with another 15 weed species, per 100 m^2 plot – consistent with high weed richness in other occurrences of this community (Gibson *et al.* 2005). After the fire, weeds increased by 10 annual species per plot on average, largely dispersed via kangaroo faeces derived from foraging in adjacent farmland. These weed species remained in the area when the community was surveyed in 2020. Weed invasion, particularly after disturbances such as fire, is regarded as one of the key processes threatening both tuart and clay pan communities (DPaW 2015; DEE 2019).

Recommendations

For *Phytophthora* and myrtle rust, there is low feasibility of eradication or prevention of all impacts on native flora once the pathogen has become established, based on currently available pathogen management options. This reality, combined with compounding impacts with fire, indicates that pathogen management in fire-prone ecosystems will be most effective through considering the following points:

- While recognising that disease and weed hygiene procedures are well embedded in many organisations involved in fire operations in natural ecosystems, the risk of inadvertent dispersal of pathogens, such as through *Phytophthora*-infested soil attached to machinery, equipment and clothing, can be minimised by continued and enduring attention to hygiene management through all stages of fire planning and operations. The same types of precautions apply to vehicles, equipment and personnel moving from established myrtle rust zones to uninfected States or areas.
- Expansion of species and population representation of pathogen-susceptible species in *ex situ* germplasm collections, and where appropriate translocations to pathogen-free locations and/or the selection and reintroduction of pathogen-resistant genotypes, will assist in buffering the risk of population extinctions in the wild.
- Adequate resourcing of pre- and post-fire pathogen management and monitoring can improve conservation outcomes, as pathogen impacts can be greater after fire but can be mitigated in some situations with appropriate controls.
- Adequate resourcing of research into novel pathogen management techniques, methods of pathogen detection and quantification of impacts on biodiversity may improve pathogen management. While significant advances in knowledge have been made over recent decades, pathogen management is currently limited by a lack of effective tools for mitigation of impacts and knowledge of the extent of impact in natural ecosystems with variation in environmental conditions.

The availability and effectiveness of tools for post-fire weed control varies. Considering the following points may lead to improved weed management after significant fire events:

- Innovative approaches are needed for resourcing post-fire weed control and developing collaborations with community groups and other partners that can assist in on-ground actions. Tools for the control of some weeds after fire in natural ecosystems are well established and effective, for example the selective control of grassy weeds by the 'fops' group of herbicides (Brown and Brooks 2002). In these cases, effective post-fire weed management is largely a question of adequate and timely deployment of resources. However, this is by no means a simple task, as human, equipment and financial resources of many land management agencies will be stretched in the aftermath of fire emergencies, due to the significant role many of these agencies play in fire emergency response and other fire-related conservation actions (see Chapter 23).
- Ongoing research is needed into chemical, ecosystem management and biological weed control approaches, and their integration, for weed-ecosystem combinations. Such enhanced knowledge may be most needed where effective weed control methods after fire are currently lacking, such as for weedy bulbs in ephemeral wetlands.
- Taking into account locations where fire retardant use intersects with important biodiversity values could assist in targeting weed control efforts. Furthermore, additional

research is warranted on the weed invasion implications of fire retardants, given preliminary research suggesting retardants can alter soil properties. Needless to say, use of fire retardants will remain crucial in fire emergencies for protection of life and property.

References

ABARES (2021) 'The National Priority List of Exotic Environmental Pests, Weeds and Diseases: Information Paper (Version 2.0)'. Department of Agriculture, Water and the Environment, Canberra.

Barrett S, Rathbone D (2018) Long-term phosphite application maintains species assemblages, richness and structure of plant communities invaded by *Phytophthora cinnamomi*. *Austral Ecology* **43**, 360–374. doi:10.1111/aec.12574

Barrett S, Yates CJ (2015) Risks to a mountain summit ecosystem with endemic biota in southwestern Australia. *Austral Ecology* **40**, 423–432. doi:10.1111/aec.12199

Barrett S, Shearer BL, Crane CE, Cochrane A (2008) An extinction-risk assessment tool for flora threatened by *Phytophthora cinnamomi*. *Australian Journal of Botany* **56**, 477–486. doi:10.1071/BT07213

Bell T, Tolhurst K, Wouters M (2005) Effects of fire retardant Phos-Chek on vegetation in eastern Australian heathlands. *International Journal of Wildland Fire* **14**, 199–211. doi:10.1071/WF04024

Brown K, Brooks K (2002) *Bushland Weeds: A Practical Guide to their Management*. Environmental Weeds Action Network, Greenwood.

Burgman MA, Keith D, Hopper SD, Widyatmoko D, Drill C (2007) Threat syndromes and conservation of the Australian flora. *Biological Conservation* **134**, 73–82. doi:10.1016/j.biocon.2006.08.005

Commonwealth of Australia (2018) *Threat Abatement Plan for Disease in Natural Ecosystems caused by* Phytophthora cinnamomi. Commonwealth of Australia, Canberra.

Csurches S, Edwards R (1998) *Potential Environmental Weeds in Australia*. Environment Australia, Canberra.

D'Antonio CM, Vitousek PM (1992) Biological invasions by exotic grasses, the grass/fire cycle, and global change. *Annual Review of Ecology and Systematics* **23**, 63–87. doi:10.1146/annurev.es.23.110192.000431

Department of Parks and Wildlife (DPaW) (2015) *Interim Recovery Plan 2015–2020 for Clay Pans of the Swan Coastal Plain (Swan Coastal Plain Community Types 7, 8, 9 and 10a) and Clay Pans with Mid Dense Shrublands of* Melaleuca lateritia *Over Herbs*. Interim Recovery Plan No. 354. DPaW, Perth.

Department of the Environment and Energy (DEE) (2019) Approved conservation advice (incorporating listing advice) for the Tuart (*Eucalyptus gomphocephala*) woodlands and forests of the Swan Coastal Plain ecological community. Department of the Environment and Energy, Canberra, <http://www.environment.gov.au/biodiversity/threatened/communities/pubs/153-conservation-advice.pdf>.

Fensham RJ, Radford-Smith J (2021) Unprecedented extinction of tree species by fungal disease. *Biological Conservation* **261**, 109276. doi:10.1016/j.biocon.2021.109276

Fensham R, Carnegie AJ, Laffineur B, Makinson RO, Pegg GS, *et al.* (2020) Imminent extinction of Australian Myrtaceae by fungal disease. *Trends in Ecology & Evolution* **35**, 554–557. doi:10.1016/j.tree.2020.03.012

Fisher JL, Veneklaas EJ, Lambers H, Loneragan WA (2006) Enhanced soil and leaf nutrient status of a Western Australian Banksia woodland community invaded by *Ehrharta calycina* and *Pelargonium capitatum*. *Plant and Soil* **284**, 253–264. doi:10.1007/s11104-006-0042-z

Fisher JL, Loneragan WA, Dixon K, Delaney J, Veneklaas EJ (2009) Altered vegetation structure and composition linked to fire frequency and plant invasion in a biodiverse woodland. *Biological Conservation* **142**, 2270–2281. doi:10.1016/j.biocon.2009.05.001

Gallagher RV, Allen SP, Mackenzie BDE, Keith DA, Nolan RH, *et al.* (in press) An integrated approach to assessing abiotic and biotic threats to post-fire plant species recovery: lessons from the 2019–20 Australian megafires. *Global Ecology and Biogeography.* doi:10.1111/geb.13478

Gibson N, Keighery GJ, Lyons MN, Keighery BJ (2005) Threatened plant communities of Western Australia. 2 The seasonal clay-based wetland communities of the South West. *Pacific Conservation Biology* **11**, 287–301. doi:10.1071/PC050287

Gosper CR, Yates CJ, Prober SM, Williams MR (2011) Fire does not facilitate invasion by alien annual grasses in an infertile Australian agricultural landscape. *Biological Invasions* **13**, 533–544. doi:10.1007/s10530-010-9847-z

Gosper CR, Prober SM, Yates CJ, Scott JK (2015) Combining asset- and species-led alien plant management priorities in the world's most intact Mediterranean-climate landscape. *Biodiversity and Conservation* **24**, 2789–2807. doi:10.1007/s10531-015-0973-x

Gosper CR, Coates DJ, Hopper SD, Byrne M, Yates CJ (2021) The role of evolutionary history in the distribution, biogeography and conservation of Threatened flora in the Southwest Australian Floristic Region. *Biological Journal of the Linnean Society* **133**, 394–410. doi:10.1093/biolinnean/blaa141

Graham M, Taylor K (2018) 'Fire, weeds and the native vegetation of New South Wales'. Report by Hotspots Fire Project, Nature Conservation Council of New South Wales and New South Wales Rural Fire Service, Sydney.

Hopper SD (2009) OCBIL theory: towards an integrated understanding of the evolution, ecology and conservation of biodiversity on old, climatically buffered, infertile landscapes. *Plant and Soil* **322**, 49–86. doi:10.1007/s11104-009-0068-0

Keighery GJ, Keighery BJ, Longman VE (2009) *A Preliminary List of the Flora and Significant Flora of Yalgorup National Park*. Department of Environment and Conservation, Perth.

Keith D (1996) Fire-driven extinction of plant populations: a synthesis of theory and review of evidence from Australian vegetation. *Proceedings of the Linnean Society of New South Wales* **116**, 37–78.

Keith DK, Allen SP, Gallagher RV, Mackenzie BDE, Auld TD, *et al.* (in press) Fire-related threats and transformational change in Australian ecosystems. *Global Ecology and Biogeography.* doi:10.1111/geb.13500

Makinson RO (2018) *Myrtle Rust Reviewed: The Impacts of the Invasive Pathogen* Austropuccinia psidii *on the Australian Environment.* Plant Biosecurity Cooperative Research Centre, Canberra, <http://www.apbsf.org.au/wp-content/uploads/2018/11/Myrtle-Rust-reviewed-June-22-2018-web.pdf>.

Makinson RO, Pegg GS, Carnegie AJ (2020) Myrtle rust in Australia – a national action plan. Australian Plant Biosecurity Science Foundation, Canberra, Australia, <http://www.apbsf.org.au/myrtle-rust/>.

Monks L, Barrett S, Beecham B, Byrne M, Chant A, *et al.* (2019) Recovery of threatened plant species and their habitats in the biodiversity hotspot of the Southwest Australian Floristic Region. *Plant Diversity* **41**, 59–74. doi:10.1016/j.pld.2018.09.006

Moore NA (2005) Role of fire on *Phytophthora cinnamomi* in the Stirling Range National Park, Western Australia. BSc (Hons) thesis. Murdoch University, Murdoch.

Moore N, Barrett S, Howard K, Craig MD, Bowen B, *et al.* (2015) Time since fire and average fire interval are the best predictors of *Phytophthora cinnamomi* activity in heathlands of south-western Australia. *Australian Journal of Botany* **62**, 587–593. doi:10.1071/BT14188

Pegg G, Taylor T, Entwistle P, Guymer G, Giblin F, *et al.* (2017) Impact of *Austropuccinia psidii* (myrtle rust) on Myrtaceae-rich wet sclerophyll forests in south east Queensland. *PLoS One* **12**, e0188058. doi:10.1371/journal.pone.0188058

Pegg GS, Lee DJ, Carnegie AC (2018) Predicting impact of *Austropuccinia psidii* on populations of broad leaved Melaleuca species in Australia. *Australasian Plant Pathology* **47**, 421–430. doi:10.1007/s13313-018-0574-8

Pegg GS, Entwistle P, Giblin FR, Carnegie AJ (2020) Fire and rust – the impact of *Austropuccinia psidii* (myrtle rust) on regeneration of Myrtaceae in coastal heath following wildfire. *Southern Forests* **82**, 280–291. doi:10.2989/20702620.2020.1819154

Shearer BL, Byrne A, Dillon M, Buehrig R (1997) Distribution of *Armillaria luteobubalina* and its impact on community diversity and structure in *Eucalyptus wandoo* woodland of southern Western Australia. *Australian Journal of Botany* **45**, 151–165. doi:10.1071/BT95083

Shearer BL, Crane CE, Cochrane A (2004) Quantification of the susceptibility of the native flora of the South-West Botanical Province, Western Australia, to *Phytophthora cinnamomi*. *Australian Journal of Botany* **52**, 435–443. doi:10.1071/BT03131

Shedley E, Burrows N, Yates CJ, Coates DJ (2018) Using bioregional variation in fire history and fire response attributes as a basis for managing threatened flora in a fire-prone Mediterranean climate biodiversity hotspot. *Australian Journal of Botany* **66**, 134–143. doi:10.1071/BT17176

Silcock JL, Fensham RJ (2018) Using evidence of decline and extinction risk to identify priority regions, habitats and threats for plant conservation in Australia. *Australian Journal of Botany* **66**, 541–555. doi:10.1071/BT18056

Yates CJ, Abbott I, Hopper SD, Coates DJ (2003) Fire as a determinant of rarity in the south-west Western Australian global biodiversity hotspot. In *Fire in Ecosystems of South-west Western Australia: Impacts and Management.* (Eds I Abbott and N Burrows) pp. 395–420. Backhuys Publishers, Leiden.

Yates CJ, McNeill A, Elith J, Midgley GF (2010) Assessing the impacts of climate change and land transformation on *Banksia* in the South West Australian Floristic Region. *Diversity and Distributions* **16**, 187–201. doi:10.1111/j.1472-4642.2009.00623.x

Yates CJ, Barrett S, Dilly M, Hopper SD, Stewart B, *et al.* (2021) Modelling the impact of canker disease and fire regimes on the population dynamics and extinction risk of the Critically Endangered and granite endemic shrub *Banksia verticillata* R.Br. *Australian Journal of Botany* **69**, 274–284. doi:10.1071/BT20156

19

Interacting and compounding impacts: fire and forestry in the 2019–20 wildfires

David B. Lindenmayer, Elle Bowd, Chris Taylor and Phil Zylstra

Summary

- Climate and weather are often the dominant drivers of fire behaviour, but the effects of forest management (including logging) also can be important and must not be ignored.
- An increasing body of evidence indicates that there is a significantly greater risk of high-severity fire in logged forests relative to undisturbed forest. This includes evidence from the 2019–20 wildfires.
- Elevated logging-induced forest flammability can last for several decades after cutting and is a particular concern in areas subject to prolonged and widespread industrial forestry.
- Logging prior to fire can simplify stand structure and reduce the potential for recovery following fire, an effect resulting from the limited prevalence for cutover areas to transmit biological legacies like large living and dead old trees from pre- to the post-fire stands.
- Logging soon after fires – sometimes termed 'salvage logging' – has negative effects on many elements of forests, from microbes and soils to plants and vertebrates.
- The 2019–20 wildfires left few and often small areas of unburnt refugia within the fire footprint. Some of these areas may be important for promoting post-fire biodiversity recovery. It will therefore be important to ensure that such areas are not further disturbed, such as by logging operations.
- The links between logging and fire and their combined effects on flammability, as well as on forest condition and biodiversity, suggest a need to cease widespread industrial logging in Australian native forests. This includes the cessation of post-fire (salvage) logging.
- Cessation of industrial logging is needed to reduce the number of stressors in many forest ecosystems which have been subject to frequent, widespread and severe wildfires in the past 25 years. Cessation of logging is also essential for

increasing the extent of old growth forest cover, where fire severity is lowest and post-fire ecological recovery is strongest.

- Protection from fire of the extensive areas of flammable forest created by past logging operations demands the development of new technologies such as drone fleets or satellite platforms, along with expanded capacity for proven techniques such as specialist remote-area fire-fighting crews for rapid detection and expeditious suppression of ignitions.

Introduction

The 2019–20 Australian wildfires had major impacts on forests (Collins *et al.* 2021), biodiversity (Ward *et al.* 2020), carbon emissions (van der Velde *et al.* 2021), human health and property (Inspector General for Emergency Management 2020). The extreme temperatures and prolonged drought, including climatic conditions associated with climate change, contributed to the extent and severity of the 2019–20 wildfires (van Oldenborgh *et al.* 2021; Mackey *et al.* 2021b). Indeed, climate and fire weather are major contributors to fire behaviour (Jones *et al.* 2020). Key attributes of the 2019–20 wildfires, such as severity, were strongly linked with fire weather (Collins *et al.* 2021; Lindenmayer *et al.* 2021). Beyond climate and weather as key drivers of fire, other factors that can influence fire behaviour include slope, aspect, fuel loads and fuel distribution (Catchpole 2002; Bradstock 2010; Sullivan *et al.* 2012). In addition, past disturbance, such as past fires and logging, can influence the severity of fire (Lindenmayer *et al.* 2020, 2021; Bowman *et al.* 2021; Taylor *et al.* 2021). Hence, how the forest is treated before the fire can have impacts on the fire itself (Lindenmayer *et al.* 2021) (see Box 19.1). Here, we briefly examine some of the evidence for the compounding and interacting effects of fire and logging, both in a pre-fire and post-fire (salvage) logging context. Many of these insights come from work in Victoria. The bulk of empirical work from which we draw examples has been conducted in the wet eucalypt forests of Victoria dominated by obligate seeder tree species such as mountain ash (*Eucalyptus regnans*) and alpine ash (*Eucalyptus delegatensis*), which are often killed by high-severity wildfires. However, some of the general patterns we have outlined are also found in drier mixed species forests (Taylor *et al.* 2020) and dry lowland forests (Lindenmayer *et al.* 2021; Taylor *et al.* 2021). We therefore suggest that some of the general principles derived from these bodies of work apply in other forest types, including those elsewhere in Australia and overseas.

Box 19.1. Does thinning reduce fire severity?

Thinning involves removing a proportion of the trees and hence some of the biomass in a forest and leaving a proportion of the overstorey intact (Helms 1998). There is some evidence that thinning has some effectiveness in North American forests and particularly when it is accompanied by prescribed burning (Kalies and Yocom Kent 2016). However, in an Australian context, thinning and prescribed burning are often not interlinked because of the damage done by the latter to sawlog value (Taylor *et al.* 2021). Indeed, empirical analyses after the 2009 wildfires and the 2019–20 wildfires have indicated that thinning often makes limited difference to subsequent fire severity and, in some cases, it can even elevate fire severity (Taylor *et al.* 2020, 2021).

We have structured our chapter around six key sections. We first outline evidence for relationships between logging and elevated fire severity. We then discuss how the post-fire recovery process can be impaired in previously logged forests (before they were then burnt). The third section of this chapter outlines evidence for yet other compounding interactions between logging and fire in a spatial and topographic context. Some of the impacts of post-fire logging are briefly described in the fourth section. Each of these negative impacts interacts with others, leading to major compounding negative effects on biodiversity and ecosystem condition. We present a new conceptual model to illustrate these effects and the overall disturbance burden faced by some forest landscapes. The final two sections focus on some overarching conclusions and associated general recommendations for forest management. We do not explore the direct effects of logging on biodiversity or key ecosystem processes, such as through the removal of key elements of stand structure (like large old trees) (Lindenmayer *et al.* 2016), altering plant species composition (which can influence animals such as through degrading foraging resources) (Au *et al.* 2019), or reducing the amount of old growth (Lindenmayer and Taylor 2020a) and other habitat attributes that provide critical habitat for a wide range of species (Lindenmayer *et al.* 1990, 2019a; Lefoe *et al.* 2022).

Key findings

Prior to the fire: logging makes forests more prone to high-severity wildfire

Several studies, encompassing a range of datasets in different jurisdictions, contain evidence that logged forests are at risk of an increased probability of high-severity wildfire (Attiwill *et al.* 2014; Bradstock and Price 2014; Taylor *et al.* 2014; Bowman *et al.* 2021; Lindenmayer *et al.* 2021). Some of these studies have found a non-linear relationship between stand age (which is often a measure of time since logging) and fire severity. This occurs because there is a very low probability of high-severity fire in very young forests soon after logging and in long-undisturbed forests (e.g. stands exceeding 100 years of age) (Taylor *et al.* 2014). However, stands aged between ~10 and 40 years of age had a seven times increased risk of high-severity wildfire relative to very young and long-undisturbed forest (Taylor *et al.* 2014) (Figs 19.1, 19.2).

Further analyses of a different dataset from the 2009 wildfires also revealed that the probability of a crown burn was significantly greater in young stands than older stands (Taylor *et al.* 2020). Evidence of such stand-age-fire-severity relationships were recently uncovered in analyses of the 2019–20 wildfires in north-eastern Victoria (Lindenmayer *et al.* 2021). Similarly, data in Bowman *et al.* (2021) show that logging has a marked effect on the probability of high-severity fire. Moreover, there was a broadly similar probability of high-severity fire in logged forests under the mildest conditions as in undisturbed forests under high and very high fire danger conditions (Bowman *et al.* 2021). Attiwill *et al.* (2014) suggested there was no relationship between logging and fire severity, but the data they presented actually showed that crown fires in the 2009 wildfires occurred twice as frequently in regrowth compared to old growth forests. The likelihood of crown fire was 30–40% lower in long-unlogged forests (exceeding 100 years old) than in regrowth forest (Attiwill *et al.* 2014). Attiwill *et al.* (2014) found that forests older than 10 years and younger than 80 years experienced the greatest fire severity. Yet other authors who have claimed there is no link between logging and fire severity (e.g. Tolhurst and McCarthy 2016) actually did not test for such effects.

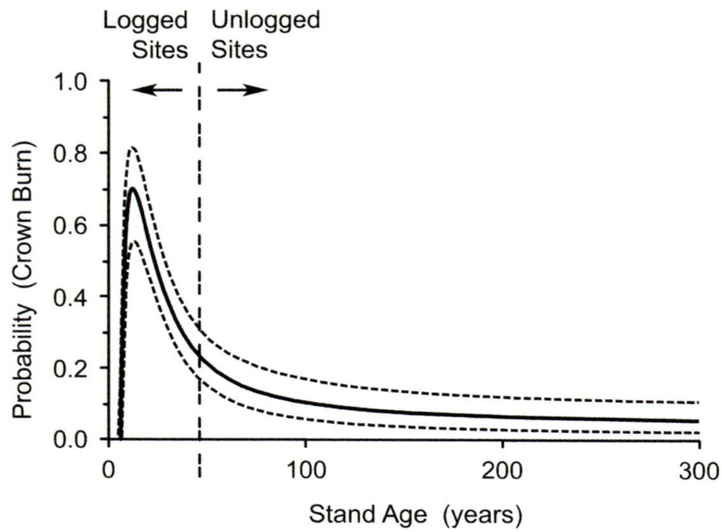

Fig. 19.1. The non-linear relationship between stand age and the probability of canopy fire or crown burn. The probability of crown burn peaks at ~40 years before declining as stands approach 80–100 years old. The solid line is the mean of the posterior prediction of the probit regression model fitted to a stratified sample of the data, and the dashed lines are 95% credible intervals. (Redrawn from Taylor *et al.* (2014))

Fig. 19.2. (A) Logged and regenerated mountain ash forests; and (B) logged and regenerated forest that was burnt at high severity in the 2009 wildfires. (Photos: Elle Bowd and David Lindenmayer)

There are several reasons why young logged and regenerated forests are at risk of elevated fire severity. These include (1) the additional volumes of logging slash that remain in cutover areas (and that are not consumed in regeneration burns), (2) the vertical structure of fuels that can move a surface fire to the canopy (Zylstra *et al.* 2016), (3) the loss of mesic elements such as the tree fern layer from logged forests (Blair *et al.* 2016), (4) the promotion of high-densities of flammable plant taxa in logged forest (Bowd *et al.* 2021), and (5) drying of soils (Bowd *et al.* 2019) as well as hotter microclimatic conditions in young logged and regenerated forests.

We argue that because forest management (including logging) is one of the factors contributing to fire severity, it is critical to mitigate that impact because it is one of the few drivers of fire severity over which we have direct control. This is especially important in areas like the Central Highlands of Victoria where there has been extensive disturbance as a result of past logging, and much of the landscape is dominated by highly flammable young and intermediate-aged forest (Taylor and Lindenmayer 2020).

Prior to the fire: logging impairs biological legacy effects and curtails post-fire recovery

There is a large body of work illustrating that disturbances such as wildfires leave behind many kinds of biological legacies (*sensu* Franklin *et al.* 2000) that can be incorporated into the regenerating forest and facilitate the recovery of biodiversity and the forest stand (Blackhall *et al.* 2017). Biological legacies include living and dead trees and other plants, carcasses, seeds, eggs and living animals. As an example of the role of biological legacies in burnt Australian forests, populations of small mammals that survived the 2009 wildfires were important for population recovery in burnt areas, including those burnt at very high severity (Banks *et al.* 2017).

The prevalence of many kinds of biological legacies after fire will be a function of forest condition at the time of a fire. Forests that are old growth at the time of a severe fire will produce many more large living and dead trees, and the regeneration of vegetation will be more advanced, than forests that are younger and previously logged at the time of a severe fire (Smith *et al.* 2014; Lindenmayer *et al.* 2019b; Bowd *et al.* 2021). For example, the multi-aged forests that are a key part of the habitat requirements of many species in Victorian mountain ash forests (Lindenmayer *et al.* 1990) can be produced where an old growth is burnt, but not when a young logged forest is burnt (Lindenmayer *et al.* 2019b).

One of the important legacies remaining after wildfires is the presence of unburnt patches within the overall footprint of a fire. These patches can be valuable refugia for some species (Berry *et al.* 2015) and in some cases help speed post-fire species recovery (Lindenmayer *et al.* 2009). Spatial analyses of the 2019–20 wildfires indicate that, within the fire footprint, there were relatively few and typically very small patches of unburnt vegetation (Mackey *et al.* 2021a). Several studies have shown that long-unburnt refugia can be important for some elements of biota (Dixon *et al.* 2018) and we argue that such places should not be subject to yet further disturbances (like logging) wherever possible.

Other compounding effects: logging on steep slopes and 'downstreaming' effects

There may be additional compounding effects of logging on fire beyond those outlined above. First, fire burns more severely in forest on steep slopes (Catchpole 2002), and the 2019–20 wildfires were no exception (Lindenmayer *et al.* 2021). A spatial analysis of areas logged under past Victorian Government timber release plans has revealed that many stands on steep slopes (exceeding 30°C) have been cut (Taylor and Lindenmayer 2021). The combination of young, flammable regrowth forest on steep terrain means that such areas will be at risk of elevated severity wildfire for many years.

A second kind of compounding effect is the potential for high-severity fires to burn beyond logged forest into adjacent unlogged areas. That is, there is potential for spatial dependence in fire behaviour with close-by areas being more likely to suffer from similar levels of fire severity than forests located far apart. There is strong evidence of spatial dependence in high-severity fire following the 2009 fires (Taylor *et al.* 2020) and the 2019–20 fires (Taylor *et al.* 2020; Lindenmayer *et al.* 2021). In extensively logged areas, there is a risk of a landscape trap developing whereby widespread young, flammable forest promotes repeated high-severity wildfires and effectively precludes stand maturation. This positive feedback loop maintains ecosystems as young, flammable forest unable to grow to a less-flammable old growth state (Lindenmayer *et al.* 2011). That is, fire and logging 'beget more fire'. Without extensive and expensive management interventions, such a landscape trap can result in regeneration failure in dominant eucalypt species and replacement by non-forest vegetation, ultimately leading to ecosystem collapse. There is increasing evidence that a landscape trap is developing in the ash-type eucalypt forests across eastern Victoria, and in some other obligate-seeder dominated forests. For example, a recent paper from Tasmanian wet forests (Furlaud *et al.* 2021) concluded that:

> *the application of the moisture model flammability function, in which fire risk decreases with stand development ... suggest[s] that widespread logging and wildfire can increase landscape-scale fire risk, as regenerating stands are more prone to high- severity fire than mature stands, which is consistent with the notion of a 'landscape trap'.*

After the fire: post-fire ('salvage') logging has major negative impacts

A controversial form of logging is post-fire logging, sometimes called 'salvage logging' (see Box 19.2). Under such operations, sawlogs and pulplogs are removed from burnt areas in an attempt to recover some of their economic value. Salvage logging is a double set of disturbances; a high-severity fire followed by intensive logging operations on the fire grounds.

Box 19.2. The effects of post-fire 'salvage' logging

Post-fire logging is occurring in southern New South Wales and north-eastern Victoria following the 2019–20 wildfires. Approximately 3500 ha of forest in Victoria will be subject to post-fire logging (VicForests 2020). The extent of post-fire logging in New South Wales remains unclear, but it is occurring in native forests in many regions throughout the state (Fig. 19.3). Post-fire logging occurred in Victoria following wildfires in 1939, 1983 and 2009 and is a form of logging that is widely practised globally, not only after fires but also after windstorms and insect attack (Thorn *et al.* 2018). There have been numerous studies of post-fire logging globally and almost all of them highlight the negative ecological effects of such operations (Thorn *et al.* 2018). In an Australian context, salvage logging has been shown to (1) deplete populations of hollow-bearing trees that are critical habitat for cavity-dependent wildlife (Lindenmayer and Ought 2006), (2) reduce bird species richness (Lindenmayer *et al.* 2018), (3) alter plant species composition and deplete populations of resprouting taxa like tree ferns and promote flammable plant species (Bowd *et al.* 2018), and (4) deplete soil organic carbon and plant-available nutrients including nitrate and phosphorus, and change the structural composition of soil (e.g. an increased sand fraction) (Bowd *et al.* 2019). Where areas subject to post-fire logging are then regenerated, the subsequent young stands are characterised by a prolonged period of elevated risk of high-severity wildfire.

Fig. 19.3. Post-fire logging following the 2019–20 fires in the Eden region of southern New South Wales. (Photo: Nina Lindenmayer)

The majority of post-fire logging operations (and associated studies) in Australian forests have been conducted in wet ash-type eucalypt ecosystems where the dominant tree species are obligate seeders that are killed by wildfires. Following the 2019–20 wildfires, other kinds of burnt ecosystems also were targeted for post-fire logging, such as those dominated by drier lowland forest in southern New South Wales. Post-fire logging operations are inappropriate in such environments because most of overstorey trees in such forests were not killed by the fires, but rather resprouted and were recovering naturally before being cut down.

Does it matter if logged forests are more flammable?

Some authors have argued that because logging occurs in limited parts of the forest estate, then relationships between logging and fire are trivial (Bowman *et al.* 2021). This argument does not hold for those forest types, landscapes and regions where logging operations are concentrated. For example, 65% of all native forest clearcut logging in Victoria occurs in Wet and Damp forest types (Taylor and Lindenmayer 2019). The elevated flammability of numerous cutblocks throughout these forest types in such regions is therefore of concern, both for the long-term persistence of the forest and its associated biota, but also for the ability to limit the spread of high-severity fires in these environments.

Lack of congruence between the impacts of logging and fire

There is an extensive literature that pivots around the hypothesis that if logging 'mimics' the impacts of wildfires, then there will be limited effects of logging on biodiversity and ecological recovery (Florence 1994). This is, in part, because species will have co-evolved with high-intensity disturbances. Such thinking has, at times, been used to justify the widespread use of silvicultural systems like clearcutting in ecosystems where the natural fire regime is high-severity, stand-replacing wildfire (Attiwill *et al.* 1994). However, a substantial body of work shows that logging and fire have markedly different impacts on many elements of biodiversity and on ecosystem structure and condition. For example, unlike logging, wildfires leave behind numerous living and dead biological legacies. These differences influence the structure and species composition of forest stands recovering after disturbance (Lindenmayer *et al.* 2019b). Indeed, studies comparing the impacts of the 2009 wildfires and ongoing clearcutting in the montane ash forests of the Central Highlands of Victoria, showed major differences in responses of birds to the two kinds of disturbance (Lindenmayer *et al.* 2019a). Moreover, the high-intensity, mechanical nature of logging operations can compact soils and destroy subterranean plant propagules, including the root systems of tree-ferns and broadleaved shrubs, which would otherwise survive and resprout under natural fire regimes (Blair *et al.* 2016; Bowd *et al.* 2018). Natural and human disturbances in these ecosystems are therefore not ecologically congruent.

Perhaps what is more important than the clear difference in the effects of fire and logging is that ecosystems and the species they contain must now deal not with one form of disturbance or the other, but with both – and, as indicated throughout this chapter, the compounding and interactive effects of both. This problem is perhaps best illustrated by the parlous state of the mountain ash forests of the Central Highlands of Victoria in which the legacy of wildfires and logging (as well as past widespread post-fire logging) has resulted in less than 1.16% of the ecosystem being old growth forest – between 1/30th and 1/60th of what it was historically (Lindenmayer and Taylor 2020a). The paucity of old growth forest resulting from both fire and logging (and their interaction such as salvage logging) matters for many species and species assemblages in these forests (Lindenmayer *et al.* 2019a).

Finally, some of the issues around the interrelationships between fire and logging have major consequences for the sustainability of timber yields from the forest industry *per se*. Notably, ~30% of the forest that was planned for logging in Victoria in the five years from 2019 under that state's Timber Release Program was burnt in the 2019–20 fires; this number approached almost 60% for individual regions such as East Gippsland (Lindenmayer and Taylor 2020b). Figures from southern New South Wales are even higher and vary from 74–80% in the Southern and Eden Regional Forest Agreement areas, respectively (Forestry Corporation 2020). The frequency of high-severity fire in several parts of Victoria means that there is a high probability (~80%) that some forest types (like those dominated by ash-type eucalypts) will be burnt before the trees reach an age where they are suitable for sawlog production (Cary *et al.* 2021). This has economic and ecological ramifications as it indicates there is both declining certainty in future sawlog supply and potentially increasing pressure to log remaining unburnt forest (which often has high conservation value for threatened forest-dependent species (Taylor and Lindenmayer 2019)) – thereby adding to the overall extent of disturbed forest in the landscape. Indeed, recent analyses (Mackey *et al.* 2021b) showed that unburnt refugia within the footprint of the 2019–20 wildfire were both rare and typically small (1–3 ha in size). Further disturbance of these areas, such as through logging operations, may well exacerbate the problems for species persistence created by the widespread, high-severity wildfires that characterised the 2019–20 fire season.

Responses to long-term elevated flammability

The evidence that logging contributes to elevated fire severity is compelling. These effects can last for several decades after forests have been cut and then regenerated, leaving a prolonged flammability risk (Taylor *et al.* 2014). What can be done to counter such problems in flammable forest environments? One step is to expand the old growth forest estate as this is where fire severity is typically lowest. Another, interrelated, approach is to invest in new technologies for early detection and rapid suppression of fires, thereby increasing the chances of forest maturing to an old growth stage. These technologies include the development or enhanced use of (1) drone fleets and satellites to quickly detect ignitions like those triggered by lightning strikes, and (2) aerial assets to dispatch targeted water bombing or specialist remote-area fire-fighting crews to rapidly suppress ignitions and limit fire spread in highly flammable environments.

Conclusions

Climate and weather are typically the dominant drivers of fire behaviour including key aspects of fire regimes such as fire severity. However, other factors contribute to fire severity, and forest management is one of them. Although its effects are generally less pronounced than those of climate and weather, that does not mean we can ignore the potential benefits of changing forest management to reduce fire severity and promote forest conservation.

The amount of disturbance in some forest types in south-eastern Australia is very high and is likely increasing because of climate change. The persistence of biodiversity and the maintenance of ecosystem condition in these environments is likely to depend on attempts to reduce the overall disturbance burden and reduce the number of drivers of decline. Removing industrial logging from threatened forest communities is a key way to reduce the number of drivers.

Logging elevates the probability of subsequent high-severity wildfire, an effect that lasts for up to 40 years after cutover stands have been regenerated. Logging simplifies stand structure and when regenerating stands are then burnt, biotic recovery is impaired. Logging and fire may interact at large spatial scales (and well beyond a logged area),

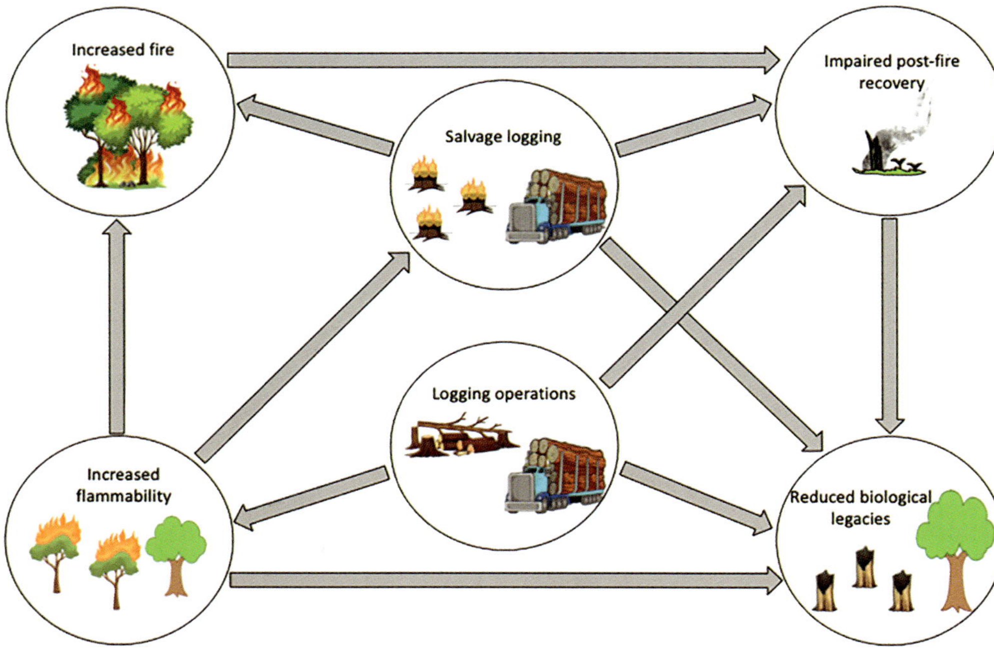

Fig. 19.4. Conceptual model showing compounding and interacting fire and logging effects in native forests. The model relates primarily to wet montane ash-type eucalypt forests where the dominant trees are obligate seeders and typically killed by high-severity stand-replacing wildfires.

potentially leading to more extensively burnt areas than would occur in the absence of logging. Post-fire logging has significant negative impacts on ecosystem condition and ecological recovery – from microbes and soils to plants and vertebrate groups such as birds and arboreal marsupials. The interactions and compounding impacts of forestry and fire described in this chapter are summarised in Fig. 19.4.

Recommendations

- The cessation of industrial logging from native forests is one of the key steps governments can take to reduce the problem of high fire severity. States such as Western Australia have begun to do this, and Victoria plans to do the same by the end of this decade (2030). This transition in Victoria should be brought forward and the exit from the native forest logging industry made far earlier, given the extent of repeated fires that have already occurred in that state over the past 25 years (Lindenmayer and Taylor 2020b), coupled with the fact that logging adds to the fire proneness of native forests.
- Forest management policies and practices should aim to ensure that as much of the forest estate is as old as possible. This is because old growth forests are where fire severity is lowest (Attiwill *et al.* 2014; Taylor *et al.* 2014; Lindenmayer *et al.* 2021) and value for many elements of biodiversity is often highest (Lindenmayer *et al.* 1990, 1999, 2019a). Efforts to boost the amount of old growth forest also will be important in the event of future fires, as such areas will more likely support more of the key kinds of

biological legacies that make regenerating stands more suitable for species to persist in and/or eventually recolonise burnt forests (Lindenmayer *et al.* 2019b).

- Land managers should focus on protecting and conserving fire refugia and limiting any additional disturbances in unburnt patches within already extensively disturbed landscapes.
- There should be a ban on all post-fire logging in native forest (i.e. no more salvage logging). The impacts of such practices are substantial, and inconsistent with principles of ecological sustainability. Hence, no resource management agency or government should countenance such destructive practices.
- The elevated levels of forest flammability that are a legacy of past widespread logging practices demand new technologies for the detection and suppression of wildfires in young forests.
- Much of the work on interactions between logging and fire severity has been in tall, wet ash-type eucalypt forests and additional research in other forest types is warranted.

Acknowledgements

The authors acknowledge the Traditional Custodians of the land they conduct their research on, along with their Elders, past and present. DBL sincerely thanks his colleagues in the NESP Threatened Species Recovery Hub, who completed an astounding amount of high-quality work immediately following the 2019–20 fires. We thank the editors of this book for asking us to write this chapter. Tabitha Boyer and Jess Williams assisted with editorial tasks.

References

Attiwill PM, Florence R, Hurditch WE, Hurditch WJ (1994) *The Burning Continent. Forest Ecosystems and Fire Management in Australia*. Institute of Public Affairs, Melbourne.

Attiwill PM, Ryan MF, Burrow N, Cheney N, McCaw L, *et al.* (2014) Timber harvesting does not increase fire risk and severity in wet eucalypt forests of southern Australia. *Conservation Letters* 7, 341–354. doi:10.1111/conl.12062

Au J, Clark RG, Allen C, Marsh KJ, Foley WJ, *et al.* (2019) A nutritional mechanism underpinning folivore occurrence in disturbed forests. *Forest Ecology and Management* **453**, 117585. doi:10.1016/j.foreco.2019.117585

Banks SC, McBurney L, Blair D, Davies ID, Lindenmayer DB (2017) Where do animals come from during post-fire population recovery? Implications for ecological and genetic patterns in post-fire landscapes. *Ecography* **40**, 1325–1338. doi:10.1111/ecog.02251

Berry L, Driscoll DA, Banks SC, Lindenmayer DB (2015) The use of topographic fire refuges by the greater glider (*Petauroides volans*) and the mountain brushtail possum (*Trichosurus cunninghami*) following a landscape-scale fire. *Australian Mammalogy* **37**, 39–45. doi:10.1071/AM14027

Blackhall M, Raffaele E, Paritsis J, Tiribelli F, Morales JM, *et al.* (2017) Effects of biological legacies and herbivory on fuels and flammability traits: a long-term experimental study of alternative stable states. *Journal of Ecology* **105**, 1309–1322. doi:10.1111/1365-2745.12796

Blair D, McBurney L, Blanchard W, Banks S, Lindenmayer DB (2016) Disturbance gradient shows logging affects plant functional groups more than fire. *Ecological Applications* **26**, 2280–2301. doi:10.1002/eap.1369

Bowd EJ, Lindenmayer DB, Banks SC, Blair DP (2018) Logging and fire regimes alter plant communities. *Ecological Applications* **28**, 826–841. doi:10.1002/eap.1693

Bowd EJ, Banks SC, Strong CL, Lindenmayer DB (2019) Long-term impacts of wildfire and logging on forest soils. *Nature Geoscience* **12**, 113–118. doi:10.1038/s41561-018-0294-2

Bowd E, Blair DP, Lindenmayer DB (2021) Prior disturbance legacy effects on plant recovery post-high severity wildfire. *Ecosphere* **12**, e03480. doi:10.1002/ecs2.3480

Bowman D, Williamson GJ, Gibson R, Bradstock RA, Keenan RJ (2021) The severity and extent of the Australia 2019–20 *Eucalyptus* forest fires are not the legacy of forest management. *Nature Ecology & Evolution* **5**, 1003–1010. doi:10.1038/s41559-021-01464-6

Bradstock RA (2010) A biogeographic model of fire regimes in Australia: current and future implications. *Global Ecology and Biogeography* **19**, 145–158. doi:10.1111/j.1466-8238.2009.00512.x

Bradstock R, Price OF (2014) Logging and fire in Australian forests: errors by Attiwill *et al.* (2014). *Conservation Letters* **7**, 419–420. doi:10.1111/conl.12086

Cary G, Blanchard W, Foster CN, Lindenmayer DB (2021) Effects of altered fire regimes on critical timber production and conservation rotations. *International Journal of Wildland Fire* **30**, 322–328. doi:10.1071/WF20129

Catchpole WR (2002) Fire properties and burn patterns in heterogenous landscapes. In *Flammable Australia. The Fire Regimes and Biodiversity of a Continent.* (Eds RA Bradstock, JE Williams and AM Gill) pp. 49–76. Cambridge University Press, Cambridge, UK.

Collins L, Bradstock R, Clarke H, Clarke MF, Nolan RH, *et al.* (2021) The 2019/2020 mega-fires exposed Australian ecosystems to an unprecedented extent of high-severity fire. *Environmental Research Letters* **16**, 044029. doi:10.1088/1748-9326/abeb9e

Dixon KM, Cary GJ, Worboys GL, Gibbons P (2018) The disproportionate importance of long-unburned forests and woodlands for reptiles. *Ecology and Evolution* **8**, 10952–10963. doi:10.1002/ece3.4561

Florence RG (1994) The ecological basis of forest fire management in New South Wales. In *The Burning Continent: Forest Ecosystems and Fire Management in Australia.* (Eds PM Attiwill, RG Florence, WE Hurditch and WJ Hurditch) pp. 15–33. Institute of Public Affairs, Jolimont.

Forestry Corporation (2020) Impact of fires 2019–20, <https://www.forestrycorporation.com.au/operations/fire-management/fire-impact-of-2019-20>.

Franklin JF, Lindenmayer DB, MacMahon JA, McKee A, Magnuson J, *et al.* (2000) Threads of continuity. *Conservation in Practice* **1**, 8–17. doi:10.1111/j.1526-4629.2000.tb00155.x

Furlaud JM, Prior LD, Williamson GJ, Bowman DMJS (2021) Fire risk and severity decline with stand development in Tasmanian giant Eucalyptus forest. *Forest Ecology and Management* **502**, 119724. doi:10.1016/j.foreco.2021.119724

Helms JA (1998) *The Dictionary of Forestry.* Society of American Foresters, Bethesda, MD.

Inspector General for Emergency Management (2020) *Inquiry into the 2019–20 Victorian Fire Season.* Victorian Government, Melbourne.

Jones MW, Smith A, Betts R, Canadell JG, Prentice IC, *et al.* (2020) Climate change increases risk of wildfires. ScienceBrief Review, <https://sciencebrief.org/briefs/wildfires>.

Kalies EL, Yocom Kent LL (2016) Tamm Review: are fuel treatments effective at achieving ecological and social objectives? A systematic review. *Forest Ecology and Management* **375**, 84–95. doi:10.1016/j.foreco.2016.05.021

Lefoe M, Rendall AR, McKinnon F, Whisson DA (2022) Logging and wildfire limit the distribution of a vulnerable arboreal mammal. *Forest Ecology and Management* **503**, 119773. doi:10.1016/j.foreco.2021.119773

Lindenmayer DB, Ought K (2006) Salvage logging in the montane ash eucalypt forests of the Central Highlands of Victoria and its potential impacts on biodiversity. *Conservation Biology* **20**, 1005–1015. doi:10.1111/j.1523-1739.2006.00501.x

Lindenmayer D, Taylor C (2020a) Extensive recent wildfires demand more stringent protection of critical old growth forest. *Pacific Conservation Biology* **26**, 384–394. doi:10.1071/PC20037

Lindenmayer DB, Taylor C (2020b) New spatial analyses of Australian wildfires highlight the need for new fire, resource, and conservation policies. *Proceedings of the National Academy of Sciences of the United States of America* **117**, 12481–12485. doi:10.1073/pnas.2002269117

Lindenmayer DB, Cunningham RB, Tanton MT, Smith AP, Nix HA (1990) The habitat requirements of the Mountain Brushtail Possum and the Greater Glider in the montane ash-type eucalypt forests of the Central Highlands of Victoria. *Australian Wildlife Research* **17**, 467–478. doi:10.1071/WR9900467

Lindenmayer DB, Cunningham RB, McCarthy MA (1999) The conservation of arboreal marsupials in the montane ash forests of the Central Highlands of Victoria, south-eastern Australia: VIII. Landscape analysis of the occurrence of arboreal marsupials. *Biological Conservation* **89**, 83–92. doi:10.1016/S0006-3207(98)00116-5

Lindenmayer DB, MacGregor C, Wood JT, Cunningham RB, Crane M, *et al.* (2009) What factors influence rapid post-fire site re-occupancy? A case study of the endangered Eastern Bristlebird in eastern Australia. *International Journal of Wildland Fire* **18**, 84–95. doi:10.1071/WF07048

Lindenmayer DB, Hobbs RJ, Likens GE, Krebs CJ, Banks SC (2011) Newly discovered landscape traps produce regime shifts in wet forests. *Proceedings of the National Academy of Sciences of the United States of America* **108**, 15887–15891. doi:10.1073/pnas.1110245108

Lindenmayer DB, Blanchard W, Blair D, McBurney L, Banks SC (2016) Environmental and human drivers influencing large old tree abundance in Australian wet forests. *Forest Ecology and Management* **372**, 226–235. doi:10.1016/j.foreco.2016.04.017

Lindenmayer DB, McBurney L, Blair D, Wood J, Banks SC (2018) From unburnt to salvage logged: quantifying bird responses to different levels of disturbance severity. *Journal of Applied Ecology* **55**, 1626–1636. doi:10.1111/1365-2664.13137

Lindenmayer DB, Blanchard W, Blair D, Westgate MJ, Scheele BC (2019a) Spatio-temporal effects of logging and fire on forest birds. *Ecological Applications* **29**, e01999. doi:10.1002/eap.1999

Lindenmayer DB, Westgate MJ, Scheele BC, Foster CN, Blair DP (2019b) Key perspectives on early successional forests subject to stand-replacing disturbances. *Forest Ecology and Management* **454**, 117656. doi:10.1016/j.foreco.2019.117656

Lindenmayer DB, Kooyman R, Taylor C, Ward M, Watson J (2020) Recent Australian wildfires made worse by logging and associated forest management. *Nature Ecology & Evolution* **4**, 898–900. doi:10.1038/s41559-020-1195-5

Lindenmayer D, Taylor C, Blanchard W (2021) Empirical analyses of the factors influencing fire severity in southeastern Australia. *Ecosphere* **12**, e03721. doi:10.1002/ecs2.3721

Mackey B, Lindenmayer DB, Norman P, Taylor C, Gould S (2021a) Are fire refugia less predictable due to climate change? *Environmental Research Letters* **16**, 114028. doi:10.1088/1748-9326/ac2e88

Mackey BG, Gould S, Lindenmayer DB, Norman P, Taylor C (2021b) 'Bushfire recovery project. Bushfire Science Report No. 1. Does climate affect bushfire risks in the native forests of eastern Australia?' Joint report of Griffith University and the Australian National University, Griffith University, Brisbane, <https://www.bushfirefacts.org/>.

Smith AL, Blair D, McBurney L, Banks S, Barton PS, *et al.* (2014) Dominant drivers of seedling establishment in a fire-dependent obligate seeder: climate or fire regimes? *Ecosystems* **17**, 258–270. doi:10.1007/s10021-013-9721-9

Sullivan AL, McCaw WL, Cruz MG, Matthews S, Ellis PF (2012) Fuel, fire weather and fire behaviour in Australian ecosystems. In *Flammable Australia: Fire Regimes, Biodiversity and Ecosystems in a Changing World*. (Eds RA Bradstock, AM Gill and RJ Williams.) pp. 51–77. Cambridge University Press, Cambridge, UK.

Taylor C, Blanchard W, Lindenmayer DB (2020) Does forest thinning reduce fire severity in Australian eucalypt forests? *Conservation Letters* **14**, e12766.

Taylor C, Blanchard W, Lindenmayer DB (2021) What are the relationships between thinning and fire severity? *Austral Ecology* **46**, 1425–1439. doi:10.1111/aec.13096

Taylor C, Lindenmayer DB (2019) The adequacy of Victoria's protected areas for conserving its forest-dependent fauna. *Austral Ecology* **44**, 1076–1091. doi:10.1111/aec.12805

Taylor C, Lindenmayer DB (2020) Temporal fragmentation of a critically endangered forest ecosystem. *Austral Ecology* **45**, 340–354. doi:10.1111/aec.12863

Taylor C, Lindenmayer DB (2021) Stakeholder engagement in a Forest Stewardship Council controlled wood assessment. *Environmental Science & Policy* **120**, 204–212. doi:10.1016/j.envsci.2021.03.014

Taylor C, McCarthy MA, Lindenmayer DB (2014) Non-linear effects of stand age on fire severity. *Conservation Letters* **7**, 355–370. doi:10.1111/conl.12122

Thorn S, Bassler C, Brandl R, Burton PJ, Cahall R, *et al.* (2018) Impacts of salvage logging on biodiversity: a meta-analysis. *Journal of Applied Ecology* **55**, 279–289. doi:10.1111/1365-2664.12945

Tolhurst KG, McCarthy G (2016) Effect of prescribed burning on wildfire severity: a landscape-scale case study from the 2003 fires in Victoria. *Australian Forestry* **79**, 1–14. doi:10.1080/00049158.2015.1127197

van der Velde IR, van der Werf GR, Houweling S, Maasakkers JD, Borsdorff T, *et al.* (2021) Vast CO_2 release from Australian fires in 2019–2020 constrained by satellite. *Nature* **597**, 366–369. doi:10.1038/s41586-021-03712-y

van Oldenborgh G-J, Krikken F, Lewis S, Leach NJ, Lehner F, *et al.* (2021) Attribution of the Australian bushfire risk to anthropogenic climate change. *Natural Hazards and Earth System Sciences* **21**, 941–960. doi:10.5194/nhess-21-941-2021

VicForests (2020) VicForests starts post-fire timber recovery, <https://www.vicforests.com.au/fire-management-1/vicforests-starts-post-fire-timber-recovery>.

Ward M, Tulloch AIT, Radford JQ, Williams BA, Reside AE, *et al.* (2020) Impact of 2019–2020 mega-fires on Australian fauna habitat. *Nature Ecology & Evolution* **4**, 1321–1326. doi:10.1038/s41559-020-1251-1

Zylstra PJ, Bradstock RA, Bedward M, Penman TD, Doherty MD, *et al.* (2016) Biophysical mechanistic modelling quantifies the effects of plant traits on fire severity: species, not surface fuel loads determine flame dimensions in eucalypt forests. *PLoS One* **11**, e0160715. doi:10.1371/journal.pone.0160715

20

Review of impacts of the 2019–20 wildfires on biodiversity

Sarah M. Legge, Libby Rumpff and John C. Z. Woinarski

Chapters 3 to 19 provide a reckoning of the 2019–20 wildfire impacts on natural and cultural values. The chapters document that these fires affected, in varying ways, much of the environmental fabric of Australia including key biodiversity sites and protected areas, soils, aquatic and marine systems, ecological communities, plants, fungi and animals, including many threatened species. Here, we review the information across these chapters, drawing out common features of impacts and opportunities.

Impacts

The fires burnt over 103 400 km^2, affecting hundreds of protected areas, including substantial proportions of World Heritage areas inscribed for their cultural and natural values, and some of Australia's key biodiversity areas. The fire-caused loss of vegetation and changes to soil properties increased erosion rates up to threefold, leading to massive sediment pulses in waterways that killed aquatic organisms, filled pools and smothered streambed substrates. Sediment was transported dozens of kilometres downstream, even affecting littoral and marine areas. Some of the effects on the abiotic scaffolding of life, such as changes to soil chemistry, nutrient cycles and stream morphology, may take decades to unwind, and may not recover before future fires further degrade the system. The fires affected tangible and intangible cultural values; burnt dozens of ecological communities, including ecosystems that rarely (if ever) experience fire; and caused population losses in hundreds of vertebrate species, and tens of thousands of invertebrate, plant and fungi species (Fig. 20.1). Impacts ranged from severe to negligible, and even beneficial for some species. Some affected species and ecological communities will never be made good: some species are now extinct or nearly so (e.g. Yalmy galaxias (*Galaxias* sp. nov. 'Yalmy'; Chapter 6), *Banksia montana* mealybug (*Pseudococcus markharveyi*; Chapter 11), and some ecological communities are unlikely to return to their pre-fire state, given projections for future climate and fire regimes.

The 2019–20 wildfires were so extensive that heavily impacted species were diverse, in terms of their pre-fire conservation status (threatened/non-threatened); habitat (e.g. rainforest, wetlands, woodlands); whether terrestrial or aquatic; and whether range-restricted or wide-ranging. However, although many wide-ranging species across the forests of eastern Australia suffered major losses, the species whose distributions most extensively

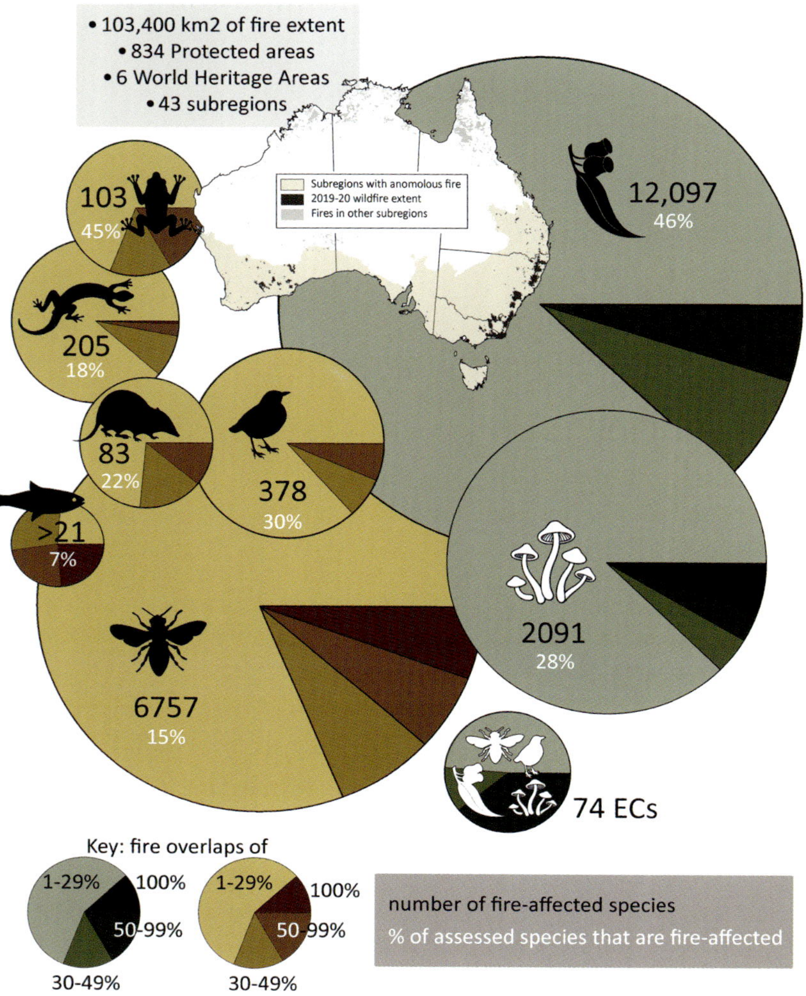

Fig. 20.1. Summary of exposure of species and ecological communities to the 2019–20 wildfires. The map shows the fire-affected subregions (beige), the fire extent within those subregions (dark grey), and areas that burnt elsewhere in Australia that season (light grey). The pies summarise fire exposure for different species groups. The number of fire-affected species with ranges that overlapped fires by 1% or more are shown in black text, and also depicted by the relative sizes of the pies (although these are not to scale). Of the fire-affected species, the relative proportions of species whose distributions overlapped with fire by 1–29%, 30–49%, 50–99% and 100% are shown by the pie segments. The number of species assessed within each species group (from top right, moving clockwise): plants (26 062), fungi (7445), ecological communities (92), invertebrates (45 428), fish (281), mammals (386), birds (1276 including ultrataxa), reptiles (1128), and frogs (227). Note that the numbers of species assessed, and therefore the numbers that are fire-affected, are substantial underestimates for invertebrates and fungi: of ~320 000 invertebrate species, only 45 428 are described and have distribution data available, suggesting the real number of fire-affected species could be at least seven times greater than the 6767 depicted here. Of the estimated 50 000 to 250 000 fungi species, 7445 are described and have distribution data, suggesting the real number of fire-affected fungi species could be considerably more than seven times the 2091 depicted here. The information to compile this figure comes from Chapters 3–19 and references therein.

overlapped with fire tended to be range-restricted. The preceding chapters indicate that 787 invertebrate, 66 vertebrate, 178 fungi and 593 plant species had at least 50% of their range burnt in these fires, an extraordinary outcome from a single event. Many of these species have now been listed as threatened, or uplisted due to these losses, with listing assessments pending for many others. Fire-affected ecological communities were similarly diverse, ranging from rainforests to peatlands, and including fire-prone communities such as heathlands as well as fire-sensitive communities.

Common features of biodiversity impacts across Chapters 3–19 include:

- For most species, estimating fire-caused population loss is challenging because of scant pre-existing monitoring and limited knowledge of species' fire response. Where evidence on fire responses is available, it is usually from studies of less extreme fires. Assessing fire impacts has been especially difficult for poorly known groups such as fungi and invertebrates.
- Fires can kill individual plants, fungi and animals directly, but (indirect) mortality after fire can exceed mortality during the fire because resource availability changes radically, and exposure to other threats may increase.
- As well as amplifying the short-term effect of fires, these interactions with other threats can suppress longer-term recovery of populations. For example:
  - Areas of lower habitat integrity, including areas that are fragmented, or used for forestry in recent decades, are more likely to burn at high severity.
  - For animals, fire may amplify predation risk by introduced predators, or disease susceptibility, or exacerbate competition from introduced herbivores.
  - For plants and ecological communities, the antecedent conditions critically modify the impacts of fire. For example, in areas of increased fire frequency, short fire intervals strongly disadvantage some species and powerfully reorganise species assemblages. Pre-fire drought also strongly moderates the fire response of ecological communities, and many plant and animal species.
- The response of populations and ecological communities to fire depends on their pre-fire conservation status; recovery is more challenging for populations that are already reduced or declining.
- Fire impacts may be most significant, and recovery most challenging, for species and environments with an evolutionary history of infrequent disturbance.
- Some species are unlikely to return to their pre-fire size over the next decade or more; their recovery may be thwarted by projected increases in fire incidence.
- Some of the 2019–20 wildfire impacts will be long-lasting and reverberating. For example, as well as changes to soil properties and the physical structure of streams and littoral areas, hollow-bearing trees and logs may be scarce in some areas for decades. Species that rely on these features will be correspondingly affected.
- Not all species were adversely affected by the fires; for example, some plant species that rely on disturbance for germination have experienced large recruitment events.

Opportunities

The review of Chapters 3 to 19 also highlights some conservation opportunities. For example, populations of introduced species were reduced by the 2019–20 wildfires, or made more vulnerable to control options. Some of these opportunities were seized by management agencies (e.g. increasing the extent or intensity of aerial baiting and shooting). Other options remain for future trials, such as capitalising on the range reductions of invasive fish species by preventing their re-expansion using instream barriers. In addition, some of the emergency rescue operations for freshwater species and birds, and seed and cutting salvage for plants, demonstrated what could be achieved in the future, with more forethought, planning and resourcing.

It is important to reflect that all knowledge available before the 2019–20 wildfires (on species distribution, status and trends, sensitivity to threats and response to management) was crucial for informing the management response; similarly, all conservation management that had been undertaken to that point, to support species and ecological communities, contributed to resilience in the populations and ecological communities affected by the wildfires. Much survey, monitoring and management action are occurring in the aftermath of the 2019–20 wildfires, and this effort will sharpen current and future assessments of impacts and management effectiveness, support recovery, and help build resilience to future fires.

21

Prescribed burning in Australian forests: characteristics, impacts and effects

Owen Price

Summary

- Two to seven per cent of the forests of southern Australia are prescribed burnt each year.
- Prescribed burning reduces the area burnt by wildfire by approximately 1 ha for every 3 ha treated, and other measures of wildfire are also reduced accordingly.
- At current levels, prescribed burning is reducing risk by a modest amount, but it is not causing substantial negative effects on biodiversity. However, there may be negative impacts in areas treated more frequently such as the interface with human populations.
- In the 2019–20 fires, recent prescribed burning within 500 m of houses reduced the rate of house loss such that areas where 50% had been treated in the last 5 years had only 41% of the loss rate of untreated areas.
- Under climate change, prescribed burning rates will need to increase markedly in order to keep wildfire risk at current levels. This is probably both impractical and detrimental to biodiversity.
- There are major knowledge gaps about the response of individual species to the full range of fire regime factors, about emissions and carbon storage under different regimes, and indirect effects of fire such as through increasing predator pressure, and interactions between prescribed burning and wildfire.

Introduction

Prescribed burning is one of the main wildfire risk reduction measures practised in Australian eucalypt forests and heathlands. But it is a contentious practice because there are concerns over its cost-effectiveness and that its widespread application reduces intervals between fires and thus may harm some components of biodiversity. Most major wildfire seasons spawn official inquiries and these have all investigated the role of prescribed burning (see Chapter 30). The modern fire management era began after the Stretton Inquiry into the 1939 wildfires in Victoria lamented that the rate of 'controlled burning' was 'ridiculously inadequate' (Stretton 1939). Of the subsequent inquiries, most have recommended improvements to the strategy and coordination of prescribed burning, but only two have recommended specific targets. These were the 2008 Victorian Parliamentary Inquiry into the management of wildfires on public land and the 2010 Victorian Royal Commission into the Black Saturday Fires (Anon 2008; Teague *et al.* 2010). The Black Saturday royal commission target that 5% of public land be prescribed burnt each year was adopted by the Victorian Government but abandoned in 2016 in favour of a more focused risk-based strategy.

Depending on where, when and how it is done prescribed burning can:

- Protect sensitive areas from burning in a wildfire. These areas may be houses or fire sensitive species or ecosystems.
- Reduce the severity of a wildfire. This will reduce the ecological impact of the wildfire, which will protect populations of some fire-sensitive species, but may be detrimental to populations of species that require high severity to complete their life cycle.
- Increase the frequency of burning (and thus decrease intervals between them). Because wildfires are rare, most prescribed burns never interact with one, and so act to increase fire frequency. This may be detrimental for fire-sensitive species or those that require long fire-free intervals to complete their life cycles.

Increases in prescribed burning are often called for after major wildfire seasons, and if prescribed burns are to be used to prevent the risk of future wildfires under climate change then the area treated will have to keep increasing (or become ineffective), so it is important to understand how prescribed burning, and increasing it, affects biodiversity.

What is prescribed burning?

Prescribed burning is the deliberate application of fire into the landscape. In contemporary times, most are 'hazard reduction burns': the removal of fuels to reduce risks to people and property (Fig. 21.1). The managers of conservation reserves may impose burns for

Fig. 21.1. A prescribed burn near houses, Stanwell Park, NSW, April 2021. (Photo: Owen Price)

ecological purposes, such as promoting the regeneration of a senescent seeding shrub species, or increasing the mix of post-fire ages within a reserve. Prescribed burning is most commonly applied in dry sclerophyll forests and heathland. Wet sclerophyll forests are more rarely treated because fire is more difficult to control there, and rainforests are generally never deliberately burnt. Sometimes dry sclerophyll forests are control burnt to protect other vegetation types.

Before European colonisation most fires were Indigenous burns or 'cultural burns' for a variety of purposes that can be grouped under the umbrella term 'caring for Country'. Colonisation brought a collapse of the practice, but there is now a resurgence across Australia, spearheaded in monsoonal Northern Territory where Indigenous savanna burning projects cover more than half of the land area (Russell-Smith *et al.* 2020). Indigenous land management is now expanding in south-eastern Australia (see Chapter 33).

While there is some degree of overlap among these three types of fires in terms of their ecological effect, they can be separated along a gradient of fire intensity. Many aspects of fire behaviour and impact are correlated: area, rate of spread, flame length, intensity and severity (the degree of impact on the vegetation). In general, cultural burns are mild, meaning all of those aspects are on the low end of the spectrum (small, trickling fire with short flames that only impact the ground and shrub layer). Wildfires are at the other extreme (severe) and hazard reduction burns are in the middle (moderate). In practice, this means that hazard reduction burns usually leave unburnt patches (ranging from 6–35% of the fire extent) (Penman *et al.* 2007; Volkova and Weston 2015; Duff *et al.* 2019). Cultural burns may leave most of the vegetation unburnt (Price *et al.* 2022). Across Australia there is a negative association between fire severity and fire frequency, which is a consequence of climate and the rate at which fuels accumulate. At one end of this gradient, tropical savannas experience high-frequency, low-severity fires and, at the other, south-eastern forests experience infrequent severe fires (Murphy *et al.* 2013). This means that the nature of both wildfire and prescribed fire varies across the nation.

Prescribed burning and wildfire risk to property

Protection of human life and property is the main purpose of hazard reduction burning. It is mostly focused on the interface between forest and houses, though there is considerable debate about the amount and locations that should be burnt. Victoria set targets of 5% of all public land from 2010–16, but shifted to a risk-based rather than area-based approach in 2016. NSW National Parks and Wildlife Service (NPWS) have targets of 135 000 ha per year (1.5% of the protected area estate and ~2.5% if only applied to forests). In Western Australia, the Department of Biodiversity, Conservation and Attractions (DBCA) aims to burn 7% of its forest estate each year (Bradshaw *et al.* 2018). In all states bar Western Australia, the annual treatment rate is much less than the area of wildfires, and has been relatively stable over recent decades (see Fig. 21.2 for trends in New South Wales).

Research evidence finds that prescribed burning reduces many aspects of wildfire risk. It reduces severity of a subsequent wildfire for ~5 years (Bradstock *et al.* 2010), but effectiveness reduces as fire weather increases. For example, recent prescribed burning had no effect under Catastrophic weather conditions in the 2009 Black Saturday fires (Price and Bradstock 2012). Under most conditions, there will be less impact on houses adjacent to the treated area and it is more likely that firefighters will be able to stop fire from reaching houses. The extent or proximity has been found to reduce wildfire impact in both the 2009 Black Saturday fires (Gibbons *et al.* 2012) and also in the 2019–20 megafires (Price and Haynes 2020). Prescribed burning also reduces the regional area burnt by wildfires by

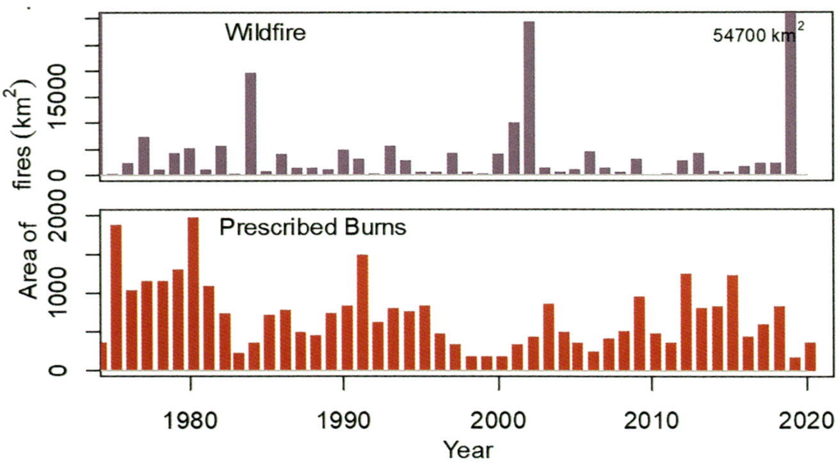

Fig. 21.2. Historical area burnt in New South Wales by wildfire and prescribed burns according to NSW DPIE fire history data (downloaded from SEED portal (https://www.seed.nsw.gov.au/). On average, wildfire area is 6.2 times prescribed burning area. Notice that the scales on the *y* axes differ.

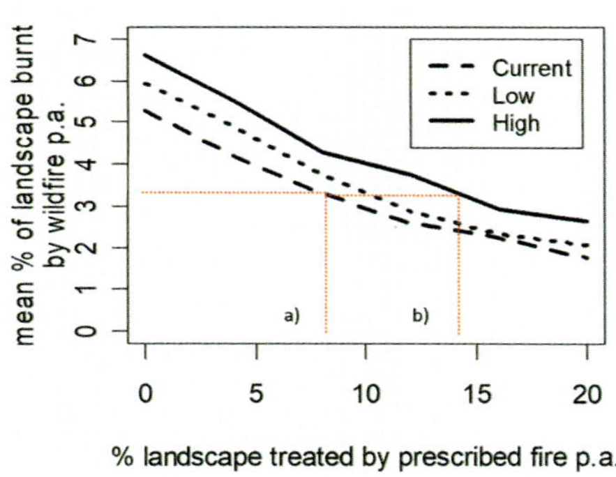

Fig. 21.3. Prescribed burning effect on the area of wildfire as predicted by fire simulations (from Bradstock *et al.* (2008)) for the Blue Mountains, showing (a) the annual treatment rate required to halve annual wildfire area and (b) the rate required under 2050 climate. (Image reproduced from Bradstock R, Davies I, Price OF, Cary G (2008) Effects of climate change on bushfire threats to biodiversity, ecosystem processes and people in the Sydney region: Final Report to the New South Wales Department of Environment and Climate Change. Centre for Environmental Risk Management of Bushfires, Faculty of Science, University of Wollongong, Wollongong)

~1 ha for every 3 ha treated over the long-term in forests (Fig. 21.3) (Bradstock *et al.* 2008, 2012b). However, prescribed burning rarely stops a fire unaided (without the addition of firefighters or a fire break such as a road) (Price and Bradstock 2010). Also, prescribed burns produce smoke, which can expose the local community and some land uses to poor air quality (up to ~5 km from the burn), and sometimes cause extensive exposure (Broome *et al.* 2016; Price and Forehead 2021).

Although protection of property is not the main reason for cultural burning, it can have that effect (Steffenson 2020), albeit to a lesser extent than hazard reduction burning due to the small scale and lower fuel reduction.

Prescribed burning and biodiversity

All Australian fauna and flora are adapted to fire to some extent, with specialised mechanisms to aid recovery. The way fire influences them depends on those adaptations and on the fire regime they experience (each individual fire, the sequence of fires, and the intervals between fires). The possible combinations are almost endless, so understanding these dynamics is complicated. For plants, there is a useful general division between species that recover from fire by surviving and resprouting (from the base, trunk or canopy depending on the severity of the fire) and those that are killed and regenerate from seeds: 'resprouters' (which include most eucalypts) and 'seeders' respectively (Keeley *et al.* 2011). Regular prescribed burning generally has the twin effects of reducing the interval between fires and their severity (because there is less time for fuel to build-up). Competition among plants is high after a mild fire because much of the vegetation is intact so those plants that survive (the resprouters) have the advantage. Seeders will decline if the interval between fires is less than the time required to mature and set another seed crop, and this effect has been shown to eliminate some plant species from certain places (Gill and Bradstock 1995). Resprouters can tolerate shorter intervals, but even they have limits (Enright *et al.* 2011): up to 5% of resprouting eucalypts are killed by low-severity fires typical of prescribed burns (Bennett *et al.* 2016; Etchells *et al.* 2020), and frequent prescribed burning may particularly impact larger trees (Watson *et al.* 2020). Also, many of the seeders require high temperatures to germinate their seeds, temperatures only experienced in severe fire (Liyanage and Ooi 2015). So generally, prescribed burning favours resprouters. This is a significant concern considering the large number of seeder species. For example, the NSW Flora Fire Response Database (V2.1, DPIE, unpublished data) lists 1113 plant species as seeders (36% of all species). Prescribed burns are usually conducted in spring or autumn, outside the typical summer wildfire season, and this will negatively affect species that are adapted to summer fires (Miller *et al.* 2019).

Set against these changes in interval, severity and season is the fact that prescribed burns tend to leave more unburnt patches than do wildfires. Estimates range from 22–35% unburnt (Penman *et al.* 2007; Volkova and Weston 2015), compared to < 5% in the 2019–20 wildfires (Price *et al.* 2022). This means that some sensitive plants will escape each prescribed burn, and the actual impacts of a regime of frequent prescribed burns may not emerge for many decades. It also means that concerns about the impact of reduced intervals may be overestimated if this patchiness is not factored into assessments. In general, the floristic composition, forest structure and carbon storage have been found to be resilient to a range of regimes (Knox and Clarke 2012; Gordon *et al.* 2018; Rahmani and Price 2021), although previously assumed resilience may be subverted by regimes characterised by an increasing extent and severity of wildfires.

The response of animals to fire is even more complicated than for plants because, depending on their size and mobility, they can escape the flames either *in situ* (in a burrow or hollow) or by fleeing. It follows that the most sensitive species are those that do not shelter (such as litter-dwelling lizards and insects or koalas (*Phascolarctos cinereus*)), have small home ranges, or both. Because prescribed burns are generally smaller, less severe and leave more internal unburnt patches than wildfires, the impact of any single prescribed fire on fauna is less than from wildfires. Most studies of animal responses to single fire events find either that populations are only slightly reduced (Jolly *et al.* 2022) by fire or that they recover to pre-fire levels in a few years, and this is especially the case for prescribed burns (Kuchinke *et al.* 2020; McHugh *et al.* 2020), although this finding is based on relatively few studies of impacts of high-severity fires. Many studies also find that the direct influence of fire is smaller than its indirect effect through modification of the habitat (such as temporarily removing the ground or shrub layer) (Monamy and Fox 2000; Swan *et al.* 2015). However, in contrast to wildfires and cultural burns, the ignition pattern often in hazard reduction burns involves a line of flame around the perimeter of the burn, spreading into the centre. This potentially leaves animals without easy escape routes, though the phenomenon has not been examined.

What we learned about prescribed burning from the 2019–20 wildfires

Previous prescribed burning affected the 2019–20 wildfires. The rates of house loss were ~30% lower where 50% of the area within 500 m of houses had been prescribed burnt within the past 5 years (Price and Haynes 2020). Fire severity was ~30% lower in areas burnt by recent prescribed burns (Hislop *et al.* 2020; Price *et al.* 2021). Of 1699 New South Wales prescribed burns in the five years before spring 2019, 30% were encountered by the wildfires, and of those 42% resulted in internal unburnt patches, and 13% were aligned with the wildfire boundary and so had a potential role in stopping the fire (Price *et al.* 2021). This shows that prescribed burning did reduce risks associated with property loss and can introduce refuges from wildfire and particularly from high-severity fire.

Prescribed burning and carbon storage

Prescribed burning in south-eastern forests decreases fire intervals. This is different to tropical savannas where spatially targeted treatments can slightly increase intervals (Evans and Russell-Smith 2020). Also, since one objective of prescribed burning programs is to keep fine fuel loads lower than they would otherwise be, a lower carbon stock may be expected under a regime dominated by prescribed burning. This means that prescribed burning may decrease forest carbon stocks. However, whether the total forest carbon stock falls depends on the comparative biomass consumption rates of prescribed and wildfires (Bradstock *et al.* 2012a), which in turn depends upon which fire type promotes the stocks of large trees where most of the carbon resides. This is a complicated and poorly understood issue, in large part because current estimates of rates of tree mortality and consumption are highly uncertain. On the one hand, a wildfire regime may kill many trees but provides the conditions for recruitment and regrowth. On the other a prescribed fire regime kills very few large trees, enabling them to grow to their maximum size, but the short inter-fire intervals may prevent recruitment when those trees senesce. The available field data suggest that differences in fire regime over several decades have a small effect on carbon stocks (Gordon *et al.* 2018) and that prescribed fires have little potential to increase carbon sequestration (Volkova *et al.* 2021).

Prescribed burning and climate change

Climate data suggest that the conditions conducive to wildfire have increased substantially over recent decades and are predicted to escalate further over coming decades (Clarke *et al.* 2013), and this is resulting in increased wildfire activity (Bradstock *et al.* 2014; Canadell *et al.* 2021). The 2019–20 wildfires were characterised by a clear climate change signal (Abram *et al.* 2021). Climate change will result in increased frequency and severity of wildfire, and this combined with probable poorer conditions for growth may lead to risk of loss of woody plants, and the ecological communities they comprise, through 'interval squeeze' (Enright *et al.* 2015). Should this occur, the reduction in biomass will eventually have a dampening effect on wildfire activity so that wildfire activity may increase over several decades, but then start to decline. Wildfire area and associated risk can be maintained at current levels by significantly increasing the annual amount of prescribed burning (probably to nearly 15% of the landscape by 2050 (Bradstock *et al.* 2012b; Fig. 21.3)), but this will be very costly; decrease inter-fire intervals even more; may be impossible to achieve; and the effectiveness of control burning in constraining wildfire may decrease under hotter, drier conditions forecast for the future, and may contribute to interval squeeze.

How much of a problem is prescribed burning?

Prescribed burning can help to protect areas from high-severity fire so it is beneficial for species whose survival is a function of severity. However, the dominant effect is to decrease intervals between fires and this is likely to be detrimental to many plant and animal species, and some ecological communities. The extent to which this is a problem depends on how much the average inter-fire interval is reduced. Current fire management policies in national parks rely on thresholds of potential concern (Kenny *et al.* 2004), which define minimum and maximum fire intervals for vegetation formations or individual species. Analysis of these threshold status in New South Wales shows that only a small area was burnt too frequently before the 2019–20 wildfires, and even those massive fires caused only a modest increase of 25% in that area burnt below threshold from 800 000 ha to 1 million ha (Williamson 2020) (Fig. 21.4). However, the entire burnt area (of 550 million ha) is now vulnerable to falling below threshold should another wildfire occur in the next 5–10 years. Prescribed burning rates are much lower than wildfire rates (< 3% of the landscape per year) so it is unlikely that they will cause significant shortening of intervals. For example, on average, a prescribed fire regime that burns 3% of forests on average each year overlaid on a 5% annual wildfire rate will reduce inter-fire intervals from ~20 years to 14 years, still considerably higher than thresholds of potential concern for most species.

However, there are some causes for concern. Prescribed burning treatments may be increased to counter increased risk from wildfires under climate change. Simulation studies suggest that if treatment rates are increased to 10% per annum in Tasmania or the Australian Capital Territory, then the area of forests burnt below thresholds of potential concern will increase by 15% relative to current levels (Cirulis *et al.* 2019). These kinds of treatment levels have been proposed to be needed to keep wildfire area constant in the future. Current treatments are unevenly distributed across the forests, and are focused at the interface between forests and human settlements (Price and Bradstock 2013). These areas may well be burnt below threshold of potential concern (i.e. leaving inter-fire intervals too short to maintain some species) and this is also likely to worsen under climate change. In some respects, these are areas managed by frequent prescribed fires as sacrificial areas, but there has been no assessment of the area involved either as hectares or as a proportion of the total forest. While a modest increase in fire frequency will have a small effect on biodiversity overall, the impact on particular threatened species and ecological communities may be much more significant. Changed fire regime is a listed threatening process for 66% of threatened species in Australia (Kearney *et al.* unpublished), and many of these will be interval sensitive. Considering that many of these species have small populations that may be completely burnt in a single fire, close attention should be paid to applying prescribed burns to threatened species habitat.

Prescribed burning may have some detrimental impacts on biodiversity through demographic factors influenced by the inter-fire interval, but it also may have more indirect impacts such as through increasing the risk and impact of compounding threats. For example, many native plants and animals may survive any fire event, but those

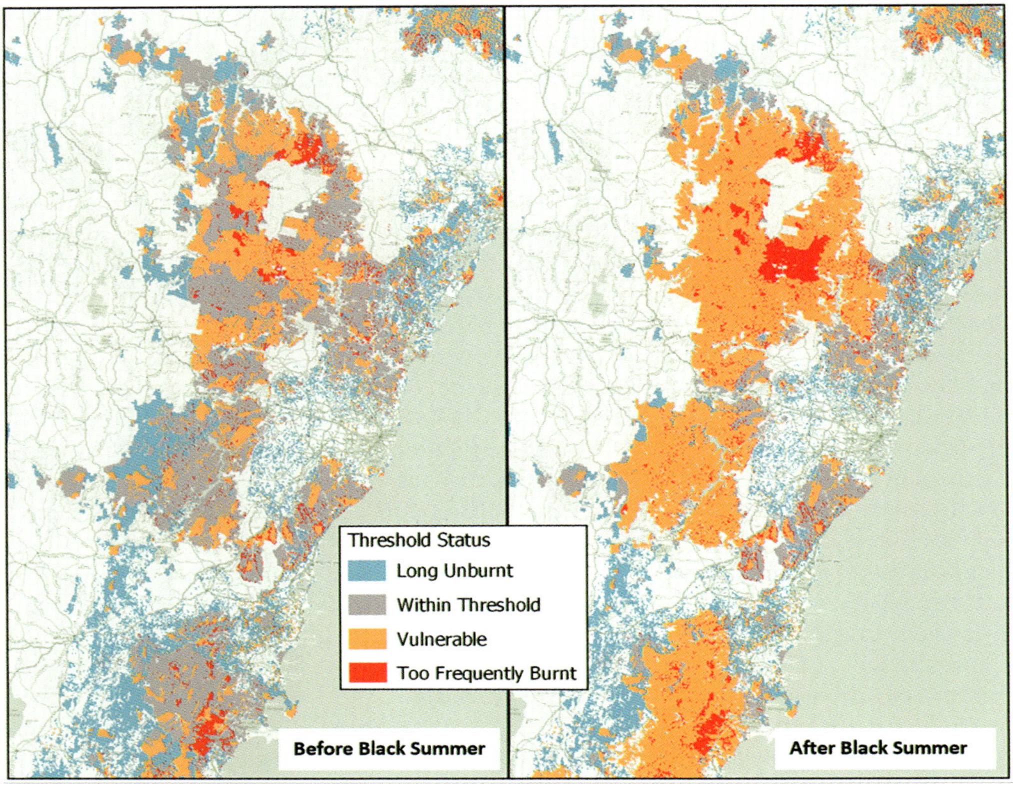

Fig. 21.4. Firetools analysis of thresholds of potential concern before and after the Black Summer fires (from Williamson (2020)) for vegetation formations in the Sydney region. The main effect has been to shift a large area into the vulnerable class (83% increase across New South Wales), with more modest 21% increase in area burnt too frequently and 30% decrease in the area long unburnt. Vulnerable means an area that would be burnt too frequently if another fire occurred within 5–7 years. (Image: Grant Williamson)

survivors may then be exposed to increased rates of consumption by introduced herbivores and predators (e.g. Hradsky 2020; Chapter 17), and these compounding impacts may escalate and become more extensive with increased application of prescribed burning.

Prescribed burns can also escape their proposed boundaries and cause damage. A notable example was the 2011 Ellenbrook burn near Margaret River, Western Australia, which destroyed 32 homes (Keelty 2012).

Knowledge gaps

Prescribed burning involves winners and losers, and finding the best settings to balance the trade-off between these, and with property risk reduction, is a very complex problem. Some issues are noted below:

- There is still a basic need to understand species-level responses to elements of the fire regime. There is a reasonable knowledge base for plant species in respect to interval

thresholds, but very little in respect to severity, seasonality and long-term regimes, and the knowledge base is notably flimsier for animals. Prescribed burning impacts can be interpreted in this context via a species' minimum interval requirement for low-severity fire. To illustrate the scale of the problem, the NSW Flora Fire Response Database (DPIE, unpublished data) has a defined fire response (seeder, sprouter, etc.) for 3086 plant species, but an estimated primary juvenile period (minimum threshold) for only 474 of these, and no information about responses to fire of different severity or season. There is no equivalent database for animals. These gaps can only be filled by field work targeting a range of species and fire regime conditions, most practically via natural experiments such as chrono-sequences (Driscoll *et al.* 2010). Studies collecting this information are ongoing, spawned in part by the opportunity presented by the 2019–20 wildfires.
- The response of trees is a high priority because they contribute significantly to carbon storage and ecosystem health. The main questions here are rates of mortality and recruitment under different combinations of fire frequency and severity and drought.
- The extent of internal patchiness of fires and its effect on survival of plants and animals, and the viability of their populations, needs much more attention, especially in the context of prescribed burns because they are patchier than wildfires.
- Fire interacts with many other processes that potentially can have synergistic impacts on biodiversity. Drought is probably the most significant because it is always associated with major wildfire seasons, but post-fire impacts of feral herbivores and predators, and of weeds, have all been identified as problems for certain species (see Chapters 17, 18). Much more needs to be understood about these effects: how pervasive they are, how they play out and how to ameliorate them (Driscoll *et al.* 2010).
- There is a need to compare different types or methods of prescribed burning. In particular, hazard reduction burns are likely to cause different effects to cultural burns because they are generally larger, more severe and are lit around the perimeter of the treated area. One knowledge gap here is the extent to which the conventional practice of hazard reduction burning can incorporate and adopt cultural burning principles and practice, and the extent to which such modification may maintain the benefits of hazard reduction burning but reduce its risks and detriments.
- Alternatives to frequent prescribed burning applications should be explored. This includes the potential that fire risk may be reduced when forests are left long unburnt, and that more focus should be placed on the human dimensions of wildfire risk, such as ignitions, property preparedness, defence and evacuation and wildfire suppression.

Recommendations

- There is much still to understand about the influence of fire regime factors on species survival, growth and reproduction, emissions and carbon. Prescribed burning and cultural burning are two of those factors that need more research.
- Climate change will squeeze species through drought and decreased fire intervals. Prescribed burning should not be applied in a way that makes this worse.
- All aspects of wildfire risk (to human, economic and environmental values) and all mitigation strategies (prescribe burning, suppression, town planning) need to be evaluated holistically to identify optimum, cost-effective strategies.

References

Abram NJ, Henley BJ, Sen Gupt A, Lippmann TJR, Clarke H, *et al.* (2021) Connections of climate change and variability to large and extreme forest fires in southeast Australia. *Communications Earth & Environment* **2**, 8. doi:10.1038/s43247-020-00065-8

Anon (2008) 'Inquiry into the impact of public land management on bushfires in Victoria'. Victorian Parliament, Environment and Natural Resources Committee, Melbourne.

Bennett LT, Bruce MJ, MacHunter J, Kohout M, Tanase MA, *et al.* (2016) Mortality and recruitment of fire-tolerant eucalypts as influenced by wildfire severity and recent prescribed fire. *Forest Ecology and Management* **380**, 107–117. doi:10.1016/j.foreco.2016.08.047

Bradshaw SD, Dixon KW, Lambers H, Cross AT, Bailey J, *et al.* (2018) Understanding the long-term impact of prescribed burning in mediterranean-climate biodiversity hotspots, with a focus on south-western Australia. *International Journal of Wildland Fire* **27**, 643–657. doi:10.1071/WF18067

Bradstock R, Davies I, Price OF, Cary G (2008) 'Effects of climate change on bushfire threats to biodiversity, ecosystem processes and people in the Sydney region: Final Report to the New South Wales Department of Environment and Climate Change'. Centre for Environmental Risk Management of Bushfires, Faculty of Science, University of Wollongong, Wollongong.

Bradstock RA, Hammill KA, Collins L, Price O (2010) Effects of weather, fuel and terrain on fire severity in topographically diverse landscapes of south-eastern Australia. *Landscape Ecology* **25**, 607–619. doi:10.1007/s10980-009-9443-8

Bradstock RA, Boer MM, Cary GJ, Price OF, Williams RJ, *et al.* (2012a) Modelling the potential for prescribed burning to mitigate emissions from fire-prone, Australian ecosystems. *International Journal of Wildland Fire* **21**, 629–639. doi:10.1071/WF11023

Bradstock RA, Cary GJ, Davies I, Lindenmayer DB, Price O, *et al.* (2012b) Wildfires, fuel treatment and risk mitigation in Australian eucalypt forests: insights from landscape-scale simulation. *Journal of Environmental Management* **105**, 66–75. doi:10.1016/j.jenvman.2012.03.050

Bradstock R, Penman T, Boer M, Price O, Clarke H (2014) Divergent responses of fire to recent warming and drying across south-eastern Australia. *Global Change Biology* **20**, 1412–1428. doi:10.1111/gcb.12449

Broome RA, Johnstone FH, Horsley J, Morgan GG (2016) A rapid assessment of the impact of hazard reduction burning around Sydney, May 2016. *The Medical Journal of Australia* **205**, 407–408. doi:10.5694/mja16.00895

Canadell JG, Meyer CP, Cook GD, Dowdy A, Briggs PR, *et al.* (2021) Multi-decadal increase of forest burned area in Australia is linked to climate change. *Nature Communications* **12**, 6921. doi:10.1038/s41467-021-27225-4

Cirulis B, Clarke H, Boer M, Penman T, Price O, *et al.* (2019) Quantification of inter-regional differences in risk mitigation from prescribed burning across multiple values. *International Journal of Wildland Fire* **29**, 414–426. doi:10.1071/WF18135

Clarke H, Lucas C, Smith P (2013) Changes in Australian fire weather between 1973 and 2010. *International Journal of Climatology* **33**, 931–944. doi:10.1002/joc.3480

Driscoll DA, Lindenmayer DB, Bennett AF, Bode M, Bradstock RA, *et al.* (2010) Fire management for biodiversity conservation: key research questions and our capacity to answer them. *Biological Conservation* **143**, 1928–1939. doi:10.1016/j.biocon.2010.05.026

Duff TJ, Cawson JG, Penman TD (2019) Determining burnability: predicting completion rates and coverage of prescribed burns for fuel management. *Forest Ecology and Management* **433**, 431–440. doi:10.1016/j.foreco.2018.11.009

Enright NJ, Fontaine JB, Westcott VC, Lade JC, Miller BP (2011) Fire interval effects on persistence of resprouter species in Mediterranean-type shrublands. *Plant Ecology* **212**, 2071–2083. doi:10.1007/s11258-011-9970-7

Enright NJ, Fontaine JB, Bowman D, Bradstock RA, Williams RJ (2015) Interval squeeze: altered fire regimes and demographic responses interact to threaten woody species persistence as climate changes. *Frontiers in Ecology and the Environment* **13**, 265–272. doi:10.1890/140231

Etchells H, O'Donnell AJ, McCaw WL, Grierson PF (2020) Fire severity impacts on tree mortality and post-fire recruitment in tall eucalypt forests of southwest Australia. *Forest Ecology and Management* **459**, 117850. doi:10.1016/j.foreco.2019.117850

Evans J, Russell-Smith J (2020) Delivering effective savanna fire management for defined biodiversity conservation outcomes: an Arnhem Land case study. *International Journal of Wildland Fire* **29**, 386–400. doi:10.1071/WF18126

Gibbons P, van Bommel L, Gill AM, Cary GJ, Driscoll DA, *et al.* (2012) Land management practices associated with house loss in wildfires. *PLoS One* **7**, e29212. doi:10.1371/journal.pone.0029212

Gill AM, Bradstock R (1995) Extinction of biota by fires. In *Conserving Biodiversity: Threats and solutions.* (Eds RA Bradstock, TD Auld, DA Keith, RT Kingsford, D Lunney and DP Sivertsen.) pp. 309–322. Surrey Beatty & Sons, Chipping Norton.

Gordon CE, Bendall ER, Stares MG, Collins L, Bradstock RA (2018) Aboveground carbon sequestration in dry temperate forests varies with climate not fire regime. *Global Change Biology* **24**, 4280–4292. doi:10.1111/gcb.14308

Hislop S, Stone C, Haywood A, Skidmore A (2020) The effectiveness of fuel reduction burning for wildfire mitigation in sclerophyll forests. *Forest Science* **83**, 255–264.

Hradsky BA (2020) Conserving Australia's threatened native mammals in predator-invaded, fire-prone landscapes. *Wildlife Research* **47**, 1–15. doi:10.1071/WR19027

Jolly CJ, Dickman CR, Doherty TS, van Eeden LM, Geary WL, *et al.* (2022) Animal mortality during fire. *Global Change Biology* **28**, 2053–2065. doi:10.1111/gcb.16044

Kearney SG, Watson JEM, Reside AE, Fisher DO, Maron M, *et al.* (unpublished) A novel threat-abatement framework confirms habitat retention and invasive species management are the key actions for avoiding further declines in Australia's threatened species.

Keeley JE, Pausas JG, Rundel PW, Bond WJ, Bradstock RA (2011) Fire as an evolutionary pressure shaping plant traits. *Trends in Plant Science* **16**, 406–411. doi:10.1016/j.tplants.2011.04.002

Keelty MJ (2012) 'Appreciating the Risk: Report of the Special Inquiry into the November 2011 Margaret River Bushfire'. Government of Western Australia, Perth.

Kenny BJ, Sutherland E, Tasker E, Bradstock R (2004) *Guidelines for Ecologically Sustainable Fire Management NSW Biodiversity Strategy.* NSW National Parks and Wildlife Service, Hurstville.

Knox KJE, Clarke PJ (2012) Fire severity, feedback effects and resilience to alternative community states in forest assemblages. *Forest Ecology and Management* **265**, 47–54. doi:10.1016/j.foreco.2011.10.025

Kuchinke D, Di Stefano J, Sitters H, Loyn R, Gell P, *et al.* (2020) Prescribed burn severity has minimal effect on common bird species in a fire-prone forest ecosystem. *Forest Ecology and Management* **475**, 118437. doi:10.1016/j.foreco.2020.118437

Liyanage GS, Ooi MKJ (2015) Intra-population level variation in thresholds for physical dormancy-breaking temperature. *Annals of Botany* **116**, 123–131. doi:10.1093/aob/mcv069

McHugh D, Goldingay RL, Parkyn J, Goodwin A, Letnic M (2020) Short-term response of threatened small macropods and their predators to prescribed burns in subtropical Australia. *Ecological Management & Restoration* **21**, 97–107. doi:10.1111/emr.12407

Miller RG, Tangney R, Enright NJ, Fontaine JB, Merritt DJ, *et al.* (2019) Mechanisms of fire seasonality effects on plant populations. *Trends in Ecology & Evolution* **34**, 1104–1117. doi:10.1016/j.tree.2019.07.009

Monamy V, Fox BJ (2000) Small mammal succession is determined by vegetation density rather than time elapsed since disturbance. *Austral Ecology* **25**, 580–587. doi:10.1111/j.1442-9993.2000.tb00063.x

Murphy BP, Bradstock RA, Boer MM, Carter J, Cary GJ, *et al.* (2013) Fire regimes of Australia: a pyrogeographic model system. *Journal of Biogeography* **40**, 1048–1058. doi:10.1111/jbi.12065

Penman TD, Kavanagh RP, Binns DL, Melick DR (2007) Patchiness of prescribed burns in dry sclerophyll eucalypt forests in South-eastern Australia. *Forest Ecology and Management* **252**, 24–32. doi:10.1016/j.foreco.2007.06.004

Price OF, Bradstock R (2010) The effect of fuel age on the spread of fire in sclerophyll forest in the Sydney region of Australia. *International Journal of Wildland Fire* **19**, 35–45. doi:10.1071/WF08167

Price OF, Bradstock R (2012) The efficacy of fuel treatment in mitigating property loss during wildfires: insights from analysis of the severity of the catastrophic fires in 2009 in Victoria, Australia. *Journal of Environmental Management* **113**, 146–157. doi:10.1016/j.jenvman.2012.08.041

Price OF, Bradstock RA (2013) The spatial domain of wildfire risk and response in the Wildland Urban Interface in Sydney, Australia. *Natural Hazards and Earth System Sciences* **13**, 3385–3393. doi:10.5194/nhess-13-3385-2013

Price OF, Forehead H (2021) Smoke patterns around prescribed fires in Australian eucalypt forests, as measured by low-cost particulate monitors. *Atmosphere* **12**, 1389. doi:10.3390/atmos12111389

Price O, Haynes K (2020) 'People and property: probability of house destruction. Report to the NSW Bushfire Inquiry'. NSW Bushfire Risk Research Hub, University of Wollongong, Wollongong.

Price O, Barker J, Rahmani S, Wilkinson C, MacDonald D (2021) *Analysis and Characterisation of Bushfire-Meets-Prescribed Burn Events from the 2019/20 Fire Season.* University of Wollongong, Wollongong.

Price O, Nolan RH, Samson SA (2022) Fuel consumption rates in eucalypt forest during hazard reduction burns, cultural burns and wildfires. *Forest Ecology and Management* **505**, 119894. doi:10.1016/j.foreco.2021.119894

Rahmani S, Price O (2021) Effects of 38 years of wildfires on tree density in the Blue Mountains, Australia. *Austral Ecology* **46**, 20–30. doi:10.1111/aec.12952

Russell-Smith J, Edwards AC, Sangha KK, Yates CP, Gardener MR (2020) Challenges for prescribed fire management in Australia's fire-prone rangelands – the example of the Northern Territory. *International Journal of Wildland Fire* **29**, 339–353. doi:10.1071/WF18127

Steffenson V (2020) *Fire Country: How Indigenous Fire Management could help save Australia.* Hardie Grant Travel, Melbourne.

Stretton LB (1939) 'Report of the Royal Commission to inquire into the causes of and measures taken to prevent the bush fires of January, 1939, and to protect life and property'. Melbourne.

Swan M, Christie F, Sitters H, York A, Di Stefano J (2015) Predicting faunal fire responses in heterogeneous landscapes: the role of habitat structure. *Ecological Applications* **25**, 2293–2305. doi:10.1890/14-1533.1

Teague B, Mcleod R, Pascoe S (2010) '2009 Victorian bushfires Royal Commission final report'. Parliament of Victoria, Melbourne.

Volkova L, Weston CJ (2015) Carbon loss from planned fires in southeastern Australian dry Eucalyptus forests. *Forest Ecology and Management* **336**, 91–98. doi:10.1016/j.foreco.2014.10.018

Volkova L, Roxburgh SH, Weston CJ (2021) Effects of prescribed fire frequency on wildfire emissions and carbon sequestration in a fire adapted ecosystem using a comprehensive carbon model. *Journal of Environmental Management* **290**, 112673. doi:10.1016/j.jenvman.2021.112673

Watson GM, French K, Collins L (2020) Timber harvest and frequent prescribed burning interact to affect the demography of Eucalypt species. *Forest Ecology and Management* **475**, 118463. doi:10.1016/j.foreco.2020.118463

Williamson G (2020) 'Report to the NSW 2019 Bushfire Inquiry: Theme 2.2 – Biodiversity and environmental impacts: threshold analysis'. NSW Bushfire Risk Research Hub, Wollongong.

22

Response by the Australian Government to the 2019–20 wildfires

Glenn Block, Fiona Fraser, Anna-Lisa Hayes, Helene Marsh, Libby Rumpff and John C. Z. Woinarski

Summary

Context and challenges

- The *Environment Protection and Biodiversity Conservation Act 1999* (EPBC Act) focuses Australian Government interests on the protection of matters of national environmental significance.
- Many of these matters were affected by the 2019–20 wildfires.
- Given the impacts of these fires, responses to prevent extinction and support recovery needed to be made urgently and targeted towards those species and ecological communities that were in most need of such management. Determining this response required rapid assessment of impacts.

Main findings

- The Australian Government committed an initial $2 billion National Bushfire Recovery Fund to support a wide range of responses to the 2019–20 wildfires; this included a $200 million investment for the recovery of native wildlife and their habitats. This funding provided the foundational support for many post-fire emergency responses to prevent extinction and support recovery of fire-affected species and ecological communities.
- The Australian Government responded rapidly to the biodiversity losses caused by the 2019–20 wildfires.
- This response was characterised by a high degree of consultation and collaboration with state/territory conservation agencies, conservation non-government organisations (NGOs), researchers and many other stakeholders.
- The response was substantially informed by an independent advisory panel established by the Minister for the Environment.
- Largely as informed by this panel, the Australian Government Department of Agriculture, Water and the Environment ('the Department') rapidly developed

priority listings of the ecological communities, plants and animals that were most affected by the 2019–20 wildfires, with such listings then used to help allocate investments in recovery efforts.

- The Australian Government undertook a series of regional consultative workshops in the most fire-affected areas to develop, fund and implement priority actions at regional scale for the recovery of fire-affected biodiversity.
- The Australian Government supported the independent Threatened Species Scientific Committee to prioritise and undertake an expedited assessment of the conservation status of a large set of fire-affected species and ecological communities most in need of conservation and management.
- The Australian Government established a Royal Commission into National Natural Disaster Arrangements, and the Department's evidence encompassed the impacts of the fires on biodiversity, measures to reduce impacts of future fires and assist recovery, the respective roles of the Commonwealth and the states in environmental management, and options for improvement around data, mapping, and species distribution modelling. These contributions to the Commission were one way in which the biodiversity impacts, issues and response were integrated with broader policy considerations following the fires. The Australian Government supported in principle the Commission's recommendation relating to biodiversity.
- The Department worked closely with the National Bushfire Recovery Agency (now National Recovery and Resilience Agency) to maintain a focus on biodiversity impacts and response among other Australian Government actions.
- Recognising that shortcomings in national coordination and availability of critical data constrained assessment of the impacts of these fires on biodiversity, the Australian Government supported a suite of projects and initiatives to improve data availability and fill some key knowledge gaps. Such evidence will increase preparedness for any future comparable event.

Introduction

The Australian Government had a leadership and coordination role in many aspects of the response to the 2019–20 wildfires. The role was backed by the Australian Government making more than $2 billion available for recovery. With respect to the natural environment, the Australian Government's role is focused on protection of matters of national environmental significance, as defined in the EPBC Act. These matters include species and ecological communities nationally listed as threatened, species listed as migratory under a range of international agreements, World Heritage properties and wetlands of international importance (listed under the Ramsar Convention on Wetlands).

From late 2019, it became evident that wildfires across large areas of eastern Australia were having an unprecedentedly severe impact on many of these matters, with reports in early January 2020 of more than 1 billion animals affected (see Chapter 12) and impacts on many threatened species and ecological communities (Woinarski *et al.* 2020). Given the exceptionality of the threat, there was no pre-existing response package. However, the Australian Government responded rapidly, consultatively and strategically: it provided national leadership, fostered collaborations, and invested substantially in research and recovery actions, guided by an evidence-based prioritisation process.

A rapid national overview of the extent and severity of impact of the wildfires on biodiversity was conducted internally by the Department of Agriculture, Water and the Environment ('the Department'). This analysis, published online on 20 January 2020 (while fires were still raging), confirmed that the impacts on biodiversity were exceptional, reporting that 327 nationally listed threatened species (nearly 20% of all Australian threatened species) had at least 10% of their modelled likely or known distributions within the fire extent, of which 49 species had more than 80% of their modelled likely or known distributions within the fire extent. The Department's rapid assessment also demonstrated significant impacts on some World Heritage properties.

The Department's initial analyses also identified that the extent, severity and impacts of the 2019–20 wildfires were exceptional across much of eastern and southern Australia and delineated these areas (the 'Preliminary Analysis Area') as a focus for concern (Chapter 1), rather than diluting management response across the nation as a whole. Most of the analyses of biodiversity impacts of the 2019–20 fires have subsequently been restricted to this area (Legge *et al.* 2022), including throughout this book.

In January and February 2020, as fires were still burning, the Minister for the Environment, the Hon Sussan Ley MP, hosted a series of emergency consultative roundtables, focused on immediate recovery actions, with more than 200 representatives from state and territory agencies, conservation and wildlife care organisations, business, industry, Traditional Owners, scientists and natural resource managers. These meetings ensured a high level of collaboration and transparency among the many respondent groups, and helped galvanise and target the rapid deployment of remedial actions, with such response supported by the urgent provision of funding (see below).

In January 2020, the Minister also established an expert advisory group, the Wildlife and Threatened Species Bushfire Recovery Expert Panel, to provide strategic advice, particularly about identifying the natural values that had been most fire-affected and prioritising management responses to support their recovery. The functions, operation and outcomes of this panel are described in more detail below.

The Australian Government established a royal commission into the 2019–20 wildfires. The commission had a broad scope, explicitly including the clarification, coordination and accountability of Commonwealth and state/territory responsibilities for fire management, the need to try to improve resilience within the context of changing climate, and legislative and policy reform; it also included consideration of biodiversity impacts. This commission is considered in more detail in Chapter 30, but we note here that many matters described in this chapter were presented by the Department as evidence to that commission. The Australian Government supported in principle the commission's recommendation relating to biodiversity.

In this chapter, we describe some main components of the Australian Government's response to the 2019–20 wildfires, focusing particularly on its national leadership and coordination role in supporting post-fire recovery actions for the most fire-affected components of biodiversity.

Funding support

The Australian Government established a $2 billion National Bushfire Recovery Fund to support a wide range of responses to the 2019–20 wildfires. Within this program, there was an explicit component to support biodiversity recovery efforts ('Wildlife and Habitat Bushfire Recovery' investment). An initial $50 million investment ('Phase 1 emergency response', for short-term projects) was established in January 2020. This comprised rapidly delivered allocations of $13 million to state and territory governments for on-ground emergency interventions and priority recovery actions, $11.9 million for a competitive grants program, $10.3 million for wildlife rescue and care services and establishment of insurance populations of fire-affected species, $7 million for natural resource management organisations for emergency actions including control of pests and weeds that may compound fire impacts and constrain recovery, $5 million to Greening Australia for revegetation efforts, and $2.5 million to Conservation Volunteers Australia to support volunteer involvement in recovery efforts.

In addition to this support for urgent on-ground management responses, a further $2 million was allocated by the Australian Government to fire-related research, through the Threatened Species Recovery Hub of the National Environmental Science Program. This funding was provided for more detailed analyses of the impacts of the 2019–20 wildfires on biodiversity, to help support survey and monitoring efforts, and to assess management effectiveness.

A subsequent Phase 2 ('Recovery and Resilience') funding stream of $150 million was announced in May 2020, for projects operating for up to 2 years from July 2020. This comprised $110 million for on-ground actions to recover biodiversity in the seven most fire-affected regions (see more detailed account below); $28 million to support the assessment and listing process for fire-affected species, and planning, coordination and monitoring of species' recovery (including $2 million for a national monitoring program for the koala *Phascolarctos cinereus*), and administration of the program (see more detailed account below); $10 million in grants for community-led conservation actions; and $2 million to support knowledge exchange on Indigenous fire management. Collectively, this support provided for recovery actions that were appropriate, and could be implemented, at national, state/territory, regional and local scales. The $200 million allocated over ~2 years by the Australian Government to recover biodiversity affected by the 2019–20 wildfires is exceptional in the context of typical resourcing for threatened species management in Australia.

A key obligation for all funded projects was for regular reporting on outcomes. Such reporting will help monitor recovery and build the evidence base about the relative effectiveness of management actions, and hence will allow for an increased preparedness for, and effectiveness in, responding to any future comparable event.

However, as the Australian Government funding focused on addressing the immediate and shorter-term impacts of the fires, all projects were initially required to be delivered and funding expended by June 2022. Due to COVID-19 restrictions affecting on-ground activities throughout 2020–21, some projects have been extended through to mid-2023 to ensure the contracted wildlife recovery outcomes can be met. For many species and ecological communities, post-fire recovery will not occur in such a short period (Legge *et al.* 2021), and much longer commitments to remedial management will be required.

Much of the assessment of impacts and management response described in this book owes at least in part to this funding program provided by the Australian Government.

The Wildlife and Threatened Species Bushfire Recovery Expert Panel

This specially established advisory panel was chaired by the Australian Government's Threatened Species Commissioner, Dr Sally Box, and comprised eight additional independent experts (including four of this book's editors) in fire ecology, threatened species management, structured decision analysis, *ex situ* conservation and Indigenous knowledge. Representatives of state and territory conservation agencies attended panel meetings as observers, to provide jurisdictional perspectives and to foster coordination in priority management responses. The panel's primary purpose was to advise on prioritising recovery actions for fire-impacted native species, ecological communities, natural assets and their Indigenous cultural value.

At its first meeting in January 2020, the panel set four broad objectives for the recovery effort:

1. Prevent extinction and limit decline of native species.
2. Reduce the immediate suffering of native animals directly impacted by the fires.
3. Maximise the chances for long-term recovery of native species and communities.
4. Ensure learning and continual improvement are at the core of the response.

The panel also identified a series of priority actions that needed to be implemented urgently to meet these objectives:

- protection of unburnt areas within or adjacent to recently burnt ground that provide refugia;
- control of feral predators and herbivores to reduce the pressure on native species where appropriate;
- emergency salvage of plant and animal species for *ex situ* conservation or wild-to-wild translocation;
- rapid on-ground assessment for species and communities of concern; and
- supplementary shelter, food and water for animals where appropriate.

These objectives and priority activities provided the subsequent framework for the Australian Government's response and the distribution of its funding support. The panel met 20 times over the period of its 9-month tenure (January to September 2020).

Overseen by the panel, a series of analyses prepared by the Department and other experts provided rapid assessments of the fire impact on biodiversity. These assessments were used to help direct urgent management response to the species and ecological communities most needing such support, and to help identify the type of management action that could be most effective. To help coordinate recovery efforts and to inform the public more generally, these prioritisation analyses were published on the Department's website, for vertebrates and a small set of invertebrates (initially in February 2020, updated in March 2020), ecological communities (in February 2020), invertebrates (in April 2020) and plants (initially in April 2020, updated in October 2020), with subsequent formal publication (Legge *et al.* 2022; Gallagher *et al.* in press). These assessments also served to indicate which species and ecological communities may have become newly eligible, because of the fires, for national listing as threatened or – in the case of already listed species or ecological communities – for up-listing to a more imperilled status.

Informed by such prioritisations, the expert panel then provided advice to the Australian Government on recovery actions to be delivered by state and territory governments as well as applicants for funding under successive tranches of the $11.9 million for a competitive grants program (the 'Wildlife and Habitat Bushfire Recovery' investment fund), to ensure funding was directed as comprehensively as possible across the set of priority fire-affected species and ecological communities, targeted to those whose recovery most needed support, and applied the management actions most likely to be effective.

Regional recovery planning and implementation

The Australian Government invested $110 million ('the Regional Fund') to build on existing recovery work and support efforts beyond the immediate post-fire recovery period, focused on seven regions recognised to have the greatest biodiversity impacts as a result of the wildfires (Fig. 22.1). These funds were allocated to priorities determined through regional co-design workshops, to strategic recovery initiatives and to recovery efforts that required coordination across regions.

Co-design workshop investments

Co-design workshops were held in August and September 2020 with the stakeholders involved in post-fire management (state governments, regional natural resource management (NRM) groups, Traditional Owners, non-government organisations (NGOs), Landcare groups, local council) in each region to (1) share and consolidate knowledge of existing efforts; (2) identify ecological priorities that required additional funding to secure their recovery (i.e. gaps in investment); and (3) outline an indicative action plan to guide the allocation of these funds. The aim was to develop plans that would deliver the greatest benefit to the priority species, ecological communities and other natural assets in each region by 2030.

For each workshop, the Department developed a regional profile including information on the distributions of priority fire-affected species and ecological communities, and the actions most likely to support their recovery. This detailed compilation was made

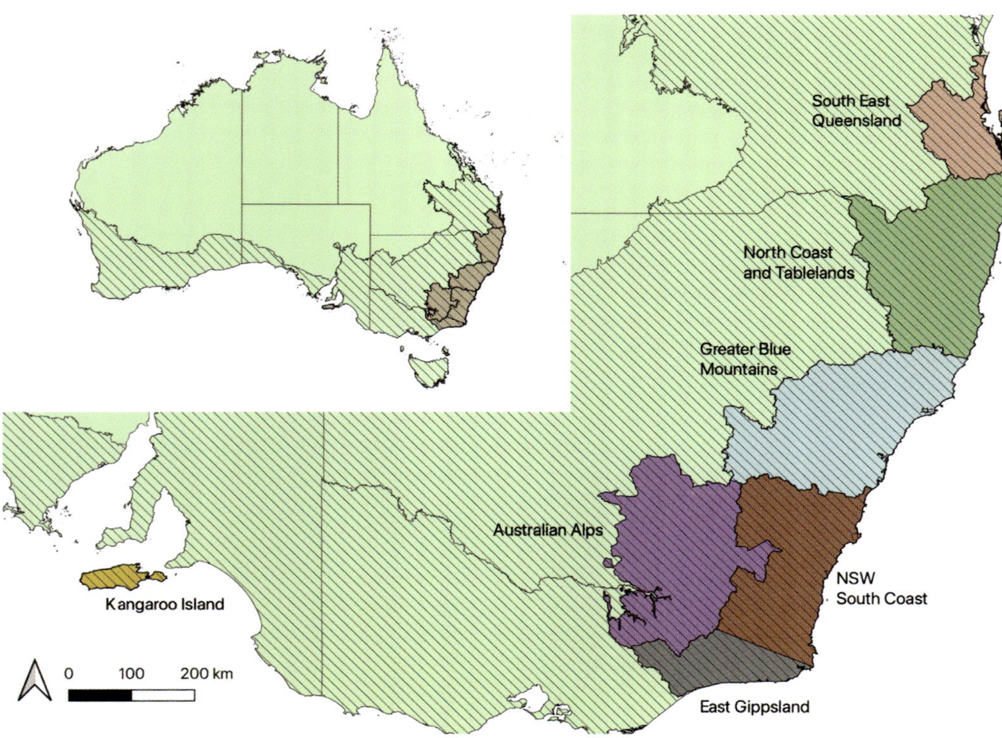

Fig. 22.1. The seven priority areas identified by the Department of Agriculture, Water and the Environment for regional scale biodiversity recovery planning and management.

possible through the development within the Department of an advanced spatial analysis tool that helped illustrate and compare benefits for priority species arising from any potential management option, and to ensure that benefits were spread widely across priority fire-affected species and ecological communities (Fig. 22.2). State agencies and NRM regions contributed to the regional profiles by providing state and regional-scale information on fire impacts and priorities for recovery.

Workshop participants shared perspectives on the impacts of the 2019–20 fires, the priorities and threats in each region, the existing management and planning underway (targeting priorities), and any perceived gaps in investment. Participants were then split into groups to develop a strategy, with detail on the actions, locations, priority matters targeted by each action, and the approximate budget allocated to each action. Individual strategies were then explored and discussed by the larger group, which was tasked with developing a consensus action plan for the region.

Table 22.1 highlights the level of investment and types of actions supported in each region. The funding was often used to consolidate or expand existing programs (e.g. pest control programs), highlighting the perceived importance of managing ongoing threats. In order to assess and hone the effectiveness of management actions, monitoring was explicitly built into the investments, and the results of this monitoring may resolve some of the uncertainty associated with the type, duration and magnitude of intervention required to achieve long-term benefits.

The additional investment was seen as an opportunity to support and build capacity for Traditional Owner-led planning and recovery of Country. Notably, moving beyond the emergency recovery period, participants also wanted to improve fire planning and management (including cultural and ecological burning) to increase the resilience of priority assets, and mitigate the effects of climate change. Likewise, focus was given to the identification, management and protection of refugial areas. In some regions (East Gippsland and

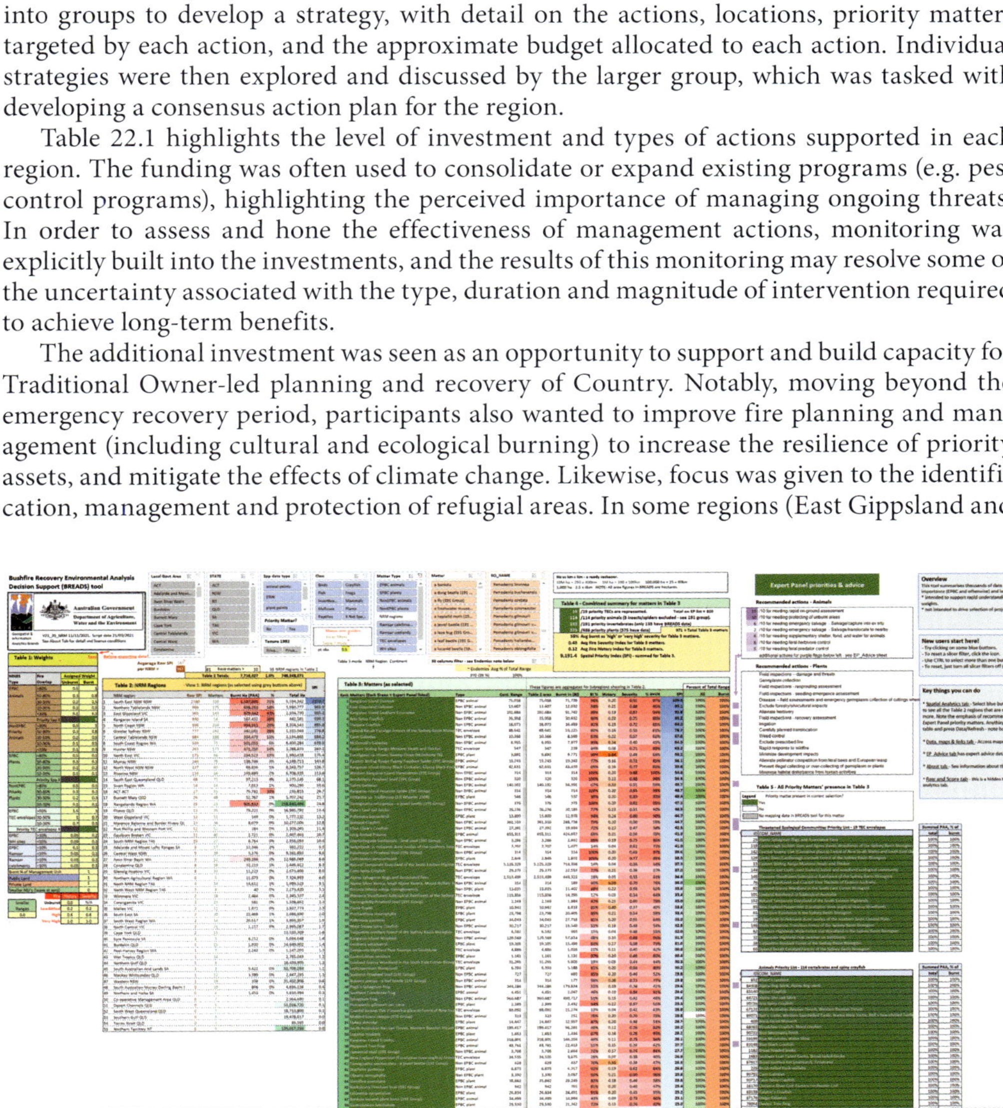

Fig. 22.2. Illustration of planning tool used in regional workshops to provide rapid assessment of the relative benefits of management options for priority fire-affected species and ecological communities.

Table 22.1. Summary of regional fire metrics and priority management themes highlighted (dark shading) for recovery funding identified through co-design workshops.

Note that much more detailed actions are given in individual regional reports, available at https://www.dcceew.gov.au/environment/biodiversity/bushfire-recovery/funding-support/regional-delivery-program.

Region	Total regional area (km^2)	% total area burnt (% of this in high to very high severity fire)	Regional Fund co-design workshop budget allocated ($ million)	Management theme										
				Traditional Owner-led planning & management	Feral herbivore control	Predator control	Fire planning and management	Weed control and vegetation restoration	Erosion control	Species-specific intervention	Refugia management	Data collation/ validation	Governance	Capacity building
Australian Alpine environment	131 143	4 (60)	8	■	■	■	■	■	■	■				
Forests and coastal ecosystems of East Gippsland	21 047	42 (59)	14	■	■	■		■		■	■			
Greater Blue Mountains and World Heritage Area	76 563	15 (50)	13	■	■	■	■	■	■	■	■	■		
Kangaroo Island	4426	38 (83)	5			■	■			■				
Rainforests of New South Wales north coast and tablelands	Region: 71 553; Gondwana Rainforests: 2688	Total region: 23 (45); Gondwana Rainforests: 46	13	■	■	■	■	■		■		■		■
Forests of the New South Wales south coast	55 543	21 (61)	15	■	■	■		■		■	■			■
Rainforests of south-east Queensland	23 731	4 (46)	4		■	■	■	■					■	

Greater Blue Mountains) this was highlighted as a particular focus of investment, while in other areas participants specified these areas as locations for implementation of strategic actions such as weed control.

Regional workshop participants were faced with difficult decisions, particularly when costs prohibited action at the scale necessary to have meaningful outcomes. This is evident, for example, in the south-east Queensland and Kangaroo Island regions, which had to target their investment in a smaller set of actions. Another example was the limited investment in erosion control, despite some regions (e.g. the Alpine region) facing serious ongoing issues with health of waterways. Participants discussed gaps in investment in targeted actions for some priority species. However, strategic actions such as herbivore control were often favoured over targeted actions, because it was likely that more enduring benefit would be realised across the list of priority assets.

These regional workshops were highly effective collaborative mechanisms to craft and deliver a carefully considered, coordinated and integrated package of much needed immediate management responses, and strategic investments for the future. They served to mesh national priorities for fire-affected biodiversity with local and regional perspectives and actions, and provided the evidence-based foundation to deliver substantial targeted investment in a remarkably short timeframe. Without doubt, they provided a critical contribution to the recovery of a high proportion of the most fire-affected species, ecological communities and other environmental assets. A more precise evaluation of that benefit will be possible in the near future, as individual actions, the overall portfolio of investments, and the recovery of species and ecological communities are monitored and reported.

Multiregional species interventions and strategic investments

A further component, the Regional Bushfire Recovery for Multiregional Species and Strategic Projects program, is investing $10 million to fund priority actions for species' recovery at a national scale, and to avoid duplication and promote complementarity with other fire recovery investments.

Species and species groups identified as priorities for this funding were determined in consultation with the Expert Panel, experts from the Department, and state and territory agencies. Principles used to assess whether fire-affected species should be a high priority for multiregional action included whether:

- the species occurs, or migrates across, multiple regions and/or jurisdictions
- actions are needed across jurisdictions to address the species' recovery
- there is a need for coordinated collaborative action
- recovery actions over 18 months will make a tangible difference to the species' survival or recovery.

The 10 priority species and species groups are the spot-tailed quoll (south-east mainland population) (*Dasyurus maculatus maculatus*), grey-headed flying-fox (*Pteropus poliocephalus*), long-nosed potoroo (south-east mainland subspecies) (*Potorous tridactylus tridactylus*), brush-tailed rock-wallaby (*Petrogale penicillata*), platypus (*Ornithorhynchus anatinus*), eastern bristlebird (*Dasyornis brachypterus*), gang gang cockatoo (*Callocephalon fimbriatum*), south-eastern glossy black-cockatoo (*Calyptorhynchus lathami lathami*), two gliders (greater glider (*Petauroides volans*) and yellow-bellied glider (*Petaurus australis*)), and alpine reptiles (including the alpine she-oak skink (*Cyclodomorphus praealtus*), alpine bog skink (*Pseudemoia cryodroma*), and Guthega skink (*Liopholis guthega*)). Under the program, national species coordinators established working groups to foster collaboration among jurisdictions, experts, NGOs and community groups, as well as maximising the value of wildfire recovery investments by sharing knowledge and reducing duplication.

A grant process was opened to project applications generated through the multi-jurisdictional working groups for each species or group of species. This ensured a strong suite of targeted and high-quality project grant applications was received that would be well placed to deliver the coordinated national-level recovery being sought under the program. Projects are focused on data, monitoring and citizen science; translocation planning or implementation; on-ground efforts; engagement and communication; and research. Funding recipients include NRM organisations, species experts, community groups, universities, state and territory governments, and zoos.

In addition to the 10 priority species described above, $12 million of the Regional Fund was directed to one iconic and wide-ranging fire impacted species, the koala, to support habitat restoration and threat mitigation in bushfire-affected regions of New South Wales and south-eastern Queensland and for coordination and delivery of more effective health research. These investments were complemented by a reassessment of the species listing status by the Threatened Species Scientific Committee and the finalisation of the national recovery plan for the koala.

Additional strategic investments were made to address gaps, further engage the community and support cross-jurisdictional coordinated activities (e.g. seed banking of priority plant species, citizen science initiatives to boost knowledge on fire impacts and recovery, and to address critical knowledge gaps for invertebrates), invasive animals coordination, and Landcare-led wildfire recovery activities.

Indigenous Fire and Land Management workshops program

The Australian Government allocated $2 million over 2020–21 and 2021–22 to support partnerships with Traditional Owners, Indigenous organisations and Indigenous enterprises to hold cultural burning and traditional land management workshops. The program design was informed by consultation with Traditional Owners, thematic experts, state cultural fire strategies and research including CSIRO's 'Climate disaster and resilience technical report' (CSIRO 2020), which identified the importance of enabling adoption of Indigenous cultural burning into contemporary land and bushfire management. The report also noted the need for targeted investment to support Indigenous-led recovery and resilience measures.

The Indigenous Fire and Land Management workshops program is also consistent with recommendation 18.2 of the 2020 Royal Commission, which stated: 'Australian, state, territory and local governments should explore further opportunities to leverage Indigenous land and fire management insights, in the development, planning and execution of public land management activities' (Royal Commission into National Natural Disaster Arrangements 2020).

The objectives of the program are to:

- support Indigenous-led knowledge sharing of Indigenous fire and land management practices;
- strengthen the knowledge and understanding of traditional Indigenous fire and land management practices; and
- support Indigenous communities to benefit from traditional fire and land management knowledge.

The intended outcomes of the program are:

- Indigenous-led fire and land management workshops that support a greater understanding of Indigenous fire and land management practices within Indigenous communities;
- supporting the inclusion of Indigenous fire and land management practices in wider conventional fire and land management arrangements; and
- empowering Indigenous communities to progress Indigenous fire and land management activities.

The program funded 13 grants. The grant guidelines also encouraged separate workshops for women and young people where appropriate. The program does not include landscape burning activities – however, demonstrating and teaching burning activities can occur as part of workshops. Originally called the Cultural Burning Workshops program, the program name was changed to ensure workshops incorporated the breadth of Traditional Owner fire and land management knowledge and experience following consultation and advice that burning on Country occurs for many reasons.

Threatened species assessment and listing

The Threatened Species Scientific Committee (TSSC) is the statutory committee appointed by the Australian Minister of Environment (henceforth 'Minister') to provide independent scientific advice to achieve biodiversity conservation outcomes, specifically for threatened species and ecological communities under the EPBC Act.

As soon as possible after the 2019–20 wildfires, the TSSC developed a bushfire response plan (the Threatened Species Scientific Committee 10-point Bushfire Response Plan), which was approved by the Minister on 2 March 2020. The plan has four key objectives, which align with those of the Wildlife and Threatened Species Bushfire Recovery Expert Panel, and 10 actions (https://www.dcceew.gov.au/environment/biodiversity/threatened/publications/threatened-species-scientific-committee-bushfire-response-plan).

The major task for the TSSC has been to make recommendations to the Minister about changes in the listing status of fire-affected species and ecological communities, so that these entities could be protected as matters of national environmental significance (for entities unlisted at the time of the fires) or transferred to a higher threat status (entities listed at the time of the fires).

With the agreement of the Minister, the TSSC accepted public nominations of fire-affected species and ecological communities at any time, rather than once per year. In addition, fire-affected entities were prioritised for assessment based on advice from the Bushfire Recovery Expert Panel and the states and territories.

This number of entities that required assessment or reassessment represented a huge increase in workload for the TSSC and the Department and was enabled by:

- the Minister allocating funding to assess the listing status of fire-affected species and ecological communities and update planning documents for fire-affected biota;
- the Department providing a temporary increase in resourcing in the Protected Species and Ecological Communities Branch, including the establishment of a new directorate and reforming quality assurance processes;
- the Minister increasing the size of the TSSC from 10 to 12 members;

- public nominations for Proposed Priority Assessment List remaining open to allow the public and experts to submit nominations as information came to light on the impact of the fires on species;
- the TSSC and Minister establishing interim proposed prioritisation processes to identify further species for assessment as the impacts of the fires were better understood. In addition to annual statutory prioritisation processes, five additional interim prioritisation processes have been established;
- the TSSC increasing its number of meetings from four to 11 per year and making operational changes such as increased use of out-of-session working groups and parallel sessions within meetings;
- the Department contracting experts from the Bushfire Recovery Expert Panel, state and territory governments, the National Environmental Science Program Threatened Species Recovery Hub, the International Union for the Conservation of Nature and other scientists to assist in the preparation of listing assessments and conservation advices;
- the TSSC collaborating with the Department to provide advice on the development and implementation of the Species Expert Assessment Plan through which scientific experts are delivering listing assessments and conservation advices for large groups of species and ecological communities – including freshwater fish, turtles, three groups of plants, lizards and snakes, terrestrial mammals, frogs, and rainforests and wet forests of south-eastern Australia – to support wildfire recovery, environmental resilience and preparedness for future fire events; and
- the TSSC and the Department working in partnership to undertake preliminary evaluations for high-profile species such as the koala and greater glider. The committee is continuing to undertake preliminary evaluations of other fire-affected animals, plants and ecological communities to maximise the efficiency of the assessment process.

As of May 2022, 169 species and nine ecological communities affected by the wildfires have been included in the 2020 and 2021 Finalised Priority Assessment Lists, the formal workplans of the TSSC for status reassessment and conservation planning, by the Minister. Of that set, assessments have been completed and listing decisions made for 51 species and four ecological communities. In addition, conservation planning documents are being revised for entities listed before the 2019–20 wildfires to include the impacts of, and management response to, that extreme event. The first set of conservation advices has been prioritised based on the level of fire impact.

The TSSC also led the rewriting of the listing assessment for the proposed Key Threatening Process 'Fire regimes that cause biodiversity decline', which came into effect in April 2022.

The scale of the 2019–20 wildfires was unprecedented and has resulted in numerous challenges for assessing impacts on species and ecological communities. Ideally, the partnerships established between the TSSC, the Department and the expert scientific community in the aftermath of the wildfires will endure and lead to more efficient and effective processes in future.

Data and information

A key finding of government inquiries reporting on the 2019–20 wildfires (e.g. Royal Commission into National Natural Disaster Arrangements 2020) was that coordination among jurisdictions in many aspects of information was suboptimal, that some critical information was not readily accessible, and that there were many critical gaps in biodiversity knowledge (see Chapter 30). One example, evident as the fires were burning, was that fire mapping was undertaken largely independently by the states and territories, with such devolution hampering national assessments of fire extent and impact. In response, from February 2020, the Department worked with states and territories to collate a national spatial coverage for fire, the National Indicative Aggregated Fire Extent Dataset (NIAFED), which provided fire extent mapping across jurisdictions. In June 2020, the Department then published the Australian Google Earth Engine Burnt Area Map (AUS GEEBAM) developed by remote sensing specialists from the NSW Government. These national datasets allowed for consistent national assessment of impacts of these fires on species and ecological communities; such approaches will be critical for rapid assessment in future comparable events.

Further recognising the constraints to the assessment of impacts of, and implementation of responses to, the 2019–20 wildfires imposed by knowledge gaps and limited accessibility of existing biodiversity data, the Australian Government has subsequently supported a range of projects relating to biodiversity data collation and knowledge acquisition. These projects have included refined modelling of the distributions of fire-affected species (Legge *et al.* 2021); rapid taxonomic appraisal of some fire-affected species (Catullo *et al.* 2021); efforts for more strategic development and coordination of biodiversity monitoring programs especially for fire-affected species (Chapter 31; Southwell *et al.* 2022); development and maintenance of databases of life history and ecological traits that are informative of species' susceptibility to fire (Gallagher *et al.* in press); digitisation of locational information for invertebrate specimens in the Australian National Insect Collection to help make such data more accessible (https://digivol.ala.org.au/project/index/192261079); and the nationally integrated collation, storage and access ('innovative digital infrastructure'), through a Bushfire Data Commons, of a suite of datasets and models relating to fire and its impacts, including biodiversity, through the Australian Research Data Commons.

Conclusions

The Australian Government rapidly recognised the unprecedented severity of the impacts of the 2019–20 wildfires on biodiversity, including on many matters of national environmental significance. It responded urgently, with this response notable for its recognition of the need for independent expertise, for being framed by clear objectives and evidence that explicitly informed prioritisation, for being highly collaborative, and by being characterised by a substantial investment that recognised and reflected the magnitude of the impacts. Consequently, a large suite of projects, mostly involving on-ground management, undertaken by conservation NGOs, landholders, Indigenous groups, NRM organisations, research bodies, and state and territory governments, were coordinated and supported by Australian Government investments. These projects have undoubtedly done much in the short term to redress the biodiversity impacts caused by the 2019–20 wildfires. Much of this response was developed *de novo* as the 2019–20 wildfires burnt, but lessons learned will help preparedness for any future comparable wildfires.

Key implications

- The scale and magnitude of impacts on biodiversity of environmental catastrophes such as the 2019–20 wildfires require a national leadership response (and substantial

targeted funding support), with such response made more effective by consultation and collaborations across all stakeholder groups.
- An urgent, but evidence-based, response is required to such environmental catastrophes, else critical opportunities to prevent extinction and enable recovery may be lost. In this case, responses were prioritised through rapid assessment that determined the most fire-affected species and ecological communities, and the management actions they most needed.
- As demonstrated in this case, the establishment of an independent expert group allows for rapid access to information and evidence-based decision making. It may be appropriate to retain such a grouping to help prepare for future cases, rather than newly commissioning such a body after the event of the next catastrophe.
- Information shortcomings constrained some assessments of fire impacts and hence priorities for management. These need to be remedied to allow for more strategic, effective and timely responses to future catastrophes. In particular, the number and range of funded management actions now being implemented to recover biodiversity provides an extraordinary opportunity to assess the relative effectiveness of management options and inform responses to future events.
- Recovery from the 2019–20 wildfires is a long-term proposition, and additional resources and carefully coordinated programs of work will be needed to consolidate efforts to date and build the resilience of native wildlife to future disasters.

References

Catullo RA, Schembri R, Tedeschi LG, Eldridge MDB, Joseph L, *et al.* (2021) Benchmarking taxonomic and genetic diversity after the fact: lessons learned from the catastrophic 2019–2020 Australian bushfires. *Frontiers in Ecology and Evolution* **9**, 645820. doi:10.3389/fevo.2021.645820

CSIRO (2020) 'Climate and disaster resilience technical report'. CSIRO, Canberra.

Gallagher RV, Allen S, Mackenzie BDE, Yates CD, Gosper CR, *et al.* (in press) High fire frequency and the impact of the 2019–2020 megafires on Australian plant diversity. *Diversity & Distributions*,. doi:10.1111/ddi.13265

Legge S, Woinarski JCZ, Garnett ST, Geyle H, Lintermans M, *et al.* (2021) 'Estimates of the impacts of the 2019–2020 fires on populations of native animal species'. NESP Threatened Species Recovery Hub, Brisbane.

Legge S, Woinarski J, Scheele B, Garnett ST, Lintermans M, *et al.* (2022) Rapid assessment of the biodiversity impacts of a megafire to guide urgent management intervention and recovery. *Diversity & Distributions* **28**, 571–591. doi:10.1111/ddi.13428

Royal Commission into National Natural Disaster Arrangements (2020) 'Royal Commission into National Natural Disaster Arrangements Report'. Commonwealth of Australia, Canberra.

Southwell D, Legge S, Woinarski J, Lindenmayer D, Lavery T, *et al.* (2022) Design considerations for rapid biodiversity reconnaissance surveys and long-term monitoring to assess the impact of wildfire. *Diversity & Distributions* **28**, 559–570. doi:10.1111/ddi.13427

Woinarski J, Wintle B, Dickman C, Bowman D, Keith D, *et al.* (2020) A season in hell: bushfires push at least 20 threatened species closer to extinction. *The Conversation*, 8 January (online).

23

State agency responses to support biodiversity recovery following the 2019–20 wildfires

Daniel J. Rogers, Rachael Alderman, Sarah Barrett, Anne Buchan, Allan H. Burbidge, Tracey Churchill, Sarah Comer, Rosemary Gales, William Geary, Carl R. Gosper, Mark S. Harvey, Jason Higham, Steve Leonard, Hannah Lloyd, Lisien Loan, Manda Page, Tania Reid and Ian S. Walker

Summary

- The wildfires that occurred in 2019–20 affected Queensland, the Australian Capital Territory, New South Wales, Victoria, Tasmania, South Australia and Western Australia. During these fires, relevant land management agencies rapidly mobilised to support management and recovery of biodiversity.
- There were some shared themes that facilitated state agencies to respond appropriately at the landscape scale to wildlife and habitat recovery. These include having collated and accessible information to support decision-making on the distribution, abundance, temporal trends, ecology, and threats to species and ecosystems; the development and maintenance of internal agency technical capacity and capability to guide and support on-ground action; and having the ability to draw on external expertise and partnerships to rapidly develop inclusive and well-informed response plans.
- Key challenges that now form lessons for the future include developing and implementing emergency response plans and strategies that more effectively coordinate across stakeholder groups; continuously improving policies and frameworks that more effectively manage conservation priorities before, during and after wildfire events; and further developing the knowledge, capacity, planning and partnerships required to meet the challenge of conserving biodiversity with the ongoing influence of climate change on wildfire events.

Introduction

The wildfires that occurred over the summer of 2019–20 affected every state and territory, except the Northern Territory (Table 23.1). The scale of these fires was nationally and globally significant, both for their impact on the human communities affected and for their impacts on the wildlife, threatened species, ecosystems and natural values within the

Table 23.1. Fire extent area per jurisdiction during 2019–20 in southern and eastern Australia (as at 28 April 2020).

This is known as the 'Preliminary Analysis Area' (Chapter 1), which enabled national analysis and evaluation of the potential impacts of the fires on wildlife, plants and ecological communities, and appropriate response (DAWE 2020). Fire extent area calculations for this national dataset can differ from fire extent areas in state and territory agency datasets as a result of using different methodological approaches to mapping fire extent and classifications of fire types (e.g. wildfires *v.* prescribed burns).

Jurisdiction	Fire area ('000 ha)
Australian Capital Territory	90
New South Wales	5682
Northern Territory	0
Queensland	574
South Australia	313
Tasmania	45
Victoria	1583
Western Australia	2044
Total	**10 331**

fire extent. What was also significant about these fires, however, was the response of the Australian and global community in wildlife recovery efforts, including that of government agencies, non-government organisations (NGOs), researcher institutions and others.

State and territory environment and land management agencies played complex and multiple roles in the response to the wildfires, and in the efforts to support short- to medium-term recovery of threatened species, ecosystems, habitats and natural values impacted by the fires. These agencies have key responsibilities in ensuring the protection and conservation of native biodiversity, are often responsible for the management of the public protected area estate (national parks and other conservation reserves), and are frequently a key (though not always primary) agency responsible for fire management and fire emergency response. This often applies personally to regional agency staff and parks rangers, who may be part of agency or volunteer fire fighting teams, and members of the affected communities themselves. The application of such complex and comprehensive roles during and after the wildfires meant that agencies were required to rapidly reprioritise their capacity and capability, and rapidly enable existing partnerships, towards the response and recovery effort required to mitigate the risks to environmental and community values.

This chapter shares the experiences of state environment agencies with respect to the impact of the wildfires on wildlife, and how we responded. This is by no means an attempt to be comprehensive, or review what worked or did not (see Chapter 29). Rather, we hope that our individual and collective experiences of the 2019–20 wildfires provided us with an opportunity to learn from one another, and to adapt our responses in the future.

State and territory accounts

Queensland

In south-east Queensland, over 7 million ha in total burnt during the 2019–20 wildfires (https://www.qra.qld.gov.au/sites/default/files/2020-08/2019_qld_bushfires_recplan_2019-20_lr.pdf; 574 000 ha classified as emergency bushfire areas in southern and eastern

Australia; Table 23.1), including important habitats for over 640 threatened species (Fig. 23.1). The Department of Environment and Science (DES), through the Queensland Parks and Wildlife Service (QPWS) had to immediately close many parks to the public until access could be made safe. Aside from visitor safety, immediate post-fire work included enabling access for post-fire ecological assessments and action to support the impacted fauna and flora.

To prioritise recovery efforts, DES led a process to analyse the spatial overlap of fire extent with the predicted distribution of biota classified as threatened under the Queensland *Nature Conservation Act 1992*. While impacts were identified state-wide, the focus was on southern Queensland, due to the preceding severe drought, and to align with the scope of emergency support from the Australian Government's bushfire recovery package for wildlife and their habitat. The information was evaluated to prioritise the threatened species and identify key threats to recovery, by a quickly assembled panel of experts from across DES (QPWS and Science and Technology), the Queensland Museum, Queensland universities, non-government conservation organisations, and other partners. The outputs were cross-checked with a broader assessment undertaken by the Australian Government for species listed as threatened under the Commonwealth *Environment Protection and Biodiversity Conservation Act 1999*. This process reinforced existing scientific networks and fostered new partnerships, which will serve to streamline ongoing efforts to protect threatened species in Queensland.

DES established a process to map fire severity in detail, helping to identify optimal locations for species' surveys and the targeted control of weed and pest animal threats. An innovative method was developed to integrate fire severity mapping with expert

Fig. 23.1. Main Range National Park in the Gondwana Rainforests of Australia World Heritage Area was extensively burnt in the 2019 fires, and included temperate rainforests at altitudes over 1350 m. (Photo: Robert Ashdown/Queensland Government)

knowledge of vegetation community fire tolerance and threats to post-fire recovery, termed the 'Potential Ecological Impact' (PEI). PEI mapping helps identify areas, within ecosystems, likely to be most severely impacted that may require increased resources (e.g. pest management), or altered management approaches (e.g. modification to planned burn program) to enhance recovery (see https://parks.des.qld.gov.au/management/programs/fire-management/post-bushfire-evaluation). This new technique will enable faster post-fire assessments in the future and deliver richer ecological insights into fire impacts to guide park management decisions.

The subsequent threatened species recovery program (The State of Queensland (Queensland Reconstruction Authority) 2020) supported the recovery of 53 threatened plant and vertebrate species, alongside priority invertebrate species, across the Queensland sections of the Gondwana Rainforests of Australia World Heritage Area, and Great Sandy, Oakview and Bulburin national parks (https://www.qld.gov.au/environment/plants-animals/conservation/bushfires-threatened-species-recovery). These species include short-range endemics representing ancient lineages that are now of increased conservation concern, particularly in the face of climate change projections including increasing fire frequency and intensity. The recovery program has enhanced the ecological data available for the priority species, and highlighted the critical need for baseline data on threatened species to meaningfully assess post-fire impacts and to track recovery. It helped establish a survey protocol for multiple species in response to an unprecedented fire event. The ready availability and collaboration of ecological experts was essential to rapidly prioritise species, undertake urgent field surveys, produce habitat models, interpret results, and guide management actions. This work has also enhanced engagement with First Nations people and built capacity to manage the ongoing threats to key biodiversity values, including culturally significant species.

New South Wales

Following the 2019–20 wildfires, ~5.4 million ha (7% of the entire state) were impacted, including damage across 37% of our national parks. The scale of the emergency response was unprecedented (Fig. 23.2). Within the NSW Government, a taskforce was established to coordinate efforts. The Saving our Species program (https://www.environment.nsw.gov.au/topics/animals-and-plants/threatened-species/saving-our-species-program) led the threatened species response. The staff from the National Parks and Wildlife Service fought fires on the front line and implemented large-scale pest and weed control actions to protect species. This involved comprehensive prioritisation across bushfire affected regions and on-ground assessments, research and monitoring into the effects of the fires on threatened species, populations and communities. NSW Government scientists undertook critical work mapping and analysing impacts on threatened species. Community groups, volunteers and NGOs, including the wildlife rehabilitation sector, provided essential support. Corporate and community sponsors also had an important role. For example, Woolworths, Foodbank and WWF-Australia donated food resources for emergency wildlife food supplementation (see Chapter 24). The Australian Government provided funding support for immediate and medium-term recovery actions.

In January 2020, the former Department of Planning, Industry and Environment (DPIE) released its Wildlife and Conservation Bushfire Recovery Immediate Response (DPIE 2020). The document included several immediate response actions including: supplementary food, water and shelter; seed banking and insurance populations; feral animal and weed control; survival, recovery and regeneration; and supporting wildlife carers.

Fig. 23.2. NSW National Parks and Wildlife Service staff work to support the recovery of species in difficult post-fire conditions. (Photo: Alex Pike/Saving our Species/DPE)

Many affected species and communities in NSW will take years to recover. To support this recovery, the ongoing work of Saving our Species aims to maximise the number of threatened species secure in the wild in NSW. The NSW Government's medium-term response plan (DPIE 2021a) outlines a suite of focus areas such as undertaking evidence-based planning, improving knowledge, undertaking intensive interventions, conducting long-term monitoring, and enhancing community collaboration, including with Aboriginal communities.

Immediately after the fires, NSW Government staff undertook a desktop assessment of all impacted threatened species through mapping and some on-ground assessments (DPIE 2020). On-ground efforts focused on providing supplementary food, water and shelter to threatened animals, undertaking feral animal control, securing insurance populations for plants and animals, and supporting wildlife carers with $1 million in emergency funding. Survey and monitoring work also commenced to help understand the impacts and species responses to fire. In 2019–20, Saving our Species developed more than 170 fire response plans for threatened species and ecological communities, and supported over 200 species and ecological communities at over 330 sites with over 470 post-fire conservation actions adopted and implemented (DPIE 2021b).

While critically important, the emergency response was just the first step in an ongoing recovery effort that requires critical actions to be implemented over longer time scales. In 2020–21, Saving our Species undertook around 200 post-fire actions for more than 90 species. Significant recovery projects are now underway, supported by the Australian Government. This includes targeted threatened species projects across priority fire-affected regions. Key actions in this phase include seed banking and captive breeding, feral animal and weed control, and habitat restoration. Insurance populations have been established for a range of impacted species including the Manning River helmeted turtle (*Myuchelys purvisi*), southern corroboree frog (*Pseudophryne corroboree*), regent honeyeater (*Anthochaera phrygia*) and Bellinger River snapping turtle (*Myuchelys georgesi*).

Now 2 years after the fires we can reflect on the factors that have been important to an effective response. Prior to the wildfires, Saving our Species had established projects to secure over 400 species and ecological communities. These existing projects put New South Wales in a good position to understand, and respond to, the needs of threatened species during and after the wildfires. The adaptive management framework and resourcing allocation process for emergencies and large environmental events has also been reviewed, with a focus on continuous improvement. Saving our Species is the centrepiece of the NSW Government's strategy to secure threatened species in the wild. Alongside the NSW Koala Strategy, it is the largest commitment to threatened species conservation in our state's history. These dedicated resources for threatened species are essential to building species resilience and enabling a rapid response to future fires.

Australian Capital Territory

The 2020 Orroral Valley wildfire burnt more than 80% of the ~100 000 ha Namadgi National Park and over 20% of Tidbinbilla Nature Reserve – both listed on Australia's National Heritage List. Namadgi National Park is also the primary water supply catchment for Canberra.

The fire affected nationally threatened bogs and fen communities in the Australian Capital Territory, resulting in losses of the threatened broad-toothed rat (*Mastacomys fuscus*), the two-spined blackfish (*Gadopsis bispnosus*) and the montane spiny crayfish (*Euastacus rieki*). The loss of vegetation and ground cover as a result of fire greatly exacerbated the impacts of two unprecedented rain events in the following year. The increased volume and intensity of run-off destroyed infrastructure and habitat, and mobilised large amounts of sediment through the Australian Capital Territory's primary water catchment.

The ACT Government, through its Bushfire and Flood Recovery Plan 2020 (Environment, Planning and Sustainable Development Directorate 2020), committed to restoring the park's natural and built environment. The ACT Government also committed increasing resilience through the 'build back better' principle, working with Canberra's communities and Traditional Custodians to restore the park's health and wellbeing. This plan involved the development of a report from the Rapid Risk Assessment Team (ACT/NSW Rapid Risk Assessment Team 2020), which identified immediate post-fire recovery priorities, focusing on mitigating immediate risk to park and rural land, removing dangerous trees, restoring access, controlling invasive species, and protecting the Australian Capital Territory's water supply.

Longer-term responses involved monitoring broader trends and drivers like climate change, as well as actions to support the recovery of the natural landscape, built infrastructure and local community. The plan supports actions to control invasive plants and animals; protect threatened species; support catchment rehabilitation efforts and protect water quality; manage and monitor erosion; manage impacts and threats to cultural heritage; consider climate change adaptation and mitigation measures; restore infrastructure including roads, fences, walking tracks, signage, bridges; and assist local landholders. The website https://www.environment.act.gov.au/ACT-parks-conservation/bushfire_management/recovering-from-the-2020-bushfires contains further information on the recovery efforts including story maps 'Namadgi Phoenix', monitoring results that document the recovery of Namadgi National Park and ecosystem recovery programs 1 year on from the Orroral Valley fire.

Recovery efforts were informed by the following guiding principles:

- Traditional Custodians, the Ngunnawal people, and the government will heal Country and walk the journey to recovery together.
- Environmental integrity and ecosystem functions will be restored through assisting natural processes.
- Heritage values will be restored through collaboration using contemporary and traditional practices.
- Canberra's communities, recreational, interest, conservation and user groups will be genuinely and meaningfully engaged throughout the recovery phase and reconnected to country.
- Recovery actions will consider ecosystem drivers and be underpinned by sound science, credible research, monitoring, and best practice techniques.
- Climate adaptation will be considered in every aspect of recovery.
- We will support our people through the recovery journey.

Several important lessons were learnt from the experience of the 2019–20 wildfires. First, protection of natural and cultural values during the fire was helped by having values officers with appropriate conservation expertise embedded in the incident management team. For example, such access to expertise ensured that retardants were not used in bog and fen communities. Second, preparation of a rapid risk assessment report while the fire was still burning set up the recovery program to respond with appropriate urgency, and set expectations with government. Third, recovery needs to plan for a different future and integrate climate resilience, rather than the status quo.

Victoria

The 2019–20 wildfires were exceptional in size and impact within Victoria. They have affected all Victorians and have meant significant change to the work many are doing. The collective biodiversity and wildlife welfare response and recovery effort was incredible. The Victorian Government worked with Traditional Owners, experts and key stakeholders to produce a report, 'Victoria's bushfire emergency: biodiversity response and recovery' (DELWP 2020), on biodiversity impacts and response actions informed by state-wide datasets and models. This report set out the approach to recover critical biodiversity values in the months following the fire season, as well as a longer-term agenda to restore natural environments and build resilience to future fire events.

In Victoria, the 2019–20 fire season affected more than half of the habitat for over 240 species, including at least 110 species that had more than half of their habitat burnt by high-severity fire. This rapid analysis of fire overlaps with species distributions helped to identify species of key concern, and the immediate and short-term actions they needed, such as emergency extraction, supplementary feeding and seed banking. The Victorian Government paired this species-specific approach with a landscape-scale spatial conservation action planning tool to identify the priority areas for invasive species control inside and outside the fire extent, to help ameliorate the intensified effect of these threats post-fire.

The ability to respond quickly and effectively is testament to the expertise and knowledge held within the Department of Environment, Land, Water and Planning and portfolio partners, and was facilitated through critical relationships and collaborative approaches with key stakeholders nurtured under Protecting Victoria's Environment – Biodiversity 2037 (DELWP 2017). This early response enabled the Victorian Government to build response and collaboration across the sector. It demonstrates the value of pre-existing capability, and foundational datasets and analyses suitable for rapid, integrated, spatial assessment of values, impacts and intervention options. At the same time, the 2019–20 wildfires highlighted the need to improve preparedness and suppression planning before the next major fire event so that biodiversity information can be targeted and fit-for-purpose.

For the first time in Victoria's history, a very senior position (called the State Controller Wildlife) was created under formal emergency response arrangements to guide delivery of wildlife welfare and emergency actions for biodiversity through the State Control Centre. This role facilitated the delivery of priority actions under a 7-day plan via the local incident management teams. The creation of the State Controller Wildlife reflected a new level of public scrutiny and expectation about the government's response to protecting not just lives and built assets from fire impacts, but also the natural environment.

The Victorian Government took a holistic, state- and species-wide approach in its response to the wildfires, one which considered long-term resilience. The response also prioritised Traditional Owner wellbeing by supporting programs to enhance connection to, and health of, Country. This included Traditional Owner-led assessments, and actions to heal Country.

Future conservation programs need to proactively account for the risk of megafires in planning, especially given climate change and associated changing fire regimes. This should include, where possible, spreading the risk to species by establishing new populations or strengthening existing populations of species vulnerable to fire. Governments will increasingly need to consider how megafire risk can be built into long-term conservation planning.

Tasmania

During the 2019–20 fire season, Tasmanian fire management agencies relied on systems for assessing and responding to wildfire threats to biodiversity, which were developed following the experience of past wildfires in Tasmania, especially the 2018–19 wildfires season. These earlier fires had threatened key occurrences of fire-sensitive natural values within the Tasmanian Wilderness World Heritage Area. Key elements of these systems include:

- the Bushfire Risk Assessment Model, which identifies natural, cultural and economic assets that are prone to wildfire impact, thereby providing strategic guidance to wildfire preparedness and response planning;
- fine-scale spatial data on the distribution of species, communities and other natural values most at risk of (local) extinction or permanent degradation if exposed to wildfire – these data guide 'tactical' wildfire response planning;
- clear lines of communication between fire managers and biodiversity specialists, with the latter 'on call' for the duration of the fire season to advise on priorities for wildfire protection and carry out rapid desktop and field impact assessments; and
- staff within the Parks and Wildlife Service Fire Management Section with fire ecology expertise tasked to liaise with specialists to feed biodiversity advice into operational planning.

These systems are brought together via the Reserve Values Protection Planning (RVPP) process, in which the natural and cultural values within key areas at risk from wildfire are identified and prioritised and response plans are formulated. While in 2019–20 these plans were formulated 'on the fly' as the fire situation evolved, the intention is to develop plans in readiness for future wildfires. The RVPP approach has been recognised for its quality and innovation (AFAC 2019) and is likely to be a useful model for other jurisdictions. These systems are subject to ongoing refinement, drawing on research and operational experience in Tasmania and beyond.

Tasmania has also progressed policies and processes to manage *in situ* supplementary feeding of wildlife and the care of injured and orphaned fauna following wildfires. Work is also in progress to develop techniques for restoration of fire-affected alpine ecosystems.

Compared to the rest of south-east Australia, Tasmania experienced a subdued wildfire season in 2019–20. Five wildfires of greater than 2500 ha occurred, with ~0.6% of the state affected by bushfire (Table 23.1). Having the systems described above in place meant that fire managers were able to readily assess threats to biodiversity as the fires progressed and direct resources accordingly. As it transpired, these wildfires mostly affected fire-adapted ecosystems, such that long-term impacts are expected to be minor.

Although the existing systems are valuable, further work is required to ensure biodiversity protection is appropriately prioritised in wildfires response. In addition, there is a need to increase the priority of planned burning programs that aim to reduce wildfire risk to biodiversity – currently most of the focus is on planned burning to protect community infrastructure. Ongoing improvement of these processes will help minimise impacts of future wildfires on Tasmania's unique natural values.

South Australia

The significance and scale of the fires of 2019–20 brought an increased scrutiny on the plight of native animals and habitats. The fires affected several areas across the state with the largest impacts in the Mount Lofty Ranges and Kangaroo Island. Widely shared images and stories highlighted those impacts and drew the attention of the state, national and global community. In South Australia there was a particular focus on Kangaroo Island, a place internationally renowned for its natural beauty and abundant iconic Australian wildlife, such as koalas (*Phascolarctos cinereus*), kangaroos and many other species. The fires on Kangaroo Island impacted ~210 000 ha (close to half of the area of the Island), including 95% of western Kangaroo Island's national park estate. The fires affected iconic species, many threatened species, and species endemic to the Island.

The fires and the interest in the wildlife led to an overwhelming desire from the community to help. This raised several challenges for the Government agency responsible. These challenges included limited data and knowledge on which species were affected and how severely; slow provision of information to the community on wildlife impacts; and coordination of activity on ground. The desire of the public to help occasionally led to perverse outcomes, such as feeding wildlife and rescuing animals before the firegrounds were safe.

In order to address some of these challenges, the SA Government set up an independent Wildlife and Habitat Bushfire Recovery Taskforce. This taskforce deliberated with experts and community and developed the Wildlife and Habitat Recovery Framework, so that next time we are in the same situation we have a template to work from to assist in the transition from emergency response to recovery. The taskforce has also released a report with eight key recommendations to address challenges and improve future preparedness.

On Kangaroo Island, a partnership with the National Environment Science Program enabled us to hold a workshop on Kangaroo Island immediately following the fires

Fig. 23.3. Workshop participants develop an action plan to guide recovery efforts for Kangaroo Island. (Photo: Nicolas Rakotopare/Threatened Species Recovery Hub)

(Fig. 23.3), and to work with experts from across government, organisations and the community to develop an action plan (Rogers *et al.* 2021) for recovery efforts. This action plan, funded by both state and Commonwealth, guided priority activities to understand the impacts, address threats and undertake recovery actions. This approach meant that effort across a range of partners was prioritised, and complementary, and has led to effective on-ground outcomes. Ongoing collaboration and coordination between partners has allowed the recovery efforts to accommodate new and emerging threats, including blue gum 'wildlings', while continuing to undertake recovery actions. There are still concerns, however, as recovery will be long and slow for many species, but the initial recovery funding has come to an end. Nevertheless, the funding and efforts to recover species after the 2019–20 wildfires are a substantial step up compared with that of past fires.

Western Australia

Wildfires over the summer of 2019–20 burnt ~1.5 million ha in relatively remote parts of the Great Western Woodlands and South Coast areas of Western Australia, but only 23 692 ha in the heavily populated south-west corner of the state. The post-fire biodiversity conservation response of the Western Australian Government, through the Department of Biodiversity, Conservation and Attractions and the Western Australian Museum, initially focused on identifying where the fires putatively had greatest impact on species and communities already recognised as being at high risk of extinction. This prioritisation was informed by pre-fire population and survey data attained through long-term monitoring of threatened flora, fauna and ecological communities, and was underpinned by an understanding that the impact of the 2019–20 fires was best understood in the context of the past fire regime and other interacting threats.

In Cape Arid National Park and Nuytsland Nature Reserve two wildfires (47 195 and 80 146 ha) impacted an estimated 18 500 ha of core habitat of the Critically Endangered western ground parrot (*Pezoporus flaviventris*) and, combined with extensive fire in 2015, over 80% of suitable habitat was burnt over a 4-year period. Slow post-fire recovery of the kwongan heath habitat resulted in significant areas of poor vegetation cover, reducing the quality of shelter and leaving the habitat accessible to foxes and feral cats, rendering such areas unsuitable for roosting or breeding by the parrots. Consequently, post-fire management has focused on maximising the effectiveness of introduced predator control through additional baiting and trapping, and implementing strategies for reducing further risk of fire in ground parrot habitat (Burbidge *et al.* 2016; Comer *et al.* 2021; https://www.dpaw.wa.gov.au/plants-and-animals/threatened-species-and-communities/threatened-animals/western-ground-parrot). Extensive acoustic monitoring of ground parrots has highlighted the importance of long-unburnt habitat in the breeding season and identified priority areas for predator management. With support from BirdLife Australia and the Friends of the Western Ground Parrot, a translocation was also initiated, with selection of the release site and threat mitigation in this area preceding the trial release of birds to the east of Albany in April 2021 (Comer *et al.* 2021).

Essentially all high-elevation areas in the eastern and central Stirling Range were impacted by wildfire over the period 2018–20, affecting a suite of threatened flora, the threatened Eastern Stirling Range Montane Heath and Thicket ecological community and many of the threatened and short-range endemic invertebrates found only in the Stirling Range National Park. The high-elevation portions of the Stirling Range support an exceptionally rich flora with high levels of endemism, which is threatened by susceptibility to *Phytophthora* dieback acting independently and in concert with short fire intervals (Barrett and Yates 2015). Management actions to support post-fire recovery consisted of (1) ongoing monitoring of post-fire recruitment and survival (see Chapters 8 and 9); (2) threat management, including phosphite application to reduce *Phytophthora* impact, and fencing and feral herbivore control to reduce browsing of post-fire recruits by mammals (see Chapters 17, 18); (3) establishment of two seed production areas with 14 species in *Phytophthora*-free locations outside of the Stirling Range to secure living collections of these species as an insurance against losses in the wild and to facilitate future seed collection; and (4) seed collections from natural populations of 11 species subsequently banked at the Western Australian Seed Centre, to support future recovery actions. Availability of pre-fire data for threatened and short-range endemic invertebrates of the Stirling Range was variable. However, for 39 of the 40 target species, 100% of known habitat was impacted by wildfires over 2018–20 (Bain *et al.* 2021); post-fire surveys documented the level of impact on these areas and provided a benchmark of population demographics and occupancy, which can be used to track recovery and future trends in these species.

Important lessons from the 2019–20 wildfires from a state agency perspective were that (1) robust pre-fire data on species, community status and susceptibility to threats supported understanding of fire impacts and development of management responses both during and after the fire; (2) pre-fire germplasm collections for those species suited to long-term storage was important for post-fire recovery actions, given fires can cover all or large portions of a species' range; and (3) having the capability and capacity for rapid post-fire monitoring and threat mitigation activities, on the scale of days to years, depending on the timeframe of post-fire recovery and elevated susceptibility to threats for different species, maximises the likelihood of population persistence.

Discussion

Australian states and territories were impacted differently by the severity and scale of the 2019–20 wildfires. The absolute and relative scale of the fires varied among the states (Table 23.1). Importantly, the ecological context of fire in these different landscapes varied enormously; the mallee–heath communities of south-west Western Australia and Kangaroo Island will experience very different recovery pathways to the rainforests of southern Queensland and northern New South Wales.

Our abilities to assess the ecological impacts, and respond appropriately, were as diverse as the range of ecological impacts experienced. While much of the response was built on existing programs and experience (e.g. invasive plant and animal control operations), the exceptional nature of the fires and the extent of their impacts required innovative responses, such as *in situ* supplementary feeding and other novel emergency-based approaches. What was important was that management agencies were required – and in most cases able – to design such responses, and have the confidence to apply them, in the rapid timeframes required.

Several key themes emerge from the narratives described in this chapter. Central to success was the development, application and maintenance of the tools and capacity to respond quickly across a broad scale. This foundation comprised:

- *knowledge and data on the distribution, abundance, temporal trends and ecology of species and ecosystems, and their drivers (including, but not only, fire regimes):* having collated and usable information available in a timely manner at the fingertips of decision-makers was instrumental;
- *internal agency technical capacity and capability*: this includes the ability to mobilise an adequate on-ground emergency response, in the context of a broader disaster situation, as well as the required technical ecological expertise from within agencies and their partners for the ongoing response and recovery; and
- *the ability to rapidly develop inclusive and well-informed response plans*: the strength of existing relationships between state, territory and Commonwealth agencies, and with external partners, and the ability to draw on expertise and support from community groups, researchers, traditional owners and NGOs, was a strong determinant of the ability to coordinate our investment and on-ground activities across diverse areas of responsibility and interest.

The development and maintenance of the required information and the human and organisational relationships over the long-term can often be a challenge for agencies as they manage the tension between long-term strategic objectives and short-term responsive demands. However, the 2019–20 wildfires acutely brought to light the critical value of maintaining this capacity: its importance is most strongly demonstrated during times of crisis, when an adequate, informed and collective response is most urgently needed.

Key implications

The novel nature of the 2019–20 wildfires revealed several challenges. To address these challenges and respond more effectively, we frame these issues as opportunities for the future:

- Developing and implementing emergency response plans and strategies that more effectively coordinate and guide efforts across stakeholder groups is critical to ensuring resources are available and targeted towards the protection and recovery of priority natural values. The enormous community concern about wildlife and threatened species, particularly in relation to the wildfires in eastern and southern Australia, was outside many of our experiences, and while this provided an enormous impetus for agencies to mobilise significant resources towards recovery, it also challenged some with how best to respond and support the desire of the community to contribute meaningfully. This was particularly the case for wildlife rescue operations.

- Agencies were also variably able to apply lessons from previous, recent fire events, particularly with respect to planning and governance. The particular lessons learnt from the 2019–20 wildfires have resulted in an increased emphasis on continuing to improve our policies and frameworks, with such ongoing monitoring and adaptation critical for ensuring that biodiversity is more effectively managed before, during and after wildfire events.
- There is a need to further build ecological and fire operational knowledge, capacity, planning and partnerships between and external to state and territory agencies to meet the challenge of a future in which fire management and biodiversity conservation is increasingly influenced by climate change. The role of anthropogenic climate change in the scale and severity of the wildfires in 2019–20 is now well established (e.g. van Oldenborgh *et al.* 2021). In an increasingly warm world, the experience, capacity and expertise held within state and territory agencies will become increasingly important for biodiversity conservation management before, during and after similar fire events.

References

ACT/NSW Rapid Risk Assessment Team (2020) Orroral Valley fire rapid risk assessment Namadgi National Park. Environment, Planning and Sustainable Development Directorate, ACT Government, Canberra, <https://www.environment.act.gov.au/__data/assets/pdf_file/0003/1495236/orroral-valley-fire-rapid-risk-assessment-namadgi-national-park.pdf>.

AFAC (2019) *A Review of the Management of the Tasmanian Fires of December 2018–March 2019.* Australasian Fire and Emergency Service Authorities Council, Melbourne.

Bain K, Harvey M, Utber D, Comer S (2021) 'Stirling Range post-fire threatened invertebrate survey report. Targeted reconnaissance survey and post-fire impact assessment for threatened and endemic invertebrates in the eastern Stirling Range National Park'. Report prepared for Department of Biodiversity Conservation and Attractions, Parks and Wildlife Service Albany. Python Ecological Services, Walpole Western Australia.

Barrett S, Yates CJ (2015) Risks to a mountain summit ecosystem with endemic biota in southwestern Australia. *Austral Ecology* **40**, 423–432. doi:10.1111/aec.12199

Burbidge AH, Comer S, Lees C, Page M, Stanley F (Eds) (2016) Creating a future for the western ground parrot: workshop report. Department of Parks and Wildlife, Perth, <http://www.cbsg.org/sites/cbsg.org/files/documents/WGP_Report_FINAL_Sept2016.pdf>.

Comer S, Burbidge A, Berryman A, Thomas A, Blythman M, *et al.* (2021) A new chapter for kyloring. *Landscope* **37**, 18–22.

DELWP (2017) Protecting Victoria's environment – biodiversity 2037. The State of Victoria Department of Environment, Land, Water and Planning, Melbourne, <https://www.environment.vic.gov.au/__data/assets/pdf_file/0022/51259/Protecting-Victorias-Environment-Biodiversity-2037.pdf>.

DELWP (2020) Victoria's bushfire emergency: biodiversity response and recovery. The State of Victoria Department of Environment, Land, Water and Planning, Melbourne, <https://www.wildlife.vic.gov.au/__data/assets/pdf_file/0030/484743/Victorias-bushfire-emergency-Biodiversity-response-and-recovery-Version-2-1.pdf>.

Department of Agriculture, Water and the Environment (2020) National indicative aggregated fire extent dataset (v20200428) current to 28 April 2020, <https://www.agriculture.gov.au/abares/forestsaustralia/forest-data-maps-and-tools/fire-data#fire-area-and-area-of-forest-in-fire-area-by-jurisdiction>.

DPIE (2020) Wildlife and Conservation Bushfire Recovery, Immediate Response January 2020. Department of Planning, Industry and Environment, Sydney, Australia, <www.environment.nsw.gov.au/research-and-publications/publications-search/wildlife-andconservation-bushfire-recovery-immediate-response>.

DPIE (2021a) NSW Wildlife and Conservation Bushfire Recovery Medium-term response plan, Department of Planning, Industry and Environment, Sydney, Australia, <https://www.environment.nsw.gov.au/research-and-publications/publications-search/wildlife-and-conservation-bushfire-recovery-medium-term-response-plan>.

DPIE (2021b) Saving our Species Year in Review 2019–20. Department of Planning, Industry and Environment, Sydney, Australia, <https://www.environment.nsw.gov.au/research-and-publications/publications-search/saving-our-species-year-in-review-2019-20>.

Environment, Planning and Sustainable Development Directorate (2020) ACT bushfire and flood recovery plan. Environment, Planning and Sustainable Development Directorate, ACT Government, Canberra, <https://www.environment.act.gov.au/__data/assets/pdf_file/0011/1622855/act-bushfire-and-flood-recovery-plan-2020.pdf>.

Rogers D, Rumpff L, Huang J (2021) Kangaroo Island wildlife and habitat recovery planning workshop – workshop summary report. NESP Threatened Species Recovery Hub, Brisbane, <https://www.nespthreatenedspecies.edu.au/media/2snbrwxc/8-0-kangaroo-island-wildlife-and-habitat-recovery-planning-workshop-report_v4.pdf>.

The State of Queensland (Queensland Reconstruction Authority) (2020) *2019 Queensland Bushfires State Recovery Plan 2019–2022*. The State of Queensland, Queensland Reconstruction Authority, Brisbane.

van Oldenborgh GJ, Krikken F, Lewis S, Leach NJ, Lehner F, *et al.* (2021) Attribution of the Australian bushfire risk to anthropogenic climate change. *Natural Hazards and Earth System Sciences* **21**, 941–960. doi:10.5194/nhess-21-941-2021

24

Immediate and longer-term responses of conservation NGOs to the 2019–20 wildfires

Tim Allard, Amelia Caddy, Brenda Duffy, Pamela Fallow, Heidi Groffen, Darren Grover, Pat Hodgens, Liana Joseph, Jenny Lau, Viyanna Leo, Rachel Lowry, Eridani Mulder, Rebecca Spindler and Sarah M. Legge

Summary

Conservation non-government organisations (NGOs) were a critical part of the emergency response to the 2019–20 wildfires, and they remain integral to ongoing post-fire conservation actions. In this chapter, five on-ground conservation NGOs outline the immediate and longer-term activities that they implemented, and the key lessons they took from the experience of the 2019–20 wildfires. Four key themes about the fire responses recur across their accounts:

1. *Collaboration*: The genuine and constructive collaboration that emerged during the 2019–20 wildfires was a signature of the overall emergency response. It was possible because Australia's conservation NGOs have spent many years developing relationships with landholders and managers, Indigenous groups, governments, natural resource management (NRM) groups, and the public, and were able to activate these networks extremely effectively, when it mattered most.
2. *Connection with the public*: NGOs have large supporter bases with existing communication lines. These engagement structures were critical for sharing information about the fires and their impacts, and what was being done to support wildlife. The engagement structures also provided immediate avenues for the public to donate funds to the recovery effort, helping them feel they were contributing to the recovery effort.
3. *Evidence*: conservation NGOs advocated an evidence-based national emergency response; they contributed data to local and national-scale prioritisations; they collected new data; and they used data and other information to communicate with their supporters and the general public.
4. *Planning*: when time is limited and stakes are high, a quality existing plan guides managers through the crucial first steps after an ecological disaster.

Introduction

Conservation non-government organisations (NGOs) are involved in a wide spectrum of activities, such as *ex situ* actions (e.g. seed storage, captive breeding), animal rescue and rehabilitation, advocacy, research, habitat restoration and on-ground conservation management delivery. In this chapter, we focus on the latter part of that spectrum: NGOs with a heavy focus on practical on-ground conservation management describe their response to the 2019–20 wildfires, including the immediate actions they undertook, how the wildfires have shaped their long-term planning and operations, and the key lessons they have taken away from the ecological disaster. The chapter profiles five organisations, but we recognise that many other conservation NGOs also contributed to the response to these wildfires.

The five NGOs vary greatly in the geographic scale at which they operate, from Kangaroo Island Land for Wildlife, which operates solely on that island, to Bush Heritage Australia and the Australian Wildlife Conservancy, both national organisations with substantial landholdings that focus on science-based conservation management, to BirdLife Australia and WWF-Australia, which are both linked to international organisations, and incorporate a mix of management, research and advocacy objectives. Despite this diversity of scale and focus, the five examples given here share recurrent messages about the importance of collaboration and public engagement, of using evidence to inform an emergency response, and of the value of being prepared.

BirdLife Australia

BirdLife Australia is a national NGO whose purpose is to conserve Australia's native birds and their habitats. We work towards this by drawing on our scientific capability, our collective knowledge and experience in bird conservation, and our extensive network of collaborators and supporters. As the 2019–20 wildfire season began to unfold, BirdLife realised the need for collaboration, for using data to prioritise the response, and for communicating with the public. The organisation rapidly mobilised its resources to achieve those immediate needs.

Immediate response

BirdLife was part of the environmental NGO network that came together in late 2019 to develop a common agenda for wildfire response. The NGO network developed a joint statement on the need for a nationally coordinated wildlife and ecosystem recovery effort. The network's collaborative efforts and urgent call for an evidence-based national response probably contributed to the formation of the Australian Government's Wildlife and Threatened Species Bushfire Recovery Expert Panel. The Expert Panel played a fundamental role in guiding a coordinated national response to the impacts of the fires on biodiversity (see Chapter 22). Its series of communiques and work to identify and prioritise both the species most at risk of long-term impacts and the management actions needed to recover them continues to influence wildfire recovery work. The transparency of the panel's work gave the conservation sector confidence that national priorities, including for funding, were based on science rather than politics. After this, government agencies, scientists, environmental NGOs and community groups came together in regional workshops to develop a shared understanding of the impacts of the fires and plan for recovery. This collaborative approach was a hallmark of the national wildfire response, which allowed organisations to identify gaps in their collective knowledge and planned response, and reduce the risk of duplication while maximising the value of limited resources.

BirdLife could rapidly estimate impacts of the 2019–20 wildfires on Australia's birds because we curate Birdata (the Atlas of Australian Birds), one of Australia's largest biological databases containing millions of records submitted over more than three decades. While the fires were still burning, our analytics team began estimating the spatial overlap between the fires and the geographical distribution of more than 371 bird taxa – a crucial first step in identifying those birds at increased risk of extinction due to the fires. In December 2019, we began communicating our concerns about fire-affected birds to our collaborators and our extensive and diverse network of supporters. We launched a wildfire appeal to raise funds that allowed us to commence wildfire response work and recruit a Bushfire Response Coordinator to secure further funding and coordinate wildfire recovery work over the longer term.

BirdLife was one of the first organisations to conduct on-ground population surveys after the fires. Kangaroo Island emerged as a high priority for BirdLife, with most of its 17 endemic bird subspecies estimated to have over 50% of their distribution impacted by the fires. While desktop analyses provided important insights into likely fire impacts, ground-truthing was needed to validate these estimates, so we conducted comprehensive surveys of both burnt and unburnt habitat in the western half of Kangaroo Island (Fig. 24.1). Establishing a network of sites for regular, standardised monitoring has provided a baseline for measuring the post-fire response of Kangaroo Island bird populations and guided ongoing recovery actions on the island.

Longer-term response

BirdLife has prioritised wildfire recovery across many of its ongoing programs to support the recovery of birds identified by the Expert Panel as in most urgent need of management

Fig. 24.1. Bird surveys by the BirdLife Australia team shortly after the 2019–20 wildfires, at Cape Du Couedic, Kangaroo Island. (Photo: Tom Hunt)

action. Our approach to recovery has been guided by existing recovery work and national recovery plans. Two examples are a project to trial the 'seeding' of gum trees with mistletoe to provide a food source for regent honeyeaters (*Anthochaera phrygia*), and a project to translocate western ground parrots (*Pezoporus wallicus flaviventris*) to establish a second population as insurance from future fires. Our ability to respond to impacts of the fires on previously secure but less-studied species, such as red-browed treecreepers (*Climacteris erythrops*) and gang-gang cockatoos (*Callocephalon fimbriatum*), has been hampered by a lack of information on the ecology, distribution and abundance of these species. Without sufficient data collected before the fires, there is no baseline against which to measure changes since the fires, nor the effectiveness of wildfire recovery actions. Expert elicitation processes have provided rough estimates in the absence of data, but impacts of the fires cannot be accurately measured for many species. This highlights the urgent and ongoing need to monitor animal populations, in order to generate data and information enabling the detection of and response to population declines, including those caused by wildfires.

Key lessons

- Pre-existing, strong linkages with collaborators in universities, governments organisation and other environmental NGOs, and the ability reach a large network of supporters, volunteers and community groups, were both important in the 2019–20 wildfire response. A national rapid response plan could facilitate and mobilise similar collaborative action on a greater scale.
- A team of experts who used science to guide nationally coordinated priorities for wildfire response was critical. Members of the next Expert Panel could be identified now to facilitate rapid establishment after the next major catastrophe.
- We urgently need more data and information on the ecology, distribution and abundance on many species. Many birds have likely become threatened as a result of impacts of this single wildfire season. However, there are insufficient data and information to guide the recovery of these species. Australia must invest in long-term monitoring of populations and research aimed at understanding the ecology of our native biodiversity.
- For threatened species, up-to-date national recovery plans enable rapid and targeted wildfire recovery actions.

Bush Heritage Australia

Bush Heritage Australia is an independent not-for-profit organisation that buys and manages land, and works in partnerships with others, to conserve our magnificent landscapes and irreplaceable native species forever. Some of our reserves are located within the regions that were heavily affected by the 2019–20 wildfires.

Immediate response

The 2019–20 fire season began for Bush Heritage as it did for almost everyone in Australia – suddenly and early. In September 2019, a fire from unknown sources spread towards the eastern boundary of Yourka Reserve in far north Queensland. It burnt through ~1000 ha (~3% of the reserve) until it was successfully suppressed.

Knowing that this would be a bad fire season, Bush Heritage had prepared its reserves for possible fires. Prescribed burns had been completed earlier in the season on our own and some neighbours' properties, strategic fire breaks put in place, and bushfire response plans reviewed. As the Yourka and nearby blazes occurred so early in the season, we were able to direct all necessary staff and resources to them and, as a result, the blazes were contained relatively quickly. But as more and more fires broke out on, or threatened, Bush Heritage reserves over the summer, personnel and resources were spread thin.

From November 2019 to late February 2020, six more Bush Heritage reserves were burnt in New South Wales, Queensland and Western Australia. Of those, Scottsdale and Burrin Burrin reserves in New South Wales were worst affected. Burrin Burrin (411 ha) is east of Canberra on Ngunawal, Ngarigo, Walbanga, Ngambri and Walgalu country. In November 2019, lightning sparked the North Black Range fire near Braidwood which burnt through the majority of Burrin Burrin, impacting the understorey through to the canopy and destroying many hollow-bearing trees. During this period, invaluable communication from our neighbours informed our initial response.

Scottsdale (1328 ha) is on the Murrumbidgee River south of Canberra, on Ngunnawal country, where we focus on restoring 300 ha of cleared grassy box woodlands and temperate grasslands. It has a dedicated reserve manager and field officers and significant infrastructure and equipment. On 1 February 2020, when the Clear Range fire breached Scottsdale's boundary, road closures prevented staff from entering the area. The Scottsdale Bushfire Response Plan proved an important tool for staff during this time. It helped decision making by outlining what actions should be taken if no staff were on site. Approximately 70% of Scottsdale was burnt, including 90% of the revegetation area (Fig. 24.2), the

Fig. 24.2. Burnt revegetation on the valley floor of Bush Heritage's Scottsdale Reserve, Ngunnawal country, in New South Wales, after the 2019–20 wildfires. (Photo: Phil Palmer/Bush Heritage Australia)

Murrumbidgee River corridor, and the steep, west-facing slopes of the high country. We also lost many large old-growth trees due to fire and fire suppression activities.

In the immediate aftermath of both the Scottsdale and Burrin Burrin blazes, our priorities were to determine fire extent and severity, and the impact on flora and fauna. Initially we used satellite fire-mapping tools, and then on-ground evaluation. Safety assessments were carried out, access tracks cleared, and supplementary feeders and watering stations deployed to support surviving wildlife such as greater gliders (*Petauroides volans*) on Burrin Burrin. Motion sensor cameras at each food and water station monitored wildlife and feral predator activity. This information informed the development of long-term recovery plans for both reserves.

There had been ~15 000 tree guards on Scottsdale; removing the melted plastic was a priority, and was achieved through the tireless work of many volunteers. Ngunnawal Traditional Owners came on Country to record and map the cultural values that were revealed where vegetation had once been. Heavy rains soon after the fires improved the survival prospects of Scottsdale's revegetation, but the rains also exacerbated erosion and promoted a flush of broad-leafed weeds. By this point, COVID-19 restrictions prevented us from calling on our volunteers to help with weed control. Instead, the Australian Association of Bush Regenerators generously carried out targeted spraying in high-quality grassland sites.

Longer-term response

Bush Heritage's long-term response has been multi-faceted. Our teams are now better resourced thanks to wildfire recovery support. We are more actively using fire for hazard reduction and conservation, and we have engaged Traditional Owners to carry out cultural burning programs across additional reserves. Wildfire response training has been expanded to staff that are not field-based, increasing the number of personnel available to support fire-fighting activities.

Prior to the 2019–20 wildfires, Bush Heritage had already begun investigating the effects of climate change on our reserves and incorporating Bureau of Meteorology long-range weather forecasts into planning. We have now shifted our focus to detecting environmental conditions that may lead to extreme events so that we can take pre-emptive action. For example, we are collaborating with researchers to develop remote sensing technology that will alert us to low soil moisture levels and develop innovative responses. We are also exploring options to improve the ecological resilience of these landscapes, so that they have a greater chance of withstanding and recovering from future disturbance.

Key lessons

The wildfires showed us that the future we have been preparing for is here, and that now is the time to be engineering solutions that will help protect our landscapes. We need to be combining our skills and knowledge with that of others, including Traditional Owners, to foresee risks, alleviate the drivers of these extreme events, and actively heal Country.

Australian Wildlife Conservancy

Australian Wildlife Conservancy (AWC) is the largest private owner and manager of land for conservation in Australia, protecting endangered wildlife across more than 6.5 million ha. AWC's hands-on approach and expertise in on-ground management and surveys enabled us to act quickly to support front-line conservation efforts and help reduce the impacts on wildlife. None of AWC's properties were impacted by the 2019–2020 wildfires, giving us full capacity to help other organisations with the recovery effort.

AWC offered emergency on-ground assistance to conservation organisations and land holders, and established long-term management to support vulnerable wildlife in the most impacted regions. The immediate response included implementing emergency management to protect and recover highly vulnerable species, and undertaking ecological assessments, providing strategic advice and regional coordination of post-fire recovery work. Longer-term projects have included research to identify critical refuges for wildfire affected species; improve designs of emergency artificial shelters to protect vulnerable species from predation; as well as develop strategic processes to predict and plan for future changes in fire regimes.

Immediate response

Emergency management

In the wake of the 2019–20 wildfires, AWC deployed teams of trained staff to conduct emergency response management. This work included reducing the impacts of introduced predators to protect vulnerable species from predation in areas where the natural protection was removed by the fires. When fires consume dense vegetation, ground-dwelling mammals, birds and reptiles are left exposed and highly vulnerable to predation (McGregor *et al.* 2014), as well as to food and water shortages (Andersen *et al.* 2012), increased competition with native and exotic species (Tuft *et al.* 2011), and other environmental factors such as adverse weather (Hale *et al.* 2016). After fire, vegetation can take months to recover, and the scale and severity of the 2019–20 wildfires meant that few unburnt refuges were available in the landscape (Collins *et al.* 2021; Legge *et al.* 2022). To address these immediate risks to wildlife, artificial refuge tunnels were deployed in fire-affected areas, including Kangaroo Island and Wollombi Valley.

On Kangaroo Island the nationally endangered Kangaroo Island dunnart (*Sminthopsis griseoventer aitkeni*) had more than 95% of its habitat burnt. AWC responded rapidly to save this at-risk species and, together with Kangaroo Island Land for Wildlife and the Doube family, completed construction of a 14 ha feral predator-free refuge to protect surviving Kangaroo Island dunnarts (Fig. 24.3). This original fence was later extended into a 370 ha area in which dunnarts can re-establish their population without the threat of feral

Fig. 24.3. The cat exclusion fence (A) built to protect 14 ha of habitat of Kangaroo Island to protect the Kangaroo Island dunnart (B) after fire. (Photos: Brad Leue/AWC)

cats. The fenced exclosure will also support other ground-dwelling threatened species, providing them with a safe area to regenerate their own populations.

Ecological assessments, strategic advice and coordination

Assessing the impact of the fire on threatened wildlife and identifying refuge areas were vital post-fire actions. AWC deployed science staff to conduct emergency post-fire ecological surveys and deliver strategic advice with several land management, conservation and environmental research organisations, including the Wollombi Landcare (16 private properties) and South Endeavour Trust (two properties), Science for Wildlife (sites in the Kanangra-Boyd National Park), Australian National University (searching for any remaining unburnt habitat of the Critically Endangered regent honeyeater through the Wolgan, Widden and Capertee Valleys of New South Wales) and Kangaroo Island Land for Wildlife (see the next section). These surveys covered birds, mammals, amphibians and introduced predators. Survey results including species inventory lists, and strategic advice was provided to all collaborators through reports and contributions to recovery plans.

Longer-term response

AWC is investing in projects to improve post-fire emergency response management and identify refuges for highly vulnerable species.

Improving emergency response

Since the 2019–20 wildfires, AWC has collaborated on research projects investigating the effectiveness of different post-fire wildlife artificial refuges and supplementary provisions. This includes examining the effectiveness of refuge islands, biodegradable habitat pods, refuge tunnels (as deployed on Kangaroo Island and Wollombi), nest boxes, and supplementary feeding and watering stations. The results from these projects will enable us to design an effective post-fire template for wildlife support.

Identifying wildlife refuge

To aid future planning, AWC is conducting research to identify priority locations for conservation projects to protect three fire-affected threatened species: koalas (*Phascolarctos cinereus*), long-nosed potoroos (*Potorous tridactylus*) and long-footed potoroos (*Potorous longipes*). Species distribution modelling is being used to map current, historical and projected future distributions under climate change. Model outputs will be combined with other relevant information (scale, land tenure, management feasibility) to assist AWC to locate areas where work may help conserve the species.

Key lessons

- AWC is developing a Climate Change Adaptation Strategy to address the projected risks of extreme fire events across AWC properties, and to prioritise future acquisition and partnership projects.
- AWC will continue to implement best practice fire management across our estate to maintain ecosystem resilience, as well as enhance planning for operational responses to extreme fire events.
- Climate refugia may be targeted for management actions as well as acquisitions.

While future ecological disasters brought about by climate change are inevitable, we hope that AWC and the Australian community are all ready to support wildlife and ecosystems, and ensure their survival into the future.

Kangaroo Island Land for Wildlife

Kangaroo Island Land for Wildlife (KI LfW) is a private conservation NGO that works with landholders and project partners to deliver long-term conservation management for Kangaroo Island's biodiversity, across private bushland properties.

KI LfW has worked with private landholders since 2017 to survey for the nationally Endangered Kangaroo Island dunnart, based largely on sampling protocols recently developed for this elusive species by Hohnen *et al.* (2019). When the presence of dunnarts is confirmed, we identify threats and prioritise management actions at those sites. KI LfW developed a 5-year Conservation Action Plan (CAP) in 2019 for the Kangaroo Island dunnart and other threatened species, identifying feral cat predation, *Phytophthora cinnamomi* infestations and catastrophic fire as the main threats.

Everyone living and working on Kangaroo Island knew that the pristine bushland we all loved was a ticking time bomb. Dry lightning in late 2019 catalysed the island's most devastating wildfires ever recorded. We witnessed first-hand the obliteration of the west end of Kangaroo Island, one of South Australia's biodiversity hotspots. During these relentless wildfires, over 95% of the range of the Kangaroo Island dunnart was burnt, most of it severely. As well as the loss of habitat, the fires took human lives, homes, livelihoods and stock and left a landscape of emptiness, despair, sadness and anger.

Immediate response

When possible, we moved ahead of the unstoppable inferno to rescue cameras and cat traps, but often all we could do was recover destroyed survey equipment from the burnt areas. We assisted landholders, many of whom were more devastated by the loss of their bushland and wildlife species than that of their own homes.

The wildfires burnt 12 of 13 sites known to support Kangaroo Island dunnarts. The fire left very few unburnt patches of vegetation in its wake. We quickly set camera traps within these few unburnt patches. The first patch (of 8 ha) was a known Kangaroo Island dunnart locality at Western River. Within 2 days, we detected a dunnart on camera, as well as other species such as southern brown bandicoot (*Isoodon obesulus obesulus*), southern emu-wren (*Stipiturus malachurus halmaturinus*), heath goanna (*Varanus rosenbergi*) and Kangaroo Island echidna (*Tachyglossus aculeatus multiaculeatus*), all competing for food and shelter in this small refuge. We also detected feral cats. With the elation of finding Kangaroo Island dunnarts came a deep sense of responsibility and urgency to protect these survivors.

We immediately commenced feral cat control using cage traps and Felixer® grooming traps, removing over 10 feral cats from this tiny patch of vegetation within a week. We knew a cat exclusion fence was the only way to mitigate rapid reinvasion by more feral cats. With our partners Australian Wildlife Conservancy (AWC) and the Doube family private landholders, within 6 weeks of the fire we built a feral cat exclusion fence around this unburnt habitat.

We searched for other unburnt patches in the western end of Kangaroo Island with helicopter surveys, funded by the Foundation for Australia's Most Endangered Species (FAME). We set up camera traps for Kangaroo Island dunnarts at many of these sites, followed up by cat control and shelter tunnel installations if Kangaroo Island dunnarts were detected.

While there was considerable attention and fundraising for the koala, an introduced and abundant species on Kangaroo Island, we sought to highlight the dire plight of all species affected by the fires. We also communicated the need to control feral cats beyond

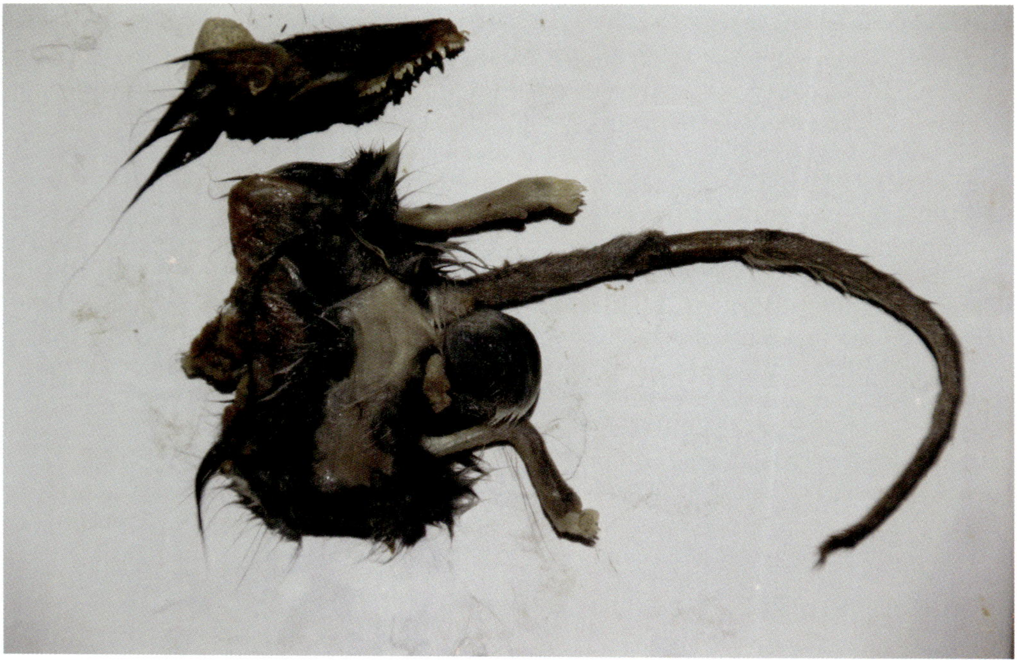

Fig. 24.4. Kangaroo Island dunnart discovered in feral cat stomach. (Photo: Georgeanna Story)

the fenced area post-fire. For example, in the months following the fire we documented that 7% of feral cats within the fire scar had preyed on Kangaroo Island dunnarts (Fig. 24.4), highlighting the need for continued effort to ensure persistence of the species on the island.

Longer-term response

We are extending our private land conservation engagement program while maintaining focused conservation effort in three existing management zones. As well as the Western River Refuge (with cat exclusion fencing), we are also working in the Church Rd Management Zone (which was severely burnt), and in the North West Conservation Alliance Zone (an unburnt 4000 ha area of 10 private properties, known to have Kangaroo Island dunnarts). This effort includes feral cat control and monitoring; surveys for threatened vertebrates, invertebrates, and flora; fire management planning; *Phytophthora* mapping, treatment and hygiene; feral pig control; and weed management. Post-fire, our conservation program has grown from two staff to 10, following investment from not just AWC and FAME, but also the Federal Government, WWF-Australia, Prague Zoo, and Zoos SA. Our next hurdle will be securing funds to continue the conservation effort into the future.

Key lessons

- Planning processes such as conservation action planning can help managers plan immediate response and management actions.
- NGOs can play a key part in emergency responses. For example, the Western River Refuge, a collaboration between NGOs and private landholders, was constructed in 6 weeks.

- We need to improve how we manage fire on Kangaroo Island, which has had two severe wildfires since 2007, with the next large fire potentially only years away. While the Kangaroo Island dunnart survived the 2019–20 wildfires, without efforts to manage fire in the landscape it is unclear whether it will survive the next wildfire event.

WWF-Australia

WWF-Australia is one of Australia's most trusted conservation organisations. For more than 40 years, WWF-Australia has worked to protect threatened species and habitats, meet the challenge of climate change, and build a world where people live in harmony with nature.

The scale of Australia's wildfires during 2019–20 was unprecedented and catastrophic. The national wildfire response framework that WWF-Australia established in January 2020, with the input of our Eminent Science Group, aimed to:

- *respond*, urgently, to the needs of fire-affected wildlife and landscapes;
- *protect and restore* what remains and intervene to mitigate the risk of future fires; and
- *future-proof* Australia's landscapes, drawing on innovative techniques, leading climate science and Indigenous knowledge.

To support this framework, WWF-Australia launched a global fundraising campaign. Supporters responded generously, giving more than $50 million to the Australian Wildlife and Nature Recovery Fund. Below, we share examples of activities under the aims listed above.

Immediate response

Due to the outstanding fundraising outcome, WWF-Australia could mobilise response teams to care for stricken wildlife. We conducted extensive searches for injured animals, utilising detection dogs, thermal imaging drones, and rapid assessment field surveys. We supported veterinary and wildlife shelters with funds for training, equipment and supplies. We also partnered with organisations to deliver food supplies to surviving animals, prioritising areas where threatened species were most at risk. A partnership with the NSW Government, Woolworths and Foodbank Australia saw tens of thousands of sweet potatoes and carrots dropped by helicopter into national parks to help threatened brush-tailed rock-wallabies (*Petrogale penicillata*) persevere until natural food sources became available (Fig. 24.5).

Overall, WWF-Australia partnered with more than 40 wildlife rescue and care organisations across wildfire zones in each fire-affected state and territory. Support was also extended to boost the capacity of seed banking and salvage initiatives to recover threatened flora.

Animals impacted report

In January 2020, WWF-Australia helped inform the world that an estimated 1.25 billion native animals had been impacted by Australia's wildfire disaster. This early estimate was led by Professor Chris Dickman (University of Sydney). By February, we had commissioned scientists to complete a more detailed and final estimate. The shocking findings generated headlines across the globe, helping people to grasp the magnitude of the crisis. Nearly 3 billion animals were killed or displaced by the fires, including 143 million mammals, 2.46 billion reptiles, 180 million birds and 51 million frogs (see Chapter 12).

Fig. 24.5. WWF-Australia partnered with Foodbank Australia and Woolworths Supermarkets to help make an immediate impact on the lives of vulnerable native wildlife in fire-affected areas by supplying fresh food as a part of the NSW Government's aerial food drop program. The food drop targeted brush-tailed rock-wallabies in the Blue Mountains, impacted by the Gospers Mountain fire in December 2019. (Photo: WWF-Australia/Veronica Joseph)

Post-fire surveys: two examples

In September 2020, WWF-Australia released a report (Phillips *et al.* 2020) that found the wildfires resulted in the loss of ~71% of koalas across six locations in northern New South Wales. The koala occupancy declines at the six fire grounds varied from 34% near Port Macquarie to a probable 100% loss in an area south of Taree. This study informed the activation of WWF- Australia's Koala's Forever Program, which has the ambitious target of doubling the number of koalas in eastern Australia by 2050.

August 2021 brought good news for one of Australia's rarest mammals, the silver-headed antechinus (*Antechinus argentus*). There were fears this endangered species, first described by science in 2013, had been pushed to the brink by the 2019–20 wildfires. Fortunately, the tiny marsupial was found surviving in burnt forest in Bulburin National Park in Queensland. With WWF-Australia support, Dr Andrew Baker and Stephane Batista (Queensland University of Technology) trapped 21 silver-headed antechinuses during field work. Despite those positive results, concerns remain for this species' survival.

Longer-term response

Landscape protection and climate-ready restoration

When developing our wildfire response framework, we wanted to ensure value was afforded to not only the trees that burnt, but also to those still standing. With this in mind, we launched a partnership with the Environmental Defenders Office called 'Defending the Unburnt'. This partnership aims to secure six unburnt areas of forest across Queensland, Victoria and New South Wales identified as highly valuable threatened species habitat.

We also joined forces with Greening Australia, launching a partnership with the goal to mainstream innovative, nature-based solutions to build nature's resilience in a changing climate. The partnership will test, validate and scale practical climate-ready restoration approaches, designed by eminent scientists from across Australia. The partnership will centre around three key areas:

1. increased resilience of Australia's flora and fauna through habitat 'renovation';
2. long-term resilience of the ecosystem services that nature provides to people; and
3. strategic actions to mitigate climate-induced risk.

Future-proofing Australia

Due to changing climate, Australia's wildfire seasons are now starting earlier, lasting longer and causing more damage. Consequently, WWF-Australia decided to embed climate mitigation, adaptation and resilience requirements into our wildfire response framework, and explicitly communicate the need to mitigate warming. Projects funded through the 'future-proofing' arm include:

- strengthening policy: working with governments to strengthen climate policy and nature protection laws;
- species adaptation and resilience: supporting long-term conservation efforts for Australia's wildlife;
- harnessing Indigenous knowledge: ensuring that Indigenous-led projects secure the greatest percentage of funds allocated via our Australian Wildlife and Nature Recovery Fund; and
- innovating: exploring and implementing innovative 'future proofing' solutions.

WWF-Australia also ran an Innovate to Regenerate Challenge, which called for innovative ideas to help regenerate Australia after the 2019–20 wildfires. One of the winners to receive funding was an Indigenous Traditional Agriculture Knowledge Hub established near Mallacoota on the farm owned by author Bruce Pascoe. The hub is reintroducing native farming practices and developing a blueprint for Aboriginal food production that returns economic benefits to Indigenous people.

Regenerate Australia

In an effort to ensure that the 2019–20 wildfires serves as a watershed moment for Australia, WWF-Australia has launched a long-term campaign called 'Regenerate Australia'. Regenerate Australia is our bold vision to rehabilitate and restore wildlife and habitats impacted by the fires, articulating the need to increase investment in climate mitigation, adaptation and resilience measures. This will require collaboration and innovation on an ambitious scale. We don't underestimate the shift needed to Regenerate Australia. Science tells us we must do more in this decade to ensure the precious landscapes and wildlife we cherish can continue to thrive. Fortunately, it is achievable.

Key lessons

- Engage widely. Working with Australia's eminent scientists to develop WWF-Australia's wildfire response framework was a key component of the success of the

Australian Wildlife and Nature Recovery Fund. Engaging directly with governments to share insights and scope response interventions was also key.
- Mobilise support. To date, WWF-Australia has committed more than $35 million dollars across more than 150 projects, in collaboration with more than 200 partners to deliver our wildfire response efforts. This would not have been possible without WWF-Australia supporters at home and abroad.
- Plan for the future. We will deepen our focus on innovation and continue to prioritise community and Indigenous-led solutions that are best positioned to drive regenerative benefits for both people and nature.

With thanks to our partners (for a full list see https://www.wwf.org.au/what-we-do/bushfire-recovery/how-your-donation-is-making-a-difference/partners).

Conclusion

The key lessons noted by each of the five NGOs share key messages about future collaborations, gathering and use of evidence, planning, communicating, and future-proofing that can be distilled into the following recommendations.

Recommendations

- **Nurture collaboration**: Relationships between universities, landholders and managers, Indigenous groups, government agencies, NRM groups and the public, need to be developed and maintained at all times, so they can be activated easily during ecological emergencies, and factored into national response plans.
- **Build evidence**: Data gaps on species' ecology, distribution and abundance hamper emergency responses. Long-term monitoring and enhanced research and survey effort are needed to inform future responses to ecological disasters.
- **Use evidence**: Be ready to engage teams of experts to help embed science into future emergency responses.
- **Plan and prepare**: Keep recovery planning and emergency response planning strategies and documents up to date; consider the forecasts of how fires and fire responses are likely to be affected by climate change.
- **Communicate widely**: Conservation NGOs are valuable conduits of information with the public, including during ecological emergencies.
- **Future-proofing**: We need to improve the ecological resilience of landscapes to prepare for a future with a greater frequency of large-scale wildfires.

References

Andersen AN, Woinarski JCZ, Parr CL (2012) Savanna burning for biodiversity: fire management for faunal conservation in Australian tropical savannas. *Austral Ecology* **37**, 658–667. doi:10.1111/j.1442-9993.2011.02334.x

Collins L, Bradstock RA, Clarke H, Clarke MF, Nolan RH, *et al.* (2021) The 2019/2020 mega-fires exposed Australian ecosystems to an unprecedented extent of high-severity fire. *Environmental Research Letters* **16**, 044029. doi:10.1088/1748-9326/abeb9e

Hale S, Nimmo DG, Cooke R, Holland G, James S, *et al.* (2016) Fire and climatic extremes shape mammal distributions in a fire-prone landscape. *Diversity & Distributions* **22**, 1127–1138. doi:10.1111/ddi.12471

Hohnen R, Murphy BP, Gates AJ, Legge S, Dickman CR, *et al.* (2019) Detecting and protecting the threatened Kangaroo Island dunnart (*Sminthopsis fuliginosus aitkeni*). *Conservation Science and Practice* **1**, e4. doi:10.1111/csp2.4

Legge S, Woinarski JCZ, Scheele BC, Garnett ST, Lintermans M, *et al.* (2022) Rapid assessment of the biodiversity impacts of the 2019–2020 Australian megafires to guide urgent management intervention and recovery and lessons for other regions. *Diversity & Distributions* **28**, 571–591. doi:10.1111/ddi.13428

McGregor HW, Legge S, Jones ME, Johnson CN (2014) Landscape management of fire and grazing regimes alters the fine-scale habitat utilisation by feral cats. *PLoS One* **9**, e109097. doi:10.1371/journal.pone.0109097

Phillips S, Wallis K, Lane A (2020) 'Quantifying the impacts of bushfire on populations of wild koalas (*Phascolarctos cinereus*): insights from the 2019/20 fire season. Final Report to WWF-Australia'. WWF-Australia and Biolink Ecological Consultants, Uki.

Tuft KD, Crowther MS, McArthur C (2011) Fire and grazing influence food resources of an endangered rock-wallaby. *Wildlife Research* **39**, 436–445. doi:10.1071/WR11208

25

Responses of natural resource management groups to biodiversity loss and recovery in the 2019–20 wildfires

Rachel Morgain, Karleah K. Berris, Leigh Blackmore, Paul Donatiu, Paul Jennings, Cherie White and Jacob White

Summary

- Natural resource management (NRM) organisations are regional bodies that span Australia, and plan, coordinate and deliver programs that support healthy and productive country, communities and regional industries.
- A significant focus of NRM involves addressing landscape-level challenges in an integrated way through fostering engagement and participation, working with and empowering communities, First Nations groups and landholders, and building regional networks and linkages to deliver ground-up responses to regionally identified needs and priorities.
- Emergency funding for the 2019–20 wildfire response has been vital to supporting NRM organisations to respond to landscape-level recovery needs across tenures and across riparian, freshwater, terrestrial and coastal systems.
- NRM organisations frequently played a pivotal role in rapid response, including helping coordinate local and regional networks, serving as a focal point for community action, supporting First Nations priorities, providing a ready conduit for resources to support rapid on-ground action, and acting as a resource for community and landholders seeking information.
- A pre-existing knowledge base supported rapid response and effective prioritisation, but was uneven across landscapes, species and ecosystems.
- Greater investment in coordination, collaboration and planning between agencies would act to strengthen on-ground responses in future. That includes coordination of leadership roles and funding, and development of pre-approved emergency response and assessment plans for species and ecosystems.
- Greater flexibility in funding would help ensure it could be better targeted in future, for example ensuring that funding for First Nations and local

partnerships meets the most important priorities of those groups, that activities are undertaken at the most appropriate times, or that priorities that currently fall through the gaps of major programs, such as protecting refuges, managing waterways, supporting non-threatened or locally threatened species and restoring ecosystem function, are met.

- Significant knowledge of ecosystems, recovery actions and effective response capabilities has been increased through the recovery phase, but needs to be consolidated and built upon to strengthen our understanding and preparedness for future events.
- There is a need for long-term investment so that the sector can manage ongoing recovery priorities and build the resilience of these populations and ecosystems for future events.
- Wildfires of this severity place significant pressures on systems in place for responding to impacts and recovering biodiversity, including on communities and NRM managers themselves, suggesting a need to embed resilience and create more surge capacity in our human and organisational systems, as well as our species and ecosystems.

Introduction

Natural resource management (NRM) organisations are regional bodies established to deliver programs to support healthy and productive country and communities. There are currently 54 regional NRM organisations across Australia, spanning the continent, ranging from statutory bodies to non-government organisations. NRM organisations receive funding from state, territory and federal governments to build local and regional approaches to manage land, water, soil and biodiversity. A significant focus of NRM organisations is on fostering resilience across landscapes, ecosystems and communities through engaging and empowering local communities, landholders and First Nations organisations to build ground-up approaches and support on-ground activities that address local landscape-level priorities.

Natural resource managers have responded to a huge diversity of needs following the 2019–20 wildfires: to care for ecosystems, soils, waterways, species, heritage, and culturally significant sites, as well as supporting landholders and community recovery across public, private and mixed-tenure landscapes. In the wake of the fires, the severe impacts on communities and organisations from the disaster placed significant pressures on systems in place for biodiversity and landscape management. A wide range of questions are still emerging from landholders, First Nations and community members seeking to understand how to respond to the severe and complex impacts of the fires, and the interactions with other threats such as drought and flooding. Despite the immense challenges, communities have sought to take action to reconnect and help heal the landscapes around them.

NRM organisations were well placed to support and develop locally appropriate responses to these diverse and complex needs. The rapid mobilisation and coordinated responses of NRM organisations was made possible by their being situated in communities and by their knowledge of local ecosystems, networks and resources. By being flexible and responsive to local needs, they have provided a critical element in the fire response. In the wake of the fires, NRM organisations were called upon to help coordinate networks and activities, assess impacts, guide and prioritise actions, mobilise communities and provide vital information, as well as to undertake assessments and recovery actions themselves for biodiversity, heritage and culturally and ecologically significant places. Taking an integrated approach to landscape management, NRM organisations have been able to address gaps between the responsibilities of other agencies, working across tenures and across terrestrial, riparian, freshwater, marine and coastal ecosystems. At the same time, being embedded in regions has allowed for an enduring focus, after the crisis response has passed, on longer-term recovery, and on preparedness for future events.

Aided by emergency funding from the Australian Government and state and territory governments, NRM organisations have helped undertake priority actions, coordinated with other response agencies and provided a focal point for wider community engagement. Forty-two regional NRM organisations were in regions affected by the fires, with 16 regions, plus Ocean Watch, receiving priority funding for urgent fire recovery actions from the Australian Government in their initial response to the fires. Thirteen of these regions were supported for longer-term recovery efforts through the Australian Government's Regional Fund (see Fig. 25.1; Chapter 22). The recovery actions coordinated by regional NRM organisations have spanned diverse activities from managing landscape-scale impacts such as weeds and overabundant and feral species, soil erosion and restoring fencing (including for grazing exclusion), to direct restoration, species-specific responses, and recovery actions for ecosystems, wetlands and river systems. Many NRM organisations have also partnered with First Nations communities to support caring for culturally significant sites and Country, and renewal of cultural fire and land management.

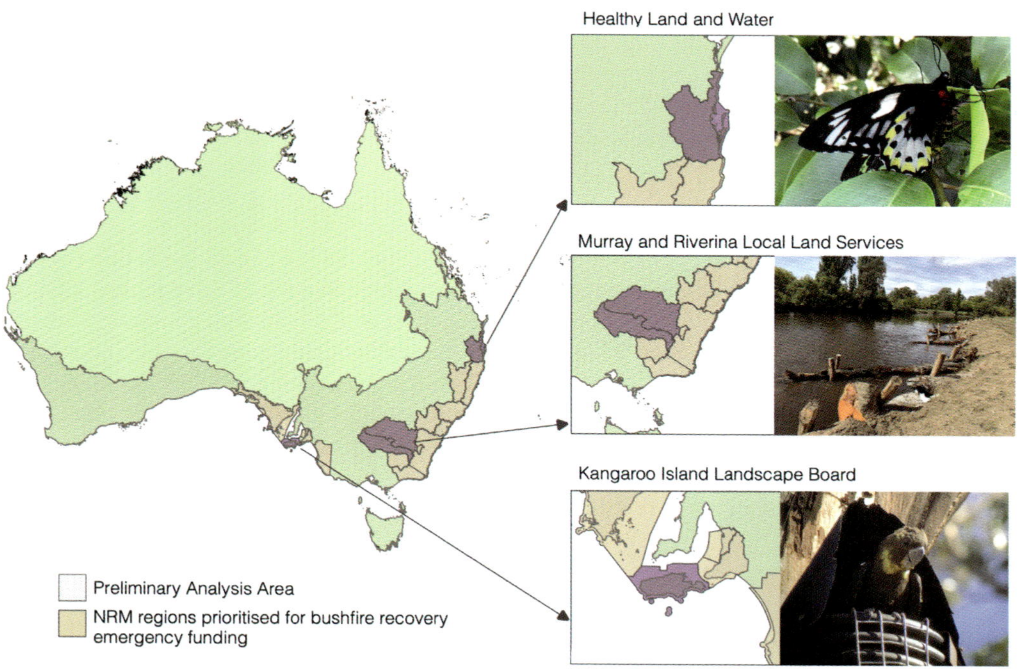

Fig. 25.1. Map of NRM regions prioritised for wildfire recovery, also showing the locations of the three case studies (Boxes 25.1 to 25.3).

In many cases, the biodiversity response of regional NRM organisations has drawn strongly on research and from the knowledge base built from years of experience. Post-fire assessments and monitoring have built further on our understanding of how species and ecosystems respond to and recover from fire, and the effectiveness of certain actions in the immediate aftermath of the wildfires. However, recovery will take many years, and will require ongoing resources for both recovery actions and monitoring to understand effectiveness over the longer term. Longer-term investment is going to be critical if we are to consolidate recovery, learn the lessons from these fires, and prepare the resilience of our ecosystems – and our response systems – for future events.

This chapter explores responses of NRM organisations across different regions in the fire footprint. These responses illustrate the breadth of activities that NRM organisations have been involved in to protect and recover biodiversity in the aftermath of the fires. They highlight the critical role NRM organisations play in coordinating regional networks to assess and prioritise recovery responses and in mobilising local communities in the response. They also show the importance of NRM organisations as repositories of local knowledge of species and ecosystems, to enable development and implementation of rapid and targeted responses to wildfire.

Box 25.1. Communication, coordination and collaboration as key ingredients to mitigate ecological impact of wildfires in Murray and Riverina NRM regions

Recovery efforts by government agencies and community groups to mitigate the ecological impact from the 2019–20 wildfires in the Murray and Riverina NRM regions were both challenging and effective. The collaborative effort demonstrates how communication, coordination and collaboration, with strong support from governments, landholders and the community, can lead to positive outcomes for the recovery of environmental health in the aftermath of large-scale wildfires.

The Dunns Road and Green Valley wildfires started on 30 December 2019, burning out of control for about 2 weeks. The fires covered an area of ~250 000 ha of private land, national park and state forest within the Murray and Riverina NRM regions, linking up with fires in neighbouring regions to form a large fire complex of over 600 000 ha.

Once the fires were under control, Murray Local Land Services (LLS; Fig. 25.1) brought together stakeholders to develop a cross-tenure landscape approach to recovery. It became evident that the scale of the ecological impact was immense and needed prompt attention, but direct impacts on people, stock and wildlife were the highest priority for many agencies. Immediate planning and coordination of the medium to long-term ecological impacts was required, but it was recognised that the wider community would not yet be ready to start these discussions.

Each organisation had its own area of interest, but no single agency held a clear mandate to provide overall leadership in coordinating ecological recovery efforts. At the initial meeting of stakeholders, the group discussed the key issues and what action would be required. The need for a landscape-scale action plan or strategy was identified as a critical component of a successful recovery.

Murray and Riverina Local Land Services negotiated with the Australian Government to allocate a portion of funding from the Wildlife and Habitat Bushfire

Recovery program (Chapter 22) to develop a landscape-scale plan to assist in identifying and prioritising actions. Holbrook Landcare Network and Murrumbidgee Landcare Inc. combined forces to develop this plan, in consultation with government agencies and the wider community. Their involvement in leading this process was an important component in capturing community interests and values, and ensuring these were well represented within the plan.

The group continued to meet on a regular basis with three key objectives: to inform each other of their recovery activities; to identify opportunities for collaboration; and to identify and plan to address priority actions that were not otherwise being undertaken. An online communication tool was set up to provide a platform for the group to communicate and track progress against the plan.

As a result of this communication, coordination, and collaboration many positive outcomes were achieved. Some examples included:

- developing a set of videos for landholders to understand the post-fire erosion process and develop low-cost techniques to reduce the impact on waterways. These were produced through a collaborative effort from North East Catchment Management Authority, Murray and Riverina LLS, Agriculture Victoria and NSW Soil Conservation Service. These were used by landholders in the recovery and are readily available for future events.
- protecting and enhancing native vegetation in key areas of private land through fencing, weed control and revegetation. Holbrook Landcare Network and Murrumbidgee Landcare Inc. played a leading role, working directly with the community and private landholders.
- creating habitat for the Endangered trout cod (*Maccullochella macquariensis*) and stabilising waterways using 65 large burnt trees that were removed from roadsides due to safety concerns (Fig. 25.2). This work was initiated by NSW Fisheries and involved Regional Roads Victoria, NSW Soil Conservation Service, Parks Victoria and Murray LLS.
- reducing sediment through a targeted erosion control program to protect key habitat for the Endangered booroolong frog (*Litoria booroolongensis*) in Gilmore Creek through a partnership with Soil Conservation Service and Riverina LLS.
- rescuing the last remaining population of the Endangered Macquarie perch (*Macquaria australasica*) in the NSW Murray system from Mannus Creek before the creek turned to sludge after the first big rain event following the fires (Fig. 25.3). Led by NSW Fisheries, this has seen the population on the road to recovery, with Charles Sturt University undertaking surveys, controlling pest fish, and developing a management plan; NSW National Parks and Wildlife Service (NPWS) controlling willow regrowth; and Murray LLS coordinating groups and funding. In the Riverina LLS region, a proportion of the Macquarie perch population was translocated into Adjungbilly Creek, part of a broader threatened species program in the creek to improve genetic diversity and restore riparian habitat.
- with reduced canopy cover in the forests, undertaking a successful coordinated aerial control program to control the numbers of pest herbivores, giving native vegetation a chance to recover. Led by NSW NPWS, this collaboration involved Murray LLS working with private landholders bordering national parks to allow the program to extend into private land.
- prioritising properties surrounding Ellerslie Nature Reserve for pest animal and plant control to support the recovery of the threatened box gum grassy woodlands

Fig. 25.2. Timber from wildfire-affected trees removed to maintain road safety used to control erosion and improve the habitat of the endangered trout cod. (Photo: David Thatcher)

ecological community. Riverina LLS led this program, engaging private contractors to assist landholders to undertake this work.

- undertaking a coordinated woody weeds control program by NSW Forestry Corporation, NSW NPWS, Snowy Valleys Council, LLS and private landholders.

The coordinated plan meant that, when funding opportunities arose, investor priorities could be matched with regional needs, and achievements could be tracked. These activities were largely funded through the Australian Government's Wildlife and Habitat Bushfire Recovery investment fund, and NSW Government funding.

Some of the key lessons and insights include the following:

- A regionally focused ecological recovery plan enabled stakeholders to address the issues in a holistic and coordinated manner.
- Collaborative action achieved significant benefits for threatened species and ecosystems.
- A lack of species information, including basic information such as species distribution, limited the ability to make informed decisions about resource allocation and prioritisation.
- A need to protect unburnt areas, or areas of low fire-intensity, as refuges for wildlife and plants was identified, but limited funds were available to implement action.
- Funding opportunities were predominantly limited to nationally threatened species or issues impacting farming production. There were limited opportunities to access funding that would assist regionally identified priorities, such as ecological recovery

of non-listed species or regionally threatened species, monitoring/surveys to establish baseline species information, waterway protection and habitat connectivity.
- Engaging Holbrook Landcare Network and Murrumbidgee Landcare Inc. to lead the development of an ecological recovery plan ensured community values and local knowledge were embedded into the plan.
- A communication and engagement plan would add value to the ecological recovery efforts, providing consistency in messaging and help bring communities along for the journey.

Communication, coordination and collaboration were the keys to success and enabled positive outcomes to be achieved. With improved clarity of leadership responsibility and additional resources to plan, coordinate and foster collaboration, even greater outcomes could be achieved, delivering significant return on a very modest investment.

Fig. 25.3. Local landholder children assisting with the introduction of captive bred Macquarie perch to boost the number and genetic diversity of the local Mannus Creek Macquarie perch population that was severely impacted by the 2019–20 wildfires. (Photo: Karin Willcox)

Box 25.2. Recovering fauna on Kangaroo Island

Between 20 December 2019 and late January 2021, lightning strikes started wildfires that burnt around 211 000 ha of land on Kangaroo Island, almost half the area of the island (Government of South Australia 2020). The wildfires impacted native vegetation and farmland, primarily on the western half and north coast of the island. Immediately, there was concern for two Kangaroo Island Endangered endemics: the South Australian glossy black-cockatoo (*Calyptorhynchus lathami halmaturinus*) and the Kangaroo Island dunnart (*Sminthopsis griseoventer aitkeni*). The regional NRM group, Kangaroo Island Landscape Board (KILB), commenced surveys for both the glossy black-cockatoo and dunnart immediately after the fires, and in the 18 months post-fire have implemented a range of management actions aimed at securing both populations from further declines after the fires.

South Australian glossy black-cockatoo

The South Australian glossy black-cockatoo became extinct in mainland South Australia in the 1970s, and has since been restricted to Kangaroo Island (Joseph 1989; Berris *et al.* 2018). A long-term recovery program commenced in 1995, focused on protecting nests from predation, and provided extra habitat through revegetation and nest box installations. Between 1995 and 2016, the population of glossy black-cockatoos on Kangaroo Island had more than doubled, thanks to this sustained conservation effort (Berris *et al.* 2018). After the 2019–20 wildfires, while surveys concluded that the glossy black-cockatoo populations were not immediately heavily impacted, loss of a large proportion of their feed trees, drooping she-oak (*Allocasuarina verticillata*), could have a long-term impact, and may already be resulting in declines, as detected in subsequent surveys (see also Chapter 15).

Monitoring by KILB indicated there was no evidence of birds relocating out of burnt western areas of the island to habitat further east, and that flocks persisted in all regions they had been recorded pre-fire. However, these flocks were now restricted to the smaller patches of unburnt she-oak remaining. Therefore, post-fire management actions commenced across all seven flock regions on the island.

KILB has focused on revegetation of feeding habitat, providing nest boxes, and predator and competitor control post-fire (Fig. 25.4). This work started immediately after the fireground was declared safe to access. The KILB Native Plant Nursery

Fig. 25.4. Many new nest boxes were established for glossy black-cockatoos after the wildfires on Kangaroo Island. (Photo: Mike Barth, Kangaroo Island Landscape Board)

propagated 6000 she-oak seedlings within a month of the fires ending. Staff built and installed nest boxes, and the Serpentine–Jarrahdale Landcare Group in Western Australia provided 'Cockatubes'. In the 18 months post-fire, more than 13 600 drooping she-oak seedlings have been planted across Kangaroo Island on private properties, and over 60 new nest boxes have been installed to replace nests destroyed in the fires.

In addition, possum exclusion collars on the base of more than 200 nest trees were repaired and canopy isolation pruning undertaken to prevent predators accessing nest trees, giving the glossy black-cockatoo population the best chance of continued high reproductive success. Trials have now commenced with closing off key nest boxes during the non-breeding season (spring) to deter self-colonised nest competitors such as galahs (*Eolophus roseicapillus*) and little corellas (*Cacatua sanguinea*). The competitors, which seem to be impacting some sites more heavily post-fire, caused several glossy black-cockatoo nest failures in 2021. Further research is also being undertaken on the post-fire recovery of drooping she-oak, with the aim to better understand the fire ecology of this tree species, and ultimately understand the fire intervals required to promote high-quality feeding habitat for glossy black-cockatoos.

One constraint has been the labour intensiveness of the work undertaken, given much of it requires tree climbing, and our working area spans the length of the island. Generous support provided by World Wide Fund for Nature and the Australian Government has funded three dedicated glossy black-cockatoo recovery staff for 2 years post-fire. However, given this species is conservation dependent and requires protected nests for adequate recruitment to sustain the population, ongoing funding for management will be needed. COVID-19 restrictions also prevented large volunteer events, limiting the ability to carry out large-scale revegetation projects.

The glossy black-cockatoo is arguably in a tenuous position post-fire, and now has access to less than half of the feeding habitat it had previously. Some flocks lost more than 90% of the available feeding habitat within their home range and are restricted to very small areas post-fire. Therefore, the glossy black-cockatoo population is extremely vulnerable to events that may affect food availability, such as further wildfires or drought. A challenge going forward will be to manage the remaining feeding habitat so that it continues to support the flocks that are reliant on it, and to prevent further habitat burning before habitat burnt in the 2019–20 wildfires regenerates.

Kangaroo Island dunnart

The Kangaroo Island dunnart had only been reported on 46 occasions from 21 sites before the fires, and it was estimated that 96% of the species' known contemporary habitat was burnt during the wildfires (see Chapter 16). At the time, there was very little understanding of how Kangaroo Island dunnarts respond to fire in the landscape; however, evidence from other small mammal species suggests that they were vulnerable not only to direct mortality from the fire but also to increased predation pressure by feral cats after the fire due to a reduction in ground cover that provides refuge. Therefore, the initial monitoring objectives were to improve understanding of the distribution of the dunnart post-fire to reduce predation through targeted and intensive feral cat control in areas where dunnarts were detected, until a better understanding of the status of the species could be ascertained.

Within a month of commencing surveys, dunnarts were detected in some of the few remaining unburnt patches of vegetation within the parks and reserves system on the western end of the island, and subsequently in severely burnt areas where dunnarts had been detected pre-fire. Survey efforts expanded across burnt and unburnt areas of suitable habitat on both private and public lands and sites of historical dunnart records across the island were re-surveyed. Dunnarts have now been detected at 66 sites from 220 surveyed and have been recorded on over 400 separate occasions across public and private property in both unburnt and burnt habitat, and in a range of habitat types, from the central lateritic plateau to coastal heath at Cape du Couedic and Cape Borda. They have also been more frequently detected from burnt sites than unburnt sites, a result not anticipated when survey efforts began. Sites where dunnarts have been detected are entirely confined to the island's western end, primarily within the protected parks and reserves, which represent some of the largest untouched areas of native vegetation on the island.

Cat management is being directed to areas where dunnarts and other high-priority species are detected (see also Chapter 17). KILB is also undertaking broad-scale control of feral cats across successive management units, which form a progressive front from west to east across the island's fire-affected landscape. This approach uses a combination of cage and padded foothold traps in each management unit. Intensive cat control over 5 or 6 weeks has seen an estimated 70–80% decline in cat activity within management units, based on detection rates on motion-activated camera arrays. Cats remaining in each management unit are then targeted using a combination of thermal night-vision techniques and trained feral cat detector dogs to reduce this number even further. This approach has seen over 750 feral cats removed since the fires.

A Curiosity® feral cat baiting trial was undertaken during the 2020 winter in the West Bay/Breakneck River area within Flinders Chase National Park to determine the efficacy of a broad-scale baiting program, so cat control efforts across fire-affected areas of western Kangaroo Island can be optimised. Curiosity® baits were deployed aerially by helicopter over an 85 km^2 baiting area, with a motion-activated camera array lying at the centre of the treatment area. This trial saw a 75% reduction in feral cat detection rates (i.e. the proportion of camera trap-nights with cat detections) following baiting. This program was expanded to cover a much larger area in the 2021 winter, targeting high-priority unburnt vegetation and the network of roads, tracks and travel ways that feral cats were known to frequent within national parks and wilderness protection areas.

COVID-19 restrictions have had a huge impact on the recruitment of staff and the availability of tools and technology to implement this project. Another constraint was that approvals were not initially in place to undertake rapid surveys and to implement management actions. Emergency response plans for threatened species recovery and rapid assessment would have been invaluable to allow activities to occur immediately and for feral cat control techniques to be pre-approved to maximise impact.

Box 25.3. Strategies to restore Critically Endangered lowland subtropical rainforest following the 2019–20 wildfires

Large areas of the Gondwana Rainforests of Australia World Heritage Area in south-east Queensland, including 70% of both Mt Barney and Main Range national parks and nearly 8% of Lamington National Park, were severely affected by high intensity wildfires in late 2019. These areas provide habitat for many threatened flora species and one nationally listed ecological community: Critically Endangered lowland subtropical rainforest (see also Chapter 8). The fires damaged rainforest and wet sclerophyll communities, opening up forest canopies and allowing weeds to become established in canopy gaps. However, post-fire there was also significant evidence of natural regeneration – both resprouting (basal, stem and root) and germination from seed (from established soil seed banks).

To maximise limited on-ground funding, the regional NRM organisation, Healthy Land and Water (HLW), sought to identify the post-fire recovery strategies of rainforest flora, and use this information as a basis for identifying the most effective approach to ecosystem restoration. This has required addressing questions about the capacity of rainforest and wet sclerophyll flora to deal with fire events of varying intensity, in particular:

1. Which native flora are regenerating post-fire?
2. What are the mechanisms used by rainforest and wet sclerophyll flora to regenerate post-fire (resprouting, by seed, or both)?
3. What successional category do regenerating species fall into?
4. How does fire intensity affect the capacity of rainforest and wet sclerophyll flora to regenerate post-fire?

Insights into these questions have supported decisions about the strategies needed to regenerate these fire-affected areas, and from this what general strategies need to be considered (such as supplementary planting of select species) when restoring fire-affected rainforest and wet sclerophyll vegetation communities.

Funding of nearly $1 million over 2 years was provided by the Australian Government through the Wildlife and Habitat Bushfire Recovery investment fund (Chapter 22), and enabled HLW to coordinate restoration activities in some of these fire-affected areas (Fig. 25.5). HLW staff have collected imagery of, and described the regeneration strategies used by, 100 rainforest and wet sclerophyll plant species in fire-affected areas of Lamington National Park, predominantly Upper and Lower Illinbah, Caves Circuit, Lower Bellbird Circuit, and Tabletop Mountain (Donatiu and White 2021). Where known, the successional status of these species (Pioneer, Early Secondary, Later Secondary or Mature Phase, as per Kooyman 1996) and their conservation status were recorded.

Fig. 25.5. Sampling impacts of the wildfires on plants in Gondwanan rainforests. (Photo: Paul Donatiu)

Very few species were killed outright by the fires, showing neither capacity to regenerate from seed nor resprout, with bangalow (*Archontophoenix cunninghamiana*) and walking stick palms (*Linospadix monostachyos*) the most notable (though the former were able to survive a trickle fire running through forest litter in the rainforest understorey). Perhaps the most unlikely finding was the high number of Mature Phase species that resprouted post-fire. Although slow-growing, and without the lateral branching structure evident in Pioneer species, their presence provides valuable assurance that large numbers of rainforest species were not lost to fire. Almost 20 Mature Phase species are clearly resprouting – probably a gross underestimate of the total number – providing compelling evidence of both rainforest recovery and resilience in the face of wildfire. For the present, and in the absence of subsequent equally intense and severe wildfire, it appears that additional restoration strategies, such as supplementary planting of select species, are not yet required to repair fire-affected rainforest and wet sclerophyll vegetation communities in places such as Lamington National Park.

Key questions on recovery and succession of species

There is an ongoing need for monitoring to assess regeneration strategies, and particularly restoration of the forest canopy cover.

In high-fire-severity areas, little mass regeneration from seed was observed, contrary to the large amounts of regeneration from seed from Pioneer species often observed following disturbance events such as treefall. It may be that soil seed banks were consumed by high severity fires. Further investigations to understand any correlation between high fire severity, damaged rainforest canopies and loss of soil seed banks would inform rainforest recovery after future wildfires, and if and when supplementary planting may be required.

While a relatively high number of Mature Phase species resprouted, their contribution to canopy formation (percentage cover) over time is unknown. Field observations to date suggest the rapid growth and lateral branching architecture of Pioneer and most Early Secondary species will see this suite of plants contribute most to restoring damaged canopies.

Ongoing need for monitoring and weed management

A priority of management has been to ensure resprouting and germinating native species are protected from weed growth that threatens to overwhelm and inhibit natural regrowth. So-called transformer weeds – such as moth vine (*Araujia sericifera*), white passionflower (*Passiflora subpeltata*), lantana (*Lantana camara*), palm grass (*Setaria palmifolia*) and devil's fig (*Solanum chrysotrichum*) – are encroaching fire-affected areas, particularly those that suffered canopy loss, threatening the habitat quality and ecological integrity of these rainforest and wet sclerophyll communities.

HLW has coordinated substantial weed management in rainforest communities in fire-affected Gondwana national parks to protect rainforest communities in these parks from being compromised by exotic species, and losing the values that make them World Heritage. This work requires continuity over a 3–4-year period to ensure canopies are sufficiently re-established to start to exclude weeds. To date almost $1 million has been invested in weed management, particularly in areas suffering severe scorching and damage. This work is essential to suppress weeds and allow native plant species to resprout and recover.

Emergency bushfire recovery funding has been essential to undertake this management activity and has provided a platform to assess and observe the regeneration strategies of rainforest flora. However, questions remain about the capacity of these species to respond to, and recover from, multiple fire events, as protected area managers continue to navigate the cumulative impacts of a warming and drying climate.

Conclusions

NRM organisations have provided a significant coordinating mechanism in the recovery from the 2019–20 wildfires, mobilising immediate responses and localised and regional networks to understand the impacts of the fires and prioritise and undertake recovery actions. Many NRM organisations were able to mobilise rapidly to respond to needs that emerged in each region. At the same time, the extent and severity of the fires, and their personal and community impacts, placed incredible pressures in some regions on the capacity to assemble the people needed to do the vital work of recovering biodiversity and ecosystems, including placing a deep toll upon NRM staff, some of whom also held emergency roles, and who were often themselves directly affected as community members. One of the critical lessons from these fires is therefore the need to have more of a buffer of organisational capacity in place to manage bushfire recovery, to create resilience in our systems and build surge capacity for emergencies.

The rapid response of NRM organisations was made possible by the very welcome emergency funding from the Australian, state and territory governments. NRM organisations provided a vital mechanism for the rapid identification of needs and deployment of this post-fire emergency response. In some cases, greater flexibility in funding, for example to support coordination, implement management actions at the most useful times, or invest in priority actions that fall through the cracks of existing programs, would have supported even better targeting of these resources.

The 2019–20 wildfires had a particularly profound impact on Country and on the First Nations owners of Country, and drew significant national attention to the need for much greater First Nations involvement in recovery responses and actions to build resilience and preparedness. It is important that the weight of resources and processes addressing these needs are guided by and go directly to First Nations communities (see Chapter 33). At the same time, NRM organisations frequently work with First Nations communities and organisations on integrated landscape management, and provide another mechanism for supporting Indigenous-led recovery actions. Flexibility in the timing and deployment of funding is also important to ensure resources can be directed to the most important, genuine needs and priorities of Indigenous people, at the time they want it.

In some cases, such as the South Australian glossy black-cockatoo, the capacity of species and systems to persist in the face of these fires has depended on vital work undertaken in the past to build up their population and resilience. This work needs to continue if these species and ecosystems are to recover and develop the resilience to withstand future events. Many of the actions needed to recover biodiversity following the 2019–20 wildfires – from weed management to habitat restoration – will take many more years of active management. Regional NRM organisations are embedded in their regions and communities in the long term, and thus are well placed to build resilience and manage the longer-term recovery needs for ecosystems and species. As critical and valuable as the emergency response has been, the long-term recovery of biodiversity will require ongoing investment and effort.

Critical actions ahead include learning the lessons from these fires and using these to build preparedness for future events within and beyond NRM organisations. Vital knowledge continues to be built, for example on the responses of species and ecosystems to the fires, the impacts on cultural sites and places, and the systems needed to coordinate priority actions. In several of the cases discussed here, knowing what actions to take and developing a rapid response was guided by a strong base of pre-existing knowledge, for example of species populations and ecology, which provided a foundation to build on. But this knowledge base, and our investments in building resilience, is uneven across our species, ecosystems and landscapes. There are gaps in our collective knowledge and management efforts which we need to fill if we are going to be prepared for future events, not least how to build greater long-term resilience in the face of increasing risks emerging from a changing climate. In addition, it will be important to embed this knowledge in pre-approved emergency response and recovery plans to ensure the right actions can be taken as rapidly as possible. There is an urgent need to consolidate the lessons that continue to be learnt through this work, to invest in legacy systems that build and share this knowledge, and to strengthen our organisational and ecological systems, in order to increase our preparedness and build resilience for the future.

Recommendations

- Prepare organisational systems and capacities, including investing in systems resilience such as surge capacity, network-building, knowledge base and legacy knowledge-sharing systems to support prioritisation and action in emergency situations.
- Develop pre-approved, knowledge-based emergency response and assessment plans for species and ecosystems.
- Invest in building capacity and flexibility of funding to support Indigenous-led recovery response within and beyond NRM organisations (see also Chapter 33).
- Improve the rapid deployment and flexibility of emergency funding to ensure funding and timing of expenditure can be tailored to local priorities, conditions and the needs of the situation.
- Ensure funding systems include the capacity to invest in planning, collaboration and coordination activities post-fire to identify priority actions and build networked capacity.
- Address current gaps in funding for recovery and protection actions between major programs, such as to protect refuges, manage waterways, support non-threatened or locally threatened species and restore ecosystem function.
- Identify clearer lines of responsibility for planning and coordination between agencies, including lead agencies and lines of funding.
- Resource recovery actions for the medium to long-term to consolidate the immediate post-fire investments.
- Invest in efforts to build up resilience of populations, ecosystems and natural resources for future catastrophic events and long-term recovery.
- Address gaps in the knowledge base of species populations and ecology, and consolidate and build upon the knowledge gained through these fires to strengthen understanding and preparedness, support rapid response and identify effective action for future events.

Acknowledgements

The projects showcased here are supported by the Australian Government's Bushfire Recovery Package for wildlife and their habitat. The NSW Government supported several initiatives in the Murray and Riverina regions. Habitat restoration for glossy black-cockatoos was also supported by the World Wildlife Fund for Nature. HLW is indebted to the professionalism and competence of the contractors involved.

References

Berris K, Barth M, Mooney T, Paton D, Kinloch M, *et al.* (2018) From the brink of extinction: successful recovery of the glossy black-cockatoo on Kangaroo Island. In *Recovering Australian Threatened Species: A Book of Hope*. (Eds S Garnett, P Latch, D Lindenmayer and J Woinarski) pp. 75–84. CSIRO Publishing, Clayton.

Donatiu P, White J (2021) *Post-fire Regeneration Mechanisms of Rainforest and Wet Sclerophyll Flora in South East Queensland*. Healthy Land and Water, Brisbane, Queensland.

Government of South Australia (2020) *Independent Review into South Australia's Bushfire Season*. Government of South Australia, Adelaide.

Joseph L (1989) The glossy black-cockatoo in the south Mount Lofty Ranges. *South Australian Ornithologist* **30**, 202–204.

Kooyman RM (1996) *Growing Rainforest: Rainforest Restoration and Regeneration: Recommendations for the Humid Subtropical Region of Northern New South Wales and Southeast Queensland*. Greening Australia, Lismore.

26

Wildlife welfare and the 2019–20 wildfires

Sally L. Sherwen, Michelle Campbell-Ward and Rupert Woods

Summary

- Given the magnitude of the 2019–20 wildfires, there were exceptional numbers of affected animals.
- There was profound community concern for and media attention directed to the welfare of fire-affected animals.
- There were marked mismatches between some media and community expectations for treatment of fire-affected animals and the reality of established triage principles and welfare considerations.
- There were major taxonomic biases in response, with a preponderance of care given to larger and iconic mammal species.
- Review of the rescue and rehabilitation activities highlighted some notable successes in outcomes, demonstrated the need for and utility of the application of ethical and welfare-based responses, and provided some important advances in the evidence base.
- However, the scale of animal welfare concerns revealed major shortcomings and inconsistencies in existing response plans and systems, and limitations on the available expertise and knowledge base. These shortcomings need to be, and are being, addressed to ensure better preparation for future comparable environmental catastrophes.

Introduction

What is animal welfare?

Enhanced scientific understanding and acceptance of sentience in many species has led to a welcome growth in the recognition of animal welfare as a core scientific discipline. This in turn, has led to the topic of animal welfare rapidly maturing as a societal issue gaining public attention (Hampton and Teh-White 2019).

Animal welfare is a characteristic of an individual animal. It describes the quality of an animal's life as it is experienced by the animal. A scientific definition that adequately captures the complexity of welfare is Broom's (1986) definition that 'animal welfare is the state of an animal as regards its attempts to cope with its environment'. This definition incorporates both the feelings as well as physical health of an animal, but it also acknowledges that welfare is a state that can be measured and can sit on a scale from negative to positive.

Each species is equipped to cope with a range of relevant stimuli through behavioural and physiological processes. However, if an animal is in an environment where it is challenged beyond its tolerable limits, there will be failure to cope and the animal will experience biological consequences such as ill-thrift, compromised immune response and, in some cases, death.

A useful framework designed to support conceptualisation of animal welfare is the Five Domains model, which describes four physical domains (nutrition, health, environment and behaviour) and one mental domain (the affective state). It was originally developed in 1994 but has since evolved through multiple iterations to incorporate the latest science (Mellor *et al.* 2020).

Wildlife welfare

The dominant drive for advancements in animal welfare science has been to guide what humans should do to protect animals in their care and give them a good life. This has been almost exclusively focused on domesticated animals and captive wildlife. The welfare of free-roaming wildlife has not been the topic of enough debate or focus in the literature, despite being of significant public interest. This interest has grown as more people become aware of the substantial impact humans have on wild animals. The growing scale and impact of anthropogenic factors on wildlife and wild places have led to an increasing sense of moral obligation to help wildlife in need in an attempt to restore or counterbalance imposed adversity (Englefield *et al.* 2019). As news about wild animal suffering across media platforms increases, this trend is expected to continue.

Humans can impact wildlife in a variety of adverse ways, including disturbance or destruction of habitats, the spread of diseases, hunting and road trauma, as well as more indirect impacts through well-intended activities such as wildlife tourism and research (Berg *et al.* 2020). When the focus is on species-level interventions, welfare of an individual animal has historically been considered a lower priority. However, as our ethical reflection deepens and scientific techniques advance, opportunities to improve both individual animal welfare and conservation outcomes have expanded. Indeed, wildlife conservation and animal welfare share the common goal of preventing harm to wild animals. Beausoleil *et al.* (2018) describe how cross-disciplinary sharing and collaborative research are needed to advance our approach to conservation and animal welfare, calling for incorporation of both 'feelings' and 'fitness' in how we approach animal and environmental interventions.

Animal welfare risk factors associated with wildfires

Although many Australian animals have evolved mechanisms to sense fire long before it arrives, giving them time to flee or seek refuge, high-intensity, fast-moving fires, or fires occurring over a large area, can overwhelm the protection afforded by these adaptive behaviours. Exposure to fire can lead to heat stress, smoke inhalation, burns and other injuries sustained in attempts to escape the immediate area. Water and food sources for all species can be impacted, resulting in dehydration and starvation, risks that are heightened if the fire occurs during drought conditions. Habitat damaged by fires can limit shelter opportunities, exacerbating the possibility of environmental exposure and predation.

For animals that are removed from the wild, physical and psychological harms can result from capture, transport, examination and temporary housing in captivity, especially if operations are poorly planned, facilities are not species appropriate, and the people involved are untrained. While some animals appear to cope better than others, the development of objective assessment tools for Australian wildlife welfare in rehabilitation settings is still in its infancy.

Animals in the wild and in care can experience distress and fear, although overt signs may be masked. The ability to exhibit species-specific behaviour may be impaired by habitat degradation or inadequate captive housing. Pain and suffering may be prolonged if veterinary care is not sought or if there are delays in decisions to euthanise or release animals.

Some of these wildfire-induced welfare impacts are difficult to mitigate. However, with advances in welfare science and the growing recognition of the significance of human factors impacting wildlife welfare during these events, there is considerable opportunity for practices to improve.

The human dimension

A range of people is likely to come into contact with wildlife during and following a wildfire, including fire-fighters, government-employed officers, park rangers, wildlife rescuers and rehabilitators, veterinary professionals, members of the public and land owners. It is critical to recognise that all interactions with wildlife, either direct or indirect, in an emergency have the potential to influence welfare outcomes.

In natural disasters, there is often a strong sense of empathy for wild animal victims. Amid devastating losses, people typically feel an intense desire to help. Community support can buoy the spirits of animal care staff and help organisations to mobilise resources quickly while working in what are often highly stressful situations.

Although interventions are invariably well intentioned, in some cases offers of help can hinder effective response and may also have adverse animal welfare impacts. For example:

- People may attempt to rescue or treat animals without the appropriate knowledge and skills.
- The logistics of dealing with unnecessary consumable donations risks diverting precious resources from the front-line.
- Emotional attachments and unrepresentative media stories and accounts can lead to unrealistic expectations that every injured animal can be saved, regardless of prognosis.
- Interpersonal conflicts, and the temptation of self or group promotion, can deflect attention away from the relief of suffering and the provision of high-quality care for rescued animals.

To help navigate these risks, it is useful to view the situation from an animal-centric perspective, to recognise the circumstances that typically compromise animal welfare and to be cognisant of judgment-impairing factors. The Five Domains model provides a useful framework to support consideration of animal welfare risks during natural disasters. Table 26.1 utilises an adapted version of the model (with 'human interaction' added as a sixth domain) applied to wildlife affected by wildfire.

Table 26.1. Examples of key welfare risk factors for consideration to assist with decision making to improve welfare outcomes for wildlife affected by fire (adapted from Mellor *et al.* 2020).

Domain	Wildlife in the wild following a fire	Wildlife in care (e.g. hospitalisation or rehabilitation) after a fire
Health	• Injury/functional impairment as a result of the fire • Presence of illness or disease • Poor physical fitness or body condition • Presence of hazards that may predispose to injury or illness • Impact of fire retardants, chemicals and ash post-fire (particularly for aquatic species)	• Injury/functional impairment as a result of the fire • Presence of illness or disease • Poor physical fitness or body condition • Impact of medical treatments or inappropriate nutrition • Biosecurity risk of housing in proximity to other species/ individuals • Fitness pre-release • Inadequate pain management • Lack of access to appropriate facilities for species-specific medical treatments
Nutrition	• Restricted access to suitable drinking water and species-appropriate natural food of appropriate quality (e.g. vegetation availability)	• Restricted access to suitable drinking water and species-appropriate natural food of appropriate quality (and ability to consume that food) (e.g. lack of browse paste for inappetant koalas)
Environment	• Weather extremes • Habitat structure that impairs retreat from threats (weather or conspecifics) • Terrain, substrate and inability to disperse to other habitats	• Inappropriate facilities for medical treatment and recovery (inability to deliver medications, isolate from other animals) • Inappropriate sensory environment (e.g. lighting, noise exposure, presence of predators) • Inappropriate enclosure furnishings and features to allow animals to build strength pre-release
Behaviour	• Restricted opportunities to express complete range of normal behaviours • Behaviours indicative of injury, illness or pain (e.g. lethargy)	• Restricted opportunities to express motivated behaviours relevant for recovery (e.g. rest) and build pre-release fitness • Behaviours indicative of injury, illness or pain (e.g. lethargy)
Human interaction	• Lack of appropriately trained people to make informed assessments and judgments • Anthropogenic activities (e.g. roads, fencing, habitat destruction, noise, disturbance)	• Lack of appropriately trained people • Excessive human interaction, human dependency • Prolonged time in care
Affective state	Examples of associated mental experiences based on physical domains above: • minimise negative states: pain, discomfort, thirst, hunger, exhaustion, isolation, fear; • maximise positive states: exploration, foraging, social, satiety, vitality, calm, content, reward.	

Key findings

Animal welfare-based decision making during the 2019–20 wildfires

As the 2019–20 wildfires burnt across Australia, an estimated 3 billion native animals were impacted (WWF 2020; Chapter 12). There is no national standard for emergency response for wildlife in such catastrophic events, with each jurisdiction deploying variable approaches according to state or territory-based wildlife legislation and emergency response plans and frameworks.

Despite variation across jurisdictions, core principles and elements in a wildlife response can be applied. For example, the skills and knowledge of experienced wildlife volunteers and veterinary professionals should be recognised and utilised where needed. For animals requiring rescue, welfare needs to be central to decision making at multiple points along the chain of events from the first sighting of an animal in the field to its medical treatment, rehabilitation and eventual release back to the wild (Box 26.1).

Box 26.1. Building an evidence base: koala post-release monitoring protocol

While thousands of sick, injured and orphaned wild koalas are rehabilitated in Australia annually, few studies have investigated the effectiveness of rehabilitation practices.

Over the 2019–20 wildfires, 27 koalas were transferred to specialised wildlife hospitals managed by Zoos Victoria (Fig. 26.1). Following initial health assessment, the severity of burns necessitated ongoing pain assessment and the provision of analgesia, regular debridement and treatment under anaesthesia, diagnostic imaging, clinical pathology and daily monitoring of demeanour, movement, food intake and faecal output. The highly specialised diet, and importance of sufficient browse to maintain healthy gastrointestinal microbiome and bodyweight in koalas, presents a major rehabilitation challenge.

While burns to the skin generally responded to treatment, a progressive deterioration in mobility and function of burnt digits was seen in some koalas, sometimes accompanied by temporary nail loss. The pathogenesis of this condition, referred to as 'progressive digital necrosis' (PDN), is believed to be initiated by thermal injury to the deeper digital tissues at the time of the original burns.

Once hospitalised care was no longer required, koalas were moved to large, semi-wild enclosures at Healesville Sanctuary and Phillip Island Nature Parks. These enclosures, containing multiple mature eucalypt trees to allow development of strength and fitness, also provided a less stressful environment with significantly reduced human contact. Koalas were held in these enclosures for a minimum of 2

Fig. 26.1. Koala being treated at Healesville Sanctuary's Australian Wildlife Health Centre for injuries caused by bushfire: close-up of burns on rear paws once bandages removed. (Photo: Zoos Victoria)

months before release, where formal behavioural assessment showed considerable improvement in fitness, foraging and climbing ability.

In late 2020, 14 *Chlamydia*-negative rehabilitated koalas were found to be in excellent health and fitness, showing wild behavioural responses, and were fitted with radio and GPS collars before being released back into regenerated forest as close as possible to their original location (Fig. 26.2). Tracking frequency gradually reduced over the life of the monitoring period, but the intensive approach provided an excellent opportunity to gain detailed understanding of each koala's behaviour and welfare status once released. A sudden change in demeanour in three released koalas facilitated rapid, in-field veterinary assessment and euthanasia on the basis of clinical findings, preventing a slow deterioration in health and welfare for these individuals and enabling detailed necropsy examination. This led to findings of acute renal disease in two animals, and clinically significant chlamydiosis in one. Further research is required to understand the development of renal disease in released koalas.

Koalas were monitored for between 4–6 months post-release to the wild, and at the end of this period all remaining koalas were in excellent health. Assessment of GPS data showed individual variation in movement, but generally good choice in habitat and spatial behaviour, and the presence of pouch young in eight of nine released females indicated successful reintegration into wild populations.

Although a small sample, this project has provided unique insight into the rehabilitation and post-release survival and health of rehabilitated koalas. The intensive approach required to ensure return to normal function in severely burnt tissues highlighted the importance of constant assessment of welfare during veterinary care and rehabilitation. There is a delicate balance between providing the level of intervention required while minimising the impact of chronic stress on healing and animal wellbeing. Knowledge gained will be used to improve approaches to the rehabilitation and release of fire impacted koalas.

Fig. 26.2. Koala released back to the wild after treatment. This release was included in a monitoring study that tracked post-release survival. (Photo: Zoos Victoria)

Prognostic thresholds for wildlife cannot be directly compared to domestic animals with similar injuries or degrees of illness. If a dog or horse sustains a severe injury and requires intensive care and/or surgery, the assumption is that the animal is likely to remain in human care for the rest of its life, can be reassessed regularly and should receive medication or other therapies as needed. With wild animals, however, veterinarians, wildlife carers and managers can only deem an animal fit for release if chronic effects are not anticipated and no further intervention or assessment is expected post-recovery. Even if the initial prognosis is deemed fair to good, long-term welfare should be continuously re-evaluated throughout the various stages of rehabilitation. If it becomes clear that the animal will not be physically and behaviourally capable of independent life in the wild – whether that is due to severity of injury or illness, progression of an earlier problem, an inability to cope with captivity or an issue that has developed in care – euthanasia is appropriate.

Figure 26.3 provides a high-level outline of the government-led, collaborative response to assessment and management of wildlife that was deployed in Victoria during the 2019–20 wildfires. It shows the various points at which an animal moves through various assessment stages and the extensive range of welfare risk factors to consider. As a result, rather than survivorship increasing, the number of animals progressing through each successive stage declines as euthanasia may be deployed at any step to prevent current or future suffering. It is usually following initial assessment that euthanasia rates are highest. For example, following the 2019–20 wildfires in Victoria, around 3000 individual wild animals were assessed in the field with only 259 (10%) progressing through to field clinics for further assessment (Parrott *et. al.* 2021) of which 33 (1% of total) were transferred for more intensive veterinary care.

It is important to recognise that triage for wildlife casualties must also take into account factors such as relevant legislation, human health and safety, and the availability of resources and facilities.

The level of veterinary and rehabilitation care required to ensure optimal outcomes for wildlife represents a significant investment in time, expertise and resources. Severely injured or debilitated wild animals, such as those suffering severe burns or smoke inhalation, require specialised intensive care and a high level of ongoing veterinary support. The expenses associated with veterinary medicines and consumables (including advanced wound dressings, sedation, anaesthesia, analgesia and antimicrobial medicines), diagnostic imaging and clinical pathology costs, and the provision of a species appropriate diet, hospital and rehabilitation facilities are significant. An estimate from the 2019–20 wildfires of cost *per week per koala* for intensive care is $16 000. Financial costs can be reduced when treating and housing wildlife, but these reductions often come at the cost of animal welfare and may negatively impact on the success of reintegration of animals back into the wild.

Compounding the challenges, the husbandry needs of wild animal patients are incredibly diverse, demanding dedicated hospital and rehabilitation facilities with the capability to cater for a wide range of species. Appropriate enclosure sizes, construction materials, management of environmental variables (temperature range, humidity, UV exposure and photoperiod) and species-specific feed are all required. While there will always be a spectrum of care needs for wildlife impacted by emergency events, and a corresponding scale in associated costs, emergency response planning must include a transparent and realistic appraisal of the investment required before the decision to undertake treatment, care and/or rehabilitation and release goes ahead.

Species presented

Survey-based research of veterinary personnel and wildlife rehabilitators involved in the 2019–20 wildfire response revealed that only a very small proportion of affected animals were rescued, with a heavy species skew towards mammals, particularly larger

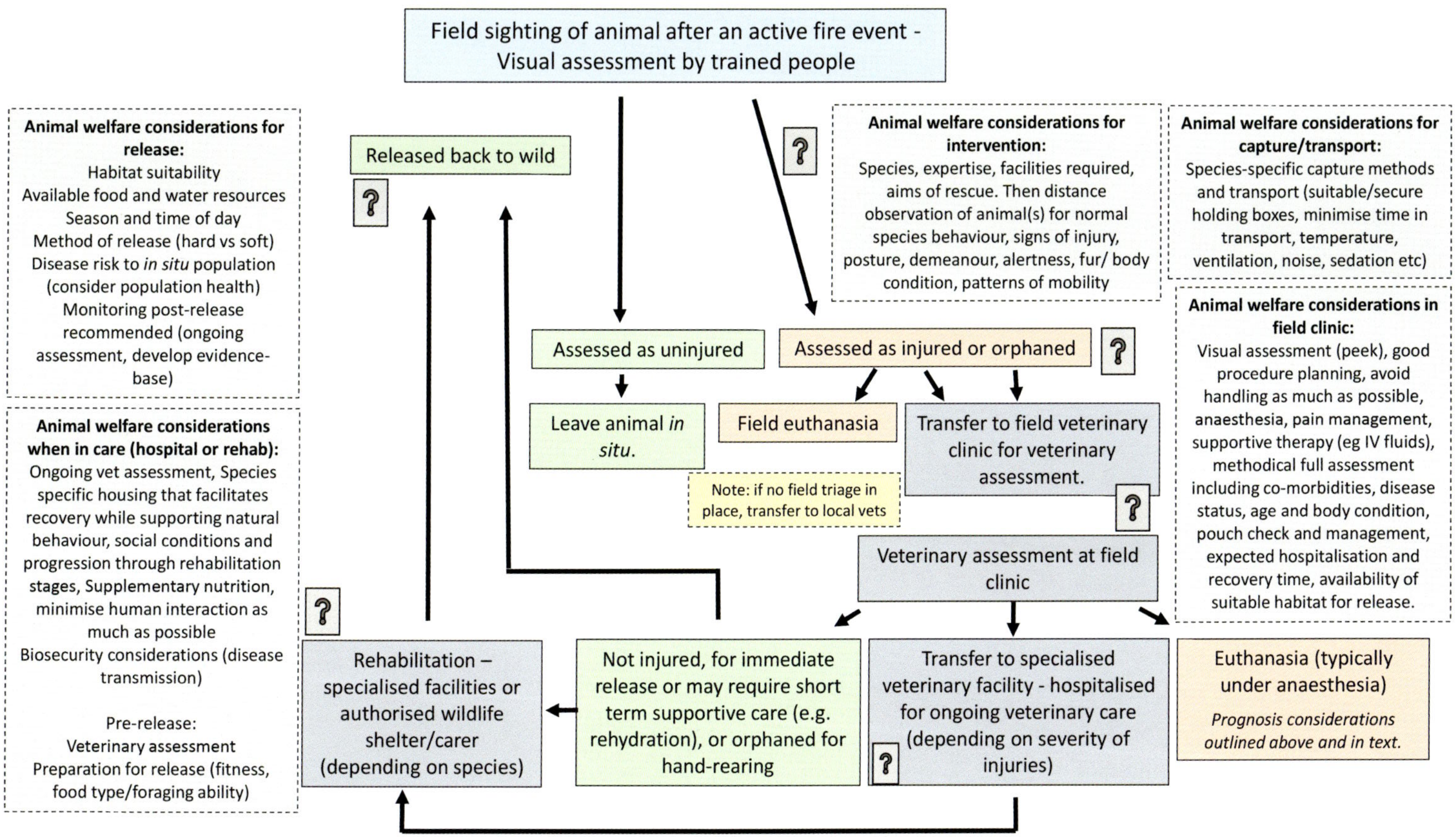

Fig. 26.3. An example of the various points of decision making for individual animal welfare and the relevant welfare risk factors considered during the 2019–20 wildfires in Victoria. '?' represents significant welfare decision points before moving to the next stage.

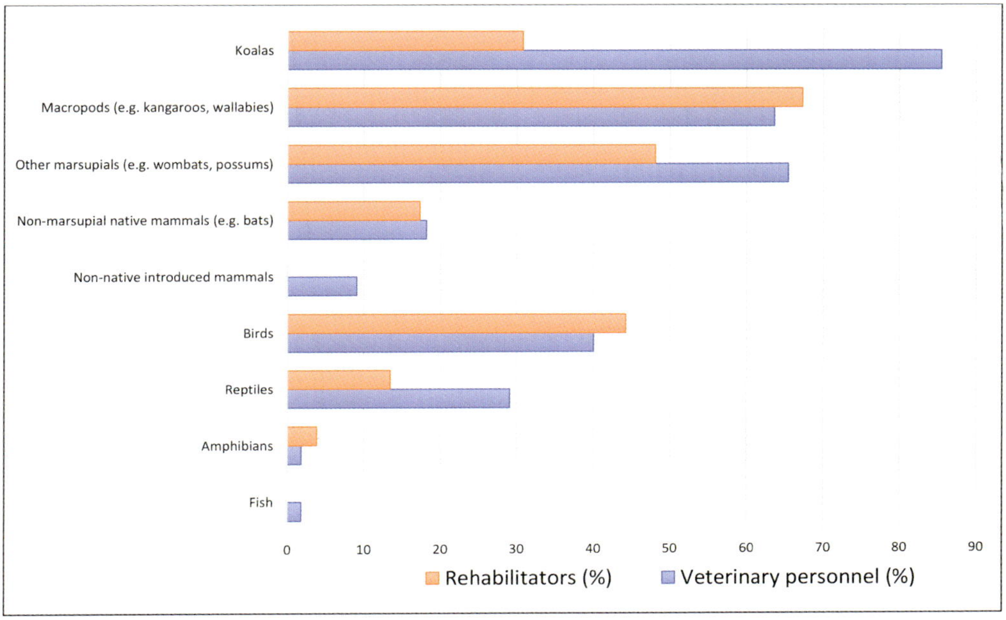

Fig. 26.4. Taxonomic caseload prevalence (%) of surveyed rehabilitators (*n* = 52) and veterinary personnel (*n* = 55) involved in the 2019–20 wildfire response (data pooled to include responses from all states and the Australian Capital Territory). Survey respondents indicated which species and or species groups were included in their caseload.

marsupials (Fig. 26.4). Rehabilitators reported caring for macropods more than any other group, whereas koalas (*Phascolarctos cinereus*) were the species most commonly assessed by veterinary personnel. Birds, reptiles and amphibians were brought into care but in much smaller numbers. Conservation status did not appear to play a role in the type of animals rescued regionally or nationally and the time in care of some victims was considerable.

There are several potential explanations for the taxonomic distribution of cases. Humans have a well-described empathy bias towards mammals (Miralles *et al.* 2019) and the fact that koalas and large macropods have high profiles is likely to have played a role. Koalas have limited options to flee a fire-affected area and may also have been easier to catch. Wild macropods are more mobile, but the severity of the fires meant that some were injured in attempts to escape. Pouch young may have survived when their mother became injured or died and, as a result, a proportion of the cases rescued were orphans.

The relative paucity of birds, reptiles and amphibians presented for assessment in comparison to the estimated numbers impacted warrants consideration. Due to their smaller body size and more cryptic nature, it is possible these species are simply not sighted by rescuers. It is also likely these smaller animals succumbed to post-fire predation due to their vulnerability without effective cover. Nonetheless, the welfare needs of these species remain poorly understood and they tend to display less recognisable signs of pain and distress. They may have been more challenging to rescue or members of the public may have been less motivated to pursue treatment for them if they were found injured.

Ethical considerations

Ethics is a philosophical discipline that attempts to discern how we ought to act, how we make decisions and what differential value we place on things. The response to wildfire-affected wildlife highlights how ethics related to animals can have a profound impact on animal welfare.

In terms of the ethical issues encountered in the 2019–20 wildfires, many are familiar to those working in the rehabilitation space, and others were novel to this event:

- There were accounts of conflict between wildlife care groups and individuals with a 'no kill' philosophy clashing with utilitarian advice that was aiming to achieve an overall 'good' outcome in situations with multiple competing interests.
- There was conflict regarding resource allocation, particularly to locally abundant species where treatment may have been of no or limited conservation value.
- There were perverse situations where eucalyptus branches for feeding rescued koalas were being cut from remnant unburnt bushland where unharmed koalas who needed this scarce and limited food resource remained (see also Box 26.2).
- In equivocal situations, there were reports that some animals were treated and response locations selected to elevate dwindling group morale rather than to address animal welfare.

The 2019–20 wildfires raised the issue of how much wild intervention is appropriate after a disaster. In terms of a 'respect for nature' ethic, how do we now define 'nature' and what is natural in the human-influenced era of the Anthropocene? For example, there was urgent public pressure for support of surviving free-living animals in wildfire affected areas through provision of food, water and shelter (see also Box 26.3). However, careful consideration must be given to how this is done to ensure there are not unintended consequences for wildlife and the environment (e.g. by encouraging the influx of introduced predators or contributing to the introduction of weeds). Multi-stakeholder groups were mobilised during the wildfires to advise government agencies and provide online resources for the community. Even among equally qualified and experienced professionals, opinions differed in terms of how much assistance was considered appropriate.

An effective response to a natural disaster relies on the coordination of many people. Individuals and groups may have to make decisions based on incomplete information in an emotionally charged environment with a high level of public interest. Moral distress and interpersonal conflict can arise from differing ethical viewpoints regarding the appropriateness of particular interventions, the relative value of species, the allocation of limited resources, and philosophies regarding animal euthanasia. Strategies for response personnel to successfully navigate these complex situations include respectful and clear communication; honest consideration and understanding of motivations influencing actions; reliance on evidence, expertise and professional judgment; and clinical debriefing.

Box 26.2. Supplementary feeding of wildlife

Throughout the 2019–20 wildfires, wildlife carers and veterinary professionals actively wishing to assist in the management of wildlife health and welfare worked in an environment of varying or absent coordination, standards and practices or supporting documentation. This affected the opportunities to provide appropriate care and to manage the health and welfare of native wildlife.

While many government agencies and wildlife organisations discourage supplementary feeding of free-ranging wildlife because of the potential adverse effects, the 2019–20 wildfires destroyed enormous areas of habitat, making it difficult for free-ranging wildlife to access the normal sources of food and water both during and after the fires. In recognition of the need for nationally harmonised advice, Wildlife Health Australia (WHA) established a small *ad hoc* Food and Water Working Group, which included ecologists, wildlife veterinarians, wildlife nutritionist, non-government agencies (zoos, universities and wildlife rehabilitation organisations) and representatives from state government to develop advice on the safe provision of food and water to free-living wildlife based on current knowledge. The working group produced several key outputs during the wildfires, including:

- general advice on the supply of water and food for free-living wildlife after natural disasters;
- detailed food and water advice for use by each state or territory environment agency as deemed appropriate for the wildlife impacted within their jurisdiction;
- bespoke advice for key threatened species, as requested; and
- advice on when supplementary food and water are no longer required by wildlife.

Despite the wealth of knowledge held by subject matter experts on the working group, there was a critical absence of evidence-based research to support and inform the advice developed. Many of the critical knowledge gaps remain. Addressing these gaps through an evidence-based, cross-disciplinary approach, while also developing a framework to facilitate dissemination of best practice guidelines to users across Australia, is required. Specifically, this research will provide the much-needed evidence to assist development of best practice guidelines for provision of food and water in advance of the next natural disaster. An added benefit from this work was the subsequent development of a Wildlife Assist app for citizen science engagement on this issue. Wildlife Assist is an app that records, monitors and advises on providing Australian wildlife with food, water and shelter in the field.

Box 26.3. Media and social media impacts on welfare of rescued wildlife

Communication is a vital component of any emergency disaster response. The media, and increasingly social media, play a vital role in facilitating communication.

Animal-related imagery, language and story can have a powerful influence on the perception of the impact of disaster on wild animals and ultimately on animal welfare. There is great potential for wild animal welfare to be enhanced via timely notification of animals at risk, dissemination of information, sharing of decision-making tools and garnering valuable financial and practical support. However, misleading information about on-ground activities and the desire to elicit a powerfully emotive connection can have unintended consequences for the wellbeing of animals.

An analysis of media outlet front pages over the period of the 2019–20 wildfires concluded that the domestic media's imagery largely portrayed a humanitarian disaster.

Conversely, international media images focused on the impacts on iconic fauna with 52% of all coverage depicting wildfire-affected landscapes or animals. In fact, international outlets featured animals more than 10 times as much as the Australian front pages, with kangaroos and koalas over-represented (Thomson 2021). Similarly, the majority of online wild animal images illustrating the disaster were of living, but seemingly helpless, koalas and macropods. In reality most victims perished, fled or were so badly injured that euthanasia was the most appropriate intervention on welfare grounds.

In the social context of immense loss, there was a desperate hunger for good news. As such, stories of compassion and heroism in animal rescue and the personalisation of animal casualties emerged. There was little discussion of the valuable and highly collaborative work being done by experienced veterinary teams around the country to relieve suffering nor the fate of severely harmed wild animals or those with injuries that would preclude release. Decision making in contemporary wildlife rehabilitation practice is often sanitised in promotional material and media stories. This false impression of the numbers of charismatic victims, their likelihood of survival and the level of available local expertise resulted in prolonged suffering due to distrust of professional euthanasia recommendations and an influx of well-meaning but untrained volunteers who had little understanding of Australian wildlife care needs, codes of practice or triage principles. User-generated social media content provided a mechanism for citizens and organisations to express concerns and values, but in an environment of celebrated heroism and high levels of donation it also, in some cases, invited self or group promotion and competition at the expense of animal welfare.

Images and videos that 'went viral' often appealed to the human–animal bond, depicting people in the process of rescuing or caring for wildlife, without acknowledging that close contact between wild animals and humans can be intensely stressful for animals. Some showed poor judgment, with people posing for media interviews with critically injured animals, delaying their transfer for urgently needed veterinary care; overtly staged images of marsupial orphans being fed in situations that did not represent established standards of care; and the postponement of release for recovered victims to meet publicity obligations. As has been demonstrated in the tourism industry, takers and viewers of wildlife 'selfies' often do not perceive the negative state experienced by the animal during the creation of such an image (Lenzi *et al.* 2020).

An improved understanding and appreciation for wild animal welfare in the context of disaster response is necessary for both the creators and consumers of media and social media content. Mechanisms to enhance the prominence and reach of accurate and ethical messages to this end remain poorly understood. The influence of digital psychological nudges such as framing, social norms, and striking visual and commitment cues have been identified as relevant to disaster communications (Mirbabaie *et al.* 2021), and their role in animal welfare impacts warrants further research.

Conclusions

The disastrous wildfire season of 2019–20 is unlikely to be a stand-alone event. The size, scale and frequency of wildfires has been steadily increasing over the last several decades. As such, we need to consider the events a learning opportunity and driver to enhance preparedness, and embrace courage and innovation in reimagining how we support both people and wildlife through future emergencies.

There are significant gains to be made if we can shape this work into a more holistic, transdisciplinary approach that embraces all knowledge systems. There is an urgent need to tackle the problems collectively as community, environmental and wildlife welfare outcomes are inextricably connected.

Recommendations

Though much good work has been done, the ability to effectively coordinate every level of the currently largely uncoordinated wildlife care sector will be crucial in the future. Partnerships and collaboration are key. Priority recommendations to enhance preparedness that will underpin effective response and recovery activities include the following:

- Develop a national framework for emergency wildlife response that integrates dedicated response roles into the emergency management arrangements of each jurisdiction, including role statements and underpinning skill prerequisites.
- Develop and set clear standards to guide decision making, including:
    - ethical guiding principles for assessment and intervention, developed in advance of an emergency to facilitate preparedness and planning (acknowledging the many factors that can cloud judgment during the event)
    - species-specific veterinary and care documentation for wildlife, including policy, standards and protocols to support an effective, consistent and timely approach.
- Build capacity and capability – identify and train a targeted cadre of on-ground wildlife responders and supporting managers for deployment in specific wildlife response roles. Training should be grounded in the latest animal welfare science. This should be a prerequisite for any involvement in wildlife response during an emergency.
- Foster a continuous improvement philosophy. As knowledge advances, so should our standards and skills for wildlife.
- Build a stronger evidence base – invest in research dedicated to better understand the effectiveness of different rehabilitation approaches, prognostic indicators and animal welfare impacts of various interventions, and build individual and population health knowledge associated with emergency response to identify potential risks.
- Engage with community – have open, transparent conversations about living with wildlife, animal welfare-based decision making and how we can foster a more productive human-nature relationship.

Acknowledgements

We acknowledge the Traditional Owners of the land on which this work occurred, and pay respects to Elders past and present, and extend that respect to all Aboriginal and Torres Strait Islander peoples. We recognise their deep connection to country and acknowledge sovereignty was never ceded. This continent always was and always will be Aboriginal land. We thank them for protecting these ecosystems since time immemorial.

For contributing their expertise: Tiggy Grillo (Box 26.2), Leanne Wicker (koala Box 26.3) and Fiona Ryan (knowledge and expertise into Fig. 26.3 development).

We express our deep gratitude to the thousands of dedicated volunteers, veterinary professionals, members of the public and others who worked exhaustively over the 2019–20 summer and beyond to prevent suffering of wildlife where they could and support both community and nature recovery.

Much of the work described above is funded through donations. We thank the state and Commonwealth governments, non-government organisations, zoos and wildlife advocacy groups, philanthropic organisations, and members of the public from around the world for their support and assistance to help wildlife.

References

Beausoleil NJ, Mellor DJ, Baker L, Baker SE, Bellio M, *et al.* (2018) 'Feelings and fitness' not 'feelings or fitness' – the raison d'être of conservation welfare, which aligns conservation and animal welfare objectives. *Frontiers in Veterinary Science* **5**, 296. doi:10.3389/fvets.2018.00296

Berg C, Lerner H, Butterworth A, Walzer C (2020) Editorial: Wildlife welfare. *Frontiers in Veterinary Science* **7**, 576095. doi:10.3389/fvets.2020.576095

Broom DM (1986) Indicators of poor welfare. *The British Veterinary Journal* **142**, 524–526. doi:10.1016/0007-1935(86)90109-0

Englefield B, Blackman SA, Starling M, McGreevy PD (2019) A review of Australian animal welfare legislation, regulation, codes of practice, and policy, and their influence on stakeholders caring for wildlife and the animals for whom they care. *Animals (Basel)* **9**, 335. doi:10.3390/ani9060335

Hampton JO, Teh-White K (2019) Animal welfare, social license, and wildlife use industries. *The Journal of Wildlife Management* **83**, 12–21. doi:10.1002/jwmg.21571

Lenzi C, Speiran S, Grasso C (2020) 'Let me take a selfie': implications of social media for public perceptions of wild animals. *Society & Animals* doi:10.1163/15685306-BJA10023.

Mellor DJ, Beausoleil NJ, Littlewood KE, McLean AN, McGreevy PD, *et al.* (2020) The 2020 five domains model: including human–animal interactions in assessments of animal welfare. *Animals (Basel)* **10**, 1870. doi:10.3390/ani10101870

Miralles A, Raymond M, Lecointre G (2019) Empathy and compassion toward other species decrease with evolutionary divergence time. *Scientific Reports* **9**, 19555. doi:10.1038/s41598-019-56006-9

Mirbabaie M, Ehnis C, Stieglitz S, Bunker D, Rose T (2021) Digital nudging in social media disaster communication. *Information Systems Frontiers* **23**, 1097–1113. doi:10.1007/s10796-020-10062-z

Parrott ML, Wicker LV, Lamont A, Banks C, Lang M, *et al.* (2021) Emergency response to Australia's Black Summer 2019–2020: the role of a zoo-based conservation organisation in wildlife triage, rescue, and resilience for the future. *Animals (Basel)* **11**, 1515. doi:10.3390/ani11061515

Thomson TJ (2021) Picturing destruction at home and abroad: a comparative visual analysis of icons and news values during disaster. *Media International Australia* **181**, 197–216. doi:10.1177/1329878X211008181

WWF (2020) Australia's 2019–2020 bushfires: the wildlife toll. World Wide Fund for Nature,, <https://www.wwf.org.au/what-we-do/bushfire-recovery/in-depth/resources/australia-s-2019-2020-bushfires-the-wildlife-toll>.

27

Ex situ responses to the 2019–20 wildfires

James R. Biggs, Jenny Gray, Judy West, Damian Wrigley, Catherine A. Offord, Heidi C. Zimmer, Berin D. E. Mackenzie, Dan Duval, Jenny Guerin, Thai Te, Marissa L. Parrott, Kim A. Miller, Michael S. McFadden and Monique Van Sluys

Summary

Context and challenges

- The International Union for the Conservation of Nature (IUCN) recognises the critically important role of *ex situ* conservation interventions and recommends that all threatened species should have *ex situ* options included in their recovery plans to allow for strategic and timely responses in times of increased risk.
- In Australia, many organisations are active in *ex situ* conservation, including botanic gardens, seed banks, zoos, aquariums, universities, research facilities and government agencies.
- With catastrophic events, *ex situ* options can be useful in securing species, by removing individuals from direct and secondary impacts or assisting species to recover *in situ*.
- Investment in *ex situ* facilities, programs and research prior to catastrophic events is a key indicator of likely success.
- Decisions to undertake *ex situ* efforts must be made strategically before, during and after catastrophic environmental events. However, if no previous effort has been made to secure *ex situ* populations or collections, the post-disaster period is the worst time to activate an *ex situ* rescue response. The numbers of individuals will be low and costs will be significantly higher.
- *Ex situ* response is not as simple as collecting and propagating. Significant time is needed to learn the species biology, care of animals and the science/art of consistently reproducing animals, as well as to unravel the science/art of germinating, propagating and/or storing plant species.

Main findings

- Many *ex situ* options were utilised for animal and plant species affected by the 2019–20 wildfires: existing *ex situ* recovery programs were boosted, new *ex situ* recovery programs were established and additional facilities created.

- Expert elicitation identified fire-affected species that would benefit most from *ex situ* support, and these were the primary focus of the *ex situ* response.
- Early indications reflect that the increased support for *ex situ* management options allowed for the protection of many fire-affected animal and plant species and is buying time for many more. In coming years, the impact of this work will bear witness to the contribution of *ex situ* options to conservation of many species.
- Partnerships and collaborations should be pursued and nurtured as an ongoing concern so that rapid, coordinated and collegiate responses to emergencies can be implemented without the need to form new relationships during times of crisis. These networks should include complementary expertise from both the *in situ* and *ex situ* community.
- Limitations on the ability to mobilise sufficient skills and facilities in the 2019–20 wildfires demonstrate the need for a more significant and strategic investment in *ex situ* capacity and skills.

Introduction

The 2019–20 Australian wildfires were unique in their scale and intensity. The long-term impact of the fires on species survival is still unknown; however, for many animal species over 80% of known habitat was affected by fire (see Chapter 11), and more than 800 or 3.5% of Australia's vascular plant species had more than 50% of their populations or ranges burnt, in vegetation communities ranging from rainforests, eucalypt forests and woodlands to heathlands (Godfree *et al.* 2021).

Effective conservation requires that a diversity of management actions is considered and tested using a cross-disciplinary and multi-institutional approach. As habitats continue to be impacted by a complexity of unabated threats, there is an increasing need for a 'one plan approach' (Byers *et al.* 2013) that integrates *in situ* and *ex situ* management processes (Schwartz *et al.* 2017; IUCN 2020; UNEP 2021).

Increasing rates of extinction, exacerbated by climate change, challenge the dependence on *in situ* strategies and necessitate increased development in *ex situ* approaches (Bowkett 2014). In some instances, *ex situ* management will be a primary strategy, while in others it will be of secondary importance, supporting other interventions (IUCN/SSC 2014).

Ex situ facilities deliver many strategies and tools that cannot be replicated solely in *in situ* settings. These include frozen zoos; seed banks and botanic gardens specialising in research geared towards maintaining germplasm and germline samples in perpetuity; hospitals and emergency response facilities for treatment and ongoing care and rehabilitation of animals; traditional zoos and botanic gardens providing education and advocacy for species and intensive genetic management; as well as larger fenced sites like sanctuaries and landscape isolation projects supporting demographic management of populations.

IUCN/SSC (2014) states that *ex situ* conservation has the potential to:

- *Address primary threats.* '*Ex situ* activities can help reduce primary threats … when specifically designed conservation research, conservation training or conservation education activities directly and effectively impact the causes of these threats.'
- *Offset the effects of threats.* 'Small populations … may require some form of intensive management of individuals and populations to improve demographic and genetic

viability and avoid extinction. Challenges faced by small populations can be counteracted by ... population management options, such as head start programmes to address high juvenile mortality, or population reinforcement to balance age and sex distribution.'

- *Buy time.* 'Establishment of a diverse and sustainable *ex situ* rescue or insurance population may be critical in preventing species extinction when wild population decline is steep and the chance of sufficiently rapid reduction of primary threats is slim or uncertain ...'
- *Restore wild populations.* 'Once the primary threats have been sufficiently addressed, *ex situ* populations can be used for population restoration ... or conservation introduction ...' (see Boxes 27.1 to 27.4 for examples)

Box 27.1. Protecting the mountain pygmy-possum and preparing for future emergencies

The total population of the Critically Endangered mountain pygmy-possum (*Burramys parvus*) (MPP) is fewer than 2000 individuals, in three remote alpine regions, plus one captive breeding and research program at Zoos Victoria's Healesville Sanctuary (DELWP 2020). While key threats to the MPP, such as historic destruction of boulder-field habitat, introduced predators and climate change, are well known (DELWP 2020), a new threat was realised in 2017. The MPP's main food source is the migratory Bogong moth (*Agrotis infusa*), which usually arrives in the alpine zone each spring in billions. However, in spring 2017 and 2018, the moths largely failed to arrive (Green *et al.* 2021). In all monitored Victorian populations, 50–95% of MPP females lost their litters of pouch young by the end of December 2018 (DELWP 2019), with starvation of young the most likely cause of mortality.

In response, Zoos Victoria, working with experts in veterinary nutrition, successfully developed a new shelf-stable nutritionally suitable supplementary food, called 'Bogong Bikkies'. The bikkies and feeders were trialled for safety and efficacy with

Fig. 27.1. Feeder station (A) established *in situ* for mountain pygmy-possums (B). (Photos: Zoos Victoria)

MPPs in the captive program and with a wild MPP population (Fig. 27.1), with work continuing under emergency services approval throughout the 2019–20 wildfire season (Parrott *et al.* 2021). The bikkies were commercially developed to human grade food safety standards through Zoos Victoria's partners, Wombaroo (Wombaroo Passwell, Australia), to be readily available in an emergency.

While the food was developed in response to low moth numbers, this preparedness proved crucial following the fires. Populations of MPPs in Mt Kosciuszko National Park were left without food following a firestorm in early 2020. Zoos Victoria sent Bogong Bikkies and feeder prototypes to staff in the NSW Department of Planning, Industry and Environment's Saving our Species team, to feed MPPs after the fires (Parrott *et al.* 2020). Due to this supplementation, MPPs in fire-affected areas were able to gain weight enough to successfully hibernate and raise their young in the following years.

Ex situ facilities exist as an extensive, cooperative network across Australia, with a reach extending globally through memberships with the World Association of Zoos and Aquariums and Botanic Gardens Conservation International.

Successful *ex situ* facilities have extensive experience in consistently and predictably breeding, propagating and caring for or storing living and frozen collections of species from hundreds of taxonomic groups. Ongoing work aimed at shifting the scope of species that are maintained *ex situ* is now mostly guided by strategic decision-making, current and emerging conservation needs, key threatening processes and expert elicitation.

Box 27.2. Dramatic growth in Bellinger River snapping turtle hatchlings

The Bellinger River snapping turtle (*Myuchelys georgesi*) is a Critically Endangered species, endemic to the Bellinger River drainage in northern New South Wales (Fig. 27.2). The species was locally abundant in the river system, numbering from 1500 to 4000 individuals, before its sudden and rapid decline in early 2015. At this time, a novel virus (Zhang *et al.* 2018) entered the river system, causing a mass mortality event and resulting in the extirpation of the adult population, leaving only juvenile turtles behind. Today, there are fewer than 150 turtles left in the river and the species is dependent on conservation intervention for its persistence.

At the time of decline, a small insurance population was collected from the upstream stretch of the river before the arrival of the virus at this site. After a year of quarantine at Western Sydney University, a population was established at Taronga Zoo. The breeding of these turtles has been successful and since 2017 over 165 turtles successfully hatched. This has allowed for the translocation of 52 zoo-bred turtles to the wild. Radio-tracking studies undertaken by NSW Department of Planning, Industry and Environment have indicated that survival of released juvenile turtles has been very high (> 95%), demonstrating the need for ongoing translocations from the zoo population as a conservation tool to recover this species.

The synergistic pressure of the 2019 drought and wildfires exerted further stress onto the remaining wild population by causing suboptimal environmental conditions

Fig. 27.2. Bellinger River snapping turtle. (Photo: Paul Fahy)

in the stream. As a result, the need for rebuilding the remaining population with zoo-bred turtles has become much more urgent.

With the support of a Commonwealth grant, holding facilities for hatchlings and juveniles are being expanded at Taronga Zoo and Symbio Wildlife Park; these will allow the increase in the size of the zoo-based population, kept both as an insurance population and for release to the wild. The expansions will allow for a 600% increase in release capacity to at least 210 individuals per year, which exceeds the number required to ensure the persistence of the species in the wild.

While there are many commonalities in operations, objectives and constraints, the existing *ex situ* efforts for animals and plants present some important differences (Table 27.1).

Table 27.1. Commonalities and differences between *ex situ* activity for Australian animals and plants.

Adapted from Commander *et al.* (2018); APC (2021); DAWE (2021); Martyn Yenson *et al.* (2021). Note that this tabulation for animals mostly focuses on vertebrates, and we recognise that *ex situ* conservation effort for most Australian invertebrate species is under-developed.

	Animals	Plants
Diversity	~8000 Australian vertebrate species, and ~320 000 invertebrate species (Chapman 2009).	~23 000 Australian vascular plant species (Chapman 2009; APC 2021).
No. threatened spp. under national legislation (pre-fire)	399 vertebrates and 67 invertebrates. At least 24% represented in living *ex situ* ZAA-accredited member populations (ZAA 2020).	1373 plants. At least 67% represented in seed banks and botanic gardens (DAWE 2021).

continued

Table 27.1. Continued

	Animals	Plants
Ex situ activities	Temporary removal from wild to protect from imminent threat. Research into biology, ecology, husbandry, behaviour, reproduction and genetic diversity. Short and long-term care of living animals in controlled environments. Demographic manipulation – 'head-starting' or removal of individuals from the wild reducing mortality during a specific life stage and subsequently returning them to the wild. Breeding for population restoration and assisted colonisation. Controlled outbreeding to improve genetics. Public engagement, education, advocacy and campaigns to influence behaviour-change. Translocation is increasingly common; listed threatened species require relevant permits.	Living plant collections on display in botanic gardens. Research into biology, ecology, parasitism, reproduction and genetic diversity. Short- and long-term storage in nurseries, seed banks and botanic gardens. Long-term storage of orthodox species (tolerant of traditional seed banking methods – sub-zero temperatures and low relative humidity). Storage in cryopreservation facilities in seed banks for orthodox and non-orthodox species. Maternal line collecting to maximise genetic integrity of seed collections. Seed production areas increasing seed supply and bolstering *ex situ* collections. Seedling establishment and propagation for restoration and translocation. Public engagement, education, advocacy and campaigns to influence behaviour-change. Translocation is common; listed threatened species require relevant permits.
Storage	Zoos and aquariums maintain demographically stable and genetically representative populations of species requiring targeted conservation actions in exhibited and non-exhibited facilities. Frozen arks maintain biological materials at ultra-low temperatures, or as freeze-dried material securing long-term integrity and relevance for future research and conservation.	Botanic gardens maintain collections of living plants for display and conservation. Some collections maintained in controlled conditions in glasshouses or nurseries. Seed banks maintain living collections of seed and other germplasm (such as cuttings, seeds, tissue culture and embryos) in controlled conditions from ambient conditions to low humidity and ultra-low temperatures.
Reproduction and constraints	Specific breeding times and environmental triggers may be hard to replicate and are often unknown. Many mammal and frog species take several years to mature to breeding age and generation lengths can vary significantly. These timeframes can impact the ability to repopulate quickly after a crisis. Need to secure release sites for offspring to avoid bottlenecks and pausing of breeding.	Seeds can be germinated readily if physical and/or chemical dormancy cues are known and can be replicated *ex situ*. Plants can be propagated rapidly, with juveniles used in botanic gardens and restoration or translocation projects. Generation times can be short for some species or significantly longer for others. Regular propagation of plant material retains skills and knowledge to maintain living populations. Restoration or translocation sites selected for conditions and ecosystem values favourable for reproductive success, including proximity to extant populations to limit inbreeding.
Survivorship	Increased survivorship compared to *in situ* for many species.	Orthodox seeds held for decades or centuries when stored in appropriate conditions. Non-orthodox seeds often short-lived in storage. Cryopreservation can prolong survivorship in storage otherwise regular replacement of collections is required.

	Animals	Plants
Emotional connections between animals and plants	High emotional connections, particularly for mammals and birds during a crisis.	Low emotional connections, allows for strategic planning and conservation actions without risk of compromising an individual plant's mental or emotional health.
Genetic diversity	Zoos use complex pedigree management and modelling software to intensively manage the genetics and demographics of animal populations, and increasingly utilise molecular genomic technologies to inform strategies that increase founder representation, maintain high genetic diversity and decrease inbreeding.	Botanic gardens often hold limited genetic diversity in historical display collections due to limited individuals on display and/or diversity of original source material. Seed banks maximise genetic diversity for target species by collecting from multiple sites across a species range, maintaining separate collections from each maternal line. Maintaining records and depositing voucher specimens with registered herbaria provides confidence in species-specific data for *ex situ* germplasm collections.
Management	Metapopulation management, whereby movement of specific individuals between known populations is managed/ augmented by humans, with a steady exchange of genes and individuals from *in situ* to *ex situ* and vice versa.	Population management through restoration or translocation of multiple individuals from different maternal lines. Introductions from multiple source populations improves genetic diversity, reduces bottlenecking and improves the adaptation potential of populations. Seed collections made from restored or translocated populations for storage and future use.

Post-fire action

Animals

Very quickly following the fires, the Wildlife and Threatened Species Bushfire Recovery (WTSBR) Expert Panel (see Chapter 22), which included one *ex situ* expert, identified 119 vertebrates as the highest priorities for urgent management intervention. Most of these species had at least 30% of their range burnt, and many had substantially more (see also Chapters 13–16).

The Zoo and Aquarium Association Australasia (ZAA) customised an expert elicitation process to identify among these fire-affected species those that would be most likely to benefit from *ex situ* support. Operational and biological factors were considered in this prioritisation process, including experience working with the species; historic success in an *ex situ* setting; diet replicability and availability; current and planned capacity of facilities; existing expertise; costs of environmental control; spatial requirements; social structures; whether or not *ex situ* action was recommended in a statutory conservation plan; and species fecundity, including the likely number of breeding events and reproductive potential.

This prioritisation process reduced the initial list of 119 priority vertebrate species to 49 species that were candidates for *ex situ* actions. Prioritisation exercises with the WTSBR Expert Panel, state representatives and *ex situ* organisations reduced the list to 11 species for which the partners had confidence and actions to deliver animals back to the wild inside a 12–18-month timeframe.

Projects supporting these 11 species were funded by the Australian Government, individual zoos and aquariums, private donors and the ZAA. While it is too early to assess the impact of this investment, all projects will be monitored and their effectiveness reported (Table 27.2). Some other important *ex situ* responses to fire are described elsewhere in this book: for example, Chapter 6 reports on the rescue, safe-keeping in custom-designed facilities, then return to the wild of some highly threatened fish and spiny crayfish species imperilled by sedimentation following the 2019–20 wildfires.

Plants

The impact of the wildfires on Australia's flora was extensive (Chapter 9). The increasing frequency and intensity of fires is of concern for the recovery and long-term survival of Australia's floral diversity (Fairman *et al.* 2016). In response, the WTSBR Expert Panel identified 486 fire-affected plant species requiring immediate interventions to enhance their prospects of recovery (Gallagher 2020). Historical records held by environment agencies, herbaria, botanic gardens and seed banks provided valuable data on the distribution and phenology of target species to inform the recovery prioritisation process.

Table 27.2. Actions and desired outcomes of post-fire response for priority animal species.

***Ex situ* management action**	**Typical actions**	**Desired outcomes**	**No. of species prioritised for *ex situ***
Expand an existing ZAA-led or ZAA member-led insurance program, to more release-to-the-wild events.	Identify stakeholders and resource requirements. Supplement genetics to bolster *ex situ* population. Continue ZAA-led intensive genetic and demographic management and rapid release to appropriate wild sites. Manage new animals to limit adaptation to *ex situ*.	Rapid reproduction of F0 and F1. Rapid release of F2+ to appropriate wild sites. Wild sites supplemented. Historic sites repopulated.	6 species
Acquire knowledge on husbandry and care, to facilitate the later (short-term) development of a new insurance population.	Identify stakeholders and resource requirement. Develop infrastructure. Salvage appropriately sized group to found *ex situ* population. Research technical methods to support consistent reproduction. Document techniques.	Optimal techniques documented. Species consistently reproduced. Advance to insurance program.	31 species
Develop a new ZAA-led insurance program, to facilitate ongoing release-to-the-wild events.	Use a genetically appropriate group to found *ex situ* population. Develop infrastructure. Analyse for disease on incoming group. Apply ZAA-led intensive genetic and demographic management. Manage to limit adaptation to *ex situ*.	Rapid reproduction of F0 and F1. Rapid release of F2+ to appropriate wild sites. Wild sites supplemented. Historic sites repopulated.	12 species

The Australian Seed Bank Partnership (the Partnership), a program of the Council of Heads of Australian Botanic Gardens, worked with herbaria, environment agencies and local experts and communities to better understand impacts and prioritise species for targeted *ex situ* interventions.

A total of 1004 conservation actions addressed many of the priority fire-affected plant species through the Partnership's projects (Table 27.3). Actions include rapid flora assessments and seed collecting; genetically diverse collections across a species range; and determining germination protocols and long-term viability of collections in seed banks (see Boxes 27.3 and 27.4). Furthermore, the use of seedlings in restoration was prioritised for local and community-led plantings where species recovery was slow and in translocation trials where recovery *in situ* was hampered by ongoing threats from pests or disease.

Table 27.3. Actions and desired outcomes of post-fire response for priority plants species (the Partnership 2021, unpublished data).

***Ex situ* management action**	**Typical actions**	**Desired outcomes**	**No. of species addressed**
Rapid flora assessments (RFA)	Identify *in situ* populations of target species. Assess target species against RFA criteria. Databasing.	Improved knowledge of fire impacts and species response. Informed species prioritisation in future fire preparedness and response. Digital reference images of seedling emergence to aid future post-fire identification.	73 species assessed 2020–2021. Up to 252 species assessed 2020–23.
Seed collecting	Determine target species' phenology. Identify optimal reconnaissance and collecting opportunities. Allocate resources for field work and laboratory processing. Coordinate volunteers and land managers for access. Databasing.	Increased representation of species diversity and within-species diversity in *ex situ* collections. Increased number of species in *ex situ* collections available for research and restoration.	128 species secured 2020–21. Up to 388 species targeted 2020–23.
Germination	Schedule trials with facility priorities and research. Do plate germination trials; monitor and score progress. Databasing.	Confirmation of physical and chemical dormancy. Development and refinement of germination protocols. *Ex situ* collections available for restoration, translocation and botanic gardens living collections.	137 species trialled 2020–21. Up to 321 species trialled 2020–23.
Propagation, Restoration, Translocation, introductions to Living Collections in Botanic Gardens	Identify and secure resource requirements. Pot germinants and cuttings. Monitor and maintain propagants. Treat for pests and/or disease. Databasing.	Propagation of threatened and priority species. Restoration of fire-impacted populations. Translocation of species impacted by wildfires. Planting of threatened and priority species in botanic gardens for *ex situ* conservation, education and research.	Up to 27 species used in restoration and translocations 2020–23. Up to 10 species planted in botanic gardens' living collections for education and research 2020–23.

Box 27.3. Addressing the threats to the Wollemi pine (*Wollemia nobilis*)

Wollemi pine is a Critically Endangered species, with only 46 mature individuals remaining in the wild and a geographical distribution of < 4 km^2 (Mackenzie and Auld in press). Discovery of the Wollemi pine in 1994 drew the attention of the world and a multifaceted recovery plan was quickly developed (NSW DEC 2006). Conservation actions are both *in situ* (Mackenzie *et al.* 2021; see also Chapter 9) and *ex situ* (Fig. 27.3), through the development of a genetically representative living collection as per the Australian guidelines (Martyn Yenson *et al.* 2021).

A propagation program was established by the Royal Botanic Gardens and Domain Trust (RBG) and, by 2005, Wollemi pines had been dispersed to many botanic gardens and became available commercially with royalties contributing to ongoing conservation.

The *ex situ* conservation of this species has enabled conservation-related research that has built knowledge of its physiology, ecology and disease susceptibility. A high priority was to explore how fire affects this species. Controlled experiments on cultivated (cutting-grown) plants confirmed that Wollemi pines could resprout after fire, with higher intensity fires reducing the regrowth (Zimmer *et al.* 2015); however, there remains some uncertainty about the response of seedlings, and of the species' response to different fire regimes.

Fig. 27.3. *Ex situ* population of Wollemi pine. (Photo: Maureen Phelan, Australian Botanic Garden)

Given the threats to Wollemi pine, including fire, recovery planning for this species has focused on translocation to suitable sites (NSW DEC 2006), with the objectives of increasing population size and distribution, and reducing risks of total loss from catastrophic fire and other threats. This ongoing process has been informed by research using *ex situ* collections, and understanding how to propagate and grow this species from seeds and cuttings, with the plant material for the translocations derived from the *ex situ* germplasm collection held by RBG (Mackenzie *et al.* 2021).

Multiple seasons following a wildfire are required to properly understand the recovery of the Australian flora. To capitalise on this opportunity, the Partnership worked collaboratively on a rapid flora assessment methodology developed by the Australian PlantBank at the Australian Institute of Botanical Science. The methodology enables seed collectors to prioritise seed collecting during field work while capturing species-specific data on whether species were heavily burnt, are resprouting or reseeding, and whether herbivory or disease are cause for immediate concern. Monitoring, including the fate of individual plants, over multiple seasons will further improve our understanding of recovery.

Collecting included targeting narrow range endemics and known fire ephemerals in the first season post fire. Additional focus was made to collect from species whose range included unburnt populations. Particular attention was made to ensure collecting efforts had minimal impact on *in situ* populations, prioritising the replenishment of the soil seed bank (Auld 2020). Other priorities include undertaking germination trials of newly secured collections to ensure scientists can break physical and chemical dormancy and achieve high germination rates for the species held *ex situ*.

Box 27.4. Restoring *Glycine latrobeana* to Lobethal Bushland Park

In December 2019, the Lobethal Bushland Park (South Australia) was extensively burnt. The fires impacted more than 23 000 ha, extirpating an entire population of the Vulnerable clover glycine (*Glycine latrobeana*) (Fig. 27.4).

While fire threatened the species' long-term survival, efforts by the South Australian Seed Conservation Centre of the Botanic Gardens and State Herbarium of South Australia (SASCC), in collaboration with ecologists from the Hills and Fleurieu Landscape Board and community volunteers, helped to return more than 230 individuals of the species to two bushland reserves.

This post-wildfire restoration actually started with a small seed collection in 2007. Following release of the recovery plan for *Glycine latrobeana* and conversations with local experts, the SASCC embarked on a mission to collect seeds of the species. More than 100 individual plants within the local population were sampled. All 1200 seeds were sent for storage at the Millennium Seed Bank, Royal Botanic Gardens, Kew, UK (MSB) as an insurance against further loss.

Following the 2019–20 fires, the SASCC requested the MSB repatriate 250 seeds from the 2007 collection to support post-fire restoration. Following a short quarantine period in Sydney, the seeds were germinated with almost 90% success. More than 230 individual plants were propagated and used in restoring the Lobethal Bushland Park

Fig. 27.4. Populations of the clover glycine (A) were severely affected by the 2019–20 wildfires, but seeds collected and stored before the fire (B) allowed for successful re-establishment. (Photos: South Australian Seed Conservation Centre)

and a nearby private bushland reserve where it had been previously recorded, but had not re-emerged since the 2019 bushfire. In 2021 a total of 730 seeds were collected from the reintroduced plants for further propagation and to duplicate collections with the MSB.

Making withdrawals from conservation seed banks to support research and restoration occurs regularly throughout Australia, but repatriation from the MSB has so far been less common. The case of *Glycine latrobeana* provides a valuable proof of concept for duplicating native seed collections in comparable facilities, illustrating how long-term *ex situ* conservation through seed banking is a viable option for future restoration and recovery efforts.

Many species will take longer to recover due to the severity of the fires, their longer reproductive cycles, or the continuing drought impacting burnt areas. Species in these groups were prioritised for germination trials of collections already held *ex situ*, aiming to ensure viability of stored seeds is effectively retained over time and made available for restoration. Several years will be needed to properly understand the recovery of various species as disease, pests and competition influence long-term survivorship.

Community engagement

The *ex situ* community galvanised its significant reach to draw public attention to the emerging wildlife crisis. They activated existing networks and engaged with new collaborators to exchange information and expertise, and share resources with the aim of enhancing the post-fire conservation response.

For fire-affected animals, numerous conservation funds were established to support three phases of action: (1) the rescue and rehabilitation of animals for return to the wild; (2) assessing the impacts on animals and habitat to understand where recovery efforts were needed; and (3) funding medium and long-term recovery efforts to return healthy animals to regenerated habitats and sustain their populations in affected areas.

Similarly for plants, the business and community responses were swift with more than \$3.2 million in grants and cash donations secured through the Partnership for collecting, research and restoration. Additional funding from grants and philanthropic sources went directly to individual facilities across the country. Other offers included seed from Australian native species in overseas plantations; however, these couldn't be accepted for use in restoration efforts due to biosecurity concerns and limitations in genetic provenance data. Local landholders and 'Friends of' groups working to protect private property, community bushlands and reserves provided local knowledge to support collecting, restoration, and translocation efforts. Many private landholders and volunteers suffered significant personal impacts, yet dedicated substantial time and effort to assist with monitoring species recovery, restoring populations, and granting access to collectors.

The impacts of COVID-19 on travel were significant with many species missed in the first-year post-fire. Furthermore, engagement with First Nations peoples was severely limited with on country collecting unable to proceed due to travel restrictions to remote communities.

Key findings

Key findings are varied and significant, incorporating constraints, prioritisation, cost and implications (Table 27.4).

Conclusions

The 2019–20 wildfires demonstrated what the *ex situ* community is able to contribute during a crisis and in the recovery. A strategic approach involving collaborative prioritisation and planning, including both the *in situ* and *ex situ* communities, will have most chance of developing and implementing conservation programs that can most effectively maintain or recover biodiversity in a future likely to be marked by many more environmental catastrophes.

Differences among taxonomic groups in the extent of knowledge base resulted in prioritisation for urgent response being developed later for plants and invertebrates than for vertebrates, with this probably leading to less investment in their recovery than for vertebrates. However, given interactions between invertebrates and plants, their recovery actions should be undertaken together, not separately.

There are many similarities between plant and animal *ex situ* activities and it is plausible that collaborative efforts may prove advantageous in future planning. Taking an ecosystem approach to *ex situ* planning, involving animals and plants, might be feasible, especially in cases where some key vegetation communities such as rainforests have been unpredictably burnt.

Recommendations

The importance of preparing ahead of crises

It is critical that a national preparedness plan for wildlife (animal and plant) recovery is established and deployed ahead of crises. Preparedness of all parties is critical to ensure that an appropriate response can be mounted in a time of crisis. Including *ex situ* actions in recovery plans of relevant threatened species enhances capacity for recovery. Plans should include support for boosting *ex situ* conservation capability.

Table 27.4. Key findings with respect to *ex situ* interventions following the 2019–20 fires.

	Key findings
Constraints	Scientific ventures to support threatened species should not be an abrupt or precipitous exercise. Learning requires time and coordinated action in advance of threats and impacts to allow experts to develop a strong understanding of species biology and care, and time to unravel the science of propagating and storing the species. Access to field sites and translocation sites was hampered by ongoing fires and safety concerns. Funding and permits are not synchronised, resulting in delays to commencing some *ex situ* programs. Recovery for animal species can only happen following the recovery of the ecosystem and many plant species can take years to recover. Expectations of short-term results can be misleading. COVID-19 was a compounding factor restricting field work, threatening supply chains including the availability of liquid nitrogen for cryopreservation, and limiting opportunities, access and permits, and the ability to properly engage and collaborate on country. Separate funding from multiple sources can be limiting if delivered in isolation. Funders that provide opportunities for flexibility and enable leveraging of complementary funding facilitate greater strategic investment to address conservation priorities.
Prioritisation	Assessing potential impacts and prioritising target species for multi-state and multi-partner projects is time-consuming and reliant on substantial collaborative efforts. The desire to catalogue the animal loss before acting resulted in delays to planning and agreeing *ex situ* interventions. A strong effective response requires strategy and preparedness before the event. The welfare impact of fire on animals is significant and resulted in the urgent prioritisation of welfare actions. Post-fire prioritisation activities considered numerous factors to inform the allocation of limited resources, and involved strong interaction with the *in situ* community, academics, community groups, and governments. Despite its thoroughness, the exercise pointed to communication weaknesses between and within organisations with many of the same species and activities offered funding from multiple sources.
Costs	The cost of responding to the wildfires is significant in the short and long term; this galvanised support for targeted actions. Multiple funding avenues were made available by governments, business and philanthropic communities, supported further by private donors. For animals, the cost of implementing *ex situ* interventions varies drastically between institutions and can be quite high in some instances, particularly for larger, state-run operations. Efficient long-term maintenance of populations that are genetically representative of wild counterparts and behaviourally competent for release demands significant infrastructure. Appropriate risk management dictates that populations managed long term must be housed at multiple facilities to limit the impact of possible stochastic events, further increasing infrastructure and care costs. For plants, the costs of implementing *ex situ* interventions to conserve biodiversity are relatively low per species. Costs of collection depend on the location and rarity of a species, number of populations and (for plants) the ability to predict fruiting based on species phenology. Long-term maintenance of *ex situ* seed collections is low cost. Regular germination trials to ensure viability is maintained, appropriate monitoring systems and back-up power generation capabilities all add to the cost of long-term storage but provide valuable safeguards. COVID-19 consequences imposed substantial costs, such as by limiting the numbers of staff and volunteers permitted in field vehicles. Furthermore, limitations on personnel numbers in laboratories and seed banks reduced the hours available for processing and testing collections.

	Key findings
Implications for *ex situ* management	The post-fire response indicated a need for a longer-term skilled and flexible work force, with ongoing resourcing that can manage growing collections. By focusing on centres of excellence and collaboration, the number of threatened species in care can be expanded. In an emergency staff can move to the place of greatest need, as was seen with the eastern bristlebird extraction (see Chapter 15). Care needs to be taken in applying a reductionist approach rather than the ecosystem approach. Both are needed. *Ex situ* tends to focus on species, while *in situ* focuses on ecosystems. Synergy between taxa exists in the wild and should be prioritised in *ex situ* management. Development and dedicated maintenance of a centrally managed database of stakeholders and projects would also assist in ensuring prioritisation activities are robust and consider all necessary stakeholders.
Implications for fire management	Translocation sites need to be protected in times of emergency. A register of translocation sites should be established and identified as critical assets. In the immediate aftermath of the fires, animal care facilities were inundated by injured animals. Fences pose a lethal barrier to livestock and wildlife in an emergency. It is recommended that operating protocols allow the creation of safe passages for animals.

Species emergency recovery capability: agency and funding

For human care and infrastructure recovery in times of crisis, a dedicated recovery agency delivers better long-term outcomes. We recommend a similar approach for wildlife recovery. A dedicated nature-based recovery agency would be able to:

- create capacity to support ecosystems and wildlife in emergencies;
- develop risk-based prioritisation frameworks to assist with decision-making and adapt to emerging threats;
- coordinate interventions and capability proactively and pre-emptively, to areas of greatest need;
- encourage development of a national inventory of coordinated *ex situ* activities;
- coordinate enabling actions and support, permits, facilities and funds matched to the intervention; and
- investigate establishing an ongoing funding model with government, non-government and philanthropic and corporate support – such as the example provided by the Future Drought Fund (https://www.agriculture.gov.au/agriculture-land/farm-food-drought/drought/future-drought-fund).

Ex situ is a long-term approach

A successful response after an emergency can take years to deliver. While there is a need for moving fast to collect samples, animals, seeds and other plant germplasm, recovery to the point of wild release/reintroduction may take significant time. The recovery process must provide long-term commitment and funding.

Pick winners

With so many species in need of help, recovery runs the risk of trying to do too much and failing to achieve real success. Ecological and biological knowledge and evolutionary factors should guide species selection. A national approach is needed to prioritise and align regional and local action.

Community recovery based on nature-based recovery

Success is improved when communities are able to contribute to the recovery effort. Especially important is early engagement with traditional owners and landowners to incorporate their knowledge and to develop long-term recovery programs with guidance and support.

Acknowledgements

We express our thanks to the scientists, carers and others who work on securing and protecting our most endangered species in *ex situ* facilities. Much of the work described above is funded through donations. We thank the state and Commonwealth governments, UK government, non-government organisations, zoos, botanic gardens, seed banks, universities, philanthropic organisations and members of the public from around the world for their support in securing species for the benefit of future generations.

References

APC (2021) The Australian Plant Census, National Species List database. Council of Heads of Australasian Herbaria, <https://biodiversity.org.au/nsl/services/search/taxonomy>.

Auld TD (2020) *How Plants Cope With Fire*. Australian Network for Plant Conservation. Canberra.

Bowkett AE (2014) *Ex situ* conservation planning is more complicated than prioritizing the keeping of threatened species in zoos. *Animal Conservation* **17**, 101–103. doi:10.1111/acv.12116

Byers O, Lees C, Wilken J, Schwitzer C (2013) The One Plan approach: the philosophy and implementation of CBSG's approach to integrated species conservation planning. *WAZA Magazine* **14**, 2–5.

Chapman AD (2009) *Numbers of Living Species in Australia and the World*. 2nd edn. Australian Biological Resources Study, Canberra.

Commander LE, Coates D, Broadhurst L, Offord CA, Makinson RO, *et al.* (2018) *Guidelines for the Translocation of Threatened Plants in Australia*. 3rd edn. Australian Network for Plant Conservation, Canberra.

DAWE (2021) 'Threatened Species Strategy 5th year progress report'. Department of Agriculture, Water and the Environment, Canberra.

DELWP (2019) 'Mountain Pygmy-Possum Operational Contingency Plan 2019–20'. Department of Environment, Land, Water and Planning, East Melbourne.

DELWP (2020) 'Mountain Pygmy-possum action statement'. Department of Environment, Land, Water and Planning, East Melbourne.

Fairman T, Nitschke C, Bennett L (2016) Too much, too soon? A review of the impacts of increasing wildfire frequency on tree demography and structure in temperate forests. *International Journal of Wildland Fire* **25**, 831–848. doi:10.1071/WF15010

Gallagher RV (2020) 'National prioritisation of Australian plants affected by the 2019–2020 bushfire season'. Report to the Commonwealth Department of Agriculture, Water and the Environment, Canberra.

Godfree RC, Knerr N, Encinas-Viso F, Albrecht DE, Bush D, *et al.* (2021) Implications of the 2019–2020 megafires for the biogeography and conservation of Australian vegetation. *Nature Communications* **12**, 1023. doi:10.1038/s41467-021-21266-5

Green K, Caley P, Baker M, Dreyer D, Wallace J, *et al.* (2021) Australian Bogong moths *Agrotis infusa* (Lepidoptera: Noctuidae), 1951–2020: decline and crash. *Austral Entomology* **60**, 66–81. doi:10.1111/aen.12517

IUCN (2020) *Motion 94 – Linking* in situ *and* ex situ *Efforts to save Threatened Species*. Gland, Switzerland.

IUCN/SSC (2014) *Guidelines on the Use of* ex situ *Management for Species Conservation. Version 2.0*. IUCN Species Survival Commission, Gland, Switzerland.

Mackenzie BDE, Auld TD (in press) *Wollemia nobilis*. The IUCN Red List of Threatened Species 2021, <https://www.iucnredlist.org>.

Mackenzie BDE, Clarke SW, Zimmer HC, Liew ECY, Phelan MT, *et al.* (2021) Ecology and conservation of a living fossil: Australia's Wollemi Pine (*Wollemia nobilis*). *EcoEvoRxiv*. doi:10.32942/osf.io/3sjwx

Martyn Yenson AJ, Offord CA, Meagher PF, Auld T, Bush D, *et al.* (2021) *Plant Germplasm Conservation in Australia: Strategies and Guidelines for Developing, Managing and Utilising* ex situ *Collections*. 3rd edn. Australian Network for Plant Conservation, Canberra.

NSW DEC (2006) *Wollemi Pine (*Wollemia nobilis*) Recovery Plan*. NSW Department of Environment and Conservation, Hurstville.

Parrott ML, Davies NE, Wicker L, Broome L, Miller K (2020) Looks like an ANZAC biscuit, tastes like a protein bar: Bogong Bikkies help mountain pygmy-possums after fire. *The Conversation*, 6 February, <https://theconversation.com/looks-like-an-anzac-biscuit-tastes-like-a-protein-bar-bogong-bikkies-help-mountain-pygmy-possums-after-fire-131045>.

Parrott ML, Wicker LV, Lamont A, Banks C, Lang M, *et al.* (2021) Emergency response to Australia's Black Summer 2019–2020: the role of a zoo-based conservation organisation in wildlife triage, rescue, and resilience for the future. *Animals (Basel)* **11**, 1515. doi:10.3390/ani11061515

Schwartz KR, Parsons EC, Rockwood L, Wood TC (2017) Integrating *in-situ* and *ex-situ* data management processes for biodiversity conservation. *Frontiers in Ecology and Evolution* **5**, 120. doi:10.3389/fevo.2017.00120

UNEP (2021) First draft of the post-2020 Global Biodiversity Framework. United Nations Environment Program, <https://www.cbd.int/doc/c/abb5/591f/2e46096d3f0330b08ce87a45/wg2020-03-03-en.pdf>.

ZAA (2020) *Regional Census and Plan*. Zoo and Aquarium Association Australasia Executive Office, Mosman.

Zhang J, Finlaison DS, Frost MJ, Gestier S, Gu X, *et al.* (2018) Identification of a novel nidovirus as a potential cause of large-scale mortalities in the endangered Bellinger River Snapping Turtle (*Myuchelys georgesi*). *PLoS One* **13**, e0205209. doi:10.1371/journal.pone.0205209

Zimmer HC, Auld TD, Hughes L, Offord CA, Baker PJ (2015) Fuel flammability and fire responses of juvenile canopy species in a temperate rainforest ecosystem. *International Journal of Wildland Fire* **24**, 349–360. doi:10.1071/WF14054

28

Shared responsibilities and collaborative responses: review of actions taken to support the recovery of biodiversity after the 2019–20 wildfires

John C. Z. Woinarski, Sarah M. Legge and Libby Rumpff

The preceding chapters (22–27) of this section described the responses made to help recover fire-affected biodiversity. An extraordinary cast of government agencies, conservation non-government organisations (NGOs), landholder groups, Indigenous groups, local councils and communities, businesses, philanthropical organisations and individuals made important and distinctive contributions to this recovery effort. Our accounts cannot hope to recognise, report on and pay tribute to all of those contributors: for example, WWF-Australia in Chapter 24 noted partnerships with more than 200 groups to implement the post-fire recovery efforts in which they were involved, and the Australian Government funded at least 240 groups to implement recovery actions. Although government agencies have primary responsibility for biodiversity management in Australia, the many groups and individuals involved in the post-fire recovery effort demonstrate the willingness of the community as a whole to share responsibilities for, or contribute to, the care of the environment, and especially so in the face of emergencies.

The collaborative effort involved a large variety of actions at local, regional, state and national level. Such an exemplary response was commensurate with the myriad impacts on biodiversity of the 2019–20 wildfires over vast areas and on many land tenures. These actions undoubtedly did much to help protect populations of many species whose existence was made tenuous because of losses in the fires. It is likely that the precarious and long pathway to recovery of many fire-affected species and ecological communities is more secure and advanced because of this response.

Several characteristics of the recovery effort are highlighted in these preceding chapters:

- Responses were initiated with extraordinary rapidity. While fires were burning, state and national governments, in collaboration with many experts and organisations, quickly identified species that were most likely affected, and contributed significant financial resources to emergency recovery actions. There were several factors that enabled such swift response, including access to biodiversity datasets and tools, the technical capacity and capability to guide and support on-ground action, and collaborative efforts between agencies, NGOs, research institutions, Traditional Owner groups and the community. For the showcased examples of Wollemi pine (*Wollemia nobilis*) and eastern bristlebird (*Dasyornis brachypterus*) (Chapters 27 and 15), existing plans facilitated heroic action during the fires when important populations were threatened. This rapid response extended to urgent post-fire action, such as for the Kangaroo Island dunnart (*Sminthopsis griseoventer aitkeni*) (Chapter 24); surveys immediately post-fire located surviving populations, and then a collaborative effort resulted in rapid establishment of a 369 ha feral predator exclusion area on Kangaroo Island, and targeted control of feral cats.

Of course, some chapters also recognised impediments to rapid action, including suboptimal biodiversity data, safety concerns for personnel in active fire zones and in recently burnt areas, the unpredicted constraints due to COVID-19, administrative requirements (i.e. having relevant plans or permits, such as for wildlife handling), availability of suitable expertise and facilities (notably for *ex situ* actions and animal shelters, some of which were swamped by the unprecedented demands) and access to some funding sources.

- Post-fire responses were generally strategic, prioritised and evidence-based. Given the potentially bewildering array of impacts of these wildfires on so many species and environments over such a large area, there was a critical need to identify and justify priorities among the fire-affected components of biodiversity and to implement priority actions for these. Such prioritisation was undertaken and reported at national and state levels, based particularly on rapid spatial assessments of fire overlap with the distributions of species and ecological communities and on advice provided rapidly by experts on those management actions most likely contribute to recovery. These conservation responses included the control of compounding threats, *ex situ* actions, and the identification and subsequent protection of important sites and populations. In many cases, rapidly implemented post-fire surveys helped to further corroborate or extend the initial assessments of impacts, and to hone management responses.

  Again, there were some impediments in this prioritisation of fire-affected biodiversity and responses: the assessments of priority species were constrained by data limitations (with one consequent bias that well-known vertebrate species received disproportionate recovery efforts) and there were uncertainties in the likely effectiveness of some management responses. To some extent, post-fire monitoring of recovery, threats and management effectiveness will fill such knowledge gaps, allowing for more confidence in responses to comparable situations in the future.
- Responses were developed and undertaken collaboratively by many stakeholders. A hallmark of the recovery effort, highlighted in most of the preceding chapters, was that the response was cooperative – to an extent that is unprecedented in conservation management in Australia – and scaled fittingly across local, regional and national levels. Inclusive workshops that focused on biodiversity values, Indigenous knowledge and values, fire impacts and management responses did much to cement this collaboration, allowing for the integration of local capability, knowledge and values with national priorities. Diverse groups displayed a much-needed willingness to work together, each contributing specific skill sets and capabilities, in recognition of the global significance of the loss and the imperative for recovery: such collaboration also helps to heal social scars imposed by the fires. The need and value of coinciding the recovery of social and environmental values is perhaps best epitomised in the Victorian nature-led community recovery program (Chapter 30).
- Financial and other resources were mobilised quickly to allow the implementation of priority recovery efforts. As described in Chapter 22, the Australian Government rapidly contributed $200 million to the recovery of biodiversity affected by the 2019–20

wildfires, and many state and territory governments provided commensurate funding. The active involvement of relevant government ministers in the fire response may have helped to ensure such ready availability of funding.

Exceptional financial contributions were also made by many organisations and individuals, including donations from across the globe. For example, the animal welfare group NSW Wildlife Information, Rescue and Education Services Inc. (WIRES) received $91 million in bushfire donations over a period of a few months (Australian Charities and Not-for-Profits Commission 2020). Many organisations and agencies had to be, and were, extraordinarily agile in responding to the fire emergency, diverting resources from other pre-existing actions and commitments and rapidly expanding their capability and expertise to address the many challenges imposed by the urgent need for response. However, this exceptional funding and need for urgent action also challenged the administrative capacity, governance and expertise of some groups.

Another concern is that most recovery funding has been invested in urgent and short-term management response, and there is likely to be far less funding available to maintain support for the long-term needs of recovery. Apparent short-term successes may be eroded without ongoing management support and without significant investment in planning and preparation for future events.

- There is a need to review, and take actions to improve, the responsiveness for future comparable events. The chapters contributed by agencies and conservation groups highlighted examples of success, but there are likely to be other cases that were less successful. These shortcomings are in part understandable given that management systems and policy settings had not previously been subject to stresses of the level experienced in the 2019–20 wildfires and the need for such wide-ranging recovery efforts. However, the fires also exposed the entirely predictable limitations of our data and knowledge, with a consequence that many highly vulnerable species and communities may have missed out on necessary emergency action – and especially so for poorly known groups and those with little public profile, such as fungi and invertebrates. These chapters included recommendations that mostly seek to address such shortcomings.

Most of the content of the preceding chapters describes the implementation of management actions. To date, understandably, there is less information on the outcomes of those actions – it is important that ongoing monitoring can assess whether these actions have led, or are leading to, the successful recovery of fire-affected species and communities.

The emphasis on management actions in these chapters also overshadows some other types of actions that will be needed for long-term improvements – such as responses in planning, policy and legislation. We visit some of these issues in the following section (Chapters 29–35).

Reference

Australian Charities and Not-for-Profits Commission (2020) *Bushfire Response 2019–20: Reviews of Three Australian Charities*' Australian Charities and Not-for-Profits Commission, Canberra.

29

What did we learn about biodiversity management, policy and operations from the 2019–20 wildfires?

Kelly de Bie, Kaye Currey and Libby Rumpff

Summary

- This chapter reports on the results of structured interviews with 32 government agency officers (across all fire-affected jurisdictions) who had responsibilities for fire management or biodiversity conservation and were active in the response to the 2019–20 wildfires. Our focus was on identifying what worked and what didn't work for the protection of biodiversity, with these lessons categorised in relation to issues before and during fire.
- *Preparedness*: Interviewees considered that significant conservation assets were more likely to be protected during fires if their locations were mapped and there were pre-existing and readily accessible plans that served to inform their protection. Some interviewees reported that management actions undertaken prior to wildfire (e.g. planned burning around their perimeter) also helped.
- *Operations*: Interviewees reported that conservation assets were more likely to be appropriately considered and protected during wildfire operations if biodiversity agency representatives were embedded in the emergency management structure (e.g. in incident management teams). This allowed for ready access to, and dissemination of, biodiversity data, prioritisation of natural assets, and connections to a broader network of biodiversity respondents. Interviewees also noted that conservation assets were more likely to be protected if there was advocacy from the managing agency or community groups (e.g. for iconic species), and if there were few competing assets for protection (e.g. where there was lower risk to human life and property). Conversely, interviewees noted that biodiversity protection during fire operations was compromised or constrained where locational data or prioritisation were inadequate, and because protection of biodiversity was almost always subordinate to protection of human life and property. Furthermore, the scale and severity of the 2019–20 wildfires overwhelmed much planning for biodiversity protection.

- If we can learn from what worked, and what did not work, in terms of preparedness and operations in the 2019–20 wildfires, biodiversity could be better protected in future large fire events.

Introduction

The scale, intensity and duration of Australia's 2019–20 fire season tested, and often overwhelmed, the capacity of fire management systems. The operational fire response focused mostly on human life and property, with less attention during fire operations to the protection of biodiversity. Actions for biodiversity mostly occurred in the post-fire period, with significant and coordinated investment into biodiversity recovery, particularly triage of animals, supplementary feeding and targeted herbivore and predator control. While some iconic species and locations were the focus of rapid and successful response efforts, many species and ecological communities were severely impacted (Chapters 4–16; Ward *et al.* 2020; Wintle *et al.* 2020). However, the fate of biodiversity in these wildfires was highly influenced by management action or inaction, taken before and during the fires. In this chapter we report on interviews undertaken with staff in government conservation and fire management agencies, reflecting on their involvement and focusing particularly on the lessons they learnt from this experience. In particular, we sought to identify and report on factors that helped conserve biodiversity in the fires as well as what did not work. This chapter complements Chapter 30, which summarises the more formal assessments on these fires by government inquiries. Some aspects described in this chapter are considered in more detail in de Bie *et al.* (2021). Note that interviewee responses have been anonymised for this chapter.

The 2019–20 fire response involved many agencies and organisations, and extensive professional and volunteer personnel across much of Australia, presenting a crucial opportunity to obtain valuable and unique insights into the management of biodiversity during the fires. Here, we report results from a set of semi-structured, online interviews with state agency biodiversity and fire operations staff, seeking to understand and explore their practices and experiences during the 2019–20 fire season (de Bie *et al.* 2021). Our focus for this research is on the periods before and during the 2019–20 wildfires, rather than on recovery actions taken after fire. Our study also considers state and territory government agencies, who have primary responsibilities for fire management. Hence, we do not consider the roles played by (and lessons that can be learnt from) Indigenous groups or conservation non-government organisations (NGOs), although we acknowledge their important roles. Furthermore, we mostly consider here policy and operational management, but again recognise that other aspects, such as legislation, resourcing and accountability, may also influence the extent to which biodiversity was protected during these fires.

Thirty-two interviews were conducted, representing 13 land management agencies across the six target states and one territory. The majority (81%) of interviewees had substantive roles with a biodiversity focus, but 57% of these also held a fire operations role during the 2019–20 fire event. Using a thematic analysis approach, we identified and interpreted key themes to understand what worked well in terms of preparedness and fire operations with respect to outcomes for biodiversity, as well as identifying key gaps to be addressed. As large-scale fires become more frequent with a changing climate (van Oldenborgh *et al.* 2021), it is critical to learn from the 2019–20 events to avoid a repeat of the large scale biodiversity impacts and to ensure the persistence of Australia's conservation assets into the future.

In the following section, we distil some of the key messages given by interviewees, in some cases using direct quotes. These are indexed to interview records (e.g. I.02) described in de Bie *et al.* (2021).

Findings

Preparedness: pre-fire planning for conservation assets

What worked?

The majority of agencies had some form of pre-fire planning, data and tools that were utilised during the wildfire event to support decision making. However, only two state agencies had specific emergency planning for threatened species or communities that outlined targeted actions to be undertaken during a fire; one of those species was the Wollemi pine (*Wollemia nobilis*) (see Chapters 9, 27). In the case of the eastern bristlebird (*Dasyornis brachypterus*) (see Chapter 15), individual birds were extracted from areas ahead of the fire front, aided by translocation planning and existing permits and resources. These few examples were successfully executed because of the presence of 'shovel ready' plans, developed before the event (Table 29.1).

Most interviewees (62%) reflected that having existing data and information products to guide decisions during the event was useful, but to varying degrees. Useful products include GIS layers and mapping that identified locations of conservation assets, as well as providing guidance on suitable suppression activities.

> *They are static maps, which are updated periodically. But on top of that, we have those GIS layers which can be pulled up at any time by fire management staff and by values officers to inform response. .... So, I think we're well positioned, in that when it comes to the point of an operation, there's reasonably good available information about where we need to be careful. (I.02)*

One agency had extensive pre-fire modelling in the form of habitat distribution models. These models helped to identify vulnerable conservation assets, and their locations, during fire. This was viewed as instrumental in helping to determine which species were highly vulnerable during the fires and informed decisions around priorities for action.

Table 29.1. Summary of existing pre-fire planning tools, their benefits, limitations, and suggested improvements.

Existing pre-fire planning tools	• Risk/bushfire management plans (prescribed burning plans) • GIS layers and mapping • Spatial modelling (e.g. habitat distribution models) • Pre-fire targeted actions (e.g. planned translocations) • Species specific fire management plans • Strategic management plans (e.g. recovery plans)
Benefits of existing pre-fire planning tools	• Provides species location data • Provides guidance regarding suitable fire suppression activities • Can initiate 'shovel ready' actions during fire • Can assist with prioritising assets for protection • Can identify possible pre-fire actions to spread risk
Limitations of existing pre-fire planning tools	• Can be lacking guidance on operational actions for protection • Little prioritisation of what assets to protect when so many may be threatened • Data may not be current

continued

Table 29.1. Continued

Improving pre-fire planning	**Adequate data and prioritisation** • Ensure priority lists are regularly updated and data layers available for all species or assets of interest (priorities). • Ensure data layers are consistent among agencies, collated and up to date. • Have a consolidated approach at a state level regarding development and maintenance of spatial layers and data. • Bring all corporate data into central, interoperable databases to avoid information being kept on individual computers or systems. • Develop habitat distribution models (noting there is limited information on population sizes for most species). **Targeted fire management plans** • Develop specific fire response plans, at species or landscape scale, that map the values and identify actions that are needed. • Employ scenario and contingency planning to identify potential actions and capacity/resources required for implementation. • Include conservation assets requirements in prescribed burning management plans so they can be used in fire events. • Adopt a nil-tenure approach to inclusion of conservation assets in strategic planning (i.e. not just those on conservation estates). **Risk/vulnerability assessments** • Develop risk management frameworks, or include conservation assets in existing risk management frameworks, including scenario analysis. • Develop landscape-scale plans that assess threats (including fire) and include spatial location of conservation assets. • Undertake meta-population analyses to identify areas that provide important refugia. • Parameterise species fire regime tolerances to establish priorities. • Review existing and new information to assess threats, and incorporate into management plans and prioritisation.
Constraints to improving pre-fire planning	• Lack of resources to acquire data (i.e. numerous assets) • Complexity of decision making regarding prioritisation (i.e. what species and actions to prioritise in what contexts)

For some agencies, mapping products were integrated with management plans. For example, the operational arm of one agency had worked closely with their biodiversity counterparts in developing strategic bushfire mitigation plans. This enabled the creation of guidelines that provided information on conservation assets, such as breeding times for key threatened species. Similarly, another agency had developed mapping of fire sensitive conservation assets and identified which of those assets were most critical to protect from fire suppression activities such as control lines and the fire itself.

> *We provide (fire agency) with a whole heap of mapping of fire sensitive values ... so they're sort of aware of what's in the path of the fire or potentially threatened, but we provide a lot of advice on things that might be in that landscape, what are the most critical things to protect. (I.19)*

What didn't work?

Although interviewees recognised the usefulness of existing planning during the fires, many also indicated that they did not consider it as being 'fit for purpose'. There was recognition that the scale and intensity of the 2019–20 wildfires were unprecedented, overwhelming some existing pre-fire planning.

> *It would really start to hit home that these fires were more intense, they were much more extensive ... from a biodiversity perspective we felt that the standard processes weren't sufficient to understand what we needed to do, and that we needed to scale up our response and understand the impacts of the bushfire. (1.21)*

The two most common concerns related to preplanning for emergency response actions for conservation assets were a lack of information on appropriate actions and no prioritisation of conservation assets (Table 29.1).

> *We do lots of fire planning, and we know where things are in the landscape. But I think what we didn't have is more specific consideration of those high-value, threatened species and other biodiversity values, to say, 'What do we do in this scenario?' (I.22)*

Existing plans were noted as having no or limited guidance on what assets to prioritise. This resulted in planners or biodiversity staff having to conduct 'on the fly' prioritisations during the fires with whatever information was available. Where particular conservation assets were identified as priorities, a lack of clarity around how and why they had been prioritised was noted.

> *It's something that we developed on the run during the fire – we prioritised the conservation assets. We need to have a plan in place for that and what we can actually do, if anything. (I.01)*

Other key issues include the availability of data and knowledge for conservation assets. This includes a general lack of knowledge on the fire requirements and appropriate fire regimes for many species and ecological communities that can lead to distorted priorities during a fire event. A lack of fit for purpose data was noted as a constraint, with data useful for informing species management plans not necessarily useful to inform actions during fire operations. Existing prescribed burning plans were primarily focused on vegetation communities or ecosystems and while spatial layers for flora exist, there was much less information on fauna, and species of concern may not be identified in planning and operations.

> *Our fire regime system is based off ecosystems, that's the primary driver in terms of how we determine our fire management zones and we don't complicate that too much with particular species management. (I.13)*

A lack of knowledge and data that are needed to inform prioritisation and risk management processes was viewed as a substantial impediment to improving pre-fire planning. State databases are not regularly updated and so using these data to inform priorities was described as problematic. A 'hybrid' approach that combines formal predictive data with updated field data, expert advice or citizen science data was recommended by some interviewees. There was recognition of a lack of capacity and resources to address this gap, given the large number of threatened or potentially threatened species and ecological communities.

> *We were overwhelmed by the sheer number of things that are a priority, or that we say (are) from our priority spatial overlays. (I.11)*

Fire operations: protection of conservation assets during the 2019–20 fires

What worked?

Having biodiversity representatives in the emergency management structure was viewed as instrumental in ensuring conservation assets were adequately considered during decision making. A few agencies created dedicated 'natural values officers' or 'natural values teams'. These roles were able to work within incident management teams to access available data and make it usable and relevant, access a network of experts or other resources, advocate for conservation assets and advise on the impact of firefighting actions. Most natural values officers were situated in incident management teams, but one agency also had a natural values officer actively working with fire teams on the ground to assist with the protection of key conservation assets. Victoria established a state-wide wildlife controller who worked at the state level, rather than the incident level. One interviewee believed that this provided much needed access to the highly structured emergency management system.

> *Once you had someone in there with a coloured vest on, access became much easier. Not to say it necessarily translated into things happening on the ground, but it certainly was critical. (I.20)*

Interviewees spoke of the ability of biodiversity representatives to 'bridge the gap' between land management and fire operations agencies, as well as being able to bring in high-level expertise. Advocating the protection of conservation assets and negotiating for resources were critical elements. Common to all the forms of biodiversity representation in the emergency management structure was the importance of the contribution of local knowledge that is often not captured in plans or mapping.

Prioritising and achieving protection were considered easier if an asset was located at a single defined site, with a restricted distribution (e.g. Wollemi pine, Table 29.2; Chapter 9). Interviewees felt that conservation assets were given greater consideration if the fire was located in national parks and the fire was primarily the responsibility of the land manager or parks agency. There were a few instances mentioned in which targeted actions were undertaken to protect broader areas, such as areas containing feeding habitat for a threatened species, or areas containing valued flora or ecological communities. For example, the

Table 29.2. Aspects that contributed to the targeted protection of assets during the 2019–20 fires.

	Wollemi pine	Bristlebird	Namadji NP
Existence of pre-fire planning	✓	✓	×
Biodiversity representative in incident management team	✓	✓	✓
Advocacy and support	✓	✓	✓
External resources	✓	✓	✓
Additional funding	✓	✓	×
Conservation tenure	✓	✓	✓
Single location	✓	✓	✓
Ecological knowledge and data	✓	✓	✓

southern section of Namadji National Park was of high value, due to being the only remaining unburnt area since the fires in 2003. When Namadji became threatened by the Orroral Valley fire, bulldozers were used to establish containment lines that successfully prevented the fire spreading into this area, preserving this important ecological community (Table 29.2; Chapter 23).

A conservation asset was more likely to be protected during the fires if it was iconic or had a pre-existing high profile, so that there was a greater knowledge of exact locations and the threats, and a higher level of community interest and political will. A few interviewees noted that more consideration was given to protection of conservation assets in areas where there was less risk to life and property. This provided opportunities for the protection of conservation assets, as resources were more readily available and there was time and space to act.

If it were a bristle worm, we would've got nothing done. (I.25)

Advocacy by highly passionate, knowledgeable individuals or stakeholder groups was able to influence the protection of particular conservation assets. Most of the assets that were discussed in the context of having been the focus of targeted actions were threatened species, but in a few instances, targeted protection actions were undertaken if the conservation asset was considered rare or unique to a particular region. Advocacy for the protection of conservation assets often occurred within incident management teams, but it was also recognised that highly passionate and knowledgeable individuals or stakeholder groups were able to influence the protection of particular species. Important elements of advocacy were that the person was motivated enough to push for action and knew who to contact to exert influence.

It comes down to one person having the passion or the drive or whatever. Without that, then the fire can be fought very differently, so it's up to single individuals and A, the relationships you develop with your colleagues, and B, how hard you're prepared to push. (I.28)

Having funding and resources available, for both pre-fire planning and during fire actions, was recognised as a vital. Pre-fire, investment in research and planning for priority conservation assets was important, to ensure suitable data and planning were available. There were two cases where the existence of a translocation plan (Chapter 27) and a detailed fire management plan (Chapter 9) enabled successful action during fire. During fire, external resources that would not compromise the firefighting effort were able to be accessed on some occasions, facilitating targeted action for priority conservation assets. This included staff from other departments, contractors, or personnel from other organisations such as zoos (see Chapters 15, 27) (Fig. 29.1).

If that money wouldn't have been provided, I guess we could have still got some things done, but it would have been more difficult, and there are some things that might not have actually occurred because they literally take money to pay people to do things (I.23)

A few interviewees were aware of guidelines around the use of retardants during planned burns, primarily near water bodies. One participant reported that they had to come up with some rules regarding retardant use during the fire. In general, interviewees

Fig. 29.1. Emergency extraction of individuals from the last known Victorian population of the endangered eastern bristlebird was facilitated by having pre-existing translocation planning to guide action, collaboration between multiple agencies, and adequate resourcing. (Photo: Mark Antos)

advised that they considered the impact of retardants and gels, but there were constraints on the knowledge about their potential detrimental impacts. Two interviewees discussed the complexity of decision making around the use of retardants and having to consider trade-offs and a 'hierarchy' of risks – for example, considering the risk of using retardants in a catchment area containing threatened species, or not using it and risking the entire catchment being burnt.

What didn't work?

A major challenge recognised by many interviewees was the unprecedented nature of the fires. The scale, intensity and often highly unpredictable behaviour of the fires had significant consequences for whether conservation assets could be considered for protection, and fire conditions often precluded the ability to plan or implement actions. There was an example where water bombers had been approved to lay lines of retardants to protect patches of a threatened ecological community and threatened flora, but by the time the machines were mobilised the fire had already burnt through the areas.

> *It was unbelievable and it was interesting because it was fire behaviour that nobody expected, it sort of defied all the predictions quite possibly because of three years of dryness and the way the wind behaves under those different topographic features it just blew out well beyond anything we expected or could have predicted. (I.32)*

Interviewees were very aware that life and property are established priorities, but the scale of the fires and limited capacity and resources further constrained the ability to be able to protect conservation assets.

> *We're just not set up for these big landscape style fires, so we have scrambled. (I.12)*

There was a high level of uncertainty as to whether resources would be available. Six interviewees spoke of having to negotiate within the incident control centre for resources, primarily aircraft. Even if aircraft had been allocated for protection of a conservation asset, it would be diverted if properties became threatened. In another instance an interviewee spoke of having to 'beg' for a highly qualified person to be included in the incident management team to do a values assessment. One interviewee felt that the lack of capacity meant that there was no action taken for many fires that caused the loss of conservation assets until fires approached property.

Resources were running around doing a whole bunch of other stuff, protecting people's houses and making sure people weren't in their houses, evacuating and things like that. So there were limited opportunity in that kind of situation to get in and protect some of those sites that we knew about even. (I.15)

A broader issue raised by a few interviewees was the need to expand the focus of always protecting life and property, to consider the equivalency of value of some natural values compared to property. One interviewee recounted a situation in which the decision needed to be made as to which direction a fire would be encouraged to go using back burning techniques. Based on existing frameworks, the decision was made to protect a small number of sheds, at the expense of 5000 ha of national park.

There's definitely a mindset shift that has to happen across the board about value. A decision has to be made about whether we're going to value environmental and cultural assets in the same way that we value built assets. (I.32)

The majority of interviewees noted shortcomings with operationalising information, due to data accessibility, relevance and currency. Data were often difficult to interpret by operations personnel, or the information was too complex and not designed for emergency management use. The ability to find and collate existing data was an issue, as it was often dispersed among databases, localities (e.g. regions) and individuals. Available data were sometimes not accessed by operational staff. One interviewee believed that the available maps and strategies were 'scantly' used in incident management teams. An interviewee who worked as a planning officer during the fires was unaware of a biodiversity risk layer that had been created by their own agency and only 'stumbled' across it. There were also concerns that data systems were not interoperable, which led to delays in information being accessed and processed.

Everyone's got their own system in terms of information management … We have our own fire management system, [the fire agency] have their fire management system, they're doing a major review of theirs and we're looking at how they can incorporate our data more effectively. (I.13)

Where information was available and accessible, there were two issues with the degree to which incident management team personnel were able to use the data in an emergency setting. Information was noted as being too broad and providing insufficient detail as to why or how to protect the asset, or available information was not current or dynamic. Biodiversity database records are often the basis for the mapping and GIS layers used in incident management teams, but the databases are not regularly updated. Some plans containing operational guidelines or fire management strategies existed only as static maps that were not regularly updated before the fire, and not able to be updated during the fire.

The other real issue for me is the accessibility and the awareness of those (fire management) strategies ... they are online, but just in a PDF version, so we didn't have live data that we could input into incident action plans. (I.05)

An overarching challenge for the inclusion of conservation assets within incident management planning was that prioritisation of conservation assets was undertaken by biodiversity departments, outside of the emergency management structure. For example, one department recognised the need for, and instituted, an intensive prioritisation process, but this process was not part of the formal decision-making structure for allocation of resources. This information was only able to be incorporated into the emergency management context when a biodiversity representative was situated in the incident management team.

A lack of personnel with biodiversity experience was noted as a common constraint. Personnel from parks agencies who may populate key roles are not always available or rostered on, and there was a limited number of adequately trained natural values officers. Protection of conservation assets was strongly influenced by who was present in the incident management teams, both at a senior level and also at local levels when access to local knowledge of key conservation assets was needed.

Because people often don't know that [conservation assets] are there. In the way that Incident Management Teams work, if you're in a normal firefighting mode, you're in these 12-hour cycles of people swapping over all the time. And people will come from out of area ... So knowledge is a big one. (I.09)

In most instances, the inclusion of a natural values officer in the incident management team was done for the first time in an emergency event setting, and there was some confusion about exactly what role they were or should be performing. One participant spoke of having to negotiate during the fire about having a natural values officer in the incident management team. In another instance interstate personnel were brought in to fill key officer roles, but they had no background with the role or the values of the area. There was a perceived difference in how incident controllers accessed advice from natural values officers. Some would actively seek advice regarding conservation asset considerations, while in other jurisdictions the onus was on the natural values officer to take the advice to the relevant incident management team personnel.

Because it is an advisory role, that's where some of that improvement can happen around how do we ensure at least that there is a clearer line of sight between the natural values officers giving the advice and the Incident Controller manager making those decisions. (I.15)

Many interviewees discussed how the relationship between biodiversity management agencies and fire agencies influenced the consideration of conservation assets. Some interviewees were very positive about interagency cooperation and willingness to assist in protecting conservation assets. However, other interviewees noted there were cultural differences between land manager and non-land manager agencies, which were demonstrated by a lack of interest in, or awareness of, the need to protect conservation assets. It was felt that both on-ground fire fighters and non-land manager operational personnel in incident control centres or incident management teams often did not have an appreciation of the importance of conservation assets, or there was a lack of training in how to protect these assets.

> *It's a tricky one because people who are in that operational role, are very hands-on people and [some] don't necessarily want to understand all of the detail around what plant is that, and where does that live, and have an appreciation of the importance of the biodiversity and other values that are there. (I.22)*

Conclusions

Based on the lessons learned during the 2019–20 wildfires, preparedness is vital for biodiversity protection in future wildfires. Being prepared involves improvements in biodiversity data availability, prioritisation of conservation assets, fire management planning to identify suitable actions during fires, and spreading risks (e.g. through translocations).

Few conservation assets were a focus of emergency fire management response in the 2019–20 fire season. When they were, several key factors contributed to their protection of these particular assets during fires (Table 29.2), such as an established priority, being highly localised and clearly circumscribed, and in conservation reserves. A key lesson from the experience of those involved in managing the 2019–20 wildfires was that the protection of biodiversity was more likely to be achieved when relevant biodiversity expertise was embedded in fire operational command structures. The review reported here demonstrates that the fate of biodiversity during wildfires may be contingent on many knowledge, policy and management factors, and that there are clear lessons that need to be learned to help improve the fate of biodiversity in future wildfires and ensure that losses of the scale in the 2019–20 wildfires are less likely to be repeated.

Recommendations

Preparedness

- Data to underpin the development of prioritisation frameworks and management plans must be collected and collated, and available in an interoperable database. This may include data on the status of species or ecological communities, the ecological needs of the species (especially their responses to fire, and to fire regimes), and the consequences of actions (including costs). A planning process that seeks to identify critical places for high priority protection would be likely to guide data collection efforts by revealing critical knowledge gaps about the distribution of priority species and ecosystems. Conservation assets need to be better defined, delineated and prioritised. These may be species that are most vulnerable and in need of protection, but may also include places of high endemism, critical habitat for multiple species, or climate refuges.
- Pre-fire planning should include targeted fire management plans for key conservation assets to assist with decision making during a fire. To be 'shovel ready', plans need to clearly identify actions, but also identify the components (e.g. skilled personnel, equipment, permits, resources) needed to assist with the rapid implementation of the plan in an emergency setting. Ideally, a combination of landscape planning and species or asset specific (targeted) plans will be required.

- Development of management plans should involve the exploration of different scenarios, to help determine the potential benefits, risks and feasibility of acting in different conditions. Importantly, it can assist in the development of different contingency management options for fire events. Contingency plans may be developed for different fire conditions, and/or for different resource availability conditions.
- A risk-based approach to acting before fire, to reduce the need to act during a fire event, was widely recognised. This may include translocations and establishing seed banks, and identifying locations for containment lines that would be enacted before a fire if a bad fire season was predicted.

Fire operations

- Conservation assets need to be better incorporated and prioritised in decision making during fire control operations. There needs to be more recognition of the value of conservation assets relative to other assets for protection during fire control, such that protection of biodiversity is not automatically considered subordinate to that of other values.
- Biodiversity representation, preferably with local knowledge, needs to be formally included in the emergency management structure, and has multiple benefits. The vulnerability and management needs of conservation assets can be clearly communicated within incident management teams to be considered for protection. Representatives can also play a role in networking with experts and other departments or organisations to provide further information, and to support and provide advice on the risks of fire-fighting actions to conservation assets.
- To be useful in supporting emergency response efforts, staff must be able to interrogate data to rapidly determine what is most important, rather than be faced with a maze of many overlapping and unprioritised assets. Data need to be current and dynamic, relevant to the decisions at hand, and located in centralised databases. Incident controllers need to be made aware of available datasets and have the training necessary to provide the skills to interpret and use the data.
- Training of fire operations staff needs to include an understanding of the value of conservation assets, and the fire control actions that may protect or jeopardise their persistence. Training should also be conducted for staff from other departments or external organisations (e.g. zoos) so that they can be ready for deployment during fires for targeted actions for conservation assets, such as species translocations
- Where required, these improvements in planning and management need to be strengthened by a supporting legislative framework.

References

de Bie K, Currey K, Woinarski J, Wintle B, Garnett S, *et al.* (2021) *Protecting Threatened Species and Ecological Communities Before and During Bushfire: Learning from the 2019–20 Fires.* National Environmental Science Program Threatened Species Recovery Hub, Brisbane.

van Oldenborgh GJ, Krikken F, Lewis S, Leach NJ, Lehner F, *et al.* (2021) Attribution of the Australian bushfire risk to anthropogenic climate change. *Natural Hazards and Earth System Sciences* **21**, 941–960. doi:10.5194/nhess-21-941-2021

Ward M, Tulloch AIT, Radford JQ, Williams BA, Reside AE, *et al.* (2020) Impact of 2019–2020 mega-fires on Australian fauna habitat. *Nature Ecology & Evolution* **4**, 1321–1326. doi:10.1038/s41559-020-1251-1

Wintle BA, Legge SM, Woinarski JCZ (2020) After the mega-fires – what next for Australian wildlife? *Trends in Ecology & Evolution* **35**, 753–757. doi:10.1016/j.tree.2020.06.009

30

Government inquiries following the 2019–20 wildfires

John C. Z. Woinarski and Libby Rumpff

Summary

- Governments have commissioned many inquiries following wildfires in Australia, mostly to assess causation, identify failings and make recommendations for improvements in order to reduce the risks and impacts of future fires.
- Such inquiries are reactive and undertaken soon after wildfire, so there has been far less scrutiny directed to longer-term post-fire recovery, and of the effectiveness of actions taken to support such recovery.
- The 2019–20 wildfires were the subject of two national inquiries and reviews in most states. Biodiversity impacts were considered most substantially in the national royal commission and inquiries in New South Wales and Victoria.
- These inquiries recognised that climate change was a fundamental driver of the severity and extent of the 2019–20 wildfires, that the future was likely to be exposed to an increasing frequency of such fires, and hence that there was a pressing need to adapt management, policy and legislation to reduce the risks associated with escalating climate change.
- Typically, biodiversity conservation has not been a primary focus of post-fire inquiries, but the 2020 royal commission collated a large evidence base that attested to the unprecedented scale of detrimental impacts on biodiversity. It identified shortcomings in policy and management practices, and exemplary practices and policy that helped reduce impacts. However, just one of its 80 recommendations focused on biodiversity conservation, and this related only to enhanced national collation of biodiversity data.
- Biodiversity was also a feature of the Victorian inquiry, which found that readily accessible spatial information on biodiversity assets was crucial for their likelihood of being protected in fire, for rapid assessment of impacts, and to help guide prioritisation of responses after fire. The Victorian inquiry also noted the importance of appropriate planning, monitoring and modelling to support preparedness for fire events, the importance of formally embedding wildlife expertise in the fire operations hierarchy, and the value of nature-led recovery for rural communities.

Introduction

Large wildfires challenge human capability; many rage beyond control. They can kill people, destroy towns, impose large economic costs, erode our sense of security, and cause major losses of biodiversity. Following such disasters, governments typically undertake or commission systematic reviews to clarify the causal factors, identify failings and propose mechanisms that could be improved to render future comparable events (and losses) less likely.

There is a long history in Australia of holding inquiries in the wake of wildfires that have a catastrophic impact on humans. The 2020 royal commission tallied 240 formal reviews into natural disasters, mostly fire, in Australia since 1927 (Royal Commission into National Natural Disaster Arrangements 2020). There are recurrent themes across most of these many inquiries: the organisational structure of fire agencies, the responses made (including coordination, communications and safety), the adequacy of equipment and infrastructure, preparedness (notably including fuel reduction burning), and research (including development and application of new technologies, spatial mapping and predictive modelling) (Kanowski *et al.* 2005) (Fig. 30.1).

Have the inquiries worked? If it is assumed the objective of the hundreds of recommendations proposed by these many inquires inquiries is to also reduce the incidence of large-scale severe fire, then they have largely failed. Furthermore, it is a difficult question

Fig. 30.1. Much of the focus of the royal commissions and other inquiries relates to human lives lost and infrastructure destroyed. Here, a bus and houses burnt in the 2019–20 wildfires at Mallacoota. (Photo: 'Mallacoota fires' by Cazz is licensed under the public domain, via Wikimedia Commons)

to answer, as there is little ongoing accounting of the effectiveness of actions implemented in response to their recommendations:

> *The lack of any overarching implementation monitoring of recommendations from previous reviews and inquiries made it challenging to understand what had already been actioned and, if not, why not. (p. 14, Government of NSW 2020)*

> *... there have been many previous reviews and inquiries into bushfires in South Australia. However, not all the recommendations accepted by government have been audited for implementation, nor did it appear during this Review that governments act upon the experiences of other jurisdictions. (p. v, Government of South Australia 2020)*

As a consequence, many of the inquiries re-evaluate the same issues, and each fire event is treated as novel:

> *In 2020, Victorians ought not be surprised by bushfire. It is a constant in our landscape. Yet every time a major bushfire event occurs, it seems that many in the community are caught unawares as if by something new and unprecedented. A demand for answers, a search for what went wrong and who to blame inevitably seem to follow. (p. 22, Inspector-General for Emergency Management 2020)*

Yet this time was different. More so than previous reviews into Australian wildfires, these inquiries recognised that the 2019–20 wildfires were not a single aberrant event, but rather part of an evolving pattern, a warning of the escalating consequences of global climate change:

> *As the events of the 2019-2020 bushfire season show, what was unprecedented is now our future. (p. 6, Royal Commission into National Natural Disaster Arrangements 2020)*

> *Our task ... required us to look to the future. A future where such events will, regrettably, be more frequent and more severe. (p. 5, Royal Commission into National Natural Disaster Arrangements 2020)*

> *... the 2019–20 fires were part of a continuum of wildfire in Victoria ... in which the impact of climate change is increasingly evident ... (p. 58, Inspector-General for Emergency Management 2020)*

The national royal commission recognised that far more proactive approaches were needed to deal with this likely future, and it called for 'a fundamental shift in strategic thinking about national natural disaster management' (p. 26, Royal Commission into National Natural Disaster Arrangements 2020).

One shortcoming of all of these inquiries is that, because of their temporal framing, they rarely, if ever, take on a longer-term and more deliberative assessment of recovery. The terms of reference frame these largely as post-disaster exercises, conducted in the immediate aftermath of wildfire, and hence focused mostly on factors leading up to and during the fire, rather than on assessment of longer-term recovery.

Our aim in this chapter is to distil the manner in which biodiversity protection was considered in the inquiries into the 2019–20 wildfires, to highlight their evaluations of actions and policy settings that influenced the outcomes for biodiversity, and to consider the recommendations they made for changes that would reduce biodiversity losses in future comparable events.

Inquiries into the 2019–20 wildfires

Post-fire inquiries were commissioned by the governments of Australia, Queensland, New South Wales, the Australian Capital Territory, Victoria, South Australia and Western Australia, with the reports of these inquiries tallying many thousands of pages. Here we consider only the most substantial of these: the Australian royal commission and the inquiries of New South Wales, Victoria and South Australia.

2020 Royal Commission into National Natural Disaster Arrangements

This royal commission had a broad scope, explicitly including the clarification and better coordination and accountability of Commonwealth and state/territory responsibilities for fire management. The intergovernmental component of the inquiry was particularly pertinent given the national scale of these fires, the uneasy partitioning imposed by federation on responsibilities for land management and emergencies, and because the 2019–20 wildfires starkly exposed some disharmony among states in control operations for fires that burnt across borders.

For the first time, a commission included a chapter on wildlife and heritage. The royal commission's framing document recognised the need to increase preparedness for the future; better protect 'wildlife management and species conservation'; and examine the extent to which 'traditional land and fire management practices of Indigenous Australians could improve Australia's resilience to natural disasters' (Commonwealth of Australia 2020).

The commission report had several key findings. The first was that biodiversity protection was more effectively considered in fire operations when relevant expertise was available in the control room, or integrated into fire control operations:

> *16.16 In some jurisdictions, emergency services are embedded in the same portfolio as agencies responsible for environmental protection, but often they are not … in many instances wildlife rescue or protection efforts were initiated outside emergency services, there was great value in leveraging emergency management incident management teams … in assisting rescue operations, for example in coordinating fire suppression and requesting air support.*

The commission report also recognised the value of relevant planning and information about biodiversity values and locations being available before wildfire, particularly where this provided a prioritised basis for rapid response:

> *16.18 States and territories had bushfire strategies and operational plans in place before the 2019–20 bushfires with a view to preventing and responding to bushfire impacts on wildlife, ecosystems and heritage values.*
>
> *16.19 These strategies and plans identify high-level arrangements, such as threatened species and heritage sites, their locations, and response and recovery strategies.*

However, this finding does not clarify whether these strategies and plans were comprehensive in their coverage of biodiversity assets, or effective in mitigating impacts (see

Chapter 29). Despite the presence of these plans, the commission reported that protection of biodiversity assets needed to be better integrated into planning and response structures. This, in turn, relies on readily available and appropriate knowledge, which is so often sub-optimal (refer to Chapter 29).

While recognising the importance of integrated emergency planning and response systems, the commission did not challenge the deeply rooted presumptive hierarchy that, in fire control operations, biodiversity protection was subordinate to other considerations:

> *16.15 In responding to disasters, ... emergency services agencies have primary responsibility for protection of people, property and the environment – they provide protection in that order.*

A blanket ranking such as this fails to recognise that the context and degree of impact is critically important in decision making. The commission did not evaluate whether there may be circumstances where the risks of significant and irreparable biodiversity loss may justify prioritisation ahead of property. For example, is directing resources to a few sheds rather than a national park justifiable (Chapter 29)?

The commission report also recognised the need for more collaboration between government agencies and non-government animal welfare groups:

> *16.34 ... governments should work together with relevant non-government organisations to establish best practice arrangements and responses relevant to emergency wildlife response and recovery.*
>
> *16.26 Including non-government wildlife organisations within emergency management arrangements can enable these groups to work in concert ... to benefit from the situational awareness of first responders, and to access the fire ground safely. Raising awareness of animal welfare, species conservation, and the capabilities of wildlife first responders can also help ensure that these groups are deployed as swiftly and safely as practicable.*

The commission noted some shortcomings in policy and practice relating to fire impacts on biodiversity. For example, it identified that the listing of species and ecological communities as threatened under the *Environment Protection and Biodiversity Conservation Act 1999* (EPBC Act) may not account for 'anticipated increasing natural hazard risks' (16.52). In light of increasing risks, the commission supported an existing nomination to list 'fire regimes' as a Key Threatening Process (KTP) under the Act (16.53). The commission did not assess if recovery plans and conservation advices adequately consider fire management and other specified actions to be taken during fires or for post-fire recovery.

Notwithstanding the substantial consideration of biodiversity conservation in this royal commission, only one of its 80 recommendations was explicitly related to its theme of 'Wildlife & Heritage':

> *Recommendation 16.1 Environmental data. Australian, state and territory governments should ensure greater consistency and collaboration in the collation, storage, access and provision of data on the distribution and conservation status of Australian flora and fauna.*

In this case, the Australian Government 'supported in principle' this recommendation, responding that:

> *The Commonwealth, with all states and territories, has agreed to establish a common method for the assessment and listing of threatened species. [This] will maintain the current high level of scientific rigour in the assessment and listing of threatened species across Australia, while promoting a more consistent, efficient and harmonised process ... (Commonwealth of Australia Department of the Prime Minister and Cabinet 2020).*

However, this collaborative arrangement between Commonwealth and state/territory governments for the listing of threatened species was already largely in place before the 2019–20 wildfires (Department of the Environment 2015). Hence, it is likely that the royal commission, although recognising the extraordinary losses of biodiversity imposed by the 2019–20 wildfires, prompted no direct response by the Australian Government to transform relevant biodiversity conservation policy or legislation to enhance preparedness for future fires or protection during such fires.

Many of the royal commission's other recommendations were more broadly focused rather than explicitly directed towards biodiversity conservation, but would be likely to result in some improved outcomes for biodiversity. To help build resilience, preparedness and allow rapid response, the commission recommended the establishment of a set of permanent collaborative entities, including a national disaster advisory body (Recommendation 3.2), a 'resilience and recovery entity ... focused on long-term disaster risk reduction' (Recommendation 3.5) and a 'national preparedness and response entity' to better develop, coordinate, test and implement planning and practice in response to natural disasters (Recommendation 3.6).

Critically, the commission also called for more engagement with Traditional Owners 'to explore the relationship between Indigenous land and fire management and natural disaster resilience' (Recommendations 18.1, 18.2). The recognition that more needs to be done to improve resilience to these events is an important step. A major focus of previous commissions, the issue of fuel load management was considered, but recommendations on this subject were largely about better reporting and accountability (Recommendation 17.1), and review of processes (Recommendation 17.2), rather than definitive and justified prescriptions. Given the varied impacts of commonly utilised fuel management practices (e.g. prescribed burning) on biodiversity (Chapter 21), the need for integrated planning (i.e. 16.20) goes beyond emergency response.

The commission also identified some improvements needed in budgeting for recovery, including better assessment of the effectiveness of post-disaster spending (Recommendation 22.1) and more emphasis on funding to build resilience (Recommendation 22.7).

State inquiries

New South Wales

The 2019–20 wildfires burnt more of New South Wales than any other state (~5.5 million ha). The New South Wales Bushfire Inquiry concluded that previously effective management responses were inadequate due to the unprecedented nature of the fires: the fires 'challenged conventional assumptions' (Government of NSW 2020). A key example is fuel reduction burning; the inquiry concluded that it 'appears to have reduced fire severity in some instances, but in others it appears to have had no effect on the severity and spread of the fire' (p. iv). Nonetheless, it subsequently recommended 'that Government re-commit to the current, regionally based approach to planning and coordinating hazard reduction activities across all tenures' (Recommendation 19), along with the collaborative application of cultural burning (Recommendations 25 and 26).

In relation to biodiversity, the inquiry considered how assets, including places and species of significant conservation value, were considered in control operations during wildfires. It recognised that there was no system in place for determining or ranking priorities 'when multiple assets of value are threatened by fire and there are insufficient resources to protect them all' and concluded that 'to avoid uninformed decisions during a fire event on what to protect, a formal mechanism is needed for working out in advance the relative value of different assets' (p. 149). It recognised that input from the relevant conservation agency into the fire operations command structure was a critical factor in protecting some biodiversity assets, notably including Wollemi pine (*Wollemia nobilis*; see Chapters 9 and 27), whose protection was also aided by the existence of a recovery plan that provided details of suitable fire management actions. However, the report posited some fundamental questions:

> *But should the Wollemi Pine be saved at the expense of human life? And at the expense of houses, farms, towns and infrastructure? And why were the Wollemi Pines saved and not, for example, some other rare botanical species? (p. 150).*

The NSW inquiry raised the issue of prioritisation and trade-offs more directly than the national commission, and recognised that a more comprehensive and systematic approach was required to attribute value to assets of varying type, in order to prioritise those assets for protection during fire. It noted that:

> *What is considered critical for saving to some people may not be considered critical to others. Nevertheless, there are assets that the community generally accepts as critical and recognises as a priority for saving in a bush fire. These assets are typically those essential to the functioning of society (and the economy), particularly infrastructure ... It is the other things of value – environmental and cultural assets and facilities such as community halls and sporting amenities – where prioritisation becomes more difficult. Therefore, trade-offs between the values must be made at the planning and preparation stage to protect these assets rather than rely on an arbitrary and ad hoc response when the bush fires are in progress.*
>
> *The Inquiry suggests that a formal mechanism is needed for working out the relative value of different assets and that the guidance provided needs to be satisfactory to the community as a whole. Further, common units of value are required across all asset classes ... For example, under such a value structure, values could be assigned as follows:*
>
> - *human life at Value Level 1;*
> - *critical infrastructure at Value Level 2;*
> - *individual properties, stock, top-level ecological and cultural assets at Value Level 3; or*
> - *general infrastructure and lower level ecological and cultural assets at Value Level 4.*
>
> *Obviously, this can be made more sophisticated, for example by applying both economic value and a value structure (and, possibly, likelihood of success if threatened by fire). (pp. 150–151)*

The report then recognised the need for better information on the location of such assets (an 'authoritative consolidated State register/repository of things of value/assets/services'), and for better information on how to protect them before and during fires.

The inquiry's terms of reference included a requirement to 'make recommendations ... on appropriate action to adapt to future bush fire risks to communities and ecosystems', and the report noted that 'more species and ecosystems are now at increased risk of decline than before the fires'. It recommended:

> *That Government invest in long-term ecosystem and land management monitoring, modelling, forecasting, research and evaluation ... This will include, among other things:*
>
> - *tracking and trying to forecast what is happening to ecosystems over decades under projected changes to climate extremes, including fire regime change*
> - *better understanding interaction of fire with other disturbances, e.g. drought, hydrological changes in the landscape*
> - *commissioning experiments and feasibility studies for ecosystem adaptation experiments – for example, facilitating shift of high conservation-value rainforest vegetation communities further south as climatic conditions change. (Recommendation 36)*

South Australia

The main considerations of the South Australian inquiry (Government of South Australia 2020) were about governance, training, resourcing and communications. The inquiry's report noted that some fires were beyond control: they 'exceeded the upper limit of direct attack with any means of firefighting, including machines, aircraft and fire tankers' (p. 32). As with other post-fire inquiries, it considered the effectiveness of fuel reduction burning, finding that: 'It is also almost certain, due to the weather and climatic conditions outlined, that the fire behaviour could not have been prevented by greater levels of hazard reduction' (p. 32).

In its reporting of impacts, the inquiry notably counterpointed tallies of human deaths and economic costs to agriculture with reporting of the extent of conservation reserves burnt, an estimate of the numbers of koalas (*Phascolarctos cinereus*) killed (claimed to be 40 000–50 000), and the numbers of threatened species affected, identifying six species as explicit priorities for urgent intervention. Noting 'the impact of bushfire on wildlife is not well understood' (p. 100), the inquiry also made one recommendation for further research on the impacts of fire on biodiversity (along with other matters):

> *Collate data and research the impact of bushfires upon communities, firefighters and animals (both native and domestic) to identify appropriate medium and long-term welfare and support requirements. (Recommendation 15)*

Victoria

The Victorian inquiry into the 2019–20 wildfires (which burnt 1.5 million ha of the state) was the most comprehensive in considering biodiversity. This reporting notably included one volume evaluating recovery to date (Inspector-General for Emergency Management 2021), including in relation to some actions taken post-fire to recover biodiversity – an assessment little considered in other inquiries. It also recognised a broader dynamic and context: that not only was climate change resulting in notably different fire regimes, but also that:

The human context is changing too. Victoria's population is growing steadily ... Some of the fastest growing parts ... are pushing settlement deeper into this increasingly active fire environment. Many Victorians who have probably never considered the possibility of confronting a bushfire ... must now adapt to living with that risk. (p. 23, Inspector-General for Emergency Management 2020)

Such human demographic changes can be expected to further complicate choices about protecting human and natural assets that need to be made urgently in fire control operations, and may increase demands for fuel reduction burning across the expanding fringes of settlement.

The inquiry's reporting included a detailed assessment of biodiversity impacts (concluding that impacts on wildlife were 'devastating' (p. 276)), and of factors that helped or hindered biodiversity conservation efforts. It noted two main factors that helped preparedness for these fires, and the protection of biodiversity during them. The first of these factors was a robust and readily accessible evidence base about biodiversity values and their location. The second factor was the inclusion of relevant experts in the fire control hierarchy, although the report recognised that such integration was not yet well established.

With respect to biodiversity information, the inquiry report noted:

the considerable research, planning and coordination to map and identify high priority biodiversity areas ... enables improved decision-making to prioritise and coordinate conservation and welfare across public and private land during emergencies ... There has been concerted work to better integrate the extensive expertise, information, decision support tools and partnerships ... into emergency preparedness for bushfires'. (pp. 88, 89, Inspector-General for Emergency Management 2020)

Detailed spatial knowledge not only helped ensure the protection of some biodiversity assets during the fires, it also allowed for a rapid assessment of the extent to which the fires burnt ecological communities and the habitat of threatened species (DELWP 2020). Complemented by decision-support tools, and informed by expert workshops, this then allowed for rapid prioritisation of emergency management responses for fire-affected biodiversity. Despite having 'no pre-existing plan for establishing an overarching program of work to protect biodiversity after a disaster' (p. 368, Inspector-General for Emergency Management 2021), the response made was mostly strategic and evidence-based. The inquiry found that, post-fire, the Department of Environment, Land, Water and Planning (DELWP) and Parks Victoria 'successfully established a biodiversity recovery program that was delivered rapidly and collaboratively' (Finding 17.5) and 'delivered actions to control pest animals and plants in a coordinated way with substantial outputs on public land' (Finding 17.3). However, the inquiry did note some failings in the biodiversity management response: 'The Department ... was not prepared for reseeding ash forests at the scale required despite it being a known gap in preparedness for post-fire forest restoration' (Finding 17.1), and '... did not have sufficient preplanning to support a strategic approach to emergency extraction of threatened species, including prioritisation, coordination and approvals' (Finding 17.4).

The inquiry recognised the significance of funding for the success of its biodiversity recovery response: 'The Victorian Government invested a large amount of funding in protecting biodiversity following the 2019–20 fires. That enabled DELWP to deliver biodiversity response and recovery actions on a scale that exceeded past events' (p. 368, Inspector-General for Emergency Management 2021). However, there were challenges: '… recovery funding arrangements were a barrier to achieving desired outcomes. In some cases, funding was not available early enough to support critical immediate interventions, while in other cases the uncertainty about long-term funding was an issue' (p. 381, Inspector-General for Emergency Management 2021).

With respect to organisational structures, the inquiry noted that 'The appointment of the Class 2 Controller – Wildlife greatly assisted in prioritising and coordinating the wildlife welfare and biodiversity response …' (Finding 7.16), and:

> *the protection of … conservation assets received greater consideration and was better managed than during past events. The standing up of a Class 2 Controller – Wildlife for the first time was an effective approach to manage the significant threat occurring to biodiversity. (p. 280, Inspector-General for Emergency Management 2020)*

However, it also found that the legislative, regulatory and planning foundation for embedding expertise in the fire operational structure was better established for animal welfare than for biodiversity conservation. So, whereas there was a pre-existing Victorian Emergency Animal Welfare Plan (which set out a pathway for efficient and effective coordination services before, during and after emergencies), there was no comparable plan for biodiversity conservation. Accordingly, the inquiry recognised:

> *A range of improvements to preparedness arrangements are required … biodiversity conservation lacks formal structure within Victoria's emergency management arrangements. This includes the absence of a biodiversity response plan, specified biodiversity roles [in the fire operation structure] and clear triggers for activation. There is also further work required to improve integration of biodiversity response roles and decision making into state emergency management arrangements. (p. 92, Inspector-General for Emergency Management 2020)*

Looking to the future, the Victorian inquiry addressed the role of adaptive management and ongoing monitoring in long-term data needs, to both assess the efficacy of recovery actions and refine modelling tools to be capable of responding to future scenarios:

> *The 2019–20 Victorian fire season showed the new potential scale and impact of bushfires on biodiversity under climate change. Additional scenario planning is required to ensure that the partnerships can be leveraged to best effect under a broad range of emergency scenarios … [and] will also enable improved readiness arrangements and response plans for the conservation of threatened species and habitats in emergencies. (p. 92, Inspector-General for Emergency Management 2020)*

Unlike the other inquires reported here, there was also recognition of the increasing need and mutual benefits of linking post-fire biodiversity recovery efforts to community, through nature-led community recovery programs.

As with other inquiries, the Victorian inquiry recognised the need for an increased role for Traditional Owners in the implementation of fire management, and reported: 'Traditional Owners [caring] for Country through cultural burning and land practices … is producing positive results. There is significant interest shown by the sector, government and community to see more cultural burning occur across the landscape. Resourcing to support mechanisms for Traditional Owners to navigate the policy requirements for planned burning would support greater initiation, delivery of and sharing of cultural burning techniques'. (Observation 4.1)

The Victorian inquiry also included a detailed consideration of fuel reduction burning (as a 'central theme of this report'), including the extent to which it helped mitigate impacts and could be further enhanced. It noted that this was a 'deeply contested question … at the centre of public debate' (p. 23), with this discord in part due to fundamental differences in values and experiences within the community. As with other inquiries, it concluded that fuel reduction burning 'is not a simple panacea' (p. 23). The inquiry report also advocated 'increased use of 'non-burning' fuel management treatments, including mechanical means' (Recommendation 6), notwithstanding that the biodiversity impact of such procedures remains poorly resolved.

Conclusions

The inquiries established in response to the 2019–20 wildfires recognised a clear signal of climate change in the severity and extent of these fires, and hence the likelihood of many more comparable fires in the future. This recognition imposes an urgent need to improve legislation, policy, planning and practice that can more effectively mitigate and adapt to climate change, to reduce the likelihood of fire, mitigate fire impacts and support post-fire recovery. Government inquiries provide a mechanism to identify and realise such enhancements.

These inquiries focused mostly on factors that led to loss of human life or property. However, the profound impacts on biodiversity of the 2019–20 wildfires forced more consideration of risks to biodiversity, and of identifying the shortcomings of existing management, planning and policy for biodiversity conservation. Across the set of these inquiries, there were key factors identified that could improve outcomes for biodiversity in the context of wildfire (Table 30.1). However, with some notable exceptions (Kanowski *et al.* 2005), governments commissioning these inquiries have generally not appointed experts in the

Table 30.1. Summary table of recommendations or observations relating most directly to biodiversity from the 2020 royal commission (RC) and state inquiries into the 2019–20 wildfires.

Area of recommendation	Before fire (pre-emptive, building resilience)	During fire (operational)	After fire (recovery)
Legislation	Include fire regimes as a Key Threatening Process under the EPBC Act (RC)		
Policy	Establish a range of collaborative bodies to build resilience (RC); improve integration of biodiversity response roles and decision making into state emergency management arrangements (Victorian inquiry)		

continued

Table 30.1. Continued

Area of recommendation	Before fire (pre-emptive, building resilience)	During fire (operational)	After fire (recovery)
Practice	Implement risk-based approach to hazard reduction burning, and implement 'increased use of "non-burning" fuel management treatments' (Victorian inquiry)	Apply pre-established prioritisation to protect significant natural assets during fire control (NSW inquiry); embed biodiversity expertise in operational control (RC, Victorian inquiry)	Rapid assessment to prioritise post-fire recovery needs (Victorian inquiry); urgent provision of adequate funding for recovery (Victorian inquiry); implement nature-led community recovery programs (Victorian inquiry)
Planning	Implement hazard reduction burning, through bush fire risk management plans (NSW inquiry); prepare bushfire strategies and operational plans that aim to protect biodiversity during wildfires (RC); ascribe values to biodiversity assets to help prioritise their protection during fire operations (RC); develop and justify valuation for natural and other assets to help prioritise their protection during fire operations (NSW inquiry); develop and implement a biodiversity response plan to coordinate actions before, during and after emergencies (Victorian inquiry); undertake scenario planning to enhance practice during emergencies (Victorian inquiry); further develop and implement integrated bushfire management area plans (SA inquiry)		
Knowledge	Better integrate biodiversity data (RC); delineate important natural assets (NSW inquiry); monitor ecosystems and their responses to fire regimes and climate change (NSW inquiry); assess compounding impacts of fire and other threats (NSW inquiry); undertake research on adaptation (NSW inquiry); delineate important natural assets and include them in register of values (Victorian inquiry); undertake research on impacts of wildfires and fire regimes on biodiversity (SA inquiry)		Monitor ecosystems and their responses to fire regimes and climate change (NSW inquiry); assess compounding impacts of fire and other threats (NSW inquiry); monitor recovery (Victorian inquiry)

fields of ecology or conservation; and while those experts may be consulted throughout the process, the inquiries have mostly been conducted by experts in emergency services, governance and auditing. As such, there is little holistic consideration of how the package of recommendations in each of the inquiries will benefit or disadvantage biodiversity.

There are some key issues evident in this chapter that will need to be addressed. First, the post-fire inquiries predominantly focus on circumstances leading up to and during fires: there is no longer-term view of recovery (Fig. 30.2). As highlighted in Chapter 29, inquiries should have incorporated both an assessment and recommendations to support integrated planning and management of biodiversity before, during and after an event. Fires are not the stand-alone threat to biodiversity, and if we continue to focus investment on improving our response to emergencies and the aftermath, without addressing and funding the mitigation of multiple threats to improve the resilience of our ecosystems, then we are both wasting resources and risking continual declines in biodiversity.

In addition, we rarely understand the effectiveness of actions arising from inquiry recommendations, or the costs of not adopting recommendations. Both the Victorian and New South Wales inquiries recognised the need for more substantial monitoring of outcomes, before and after fires. However, monitoring needs to be tied back to the original objectives (of recommendations), in order to gauge success.

Last, the inquiries have not provided substantial recommendations to amend legislation to enhance protection for biodiversity. The inquiries have also generally been reluctant to address and better resolve the fundamental concern of valuing and prioritising biodiversity protection during wildfire relative to that of human life or property. Without such questioning and legislative protection, biodiversity conservation will continue to be marginalised during fire operations, and irreplaceable conservation assets are likely to be diminished and lost over the course of successive fires.

Fig. 30.2. The long road to recovery after fires has not been a major focus of royal commissions and inquiries. Here, an echidna (*Tachyglossus aculeatus*) moves in an extensively burnt landscape, Kangaroo Island. (Photo: WWF-Australia/Douglas Thron)

Recommendations

- Inquiries to date have been undertaken soon after wildfire and have focused on factors preceding, during and immediately after fire. More focus is warranted on the adequacy and effectiveness of longer-term post-fire responses.
- Reflecting their primary focus on wildfire, the inquiries have not provided a broad context for biodiversity impacts. There needs to be recognition that an integrated and well-resourced strategy that deals with managing multiple threats to biodiversity is required to improve resilience, rather than episodic consideration and funding that deals only with the aftermath of catastrophe.
- Most recent inquiries recognise that biodiversity assets need to be more effectively delineated and prioritised, and their protection better incorporated into fire management, planning and control operations alongside other values. To be effective, such a recommendation also requires the following suite of changes:
    - improvement in biodiversity databases, and their access and interpretability in emergency settings;
    - enhanced knowledge of the fire and other management needs of these biodiversity assets, with such management requirements embedded in appropriate planning;
    - a clear prioritisation of biodiversity assets, that can readily inform a hierarchy of protection during fire operations, and with the need for such protection formalised in policy and legislation;
    - an engagement with the community about the integrated prioritisation of natural assets with potentially competing needs for protection during fires – noting that biodiversity assets will continue to be lost if the long-established convention that protection during fire operations always prioritises infrastructure and property over biodiversity.
- Inquiries into the 2019–20 wildfires recognised climate change as a key causal factor, but largely avoided making recommendations about managing the fundamental driver of climate change. It is insufficient to simply recognise the trend of increased incidence of severe fires with climate change, and more attention needs to be directed at urgently curtailing greenhouse gas emissions to reduce the risks of future comparable catastrophes.

References

Commonwealth of Australia (2020) 'Commonwealth Letters Patent 55'. Commonwealth of Australia, Canberra.

Commonwealth of Australia Department of the Prime Minister and Cabinet (2020) *A National Approach to National Disasters*. Commonwealth of Australia, Canberra.

DELWP (2020) 'Victoria's bushfire emergency: biodiversity response and recovery. Version 2'. Department of Environment Land Water and Planning, East Melbourne.

Department of the Environment (2015) *Intergovernmental Memorandum of Understanding – Agreement on a Common Assessment Method for Listing of Threatened Species and Threatened Ecological Communities*. Department of the Environment, Canberra.

Government of NSW (2020) 'Final report of the NSW Bushfire Inquiry.' Government of NSW, Sydney.

Government of South Australia (2020) *Independent Review into South Australia's Bushfire Season*. Government of South Australia, Adelaide.

Inspector-General for Emergency Management (2020) *Inquiry into the 2019–20 Victorian Fire Season. Phase 1. Community and Sector Preparedness for and Response to the 2019–20 Fire Season*. Government of Victoria, Melbourne.

Inspector-General for Emergency Management (2021) *Inquiry into the 2019–20 Victorian Fire Season. Phase 2. Progress and Effectiveness of Victoria's Immediate Relief and Recovery Arrangements*. Government of Victoria, Melbourne.

Kanowski PJ, Whelan RJ, Ellis S (2005) Inquiries following the 2002–2003 Australian bushfires: common themes and future directions for Australian bushfire mitigation and management. *Australian Forestry* **68**, 76–86. doi:10.1080/00049158.2005.10674950

Royal Commission into National Natural Disaster Arrangements (2020) 'Royal Commission into National Natural Disaster Arrangements Report'. Commonwealth of Australia, Canberra.

31

Monitoring impacts and recovery

Brendan A. Wintle, Darren Southwell, Ella Loeffler, Heloise Gibb, Emma Spencer, Adam Fawcett, Pamela Fallow, Elizabeth M. Tasker, Brad Law, Craig Nitschke, Ella Plumanns Pouton, Luke Kelly, Don Driscoll and John C. Z. Woinarski

Summary

Context and challenges

- Monitoring provides crucial information to help understand species distribution and abundance, the condition of ecosystems and changes induced by disturbances such as fire, and to evaluate and enhance actions to conserve and recover species and ecological communities.
- However, biodiversity monitoring in Australia is patchy, has poor taxonomic coverage and is generally of low quality.

Main findings

- Lack of baseline monitoring data on species range and abundance, and ecological community condition, limited our understanding of the impacts of the 2019–20 wildfires, making it difficult to efficiently allocate urgent conservation actions, and track recovery.
- For most species impacted by the 2019–20 wildfires, impacts were mostly inferred by estimating the overlap between the fire extent and species' range. Species distribution models and expert knowledge of species vulnerability to fire were used to refine impact estimates for vertebrates, invertebrates, plants and ecological communities.
- For around 300 species and ecological communities, reconnaissance survey programs were established to improve our understanding of impacts and to ascertain whether recovery was occurring. Over 1500 targeted surveys of flora, fauna and ecological communities were undertaken within 18 months of the wildfires.
- For a small subset of species, survey observation data or monitoring programs pre-dating the fires were available, allowing direct estimation of impacts.
- Surveys and post-fire observation data were collected by groups including agencies, non-government organisations (NGOs), citizen scientists and university researchers.

- Wildfire impacts on species and ecological communities inferred from reconnaissance surveys ranged from catastrophic (no individuals of a species found in previously occupied sites) to benign (unchanged occupancy).
- As a condition of the Australian Government's bushfire recovery program, monitoring will measure the effectiveness of post-fire recovery actions for some species and ecological communities.
- There is an urgent need to dramatically increase our understanding of the extent of occurrence, abundance, and condition of species through properly resourced, carefully designed, systematic monitoring programs. Such programs would enhance planning and capacity to respond decisively to future fires, improving conservation outcomes.

Introduction

Monitoring is a key to understanding the state of the environment, to evaluate the success of actions, to learn about how environments work and to trigger actions when unacceptably negative changes have occurred. Monitoring is particularly important in the case of understanding the impacts on species and ecosystems of stressors such as climate change, harvesting, invasive species, drought and fire. Indeed, monitoring is the cornerstone of most impact assessment.

Unfortunately, the investment in monitoring by Australian agencies and organisations is mostly insufficient to report accurately on the state and trends in biodiversity, or the effects of environmental management programs (e.g. Keith *et al.* 2018; Scheele *et al.* 2019). There is also little monitoring of many other components of the natural environment, including of soils (Chapter 5), ecosystem processes, aquatic health, and sites of cultural and biodiversity significance (e.g. Ramsar sites and World Heritage sites). Such shortcomings have proven a significant constraint on accurately assessing the impacts of the 2019–20 wildfires, prioritising management responses and charting subsequent recovery.

Inquiries into previous wildfires in Australia have generally not focused on biodiversity impacts. However, the immense area and diversity of habitats burnt in 2019–20, and widespread media coverage, made biodiversity impacts a stronger focus for the public and policy makers (Chapter 30). This focus on biodiversity during and immediately after the fires generated an instant demand for data and evidence about impacts on biodiversity, shining a spotlight on our biodiversity data infrastructure and, for the first time in decades, injecting some resources into national-scale data synthesis, gap analysis, field reconnaissance and post-fire monitoring. In this chapter, we provide an overview of post-fire field reconnaissance efforts, describe some of the few ongoing monitoring programs that are beginning to paint a picture of post-fire recovery and loss, and discuss opportunities to build on the pulse of survey data that has been generated following the fires.

Findings

Overall

Our ability to understand the true impacts of the 2019–20 wildfires has been hampered by a lack of measured baseline for most species. For most species the lack of (recent) pre-fire occurrence and monitoring data also made prioritisation of where and what actions should be undertaken more difficult. Population strongholds and the status of those strongholds before and after the fires could not be known. This was particularly acute for invertebrates, fungi and many plants that have to date received scant attention and data collection (Chapters 9, 10 and 11).

The use of post-fire reconnaissance surveys helped to refine understanding of impacts, especially where surveys took place at locations previously known to be recently occupied by species, or for which ecological community condition had previously been measured. In rare cases, existing long-term monitoring sites fell across burnt and unburnt locations, making strong inference about declines possible and serving as a basis for ongoing monitoring. On reviewing the post-fire reconnaissance and monitoring data collection and synthesis across the Commonwealth, state agencies (often several within a single state), non-government organisations (NGOs), Traditional Owners, universities and citizen scientists, the weaknesses in current national biodiversity data infrastructure are apparent. Lack of coordination, communication and sharing is muddying the signal from the small amount of data that does exist. A nation-wide initiative to synthesise, organise, streamline and support systematic biodiversity monitoring and analysis is urgently needed.

Post-fire reconnaissance surveys

Reconnaissance surveys are field surveys to rapidly assess the immediate impacts of fires on species or ecological communities. Reconnaissance programs were established in all fire-affected states to help refine the assessment of fire impacts and set post-fire recovery priorities. Surveys were supported by Commonwealth and state funding, public donations and citizen science. The speed of the establishment of reconnaissance surveys, sometimes part of emergency recovery work, was impressive (https://www.ari.vic.gov.au/research/fire/bushfire-response-2020-aquatic-rescues).

At the time of writing (March 2022), based on available national, state and NGO reports, the combined national reconnaissance effort has resulted in the collection of ~1500 new survey datasets across much of the country, using a variety of survey methods and approaches. The actual number is likely to be higher given that some data collections were not directly supported by state or Commonwealth funding; those have generally not been accounted for here. Approximately 850 animal surveys and 271 plant surveys have been undertaken with the support of Australian Government funding. Surveys assessed on-ground impacts and in some instances have provided evidence of early recovery trajectories. Surveys focused on species and ecological communities that were identified as national priorities for management intervention. Additional reconnaissance surveys were supported by the states and some NGOs for national, state or their own priority species and ecological communities.

In Victoria, surveys were conducted for 126 plant and 34 animal species across 1177 individual site surveys (DELWP 2022). A third of these surveys were supported under the Commonwealth bushfire recovery program reported above; the remainder were supported by the Victorian Government. Surveys were prioritised towards species with high apparent levels of loss based on spatial overlap of the fire footprint and previous records, habitat distribution models and high sensitivity to threats exacerbated by fire. Two hundred and eighty-seven sites known previously to contain threatened plants impacted by the fires were searched. Of the 126 surveyed threatened plant species, 108 species were regenerating at impacted sites, while 18 were not. Of the 520 fauna surveys (sites) conducted for 34 species at previously known locations, 175 surveys yielded positive observations.

The impacts of the fires on most fauna appeared stark, even after considering imperfect detection and spatial inaccuracies. For example, 26% of the greater glider's (*Petauroides volans*) Victorian range was burnt based on habitat overlay analysis. Forty-four per cent of the sites at which the glider had previously been observed in the East Gippsland region yielded no detections during post-fire reconnaissance surveys at those sites, even when suitable habitat remained. Results were similar for the yellow-bellied glider (*Petaurus australis*) (Chapter 16). The abundance of endangered galaxid species dropped by around 30%, and some populations of threatened frog species, including four populations of the endangered spotted tree frog (*Litoria spenceri*), could not be detected, suggesting possible local extirpation (DELWP 2022).

Aerial surveys of alpine bog ecological communities indicate that around 12% of Victoria's alpine bog area burnt at varying degrees of severity. On-ground surveys to ascertain the intensity of the damage are pending.

Within 12 months of the fire, on-ground and aerial surveys had been conducted in New South Wales for 200 plant and animal species and ecological communities rated at high risk, based on range overlap with the fires extent and vulnerability assessment (Auld *et al.* 2020). Additional on-ground surveys for 100 high-risk rare or range-restricted plant species are nearing completion. Longer-term monitoring programs for a subset of NSW *Saving Our Species* (SoS) program priority species provides a solid foundation on which impacts for a portion of the state's threatened species and ecological communities will be assessed.

In the Border Ranges region of south-eastern Queensland, post-fire surveys were conducted to detect national priority threatened species in burnt habitats including previously known locations of threatened species such as the brush-tailed rock-wallaby (*Petrogale penicillata*), Hastings River and New Holland mice (*Pseudomys oralis*, *Pseudomys novaehollandiae*), and long-nosed potoroo (*Potorous tridactylus tridactylus*). Surveys generally indicated that species persisted at burnt sites, though population-level impacts were not possible to ascertain given lack of pre-fire population data.

Pre-existing surveys for the broad-toothed rat (*Mastacomys fuscus*) in Namadgi National Park (Milner *et al.* 2016) provided a robust baseline against which post-fire sampling revealed declines in occupancy. Existing survey data informed spatial prioritisation of predator control efforts. Similarly, robust population data for several well-monitored Western Australian species, including the western ground parrot (*Pezoporus wallicus flaviventris*), allowed swift post-fire predator control efforts to help reduce fire impacts on the species (Box 31.1; Chapter 17).

Box 31.1. Coordinated monitoring and recovery planning for the eastern bristle bird and eastern ground parrot

BirdLife Australia has commenced recovery projects, including monitoring, for 12 of the 17 birds identified nationally as priorities for urgent post-fire management (Chapter 15). For example, the eastern bristlebird (*Dasyornis brachypterus*: Fig. 31.1) and eastern ground parrot (*Pezoporus wallicus wallicus*) have populations distributed across three states. Monitoring frequency and methodology vary among states and populations.

In late 2020, BirdLife Australia brought together eastern bristlebird and eastern ground parrot experts from multiple institutions to assess the nation-wide impacts of the 2019–20 wildfires and guide bushfire recovery action and monitoring. The collaboration generated a comprehensive picture of the post-fire distribution of both species, to understand recent range retractions. This information was used to develop eastern bristlebird recovery advice that was ultimately included in a national recovery plan due for publication in 2022.

Fig. 31.1. Eastern bristlebird. (Photo: Duade Paton)

Individual species experts and citizen scientists from across a range of institutions, and large NGO-led post-fire reconnaissance efforts, have been contributing post-fire studies and data into journals and data repositories since the fires. In Victoria, the SWIFFT network (https://www.swifft.net.au/cb_pages/team_bushfire_recovery_-_community_wildlife_reporting.php) was used to harness citizen science in heavily impacted areas for species of concern such as glossy black-cockatoos (*Calyptorhynchus lathami*) and diamond pythons (*Morelia spilota spilota*), which are difficult to detect in systematic survey. Using the national citizen science project, FrogID, Rowley *et al.* (2020) reported widespread persistence, at least in the short term, of many frog species with previous records within the firegrounds (Chapter 13). By mining their citizen science repository, they found that of the 66 frog species detected in the firegrounds before the fire, 45 were detected inside the fire grounds within 125 days after the fires. All 33 frog species with more than five records that were detected in the months of December–March pre-fire were detected post-fire. However, although valuable for reporting on persistence, this dataset is unsuitable for estimating the impacts of fire on the abundance of species.

In a creative analysis of bird species observation changes following the fires, Lee *et al.* (in press) analysed citizen science e-Bird data to test for systematic patterns in species use of the post-fire environment. Analysing observation change data for 76 species, they found that 22 species of smaller birds with smaller ranges and high dietary specialisation were less likely to be found in burnt areas, post-fire. In contrast, 30 larger bird species with larger ranges were more likely to be observed in burnt areas. The remaining 24 bird species showed no statistically significant change.

Carefully designed post-fire reconnaissance studies can reveal significant insights into species-level impacts when enough burnt and unburnt locations can be surveyed, such as in the 'space-for-time' study of fire impacts on the golden-tipped bat (*Phoniscus papuensis*) (Chapter 16). Similar in intent was a study looking at the impacts of the fires on native bees, led by the Arthur Rylah Institute and the Melbourne Museum, in which bee populations within burnt areas were compared with those in adjacent unburnt locations (https://www.ari.vic.gov.au/research/fire/bushfire-response-2020-impacts-on-native-bees). After collecting more than 1100 native bees representing 67 species, they found less than half the number of species in the burnt areas (25) compared to the unburnt areas (54). Bee abundance was also higher at unburnt locations. Very few above-ground nesting bees were found in the burnt areas, possibly because nesting locations were destroyed by the fire or because their preferred food plants were yet to regenerate.

Overall, reconnaissance surveys after the 2019–20 wildfires provided a foundation on which to understand ongoing impact for some species, to prioritise recovery efforts, and in some cases to serve as a baseline for ongoing monitoring efforts. For some species, they represented a much-needed injection of data on their current distribution and conservation status. For other species, such as the greater glider, the surveys revealed an even more urgent conservation status, with declines in observation rates recorded, even at sites with relatively low fire impacts.

Rare and precious gems: learning from pre-existing monitoring

For the species for which regular monitoring had been undertaken before the 2019–20 wildfires, the opportunity existed to gain significant insights into fire impacts and recovery. The examples below highlight the benefits of strong existing monitoring data.

In the Gondwana Rainforests of Australia World Heritage Area, burnt for the first time in recorded history following years of drought, surveys of mountain frogs (*Philoria kundagungan*) and Richmond mountain frogs (*Philoria richmondensis*) had been undertaken by a variety of groups since 2013 (Willacy 2014; Bolitho *et al.* 2021). These studies established sites across the species' ranges, and used a robust methodology that provided a solid basis for estimating changes in occupancy and abundance over time. Following the 2019–20 wildfires, Heard *et al.* (2021) revisited the previously surveyed locations, deploying similarly robust occupancy and abundance surveys. Occupancy rates for these two species at sites burnt in moderate to high severity varied from 14 to 23% compared with occupancies of between 54 and 64% at unburnt sites. The robust sampling design and pre-fire sampling had allowed analyses to estimate abundance change and start to disentangle the role of drought and fire.

The NSW SoS monitoring, evaluation and reporting plans have emerged as a useful tool for helping to target post-fire surveys and to understand impacts on some plant species. Metcalfe's greenhood (*Pterostylis metcalfei*) is a narrow-range endemic orchid that was not initially considered highly sensitive to fire. However, as an SoS program species threatened by grazing, trampling and weed invasion, it was closely managed and monitored. Post-fire monitoring of two populations has indicated the species is yet to recover from the fires, despite strong rains. This may be due to the very high fire intensity it experienced (Chapter 9). Similarly, existing monitoring for Guthrie's grevillea (*Grevillea guthrieana*) on the Carrai Plateau in central New South Wales indicated the sensitivity of the species to fire, and this allowed rapid deployment of post-fire mitigation measures to ensure threats to remaining populations are minimised. A third example is described in Box 31.2.

Box 31.2. The beauty of existing data: SoS priority species

Pygmy cypress pine (*Callitris oblonga*) is patchily distributed through northern New South Wales and is listed as Vulnerable in New South Wales and nationally. Twenty years of monitoring and abundance estimates across most populations provided foundational data on which to base post-fire surveys where it was identified that a large portion of its range had burnt in 2019–20. Immediately following the fire,

Fig. 31.2. A stand of dead adults of *Callitris oblonga parva* at Cathedral Rock National Park (A) living adults at Cathedral Rock National Park pre-2019 (B). (Photos: (A) John T Hunter, (B) Adam Fawcett).

standing dead stems were counted, enabling estimation of the pre-fire population (Fig. 31.2). Following the fire, damage and adult mortality were very high (83–97%) across four of five key populations, amounting to a range-wide 88% loss of adult stems. Immediately following the fire, some germinants were observed, but most have since died, probably due to grazing. This species will require intensive management and fire protection in the remaining population to avoid state-wide extinction.

Across the country, many individual monitoring and research studies will continue to provide crucial data and knowledge to allow us to piece together the evidence about the combined impacts of fire and other stressors, helping to prioritise and support recovery efforts. But despite some rare exceptions, financial and institutional support for such studies is low. With the loss of key long-term resources provided by programs such as the recently terminated LTERN, many monitoring programs rely on good will and opportunism for their ongoing existence (Keith *et al.* 2018). Furthermore, recovery of many species after these wildfires is likely to take many years, even decades, and documenting the pace and extent of recovery, and the effectiveness of supporting conservation actions, is likely to be a similarly long-term proposition. Most post-fire investments in recovery and monitoring, however, are currently constrained to the short-term, with little assurance of longer-term support.

Looking forward: opportunities to improve monitoring and preparedness

The Australian Government's Phase 1 and 2 bushfire recovery programs have prompted the development of short-term monitoring programs to document the outcomes of recovery projects. The MERIT tool (https://www.nrm.gov.au/my-project/monitoring-and-reporting-plan/merit) is being used by proponents of all projects supported by the Australian Government to establish project monitoring. The opportunity to bring monitoring projects together under a nationally coordinated monitoring program could be explored as one means of building a path to improved national biodiversity monitoring systems.

Following the fires several other initiatives that indicate significant opportunities to enhance the amount and quality of ongoing biodiversity monitoring have arisen. For example, a collaborative on-ground monitoring program led by WWF-Australia has provided over 1000 camera traps to researchers and land managers across the country, and is using innovative AI species identification technology to improve ongoing biodiversity monitoring (Box 31.3).

Box 31.3. Eyes On Recovery: a camera trap partnership

Eyes On Recovery is a collaborative venture between WWF, Conservation International, and Google that is using camera traps and AI to track the post-fire recovery of wildlife in Australia. Through partnerships with a range of on-ground land managers and researchers, EOR has deployed over 1000 remote cameras across the country. The initiative is also supporting camera data processing, analysis and sharing through an online photo management platform called 'Wildlife Insights' (WI). WI uses AI to automatically identify species and efficiently process the millions of images generated, providing ready-to-interpret datasets that inform important management decisions. For example, EOR is currently supporting a project led by the Kangaroo Island Landscape Board, to assess the impacts of fire and feral cats on Kangaroo Island dunnart (*Sminthopsis fuliginosus aitkeni*) populations across the island (Fig. 31.3). These data are contributing to an understanding of the recovery of this endangered species and are informing future management of the dunnart, invasive species and fire.

Fig. 31.3. Installation of an Eyes On Recovery camera trap to monitor Kangaroo Island dunnarts and feral cats on Kangaroo Island, South Australia. (Photo: Hannah Byrne-Willey)

To assess the effectiveness of fire management on public land, the Victorian Government is investing in the collection of biodiversity data in key vegetation types across the state. The proposed monitoring design includes ongoing surveys of biodiversity at ~200 locations within each of 11 priority vegetation types (a total of 2200 monitoring sites). Surveys in the sandy heathlands of western Victoria have commenced at 192 sites in fire-prone heathlands, stratified by time since the last fire, fire interval and fire severity. Sampling includes field surveys of birds, ground-dwelling mammals, plants and habitat structure at each site, providing a strong foundation for learning, including about species relationships to fire regimes.

Increasing awareness of invertebrate declines and what these mean for ecological processes and ecosystem services (Braby 2019) is driving a renewed interest in invertebrate monitoring. This is particularly important because of the enormous number of poorly documented, narrow-range endemic species (Box 31.4; Chapter 11).

Box 31.4. Impacts on litter invertebrates in south-eastern rainforests

Australia's invertebrates are incredibly diverse and functionally important, yet only 30% of Australia's insects have been described and there have been limited efforts to document their responses to fire. A set of long-term monitoring sites was established to test the short-term impacts of the 2019–20 wildfires on litter invertebrates of

Fig. 31.4. (A) Sites sampled in post-fire survey of litter invertebrates in temperate rainforests; and leaf litter species (mostly undescribed) collected during the fieldwork: (B) velvet worm (Onychophora), (C) cryptorhynchine weevil (Coleoptera: Curculionidae), and (D) pill woodlouse (Isopoda: Armadillidae). (Photos: Joshua Grubb)

Australia's south-eastern rainforests (Fig. 31.4). Isolated and relictual rainforests were targeted because they are likely to support many endemic species. Species were identified using morphology and DNA barcoding. Sixty-eight sites were sampled across a fire severity gradient, 1 year after the fires (Fig. 31.4). Over 50 000 litter invertebrates were collected, 19 000 of which were beetles. Abundances declined with increasing fire severity in rainforest patches. Most species collected were undescribed (e.g. eight out of 11 velvet worm species) and occupied small ranges, suggesting high levels of endemism. The range restriction of most species suggests that large-scale severe fires could have caused the extinction of as-yet-undescribed litter invertebrates in this ecosystem. This systematic and large-scale survey design will make an excellent baseline dataset for future monitoring.

Much of the limited monitoring effort for Australian biodiversity is focused on trends in species' abundance or occupancy. But many other components of our environments are also affected by wildfire and merit consideration in monitoring programs, preferably in integrated programs that can help us understand the links among the elements of biodiversity (Box 31.5). For example, Chapter 5 notes that soils can be severely depleted and changed by wildfire, but that our understanding of that process and its consequences for biodiversity is currently constrained by limited knowledge of impact and recovery; hence more targeted monitoring is required. Chapter 4 reports substantial overlaps of the

2019–20 wildfires with sites of biodiversity significance, including wetlands of international importance and World Heritage areas. In contrast to species abundance and monitoring, such sites commonly have formally accepted definitions of target states, indicators and limits of acceptable change, with ongoing monitoring reporting on the extent to which the condition of these sites is acceptable (BMT WBM 2011). This provides a ready-made framework for monitoring and contextualising the impacts of fires or other disturbance events.

Box 31.5. Integrated ecosystem monitoring: alpine ash forests on the edge

Australia supports few integrated ecosystem monitoring programs that link multiple species and ecological monitoring activities together. Integrated monitoring programs present the opportunity to learn about interactions among species, ecological processes and management in a way that single species monitoring programs cannot. For example, in Victoria's fire-sensitive alpine ash (*Eucalyptus delegatensis*) forests, there is a network of 120 plots across a gradient of fire histories, with variable histories of aerial sowing (to assist recovery from recent fires) (Wagner and Nitschke 2021). The monitoring regime includes a range of structural and compositional components of the ecosystem and soils (Fig. 31.5). Remote camera traps and audio detectors record fauna at sites. The plot network provides a baseline dataset to evaluate changes in biodiversity and ecosystem services as forest structure and composition change over time with disturbance and growth, allowing a clearer understanding of how transitions (e.g. away from dominance by alpine ash) influence fauna and other vegetation dynamics.

Fig. 31.5. Regenerating alpine ash following a single high severity fire in 2007. (Photo: Benjamin Wagner)

The impact of any given fire depends critically on the preceding sequence of fires and environmental conditions, and the environmental conditions after the fire. To maximise our learning from monitoring data, it is important to consider contextual variables such as drought before or after the fire, and fire history (the spatial and temporal dimensions of successive fires). For example, NSW vegetation types have been characterised by their responses to fire, particularly with reference to fire thresholds (mostly defined by successive fires at intervals too short or too long to allow for recruitment and persistence of key plant species). Spatial analysis of fire histories showed that many vegetation types that were largely within the optimal threshold range before the 2019–20 wildfires are now vulnerable because these fires occurred too soon after previous fires, and that the extent of long unburnt vegetation has diminished considerably (Williamson 2020). Such spatial information (e.g. remaining long unburnt patches) can also be critical for management (e.g. as refuges needing protection from future fires and logging).

Conclusions

The efforts to respond to the 2019–20 wildfires highlighted the critical importance of monitoring data. The need to urgently assess the biodiversity impacts of the fires created a spike in biodiversity survey and data synthesis. However, it is suboptimal that data collection and syntheses of monitoring datasets are completed by agencies only after a major disaster. Such an arrangement hinders rapid and accurate understanding of species most in need of rapid recovery actions, and prevents use of biodiversity data in state-wide and national disaster planning and preparation. The acknowledgement by agencies and policy makers of the important role that monitoring data play before, during and immediately following fire provides a unique policy opportunity to substantially and strategically increase investment in more extensive, taxonomically representative, and better coordinated biodiversity monitoring. The opportunity to grow a stronger biodiversity monitoring culture and systems in Australia is enhanced by the number of recent surveys and the compilation of existing monitoring data (Southwell *et al.* 2022). Together these establish a new, post-fire baseline on which a properly resourced ongoing monitoring effort could be founded. However, without effective investment in coordination and infrastructure, the data synthesis efforts of the past 2 years will be largely wasted. Now is the time for a national biodiversity data and monitoring plan and institution that will empower biodiversity management with data and evidence.

Recommendations

- Immediately invest in well-designed and strategic long-term biodiversity data and monitoring programs and infrastructure so that management decisions are routinely based on an accurate understanding of the state of biodiversity and evidence about the effectiveness of management.
- Design new monitoring programs that recognise and redress biases in the existing monitoring effort, to include more representation of the taxonomic groups currently subject to least monitoring (e.g. invertebrates, fungi and many plants).
- Invest in systems of pre-fire preparedness that specify how monitoring data will be used to inform impact avoidance (such as strategic fuel-reduction burning), strategic protection during a fire, and rapid and more accurate post-fire assessment of impacts to inform emergency recovery.

Acknowledgements

Thanks to Anne Buchan, Adrian Moorrees, Hannah Lloyd, Gregory Summerell, Alison Curtin, Dan Rogers and Matt Heard for providing insights into post-fire monitoring actions.

References

Auld TD, Mackenzie BDE, Le Breton T, Keith DA, Ooi MKJ, *et al.* (2020) 'A preliminary assessment of the impact of the 2019/2020 fires on NSW plants of national significance'. Unpublished report to NSW DPIE, Parramatta.

BMT WBM (2011) 'Ecological character description of the Gippsland Lakes Ramsar site – final report'. Prepared for the Australian Government Department of Sustainability, Environment, Water, Population and Communities, Canberra.

Bolitho LJ, Rowley JJL, Hines HB, Newell D (2021) Occupancy modelling reveals a highly restricted and fragmented distribution in a threatened montane frog (*Philoria kundagungan*) in subtropical Australian rainforests. *Australian Journal of Zoology* **67**, 231–240. doi:10.1071/ZO20037

Braby MF (2019) Are insects and other invertebrates in decline in Australia? *Austral Entomology* **58**, 471–477. doi:10.1111/aen.12411

DELWP (2022) 'Biodiversity response and recovery supplementary report: bushfire impacts on species in Victoria'. Department of Environment, Land, Water and Planning, Melbourne.

Heard G, Bolitho L, Newell D, Hines H, McCall H, *et al.* (2021) 'Post-fire impact assessment for priority frogs: northern *Philoria*'. NESP Threatened Species Recovery Hub Project 8.1.3 report, Brisbane.

Keith DA, Pellow BJ, Appleby M (2018) Monitoring threatened ecosystems and ecological communities. In *Monitoring Threatened Species and Ecological Communities.* (Eds S Legge, D Lindenmayer, NM Robinson, BC Scheele, DM Southwell and BA Wintle) pp. 101–126. CSIRO Publishing, Melbourne.

Lee JS, Callaghan CT, Cornwell WK (in press) Using citizen science to measure recolonisation of birds after the Australian 2019–2020 mega-fires. *Austral Ecology* doi:10.1111/aec.13105.

Milner R, Starrs D, Hayes G, Evans M (2016) 'Distribution and ecology of the broad-toothed rat in the ACT'. Technical Report 35. Environment and Planning Directorate, ACT Government, Canberra.

Rowley JJL, Callaghan CT, Cornwell WK (2020) Widespread short-term persistence of frog species after the 2019–2020 bushfires in eastern Australia revealed by citizen science. *Conservation Science and Practice* **2**, e287. doi:10.1111/csp2.287

Scheele BC, Legge S, Blanchard W, Garnett S, Geyle H, *et al.* (2019) Continental-scale assessment reveals inadequate monitoring for threatened vertebrates in a megadiverse country. *Biological Conservation* **235**, 273–278. doi:10.1016/j.biocon.2019.04.023

Southwell D, Legge S, Woinarski J, Lindenmayer D, Lavery T, *et al.* (2022) Design considerations for rapid biodiversity reconnaissance surveys and long-term monitoring to assess the impact of wildfire. *Diversity & Distributions* **28**, 559–570. doi:10.1111/ddi.13427

Wagner B, Nitschke CR (2021) 'Maintaining biodiversity and ecosystem functioning in a changing climate: integrating planned and autonomous adaptation in Victoria's montane eucalypt forests. Part 1: Quantifying biodiversity and other values across vegetation types'. Report to Victorian Department of Environment, Land, Water, and Planning, Melbourne.

Willacy R (2014) Baseline data for detecting the response of *Philoria richmondensis* to climate change. BSc (Hons) thesis. Southern Cross University, Lismore.

Williamson G (2020) 'Report to the NSW 2019 Bushfire Inquiry: Theme 2.2 – Biodiversity and environmental impacts: threshold analysis'. NSW Bushfire Risk Research Hub, Wollongong.

32

A precautionary tale: the consequences of, and remedies for, data deficiencies and uncertainty in conservation decisions related to the 2019–20 wildfires

Libby Rumpff, Sarah M. Legge, Jess R. Marsh, Hannah Fraser and John C. Z. Woinarski

Summary

- Biodiversity is in decline; it is imperative we take a precautionary (risk averse) approach to uncertainty in management decisions.
- Uncertainty and data deficiencies cloud almost all decisions made about biodiversity conservation, and introduce biases against the most poorly known species. However, this problem was particularly pronounced in relation to the 2019–20 wildfires, because so many species were potentially so affected and because choices about conservation responses needed to be made so urgently.
- Uncertainty is likely to be magnified when considering the fate of biodiversity in, and making management decisions for, future fire regimes rather than responses to a single fire.
- Uncertainty can be overwhelming but need not be crippling; specifying uncertainty can help identify the need for, and help target, monitoring and research (e.g. through adaptive management) to resolve critical knowledge gaps. These knowledge gaps are those that, if resolved, will result in changed and more effective management.
- Much of the research and management effort made in response to the 2019–20 wildfires will enhance the knowledge base for comparable future events.
- Lack of empirical data does not always translate to lack of knowledge – Traditional Ecological Knowledge and those that hold it have been largely ignored in management decisions.

Introduction

Ecology is plagued by epistemic uncertainty, which is a lack of (or incomplete) knowledge or certainty about the relationships between organisms and their environment (Regan *et al.* 2002). The likelihood of major biodiversity losses in, and the need for urgent conservation responses to, the 2019–20 wildfires highlighted the extent of major knowledge (or information) gaps, which we specifically refer to as those that need to be factored into conservation decision making (Kujala *et al.* 2013). During and in the aftermath of the fires, managers and decision makers were working in the face of critical uncertainty; these gaps impeded the ability to make decisions about what management strategies were best (Runge *et al.* 2011), where, when, and for what duration, to ensure the response was as effective as possible. But the critical uncertainties exist not only in relation to decisions made about the type and efficacy of actions taken in and after fires – there is also very limited knowledge about the more pervasive strategic issue of how to build more resilience for biodiversity conservation in a landscape and future likely to be subject to increasing frequency and magnitude of wildfires and other threats.

The compilation of accounts in this book highlights that knowledge is critical for effective biodiversity management, before, during and after the fires. However, there are many knowledge gaps that constrain or potentially bias decisions about such management, and the form of its implementation. Many of these knowledge gaps affect most conservation decision making (Victor 2006; Bland *et al.* 2015, 2017), but were particularly important in response to the 2019–20 wildfires. These data deficiencies and uncertainties include:

- limited knowledge about the distribution of species and ecological communities (and hence of their overlap with fire);
- limited knowledge of the responses of species and ecological communities to fire, fire severity and fire regimes and their pathway to recovery post-fire (with this uncertainty magnified in this case because the available evidence was largely derived from studies of fires of far less magnitude than the 2019–20 wildfires);
- limited knowledge of the cocktail of other threats that may be imposing pressures on species and ecological communities, and of the manner in which these threats are affected by or compound the impacts of fire;
- limited knowledge of the effectiveness of management options for these threats, and of the conservation responses that can be made to support post-fire recovery;
- limited knowledge of the population size and trends of species (and hence their likely resilience or precariousness to fire impact);
- limited knowledge of risks and efficacy of actions that can be taken prior to, and in the interval between, fires to build the resilience of ecological systems in order to reduce the susceptibility of biodiversity to extreme events and chronic threats; and
- for some groups, marked taxonomic uncertainties (and comparably for ecological communities, some marked definitional uncertainties).

These uncertainties are uneven across taxonomic groups; such bias can be further reinforced when prioritising and implementing conservation responses after catastrophes such as the 2019–20 wildfires – for example, robust evidence is required for listing species as threatened, and this requirement disadvantages the conservation of poorly known species. In recognising these critical uncertainties, we do not seek to derogate the extraordinary research effort undertaken over many years by many researchers on the responses of Australian biodiversity to fire (e.g. see Bradstock *et al.* 2012). The evidence base accruing from that research was a critical foundation for understanding impacts of the 2019–20 wildfires and for the management decisions made in their wake – but it still has many critical gaps.

Biodiversity knowledge gaps (or at least ready accessibility to relevant information) were highlighted in post-fire inquiries as shortcomings that constrained decision making (Chapter 30), with some recommendations made to redress them. But those recommendations made by the inquiries were vague and aspirational, and did not describe specifically how to resolve these gaps. So, in this chapter we attempt to synthesise the knowledge gaps that contributed to epistemic uncertainty, describe the likely consequences of this uncertainty on fire management decisions, and provide a series of recommendations to aid targeted research and treatment of uncertainty in decision making. We do not delve into the assortment of epistemic uncertainties or linguistic uncertainties (see Regan *et al.* 2002). However, we do attempt to highlight the uncertainty that is fundamental to seeking improvements for the future. Accordingly, some of the uncertainties documented in this chapter form the basis of responses provided in the 'Recommendations' chapter (Chapter 35).

A synthesis of knowledge gaps

A synthesis of key biodiversity knowledge gaps, with their relevance to the 2019–20 wildfires and suggestions to remedy associated uncertainty, is provided in Table 32.1. In Table 32.2 we synthesise which of these gaps were considered to be a concern in the chapters describing impacts to biodiversity. These knowledge gaps do not constitute a systematic account of uncertainty, nor do they tell us definitively which knowledge gaps are most important to resolve; they are those highlighted by the contributors to this book. However, Table 32.2 provides a useful capture of the different ways in which our knowledge, which varies across taxa and systems, is lacking, and provides a useful basis for targeted and critical research. Some knowledge gaps may take decades and many resources to redress, so it is important to identify which are most consequential, and then to adopt a longer-term strategic approach to resolving them.

There are some striking discrepancies in knowledge highlighted throughout this book. The information base/uncertainty is systematically biased; for taxa, our knowledge base is thinnest for invertebrates and fungi, starting with large-scale gaps regarding the identification of species and their distribution. For ecological systems our knowledge of how soils (Chapter 5) and ecological processes (Chapter 8) respond to fire is generally poor, with such ignorance most profound for marine and freshwater systems (Chapters 6 and 7), and rarely burnt vegetation types (Chapter 4 and 8). This is critical, because an understanding of how (and if) we can manage species, communities and entire systems in the long term requires an understanding of how systems and processes respond to fire regimes, against a backdrop of other threats, including climate change.

These knowledge gaps and uncertainties probably compound in their consequences for decision making. For instance, if a species has uncertain distribution (and thus uncertain fire overlap) *and* we know little of its response to fire, then we know almost nothing about fire impacts on it. It is likely the unprecedented wildfire conditions, plus this compounding of different forms of uncertainty, leads to a commonly recognised shortfall: a lack of knowledge of what management is effective (Table 32.1).

Of course, there is much more nuance required in the categorisation of management uncertainty. For instance, many chapters refer to a need to build resilience for biodiversity

Table 32.1. Different epistemic uncertainties, data deficiencies and biases affecting conservation of potentially fire-affected species, ecological communities (EC) and ecological function (adapted from compilation in Woinarski *et al.* 2021)

Type of knowledge gap	Description	Relevance to 2019–20 wildfires	Action
Linnean shortfall	Species/EC not identified or described	There may have been major losses of species that are not yet described ('dark extinctions'), or cases where taxonomic review will partition species (with some suffering major impacts not detected in the taxon considered before splitting) (e.g. Jolly *et al.* 2022). Linnean shortfall and subsequent under-reporting of actual impacts will be most likely for poorly known taxonomic groups.	Surveys and taxonomic research (including eDNA and other developing technologies).
Darwinian shortfall	Phylogenetic information (evolutionary history and potential, species classification) poorly understood	Probably limited immediate consequence, but in the longer term, important for defining (and hence protecting) sites of biodiversity significance (e.g. where many old endemic species co-occur), and possibly for sites encompassing critical genetic diversity that may be important for ongoing evolutionary potential.	Phylogenetic analysis. Assessments of sites of biodiversity significance that incorporate some measure of phylogenetic history.
Wallacean shortfall	Uncertainty about species/EC distributions	Results in major uncertainty about assessments of the extent of fire overlap and the likelihood of finding important surviving populations post-fire. Introduces biases in favour of better-known species. Also, this shortfall renders it more challenging to identify, delineate and prioritise sites of biodiversity significance that should be priorities for protection during and after fires.	Ideal: systematic landscape-wide surveys, especially for poorly known species. Short-term: structured expert elicitation, followed by validation (surveys); better harmonisation across jurisdictions in EC classification and mapping.
Movement shortfall	Uncertainty about species dispersal	Limited information about species' capability to flee fires and for species moving from burnt areas to survive in new areas. For highly dispersive species, introduces uncertainty about assessment of fire impact (because of uncertainty about whether or not populations were in burnt areas at the time of fire) (e.g. Crates *et al.* 2022); limited knowledge of responses of how fires impact at landscape scale the resources (e.g. nectar) that govern the movements of dispersive species, and on which they depend (Baranowski *et al.* 2021).	Landscape-wide, repeated surveys of dispersive species, the resources that govern their movement patterns and the responses of these resources to fire. Short-term: structured expert elicitation, followed by validation (surveys).

Type of knowledge gap	Description	Relevance to 2019–20 wildfires	Action
Prestonian shortfall	Uncertainty about abundance of species and its variation in time and space	Lack of pre-fire population data means that it is almost impossible to determine proportion of population lost in fires, and hence also target levels for recovery. Such uncertainty may also make it difficult to assess priority of fire management relative to management of other threats that may be having greater impact.	Ideal: landscape-wide monitoring. Short-term: structured expert elicitation, followed by validation (surveys).
Extinction deficit shortfall	Extinction rate or conservation status of species/ECs uncertain	Rates of loss in wildfire may be difficult to contextualise in relation to other factors; this makes modelling of impacts (and extinction rates) of future fire regimes challenging.	Ideal: targeted surveys especially of poorly known species that may have suffered major fire impacts; modelling of impacts across potential future fire regimes. Short-term: structured expert elicitation, followed by validation (surveys).
Hutchinsonian shortfall	Species ecology and sensitivity or response to disturbance poorly known	Limited knowledge of response of biodiversity and ecological function (e.g. soils) to fire reduces capability to evaluate impact of individual fires, and of fire regimes. Limited knowledge also constrains ability to impose preferred fire management, and preferred post-fire recovery actions. Major bias among taxonomic groups – there is far more knowledge of responses to fire for birds, mammals and plants than for other groups.	Ideal: long-term monitoring, research on impacts of fires on individuals, populations and ecological function, including use of experimental manipulation of fires. Modelling indicating plausible ranges of responses to future fire regimes. Short-term: inference from similar species/traits.
Raunkiaeran shortfall	Uncertainty around ecological traits and functions or processes (of ECs)	Limited knowledge of life history and ecological traits constrains our ability to understand why species may be variably affected by (and recover from) fires of varying severity and regimes. Major bias among taxonomic groups. Also, limited knowledge of the responses to fire of many ecological processes, and of how biota recovery post-fire is affected by and affects recovery of ecological processes. Critical uncertainties relating to capability to build resilience in biodiversity (e.g. systematic program of translocations) to better enable it to withstand future pressures.	Ideal: fundamental research on species and traits, and/or ecological processes. Short-term: structured expert elicitation; inference from similar species/traits or ECs.
Poirot shortfall	Causes of decline (including compounding threats) and relevant effective actions poorly known	Limited knowledge of how threats (other than fire) are affected by fire, and how they may compound impacts of fire. Limited knowledge of the efficacy of post-fire recovery actions or how long they will need to be implemented to achieve recovery. As a consequence, management responses post-fire may be poorly directed or prioritised. Also, limited knowledge of other threats (and their management) constrains capability to build resilience in the system before and between fires.	Ideal: long-term monitoring especially in relation to threat management. Short-term: structured expert elicitation; inference from similar species/traits or ECs.

Table 32.2. A synthesis of knowledge gaps described by the chapter authors, leading to uncertainty relating to the 2019–20 fires.

This is not intended to be a systematic review, and it is acknowledged that some gaps may be missing. EC = Ecological communities.

Uncertainty	Icons	Soils	Aquatics	Marine	ECs	Plants	Fungi	Inverts	Frogs	Reptiles	Birds	Mammals
Identification and description (Linnean shortfall)					+		+	+	+	+		
Phylogeny (Darwinian shortfall)												
Distributions (Wallacean shortfall)			+		+	+	+	+	+	+	+	+
Movement (movement shortfall)										+	+	
Abundance and variation in time and space (Prestonian shortfall)			+			+		+	+	+	+	+
Extinction rate or conservation status (extinction deficit shortfall)			+		+	+	+	+	+	+	+	+
Ecology and sensitivity or response to disturbance (Hutchinsonian shortfall)	+	+	+	+	+	+	+	+	+	+	+	+
Ecological traits and functions or processes (Raunkiaeran shortfall)						+	+	+				
Causes of decline and efficacy of actions (Poirot shortfall)	+	+	+	+	+	+	+	+	+	+	+	+

in relation to fire, which is likely to involve a strategy of actions such as translocation, *ex situ* population establishment, threat (pest animal and weed) management, and fire management. If there are no precursors to act as a guide, resolving uncertainty may involve trials of risk-informed programs of translocations of insecure species, and trials relating to establishment of *ex situ* populations (or germplasm material). There are risks involved due to the precarious state of species requiring protection, and planning and research that integrate multiple management options and scenario planning (e.g. to account for climate risks) are important. Threat management is undeniably part of this suite of actions, but arguably the benefits of intensive and broad-ranging ongoing management of other threats (before and after fire), such as weed and feral animals, are relatively well known. Yet there are always questions of efficiency (such as intensity, timing, duration and location), particularly when resources are limited.

A path forward for management, underpinned by greater recognition and incorporation of Indigenous fire practices, has been outlined in Chapter 33. In a discussion of uncertainty, it is necessary to acknowledge that the approaches and resulting evidence base used to inform management has to date largely ignored Traditional Ecological Knowledge held by Indigenous people. The reliance on empirical data derived from current scientific practice to justify management may be misguided, given this bias (Fletcher *et al.* 2021). While the underpinning knowledge held by Indigenous people is critical to guide management of biodiversity into the future, approaches to research and monitoring require collaboration at all phases of knowledge acquisition (Fletcher *et al.* 2021) and of implementation of the actions informed by that knowledge. Importantly, it is here that the barriers to implementation of actions (the 'implementation gap') are more critical than any knowledge gaps. These barriers need to be resolved (Chapter 33) so that Indigenous people are able to fulfil their right to self-determination, and manage Country at scale.

When does uncertainty matter?

Uncertainty matters most when lack of knowledge leads to failure to prioritise effective conservation responses where they are most needed. Following the fires, impact assessments at a national scale were undertaken for vertebrates, invertebrates, plants and ecological communities to guide allocation of resources to action (https://www.dcceew.gov.au/environment/biodiversity/bushfire-recovery/bushfire-impacts/expert-panel). With the fires either still burning or just passed, it was critical to use whatever information was available to assess potential impacts to species and communities, for inaction or misplaced actions might have compounded losses and subverted possible recovery. Given the uncertainties described above, decisions needed to be taken on the basis of incomplete information. As a result, expert elicitation was used to redress some knowledge gaps.

A key question that should be asked is: were those assessments of impacts and (hence) prioritisation of species and recovery actions right? This is an entirely valid question about how data quality impacts decision making. However, it requires more nuance. The question is more usefully refined to: which species (or ecological communities) were erroneously not classified as a priority despite being highly impacted and needing recovery support following fire, but received little or no management attention? This is a false negative or Type II error (Fig. 32.1) and is the error most environmental managers seek to avoid (Burgman 2005). False positives, where it is falsely assumed a species or community is impacted and needs conservation support, and such action is taken, have far less consequential outcomes (except perhaps to those managing inadequate budgets). As an example of avoidance of Type II error, the impact of the fires on, and susceptibility of, the slender myoporum (*Myoporum floribundum*) was difficult to assess, due to limited records and understanding of the species. However, the plant was prioritised for survey effort, and a new population was uncovered in a severely burnt area, enabling cuttings to be taken for *ex situ* conservation.

Type II errors are most likely to be made, and hence to have adverse consequences, for poorly known species. For those species about which we know next to nothing, or indeed nothing (i.e. they are data deficient), their fate is mostly left to chance and possible benefits arising from actions taken for other co-occurring species, where management

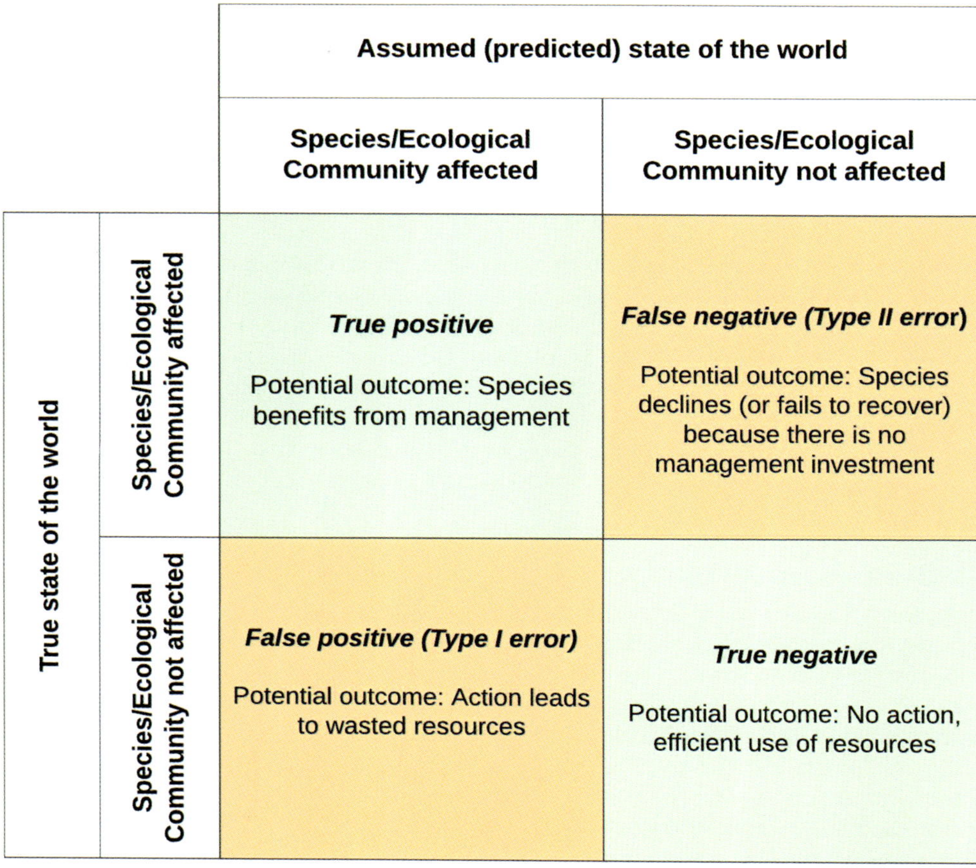

Fig. 32.1. An outline of the potential outcomes of Type I and Type II errors related to susceptibility assessments after the fires.

intervention happens to be applied where they occur and has some benefit to assist recovery. For those that are poorly known but imperilled, the potential consequences of a Type II error are dire, especially if they require targeted intervention (i.e. translocation or *ex situ* management). For example, ~400 described invertebrate and ~100 fungi species had all of their known ranges burnt in the 2019–20 fires (Chapters 11 and 10), but all of these are known from only one or two records. Some of these species may have been rendered extinct by these fires; the continuing existence of others may now be extremely tenuous, and there is some urgency in conducting on-ground surveys to assess whether any populations have survived the fires and, if so, to implement recovery actions. Given the uncertainty about the distribution and fire response of such species, there is major uncertainty about how many extinctions may have been caused by these fires: it is somewhere in the unacceptably wide range between one (the only 'confirmed' extinction of the *Banksia montana* mealybug: see Chapters 11 and 34) and perhaps 500 species. Even this larger figure may be an underestimate given that there are many more undescribed than described invertebrate species, and many of these undescribed species may have had complete overlap with fire.

Accounting for uncertainty and risk is perhaps embodied by the substantial federal and state government investment in strategic landscape-scale interventions (Chapters 22 and 23 respectively), such as feral pest and herbivore control. For researchers (and their funders), there is a need to focus effort on resolving uncertainty for those species for which the risk of imperilment is greatest, and which a change in management would benefit most.

Incorporating uncertainty to support decisions

For the 2019–20 fires, some assessment and response approaches explicitly recognised uncertainty and implemented mechanisms to deal with it. Often, these approaches required use of expert judgements and assumptions about system dynamics, whether that be through expert elicitation (Marsh *et al.* 2021; Legge *et al.* in press), or modelling (Gallagher 2020; Gallagher *et al.* in press). This is inevitable and – at least until the knowledge gaps are filled – appropriate; existing data are spatially and temporally piecemeal, biased information derived from previous research on fire may not necessarily be extrapolatable to fires of the unprecedented magnitude of those in 2019–20, and the data we have now might not tell us much about a changing future.

A specific post-fire vulnerability assessment approach was developed for invertebrates (Marsh *et al.* 2021; Chapter 11) to guide decisions about management responses and conservation listing while accounting for an overwhelming lack of knowledge about the majority of species. In this work, a trait database relevant to fire susceptibility and recovery potential was compiled using expert input on life history characteristics, threats and management needs. Critically, this trait compilation (and hence assessment of susceptibility) was not restricted to well-studied species, but experts also assigned likely traits based on knowledge of species in the same genus or family, with explicitly stated level of confidence to their assessments. This information was collated with fire overlap information, for > 1000 species, to identify species likely to be most affected by the 2019–20 fires and most needing recovery support. Although this process allowed prioritisations to be made for many poorly known species, it could not be used for the majority of invertebrate species, which are not even described. However, this structured approach provides a path forward; it will be an accessible database that can be updated and, importantly, the type of uncertainty (Table 32.2) is attributed to each taxon to inform future research needs.

Legge *et al.* (in press) expanded on the initial post-fire vulnerability assessment for vertebrates and spiny crayfish using the IDEA protocol (Hemming *et al.* 2018): a structured, repeatable, consistent approach to eliciting information that captures uncertainty and minimises bias for a large number of species. Experts were asked to predict post-fire population change over time, providing realistic upper (best case) and lower (worst case) bounds. Uncertainty from expert judgments was used to assess if population change over time crossed the population reduction threshold for a species to be considered Vulnerable, Endangered or Critically Endangered. This is illustrated in Fig. 32.2. For example, due to the impacts of the 2019–20 fires and other threats, the gang-gang cockatoo (*Callocephalon fimbriatum*) will potentially cross the Vulnerable threshold over time, when accounting for uncertainty (i.e. when using confidence limits around projected future population). Other species were predicted more unequivocally to have declined as a result of the fires such that they meet Endangered criteria (e.g. Kangaroo Island southern emu-wren (*Stipiturus malachurus halmaturinus*)). Likewise, some species do not meet listing criteria, even when factoring in uncertainty (e.g. black-faced monarch (*Monarcha melanopsis*)). This

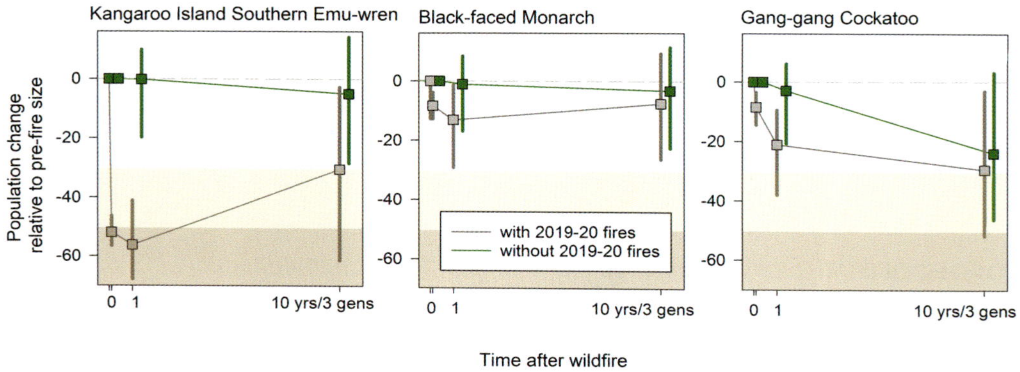

Fig. 32.2. An excerpt from the Legge *et al.* (2021) assessment for birds, highlighting how uncertainty was used to assess whether population changes (at the national scale) due to the 2019–20 wildfires met thresholds for listing as Vulnerable (light brown) or Endangered (darker brown).

structured elicitation process was used to inform national conservation assessments for threatened species listing under the *Environment Protection and Biodiversity Conservation Act* 1999.

The approach shows how the precautionary principle can provide a framework for dealing with uncertainty to inform policy. Currently, it is little applied but, given the rate of biodiversity declines in Australia and the current paucity of data, this needs to be rectified. Not incorporating and accounting for uncertainty means ignoring risk (Burgman 2005). Of course, the data can, and should be, validated, but the approach allows us to make urgently required decisions in the face of critical uncertainty, and then target knowledge acquisition where uncertainty is most critical.

Many uncertainties remain unresolved. Two metrics were pivotal in the assessment of the conservation impacts of the 2019–20 wildfires on wildlife – the extent of species' distribution that overlaps with fire, and the likely proportional mortality rate in burnt areas. However, such an evaluation is based on the presumption that animals in unburnt terrestrial areas are largely unaffected by fire – indeed, unburnt areas within or near the burnt areas are widely assumed to form crucial refuges and to be critical source areas for re-population of burnt areas. There is at least one major uncertainty in this assessment, relating to the extent to which impacts of fire on terrestrial wildlife species extend beyond the burnt area, notably due to the impacts of smoke (Chapter 16).

Uncertainty can be overwhelming, but it need not be crippling. There are existing frameworks that help us understand whether uncertainty is critical to resolve before a management decision is made, such as structured decision making (Gregory *et al.* 2012; Hemming *et al.* 2022). These approaches require clear specification of objectives, alternative management options, and evaluation of the consequences of these options with available data (with uncertainty) to explore whether a decision can be made. If the preferred (or optimal) alternative does not change after consideration of uncertainty (i.e. would you make the same decision when considering the worst case scenario?), then uncertainty is not critical to resolve in that context. A relevant example of such a process can be found in the PACES tool (NESP Threatened Species Recovery Hub 2021), which guides users through a structured process to help guide decisions about whether or not to initiate or continue an *ex situ* management plan for a species. Of course, research may be required to refine knowledge on management effectiveness, and monitoring to verify the predicted outcomes of the chosen strategy.

Conclusions

Uncertainty clouded assessments of biodiversity impacts of the 2019–20 wildfires. In some cases, these uncertainties may not have been knowledge gaps, but simply reflect lack of appropriate transfer of knowledge from experts (or data custodians) to those who needed such information (e.g. fire control centres). In either case, these gaps may have resulted in some suboptimal conservation responses. Whether that is true will only be revealed with monitoring, and time, but the 2019–20 wildfires catalysed an extraordinary spate of conservation management responses, establishment of monitoring programs, and research; and this effort should result in major knowledge gains and resolution of some key uncertainties.

Even more challenging will be uncertainties relating to future *fire regimes*, rather than individual fires, as regimes and their impacts are far more complex than single fires. In addition, many of the issues concerning uncertainty raised in this chapter are likely to be transferable: many of the same uncertainties constrain or subvert other decisions about conservation, and many of the approaches to resolving uncertainties will be applicable to other situations.

Recognition of uncertainty, and making an explicit attempt to factor it into responses, is likely to provide better, more robust, more flexible and more justified decisions and outcomes, and ready alignment with adaptive management. Critically, it allows decision makers to exercise their risk attitude; given our current declining baselines for biodiversity and an uncertain future, a precautionary approach is warranted. Approaches developed and applied after the 2019–20 wildfires demonstrate how uncertainty can be integrated into risk assessments to inform management. Yet we need to be much more proactive. The scale of the 2019–20 wildfires showed that we shouldn't have been so accepting and complacent about knowledge gaps: it should not need such catastrophic triggers to recognise and redress uncertainties.

Recommendations

- Explicitly recognise and incorporate uncertainty into risk assessments and planning, to enable decision makers to understand and moderate the risk associated with management options. Apply options and scenario analysis in planning for management, to explore the risks of an uncertain future associated with climate change and more frequent fire.
- Enable Traditional Owners to take a leading role in planning, management and research; Traditional Ecological knowledge is so often missing from decision making.
- Be specific about which type of uncertainty is critical to resolve to improve management decisions.
- Where data are lacking (particularly about the future), structured expert elicitation should be used to supplement existing data, to understand uncertainty, and to target research and monitoring to where they most affect decision making.
- Overcome some of the bias in the knowledge of biodiversity, by resolving substantial knowledge gaps for invertebrates, fungi, and ecological processes.
- Improve understanding and research about the impacts of repeat fires (fire regimes) on biodiversity.

References

Baranowski K, Faust CL, Eby P, Bharti N (2021) Quantifying the impacts of Australian bushfires on native forests and gray-headed flying foxes. *Global Ecology and Conservation* **27**, e01566. doi:10.1016/j.gecco.2021.e01566

Bland LM, Collen B, Orme CDL, Bielby J (2015) Predicting the conservation status of data-deficient species. *Conservation Biology* **29**, 250–259. doi:10.1111/cobi.12372

Bland LM, Bielby J, Kearney S, Orme CDL, Watson JEM, *et al.* (2017) Toward reassessing data-deficient species. *Conservation Biology* **31**, 531–539. doi:10.1111/cobi.12850

Bradstock RA, Gill AM, Williams RJ (Eds) (2012) *Flammable Australia: Fire Regimes, Biodiversity and Ecosystems in a Changing World*. CSIRO Publishing, Melbourne.

Burgman MA (2005) *Risks and Decisions for Conservation and Environmental Management*. Cambridge University Press, Cambridge, UK.

Crates R, Rayner L, Stojanovic D, Scheele BC, Roff AD, *et al.* (2022) Poor-quality monitoring data underestimate the impact of Australia's megafires on a critically endangered songbird. *Diversity & Distributions* **28**, 506–514. doi:10.1111/ddi.13385

Fletcher MS, Romano A, Connor S, Mariani M, Maezumi SY (2021) Catastrophic bushfires, indigenous fire knowledge and reframing science in southeast Australia. *Fire (Basel, Switzerland)* **4**, 61. doi:10.3390/fire4030061

Gallagher RV (2020) Final national prioritisation of Australian plants affected by the 2019–2020 bushfire season. Commonwealth Department of Agriculture, Water and Environment, <https://www.dcceew.gov.au/sites/default/files/env/pages/289205b6-83c5-480c-9a7d-3fdf3cde2f68/files/final-national-prioritisation-australian-plants-affected-2019-2020-bushfire-season.pdf>.

Gallagher RV, Allen SP, Mackenzie BDE, Keith DA, Nolan RH, *et al.* (in press) An integrated approach to assessing abiotic and biotic threats to post-fire plant species recovery: lessons from the 2019–20 Australian fire season. *Global Ecology and Biogeography* doi:10.1111/geb.13478.

Gregory R, Failing L, Harstone M, Long G, McDaniels T, *et al.* (2012) *Structured Decision Making: A Practical Guide to Environmental Management Choices*. John Wiley & Sons, West Sussex.

Hemming V, Burgman MA, Hanea AM, McBride MF, Wintle BC (2018) A practical guide to structured expert elicitation using the IDEA protocol. *Methods in Ecology and Evolution* **9**, 169–180. doi:10.1111/2041-210X.12857

Hemming V, Camaclang AE, Adams MS, Burgman M, Carbeck K, *et al.* (2022) An introduction to decision science for conservation. *Conservation Biology* **36**, e13868. doi:10.1111/cobi.13868

Jolly CJ, Moore HA, Cowan MA, Cremona T, Dunlop JA, *et al.* (2022) Taxonomic revision reveals potential impacts of Black Summer megafires on a cryptic species. *Pacific Conservation Biology* doi:10.1071/PC21045.

Kujala H, Burgman MA, Moilanen A (2013) Treatment of uncertainty in conservation under climate change. *Conservation Letters* **6**, 73–85. doi:10.1111/j.1755-263X.2012.00299.x

Legge S, Woinarski JCZ, Garnett ST, Geyle H, Lintermans M, *et al.* (2021) *Estimates of the Impacts of the 2019-2020 Fires on Populations of Native Animal Species*. NESP Threatened Species Recovery Hub, Brisbane.

Legge SM, Rumpff L, Woinarski JCZ, Whiterod NS, Ward M, *et al.* (in press) The conservation impacts of ecological disturbance: time-bound estimates of population loss and recovery for fauna affected by the 2019–20 Australian megafires. *Global Ecology and Biogeography* doi:10.1111/geb.13473.

Marsh J, Bal P, Fraser H, Umbers K, Greenville A, *et al.* (2021) *Assessment of the Impacts of the 2019–20 Wildfires of Southern and Eastern Australia on Invertebrate Species*. Threatened Species Recovery Hub of the National Environmental Science Program, Brisbane.

NESP Threatened Species Recovery Hub (2021) *A User Guide to the PACES Tool: Planning and Assessment for Conservation through ex-situ Management*. Threatened Species Recovery Hub of the National Environmental Science Program, Brisbane.

Regan HM, Colyvan M, Burgman MA (2002) A taxonomy and treatment of uncertainty for ecology and conservation biology. *Ecological Applications* **12**, 618–628. doi:10.1890/1051-0761(2002)012[0618:ATATOU]2.0.CO;2

Runge MC, Converse SJ, Lyons JE (2011) Which uncertainty? Using expert elicitation and expected value of information to design an adaptive program. *Biological Conservation* **144**, 1214–1223. doi:10.1016/j.biocon.2010.12.020

Victor JE (2006) Data deficient flags for use in the Red List of South African plants. *Bothalia* **36**, 85–86. doi:10.4102/abc.v36i1.346

Woinarski JCZ, Crase B, Garnett S, Rumpff L (2021) 'Addressing issues relating to the conservation of data-poor species, and options for their resolution'. NESP Threatened Species Recovery Hub Project 5.2 report, Brisbane.

33

Empowering Indigenous leadership and participation in wildfire recovery, cultural burning and land management

Cathy J. Robinson, Oliver Costello, Michele Lockwood, Petina L. Pert and Stephen T. Garnett

Summary

- The 2019–20 wildfires had a significant impact on Indigenous communities, estates, cultural heritage, landscapes and significant species.
- This chapter draws on material collected from submissions, workshops and interviews to describe the ways in which Indigenous cultural fire management currently reflect fire practices, relationships, and knowledge that are integral parts of Indigenous identity and governance systems.
- We summarise the persistent calls from Indigenous people to undertake cultural fire management and to support Indigenous leaders and fire practitioners to be involved in all aspects of landscape fire management, including planned burning and wildfire prevention, mitigation, response and recovery.
- Indigenous communities aspire to lead cultural burning and elements of wildfire recovery and related management activities, but continue to be hindered in their efforts. Reasons for this include inadequate decision making and resourcing, current regulatory and legal frameworks, disconnection with and lack of access to Country, conflicting views around fire regimes, fragmented partnerships, and a lack of information about Indigenous wildfire management.
- A key step in overcoming these constraints is to identify short-term goals to increase Indigenous leadership and capacity in fire management, as well as longer-term objectives that will establish a broader framework for empowering Indigenous decision making and involvement in all aspects of landscape fire management, including planned burning and wildfire prevention, mitigation, response and recovery.

Introduction

Indigenous people across Australia recognise that fire is both a threat to, and a healer of, Country. The specific effects of fire, whether wildfire or planned burning, on flora, fauna and ecological communities depend on the attributes of the fire and the ecosystem being burnt. Intense, frequent fires can have a negative impact on native species (Woinarski and Legge 2013; Connell *et al.* 2019) including priority threatened species identified in the Australian Government's Threatened Species Strategy (https://www.environment.gov.au/biodiversity/threatened/publications/strategy-home) (Connell *et al.* 2019; Wintle *et al.* 2020) and areas and species that are significant to Indigenous people and communities across Australia (Robinson *et al.* 2016b). At the same time, Indigenous fire and land management practitioners and ecologists recognise that fires can also be too infrequent, too mild or too small, and that some form of fire is critical to the health of ecosystems and biota across the country (Yibarbuk 1988). However, there is still much to learn about how the frequency, extent, seasonality and intensity of fire should be managed to support local habitats and species in contemporary and cross-cultural contexts (Cowley *et al.* 2014; Prober *et al.* 2016; Dickman *et al.* 2020).

Although there have been repeated calls to consider the particular circumstances and standing of Indigenous peoples, contemporary fire management activities and wildfire recovery efforts in eastern and southern Australia have only recently incorporated Indigenous fire and landscape management perspectives (e.g. Aboriginal Affairs NSW 2017; Aboriginal Victoria 2019). Recent investigations highlight that emergency events such as wildfires disproportionately affect Indigenous communities because of their relative remoteness in fire-prone areas and limited access to emergency services; as well as the deep trauma that many face due to the legacies of colonisation and discrimination that expose Indigenous people to ongoing vulnerabilities (Fogarty *et al.* 2018; Inspector-General Emergency Management 2020; Williamson *et al.* 2020).

This chapter aims to increase our understanding of what needs to be done to overcome obstacles to Indigenous involvement in cultural burning and fire management. We then use this understanding to identify a series of short- and long-term practical measures that can be implemented to enable Indigenous leadership in future fire management efforts. To do so we draw on research that sought to summarise the aspirations of Indigenous people to care for culturally significant and threatened species in fire-affected regions; and to assist with on-ground post-fire surveys, the monitoring of species recovery, and reviews of species and ecosystem responses to fire and management actions (Robinson *et al.* 2021).

Our research focused on the bioregions that were most heavily affected by the 2019–20 wildfires, to facilitate comparison with other chapters in this book (Department of Agriculture Water and the Environment 2020). However, this study region does not consider the geographical extent of Indigenous peoples' Country, the cultural heritage or song lines that connect species, kin and Country, and the geographical scope of the findings extend well beyond the 'fire-affected' regions (Williamson *et al.* 2020; Robinson *et al.* 2021).

As part of this effort, perspectives offered by Indigenous and non-Indigenous fire experts and partners in submissions to wildfire inquiries, interviews, small discussion groups and workshops were collected, collated and analysed. Excerpts from some interviews are given below: in some cases, with the interviewee being anonymised. Participants were chosen using the following process. First, the study team conducted an online search of Indigenous Traditional Owners, organisations and corporations that have the potential to be involved in fire management and/or cultural burning (e.g. National Indigenous Australians Agency rangers, and Indigenous Protected Area rangers) or had recently received funding (e.g. https://www.environment.gov.au/biodiversity/bushfire-recovery/activities-and-outcomes#a10). Next, for each state in the study region (Western Australia, Queensland, South Australia, Victoria, New South Wales, Australian Capital Territory and Tasmania), the study team searched for fire authorities, government and non-government agencies and Indigenous corporations that have been involved in fire management and/or support cultural burning (e.g. Supply Nation, National Parks and Wildlife Service, Rural Fire Service, Country Fire Authority, Country Fire Service and government environmental departments).

Key findings

The impact of wildfires on significant species and cultural heritage has been challenging for many Indigenous groups in the study region. Cathy Thomas, Women's Cultural Education Officer from the Gunaikurnai Land and Waters Aboriginal Corporation (2020), described the effects of wildfires on species and the environment:

> *Impacts on flora and fauna, it's part of our environment and our ecosystems and it's quite significant to Aboriginal people and not to mention our totems. Also, our waterways include our water species and food source in lakes, rivers and the ocean, our food source is almost scarce from fires upstream, our beaches covered in debris from fires and affecting our fish and other species in the waters.*

Wildfires can also cause considerable damage to Indigenous cultural heritage. Indigenous leaders, land managers and fire practitioners note that, while it is important to protect tangible sites that reflect cultural heritage, intangible heritage values exist across the landscape, connecting places, species and kin to other places, species, and kin. As an Indigenous representative from the Victorian Department of Primary Industries, Parks and Environment explained:

> *Heritage is the whole landscape; this is not just about fire. Fire practice is also about caring for heritage—not just about places and species. It's about why we do fire and how we do fire and this needs to be cared for. (interview, March 2021)*

In addition to the effects of wildfires themselves, submissions from Indigenous organisations reported that wildfire suppression efforts have resulted in damage to cultural heritage. Recognising that 'mechanical suppression activities generally have the most devastating impacts on sites', the Victorian Post Wildfire Indigenous Heritage Survey report cautions that 'it should never be assumed that, because a place or landscape is a known Aboriginal spiritual or named place, that it is not associated with material remains which could be damaged'. The report adds that even 'if a place has no remains, damage to the "sense of place" may occur anyway when disturbed by fire or suppression activities. Only Aboriginal people, particularly those with knowledge about a particular place, can assess whether damage has occurred', regardless of whether that damage is 'physical or less tangible' (Freslov 2004).

Managing the impacts of wildfire and wildfire suppression activities can be particularly complex in places that contain sites that may be unknown to local Indigenous communities (Tasmania Fire Service interview, February 2021; NSW NPWS interview, March 2021; National Bushfire Recovery Agency interview, February 2021) or where cultural protocols and/or lack of trust make it difficult for Indigenous communities to share the locations of sites with non-Indigenous people. Ensuring that wildfire suppression and firefighting activities provide a culturally safe space in which local Indigenous Elders can share the location of heritage sites has been raised as a key issue.

Core to this effort is recognition that fire has influenced the ways in which Australian Indigenous people live on, with and through their land for millennia. Indigenous leaders and fire practitioners are aware of this significance, and this has underpinned their advocacy to direct and practise cultural burning and land management activities in order to protect, heal and nurture Country (see Box 33.1).

Box 33.1. Culturally significant species: koalas

The effects of wildfires on koalas (*Phascolarctos cinereus*) provide a useful example of the impacts of wildfires on culturally significant species. In addition to being recognised as a national icon (Environment and Communications References Committee 2011) and listed as a species of national environmental significance under the federal *Environment Protection and Biodiversity Conservation Act* 1999 (DAWE 2020), the cultural and historical importance of koalas for many Indigenous people is well documented (Schlagloth *et al.* 2018; Cahir *et al.* 2020). In submissions relating to various wildfire disasters, it has been noted that koalas form an integral part of spiritual and cultural life and are central to many Dreamtime stories and cultural landscape features of importance to Indigenous people (see Costello 2019).

Co-author Oliver Costello notes that while wildfires are devastating to koala populations and habitats, these animals are also part of the Indigenous biocultural landscape in coastal regions of New South Wales that needs appropriate cultural burning and land management practices in order to flourish:

> *[K]oalas – we call them* budabe *or* burbi *in our language – are a really important dreaming story for our Country. A koala is an important totem for our Country. We have many different values. … Koalas are just one of many culturally significant species that we burn for. We need to make sure the canopy is healthy and safe. The pathways are also important. At home there are stories about the koalas and their songline pathways. They are pathways that we share as well. We burn to keep the pathways open.* (Costello 2019)

In 2019–20, the wildfires affected koala populations and their habitats, including areas where koala populations have been introduced (Fig. 33.1). Legge *et al.* (in press) estimated that the overall koala population suffered a population loss of 7% immediately after the 2019–20 wildfires, increasing to 26% loss three generations later (see also Chapter 16). This decline requires urgent attention (DAWE 2020). The National Recovery Plan for the Koala recognises the importance of supporting Indigenous land managers and respecting Indigenous knowledge as part of koala habitat recovery efforts (DAWE 2022). This will require locally specific approaches. On Kangaroo Island, for example, cross-cultural and collaborative approaches to koala recovery efforts are challenging in a region where there is no active Indigenous ranger group and koala populations are introduced (interview, Local Council of Kangaroo Island, May 2021). In New South Wales and parts of Queensland, Indigenous leaders are progressing pathways to provide and resource a culturally safe space for Indigenous leaders to engage in cultural burning, koala monitoring and caring for Country efforts that seek to reconcile the ways in which significant plants, animals, sites and ecosystems are identified and conserved.

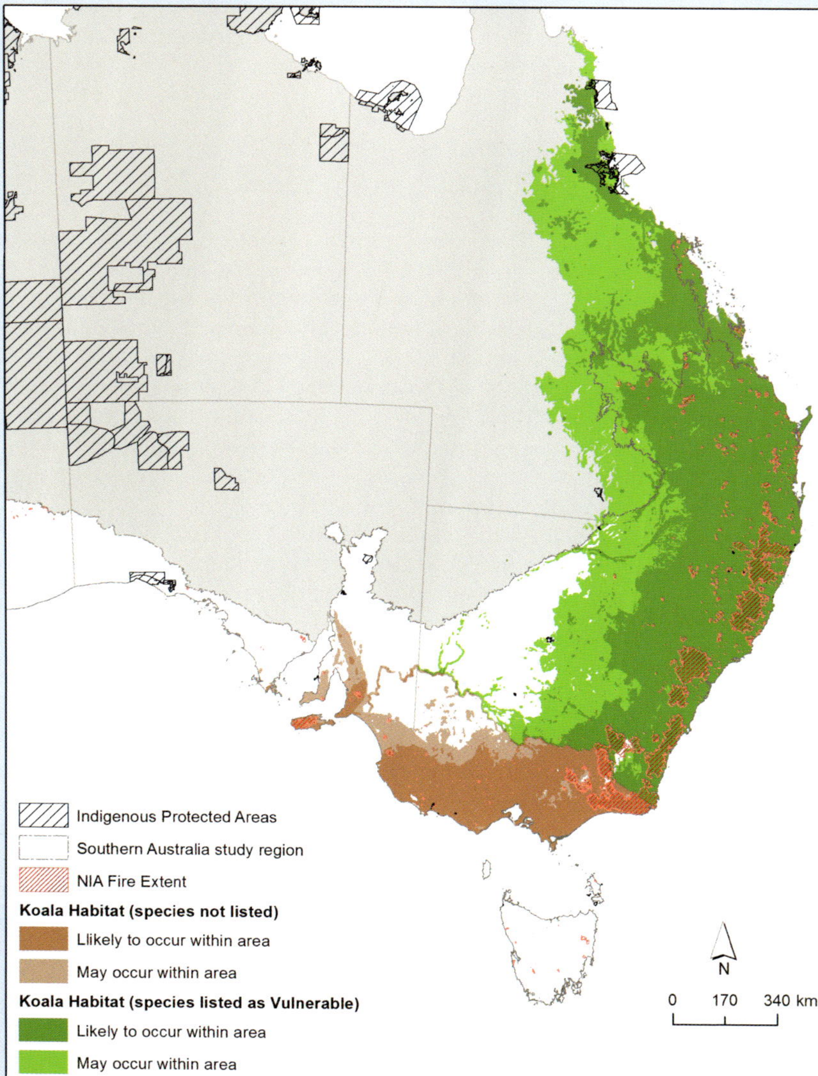

Fig. 33.1. Koalas are one of the many culturally significant species that were impacted by Australia's 2019–20 wildfires. The portion of the koala population that is in New South Wales, Australian Capital Territory and Queensland was listed as nationally Vulnerable before the fires, but was up-listed to Endangered in February 2022. Koalas in South Australia and Victoria are not listed as threatened; koalas have been introduced to some places, including Kangaroo Island.

However, several factors continue to frustrate efforts to translate these aspirations into reality, at least in some parts of Australia. These include the need for appropriate governance arrangements to support Indigenous engagement in national, state, regional and local fire management decision making. To date, this has been given limited material support. Federal and state agencies and non-government organisations (NGOs) have increasingly resourced and supported cultural burning workshops, cultural burning activities and networking opportunities to enable peer-to-peer learning. Interviewees welcomed this growing support, but they also pointed to an ongoing challenge that was succinctly summarised in an interview with a representative from the Taungurung Land and Waters Council Aboriginal Corporation (Victoria):

> *No one is employed to apply cultural burning. We need this. We need them [government] to give us time, space, and empowerment. We need Aboriginal communities to come together to answer these questions. Give us the time to flesh it out. (interview, May 2021)*

Many of the cultural burning workshops and gatherings that have taken place across Australia rely on Indigenous volunteer or day-labour funding models. While there are some notable examples of the efficacy of this approach, particularly during the early stages of re-establishing cultural authority, knowledge and support for cultural burning activities, partnerships and training are needed (see Maclean *et al.* 2018). Indigenous group interviewees noted that it is difficult to sustain community interest in cultural burning, and to grow capacity to undertake and lead landscape burning activities, without employment and training. This contrasts with some northern Australian and desert Indigenous ranger groups who have negotiated funded positions to deliver fire management programs that support locally led cultural burning and land management practices (Robinson *et al.* 2020).

The plethora of regulatory and legal hurdles that are part of contemporary fire management is also a barrier. These hurdles are ubiquitous for all local fire management and hazard reduction groups, but Traditional Owner groups may need extra support to navigate them. Wildfire codes of practice require a high level of management training and qualification, and the need for accredited training and insurance disadvantages Indigenous cultural burning (10 Deserts Project, interview, February 2021). The Victorian Traditional Owner Cultural Fire Knowledge Group (2019) reflected on these barriers, noting that 'despite the existence of supporting statements and mutual objectives in current government policies, there are numerous policy and regulatory constraints'. This means that Traditional Owners have limited authority, resources and capacity to develop and apply cultural practices on Country (Dja Dja Wurrung Clans Aboriginal Corporation 2020). Indigenous groups who practise cultural burning also lack necessary legal protections, including insurance, particularly where structures and processes rely on the volunteer model (NSW Banbai Rangers, interview, May 2021).

Input from representatives of Indigenous organisations to recent fire reports and commission inquiries has also reflected on the barriers and obstacles imposed by fragmented partnerships and burning regimes that now exist across tenures and government agencies (Firesticks 2020; Robinson *et al.* 2020). Many Indigenous groups in the study region also do not have access to their Country and have limited mechanisms available to them to engage in and lead cultural burning and land management. As a Banbai Ranger reflected:

> *Not everyone can burn Country, or they might not have Country to look after. They are missing that. Get rangers and IPAs up and running to get those people back looking*

> *after Country, to hear stories from the old people to the young people—language, dance, this all needs to be done to avoid disconnection. (interview, May 2021)*

As a result, Indigenous rangers in some regions noted that they have not had first-hand experience of burning. In other regions, Indigenous fire practitioners had not had the necessary access to land to undertake burning. Fragmented fire management regimes, resourcing opportunities and partnership arrangements also mean that Indigenous fire crews in the study region have been confined to working in patches where they are allowed to burn; during times deemed appropriate by prescribed burning regulations and protocols; and only in association with government agency, NGO or private landholder partners, who may have different (conservation or hazard reduction) fire management capabilities and agendas.

Significant differences between Indigenous, historical, academic and non-Indigenous fire practitioner understandings of fire knowledge and landscape burning practices are also a challenge (see review in Robinson *et al.* 2016a; Box 33.2). Local Indigenous and non-Indigenous fire practitioners who were interviewed as part of this study also discussed the cross-cultural challenges of working with fire officers who have limited appreciation or knowledge of Indigenous cultural burning protocols and practices, including those who are brought in from other areas and are not knowledgeable about local species or weather patterns. A Ngadju ranger in Western Australia described the consequences of these challenges:

> *Back burning is done in an efficient manner that does not consider local values and details of the local areas. This issue was amplified last year when back burning was being done quickly, with fire crews who didn't know much about the area, and which resulted in some important areas and species being damaged through fire. (interview, February 2021)*

Box 33.2. Cultural burning and biocultural landscape management

Local groups and regional alliances have articulated various strategies to empower Traditional Owners and Indigenous fire practitioners to rejuvenate or expand cultural burning as part of Indigenous people's land management activities and responsibilities. Numerous submissions, reports and interviews that were collected as part of this study described cultural burning, highlighting the intimate connections that exist between cultural burning and Indigenous landscape management (Elone *et al.* 2018).

Three key features that characterise the purpose, practice and governance of cultural burning were illuminated. First, cultural burning is a means of expressing Indigenous governance and Indigenous rights to, and responsibilities for, Country. Second, cultural burning represents a holistic approach to managing the Australian landscape that draws on the cultural authority, knowledge and customs of local Indigenous communities, including their knowledge and understanding of the nuances and needs of local landscapes. Third, cultural burning is an evidence-based approach to land management. Indigenous people manage fire using techniques that have been developed over long periods, based on observations, experience and an understanding of whether, when and how fire can maintain, benefit and heal different landscapes.

This evidence-based approach continues in contemporary fire management. This includes finding culturally safe ways to rejuvenate, share and build new knowledge to guide cultural burning in contemporary landscapes.

Non-Indigenous partners can facilitate collaborative knowledge-sharing with local communities that recognises the value of Indigenous knowledge and the practices that sustain this knowledge. For example, partners can adopt landscape burning decision and evaluation approaches that value and respect Indigenous knowledge, and recognise the biocultural features of Australia's local landscapes. Knowledge sensitivity and privacy practices should be maintained as part of this effort, as different perspectives regarding the significance of a species or habitat – and the role of fire in damaging or healing Country – can be challenging to negotiate in cross-cultural settings. Non-Indigenous partners and on-ground managers need to be mindful of ethical issues when navigating how to heal Country, so that culturally significant species can survive and thrive with fire. It is vital to negotiate with Indigenous communities regarding any new knowledge that is needed, as well as the practices that are deemed helpful or harmful for significant species, sites and habitats (Robinson *et al.* 2021).

Many Indigenous fire practitioners agree that landscape burning or fire suppression efforts that are conducted without local knowledge and appropriate Indigenous engagement can lead to poor decisions. Indigenous fire practitioners who were interviewed for this study reported that non-Indigenous fire crew assumptions about Indigenous people and their relationships with fire also create tensions in on-ground efforts. Interviewees described some of the challenges facing local Indigenous rangers, including tremendous pressure to perform cultural burning activities when they may not have sufficient capabilities, resources or capacity to undertake cultural burning in a safe or appropriate way (10 Deserts Project interview, February 2021; NSW NPWS interview, April 2021). In other cases, Indigenous rangers and fire practitioners described not feeling 'culturally safe' to participate in mainstream fire crews or fire management activities with long-established fire management decision making protocols and practices.

Conclusion

The holistic and long-term impacts of wildfires on Indigenous communities and Country provide important context for recommendations to empower Indigenous leadership in fire management decision making and activities, both in the short term and in the longer term. These efforts have primarily been driven across Australia by Traditional Owners, who assert that cultural burning is part of Indigenous people's rights and responsibilities to care for Country. With this comes the requirement to be informed about and involved in cultural burning and wildfire recovery protocols, as well as mainstream decision making processes. It is important to recognise that Australian Indigenous culture is not homogeneous and there may not be cultural consensus regarding what cultural burning entails, even within the same community group. Each region across Australia is unique, and non-Indigenous fire partners need to recognise local diversity in Indigenous fire management knowledge, capacity, priorities and techniques (Fig. 33.2). This requires careful cross-cultural collaboration based on trust, as well as recognition that the journey for Indigenous-led cultural burning across the country is far from complete.

Fig. 33.2. Widjabul Wia-Bal Country, Dorrobbee Grass Reserve, Dunoon, New South Wales. (Photo: M. Lockwood)

Local cultural burning regimes now practised across the nation are dynamic, adaptive and subject to change over time, reflecting the needs of Country and its local Indigenous culture. In some cases, cultural burning may persist, become dormant or be displaced, or it may develop into new landscape burning regimes that replace former practices that are no longer present or relevant. Over the past few hundred years, Indigenous landscape burning has been suppressed in many parts of Australia. The history and contexts of local communities, and the nature of Indigenous legal rights and interests, mean that these recent wildfires had impacts and consequences that differed among Indigenous rights-holders, non-Indigenous landowners and Indigenous community members. As a result, there are now highly variable levels of interest, capacity and cultural authority to support and reintroduce cultural burning within these areas.

Across Australia, there is a diversity of skill, expertise and capacity among Indigenous people who wish to conduct cultural burning and undertake land management in local regions. Efforts to support and sustain this interest and capacity can build on existing land and ranger activities and partnerships led by and involving Indigenous managers and communities. To ensure long-term success, collaborative partnerships must support Indigenous-led pathways to build and assert cultural authority and the capacity to direct and implement cultural burning regimes.

Recommendations

- *Empower Indigenous rights and authority,* by supporting local Traditional Owner groups to maintain, re-learn or resurrect their cultural fire knowledge, and practices. In the longer term, formal recognition of the rights and interests of Traditional Owners in land and fire management – regardless of current land tenure – is a necessary step to ensure cultural burning is central to Australia's landscape burning regimes.

- *Support collaboration and reconciliation.* With additional resourcing, Indigenous engagement and collaboration strategies and protocols can be further developed, based on initiatives that are already in place in some agencies and jurisdictions. In the longer term, existing fire management operating procedures need to incorporate locally informed Indigenous engagement processes to employ Traditional Owners in the preparation, response and recovery phases of wildfire management. These processes should be routinely reviewed to ensure communication remains appropriate and effective.
- *Empower and support Indigenous-led wildfire planning and recovery.* Indigenous leaders need to be empowered and supported to be more involved in local, regional and national wildfire recovery strategies. This includes involvement in all emergency disaster planning and strategy development, habitat and species restoration activities, rehabilitation activities, and efforts to identify and protect cultural heritage sites. At the policy level, this could include listing culturally significant species as matters of national environmental significance; at the local level, this requires ensuring that wildfire recovery practices and priorities recognise and respond to culturally significant sites, species and landscapes affected by wildfires. Cultural heritage site and asset information should also be included in wildfire management plans and, where possible, treatments should be identified to mitigate against wildfires.
- *Empower Indigenous rights and authority to care for Country through fire.* With increased investment and support, the incorporation of cultural burning into Australia's disaster risk reduction and conservation regimes has the potential to improve fire management. To achieve this, fire management partnerships must recognise the complexity of cross-cultural engagement and interactions, and respect Indigenous knowledge, know-how and protocols. These partnerships also need to provide space and opportunities to support Indigenous and non-Indigenous people to develop new knowledge and skills in their own timeframes, and they facilitate two-way knowledge exchange. Finally, testing the impact and effectiveness of cultural burning under different governance arrangements and fire management settings can help to identify what best enables Traditional Owners to lead the practice and timing of cultural burns in different contemporary landscapes, further empowering their rights and authority.

These recommendations focus on recognising Indigenous rights and interests; recognising and supporting Indigenous knowledge rights and governance; supporting Indigenous-preferred approaches to learning, sharing and passing on relevant fire knowledge; establishing place-based partnerships that enable holistic landscape burning regimes across tenures and between institutions; ensuring that Indigenous fire knowledge and management activities can safely and legally work within contemporary institutional land governance arrangements; and delivering environmental, social, cultural and economic benefits for local Indigenous people. Translating these recommendations into practice will require the resolution of institutional and cross-cultural challenges that hinder Indigenous-led efforts to re-build or expand cultural landscape regimes that reflect the contemporary cultural and biophysical landscape of Australia.

Acknowledgements

We acknowledge and are thankful to all those Indigenous and non-Indigenous fire managers and researchers who shared insights about their important work. This research would not have been possible without their input. We would also like to thank the NESP Threatened Species Recovery Hub Indigenous Reference Group and the Department of Agriculture, Water and Environment (DAWE), who supported this research. This chapter draws on material outlined in our detailed report at https://www.nespthreatenedspecies.edu.au/projects/indigenous-aspirations-and-capacity-for-bushfire-response

References

Aboriginal Affairs NSW (2017) *Transforming the Relationship between Aboriginal Peoples and the NSW Government: Aboriginal Affairs NSW Research Agenda 2018–2023*. Aboriginal Affairs NSW, Sydney.

Aboriginal Victoria (2019) 'Victorian Government Aboriginal affairs report 2019'. Aboriginal Victoria, <https://www.aboriginalvictoria.vic.gov.au/sites/default/files/2019-11/Victorian%20Government%20Aboriginal%20Affairs%20Report%202019_0.pdf>.

Cahir F, Schlagloth R, Clark ID (2020) The historic importance of the koala in Aboriginal society in New South Wales, Australia: an exploration of the archival record. *ab-Original* **3**, 172–191. doi:10.5325/aboriginal.3.2.172

Connell J, Watson SJ, Taylor RS, Avitabile SC, Schedvin N, *et al.* (2019) Future fire scenarios: predicting the effect of fire management strategies on the trajectory of high-quality habitat for threatened species. *Biological Conservation* **232**, 131–141. doi:10.1016/j.biocon.2019.02.004

Costello O (2019) 'Inquiry into Koala populations and habitat in New South Wales'. Portfolio Committee No. 7 – Planning and Environment.

Cowley RA, Hearnden MH, Joyce KE, Tovar-Valencia M, Cowley TM, *et al.* (2014) How hot? How often? Getting the fire frequency and timing right for optimal management of woody cover and pasture composition in northern Australian grazed tropical savannas. Kidman Springs Fire Experiment 1993–2013. *The Rangeland Journal* **36**, 323–345. doi:10.1071/RJ14030

DAWE (2020) 'Provisional list of animals requiring urgent management intervention. Report prepared for the Wildlife and Threatened Species Bushfire Recovery Expert Panel'. Department of Agriculture, Water and the Environment, Canberra, <http://www.environment.gov.au/system/files/pages/ef3f5ebd-faec-4c0c-9ea9-b7dfd9446cb1/files/provisional-list-animals-requiring-urgent-management-intervention-20032020.pdf>.

DAWE (2022) 'National Recovery Plan for the Koala Phascolarctos cinereus (combined populations of Queensland, New South Wales and the Australian Capital Territory'. Department of Agriculture, Water and the Environment, Canberra.

Dickman C, Driscoll D, Garnett S, Keith D, Legge S, *et al.* (2020) *After the Catastrophe: A Blueprint for a Conservation Response to Large-scale Ecological Disaster.* NESP Threatened Species Recovery Hub, Brisbane.

Dja Dja Wurrung Clans Aboriginal Corporation (2020) 'Submission to the Royal Commission into National Disaster Arrangements'. Submission Number: NND.001.00160. Dja Dja Wurrung Clans Aboriginal Corporation, Bendigo.

Elone T, White H, McKemey M, Costello O (2018) Banbai Enterprise Development Aboriginal Corporation on-country fire management. In *A National Framework to Report on the Benefits of Indigenous Cultural Fire Management.* (Eds K Maclean, CJ Robinson, O Costello) pp. 23–28. CSIRO Publishing, Melbourne.

Environment and Communications References Committee (2011) 'The koala – saving our national icon'. Senate Committee report. Environment and Communications References Committee, Canberra.

Firesticks (2020) 'Submission to the NSW Independent Bushfire Inquiry – submission number 0906. Firesticks', <https://www.dpc.nsw.gov.au/assets/bushfire-inquiry-submissions/0906-Firesticks-Alliance-Indigenous-Corporation-20200522.pdf>.

Fogarty W, Bulloch H, McDonnell S, Davis M (2018) *Deficit Discourse and Indigenous Health: How Narrative Framings of Aboriginal and Torres Strait Islander People are Reproduced in Policy.* The Lowitja Institute, Melbourne.

Freslov J (2004) 'Post wildfire Indigenous heritage survey, Volume 2: Management and impacts from wildfire and suppression activities. A report to Parks Victoria'. AAV Project No. 2833. The Department of Sustainability and Environment and Aboriginal Affairs Victoria, Melbourne.

Gunaikurnai Land and Waters Aboriginal Corporation (2020) 'Submission to the Royal Commission into National Disaster Arrangements'. Submission number: NND.001.01091'. Gunaikurnai Land and Waters Aboriginal Corporation, Victoria.

Inspector-General Emergency Management (2020) 'Queensland Bushfires Review Report 2: 2019–2020'. Office of the Inspector-General Emergency Management, Brisbane.

Legge S, Rumpff L, Woinarski JCZ, Whiterod NS, Ward M, *et al.* (in press) The conservation impacts of ecological disturbance: time-bound estimates of population loss and recovery for fauna affected by the 2019–20 Australian megafires. *Global Ecology Biogeography.* doi:10.1111/geb.13473

Maclean K, Robinson CJ, Costello O (2018) *A National Framework to Report on the Benefits of Indigenous Cultural Fire Management.* CSIRO, Brisbane, <https://www.csiro.au/en/Research/LWF/Areas/Pathways/Indigenous-futures/Collaborative-indigenous/Cultural-burning-partnerships>.

Prober SM, Yuen E, O'Connor MH, Schultz L (2016) Ngadju kala: Australian Aboriginal fire knowledge in the Great Western Woodlands. *Austral Ecology* **41**, 716–732. doi:10.1111/aec.12377

Robinson CJ, Barber M, Hill R, Gerrard E, James G (2016a) 'Protocols for Indigenous fire management partnerships. Final report'. CSIRO, Brisbane.

Robinson CJ, McKaige B, Barber M, Hill R (2016b) 'Report on the National Indigenous Fire Knowledge and Fire Management Forum, Building Protocols from Practical Experiences Darwin, Northern Territory'. CSIRO and the National Environmental Science Program's Northern Australia Environmental Resources Hub, Darwin.

Robinson CJ, Metcalfe D, Collins K, Williams L, Bohensky E (2020) Practical relief measures to build community-level resilience following a natural disaster. In *Climate and Disaster Resilience: Technical Report.* (Ed. CSIRO) pp. 195–229. CSIRO Publishing, Melbourne, Australia.

Robinson CJ, Costello O, Lockwood M, Pert PL, Garnett ST (2021) 'Empowering Indigenous leadership and participation in bushfire recovery, cultural burning and land management'. NESP Threatened Species Recovery Hub, Brisbane, Australia.

Schlagloth R, Cahir F, Clark I (2018) The importance of the koala in Aboriginal society in nineteenth-century Victoria (Australia): A reconsideration of the archival record. *Anthrozoos* **31**, 433–441. doi:10.1080/08927936.2018.1482115

The Victorian Traditional Owner Cultural Fire Knowledge Group (2019) 'The Victorian Traditional Owner cultural fire strategy'. Department of Environment, Land, Water and Planning (DELWP), <https://knowledge.aidr.org.au/media/6817/fireplusstrategyplusfinal.pdf>.

Williamson B, Markham F, Weir JK (2020) 'Aboriginal peoples and the response to the 2019–2020 bushfires, Working Paper No. 134/2020'. Centre for Aboriginal Economic Policy Research, Australian National University, Canberra, <https://caepr.cass.anu.edu.au/research/publications/aboriginalpeoples-and-response-2019–2020-bushfires>.

Wintle BA, Legge S, Woinarski JCZ (2020) After the megafires: what next for Australian wildlife? *Trends in Ecology & Evolution* **35**, 753–757. doi:10.1016/j.tree.2020.06.009

Woinarski JC, Legge S (2013) The impacts of fire on birds in Australia's tropical savannas. *Emu-Austral Ornithology* **113**, 319–352. doi:10.1071/MU12109

Yibarbuk D (1988) Notes on Traditional use of fire on upper Cadell River. In *Burning Questions: Emerging Environmental Issues for Indigenous Peoples in Northern Australia.* (Ed. M Langton) pp. 1–16. CINCRM, NTU, Darwin.

34

Of grief, spirit and hope: personal reflections of the 2019–20 wildfires

Noel Webster (Uncle Nook), Lucas Bluff, Melinda L. Moir, Jess R. Marsh, Daniella Teixeira, Jenny Gray, Libby Rumpff, Daniel J. Rogers, Fern Hames, Bruce Pascoe and John C. Z. Woinarski

Introduction

Our lives are marked by love and grief and hope. The chapters in this book are largely dispassionate accounts, tallying losses of nature, and describing actions taken for its recovery. However, although such knowledge is necessary, it is not sufficient. We could not complete this book without recognising that those losses of biodiversity – the damage to country – also deeply affected our society. Many of the contributors to this book knew that country intimately, valued it immeasurably and had spent much of their lives trying to understand and nurture it, to live within it. We grieved for what was lost, and for what this loss may foretell of our future.

The exceptionalism of the 2019–20 wildfires may have caused loss at a scale not previously witnessed in our lives. But equally striking was the extraordinary response, of communities bonding together to do whatever was possible to heal the wounds, to volunteer, to collaborate, to donate. Counterpointing the calamitous losses, this community spirit was inspiring – a recognition that our lives would be diminished by, could not afford, and most importantly would not numbly accept, loss at such a scale. We would do whatever was possible to support recovery, to restore country, and to act to reduce the likelihood of comparable devastation in the future.

So, in this chapter, we seek to do something different. We invited personal reflections of what the consequences of these fires meant to people, and of the emotions that drove them to help recover nature after these fires. We collate and interweave these perceptions around three linked themes: loss, spirit and hope. Because hope without acknowledging loss is an illusion, and spirit is necessary to transform hope into change.

Personal reflections

Noel Webster (Uncle Nook)

Descendant from Walbunja and Murramarang Peoples of the New South Wales South Coast

Sandstone emotions

My legs are shaking.

The 2019–20 wildfires may appear gone, but an eternal recollection is etched deep within my veins, blood memory of an account that will traumatise me until the day my eyes lay at rest. The guilt, the shame, the hurt and anger. I take blame contemplating my actions: could I have done more to look after our beautiful Country.

Where to for now? The recovery is it for me, for community or Country; it's all the above but where to start?

To be honest, I'm frozen with fear and have been sitting on the peripheral, evaluating the harm from the outskirts, not deep viewing only seeing the environmental aftermath, but I live in a cultural landscape where there is the holistic connection between, plant, animals, people, elements and earth.

Sandstone Country between the Shoalhaven River (Bugelli) and the Clyde River (Bhundoo), a beautiful place enriched with Cultural Values, evidenced through rock art assemblages that span across our river systems and connect the coast to the hinterland.

Are they intact, who knows, too scared to look, why, because of the anger and hurt I have experienced through the peripheral view? My shaky legs – is that an excuse not to look, not even.

Whose job is it, who's going to step up to the plate or do we do nothing and hope for the best, don't even anticipate it. It's our job, needs to be done, and there is no justification to sit back and hope.

The anxiety of the unknown, we need to know the impacts.

Ok, I've bitten the bullet and am planning to look, alone, no way, but is it fair to drag family, friends and community along for my safeguard, is that just as bad as doing nothing, what about their hurt and the way they feel. Will I be triggering their emotion and making them as angry as me, taking them back and generating their blood memory.

Time will tell, we take our first footstep next week.

Lucas Bluff

Lucas is a zoologist with responsibilities for biodiversity management in Gippsland

Notes on grief and hope

The words 'grief' and 'hope' are too simple and too strong to describe my emotions around the 2019–20 bushfires. My response has been shaped by more than a decade of biodiversity management in Gippsland, including leading fire ecology projects, visiting hundreds of sites with different fire histories, conducting landscape-scale fire-management planning, and being involved in emergency response to some half-dozen major bushfires.

Instead of grief I have experienced sadness and frustration. Sadness – at the loss, at least for a while, of aesthetic pleasure in some places that I enjoyed in the past. Open rainforest understoreys; stands of big wet forest; sites occupied by rare species … some earmarked pages have been torn from my mental catalogue. Frustration – because of lost opportunities for future study. Our ability to understand the composition of areas unburnt for decades has now been seriously depleted. Do long unburnt sites *really* have higher biodiversity value? Much harder to answer now. But grief – no, that would only come with knowledge of permanent or absolute loss.

Even events as large as the 2019–20 bushfires are not monolithic, nor, in isolation, are they likely to cause permanent loss. But by invoking this broader perspective – fire *regime* over fire *events* and *combinations* of threats over *single* threats – to deny grief, I must also accept that it erodes hope.

Like others I have treasured the moments of post-fire hope found in epicormic shoots emerging from charred bark, delicate fronds unfurling in a bare landscape, a carpet of orchids, the delight when birdsong returns. But if one interprets the 2019–20 bushfire season as a card drawn from the progressively stacked deck of climate change, long-term hope is harder to summon. My pessimism is twofold: that an increasing frequency of million-hectare fire seasons in Victoria will cause cumulative loss of natural values; and that our societal reaction may pressure fire management agencies in ways that further constrain our options. My main hope is of being proved wrong.

Melinda L. Moir

Melinda is an entomologist whose research included studies of an insect species made extinct by the 2019–20 wildfires.

One documented invertebrate extinction, how many more were lost?

The Australian wildfires of 2019–20 were the final death knell for a tiny pinkish blob of an insect, *Pseudococcus markharveyi* or the *Banksia montana* mealybug. The loss of such a small inconspicuous insect that only a handful of people ever saw alive may appear meaningless, but to me it represents the decline and loss of countless unseen invertebrate species. This species was 'fortunate' enough to get discovered, lodged in collections and given a name. But how many other invertebrates are out there, lingering on the edge of extinction, as yet undiscovered? How many will never be discovered because they were lost in the massive wildfires?

It was pure luck that *P. markharveyi* was discovered in 2007; even though its host plant was included in my postdoctoral study of herbivorous insects on threatened plants, 'vacuuming' plants did not detect the bug like it did other insects. The mealybugs cling tightly to the underside of curled leaves or within the very hairy flowerheads of the montane banksia, ensuring that their removal is difficult. Instead, the mass feasting of hungry ladybird beetles on unfortunate *P. markharveyi* gave them away. By the time I stumbled across these innocuous creatures, there were very few host plants left in the Stirling Range National Park of south-west Western Australia, and *P. markharveyi* only resided on the very oldest plants that had escaped a fire in 2000.

Despite the state of decline, I hoped that the mealybug could ride off the back of the fantastic work of the flora conservation teams. The botanists had worked hard for decades on recovering *B. montana* both *in situ* and *ex situ*. For over 8 years I was involved in different conservation actions for the mealybug (nominating for state and International Union of Conservation of Nature (IUCN) threatened species lists; monitoring; *ex situ* translocations; searching for other populations; and working with experts on the genetics). The next step was apparent; seeding young *in situ B. montana* populations with the mealybug. We just needed to wait until the young seedlings were big enough to support a few mealybugs.

Little did I know that time was very short.

Two massive wildfires, 19 months apart, finished off the only confirmed populations on the summits of the Stirling Ranges. This leads me to wonder what happened to the rest of the diverse and threatened montane invertebrate community of which *P. markharveyi* was a part, including a stunning peacock spider (*Maratus sarahae*), a series of Gondwanan spiders (*Perissopmeros darwini, Calcarsynotaxus benrobertsi, Bertmainius colonus, Zephyrarchaea robinsi*), a carnivorous undescribed snail (Rhytididae snail), a tiger millipede

(*Atelomastix tigrina*), an undescribed keeled millipede (*Antichiropus* sp.), a plant louse (*Trioza barrettae*), and an undescribed genus of a strange flightless Dictyopharidae bug. To say that these fires depress biologists is an understatement – to me it is undisputable proof of the accelerating sixth mass extinction event.

I haven't been able to bring myself to revisit the Stirling Ranges since the fire that killed the last of the *P. markharveyi*, over 2 years ago.

Jess Marsh

Jess is an arachnologist and a Kangaroo Island resident.

Life hanging on a thread

I live and work on Kangaroo Island, researching spiders and other invertebrates. One of the species I work on is the Kangaroo Island micro trapdoor spider (*Moggridgea rainbowi*) and I don't think I'll ever forget the first time after the fires that I got out there to find out how this species had fared. It was probably a changing point for me; before then I was of the belief that, yes, these fires were big and the human impacts were huge – friends lost houses and farms, everyone was so busy fighting fires or helping out – but fire is natural and species are resilient. But when I saw first-hand the severity and extent of the fires, that all that was left of the trapdoor spider colonies I had known were burnt empty holes, entire habitats reduced to blackened shells, the scale of the fires and the impact on our ecosystems really hit me.

Since then, I have been out most months, surveying the fireground, and it has been a privilege to see the land in this state of transition and regrowth. The positive finds make the effort worth it: from finding surviving Kangaroo Island micro trapdoor spiders in small unburnt remnants (Fig. 34.1) to the feeling when, after months of surveying, I found a single surviving Kangaroo Island assassin spider, a species we feared could have been

Fig. 34.1. Many known populations of the Kangaroo Island micro trapdoor spider were destroyed by fire, but here a spiderling was found after the fires. (Photo: Jess Marsh)

driven extinct by the fires. Many species are likely now imperilled, and many face ongoing threats in to the future. These fires were big, but they are unlikely to be the last, and I believe it's our job now to learn from them and to be better prepared for the next ones.

Daniella Teixeira

Daniella was completing her PhD research on glossy black-cockatoos when the 2019–20 wildfires ravaged Kangaroo Island.

More than just glossy black-cockatoos: Black Summer and my grief at a changing world

There is a word that describes feeling homesick when you're still at home: solastalgia. It captures the grief we feel when our home is threatened or changed. The fires of 2019–20 ignited a deep sense of solastalgia as they tore through places I know intimately. Kangaroo Island was one such place; I'd spent the previous 4 years researching the nesting behaviour of the island's endangered glossy black-cockatoos (*Calyptorhynchus lathami halmaturinus*).

I remember my despair as maps showed the footprint of fire spreading east across Kangaroo Island, burning one nesting area after another. On returning to the island, I found landscapes that were previously rich in life transformed to ashy 'moonscapes' (Fig. 34.2). Plastic nest hollows were melted, and iron collars, normally fixed to tree trunks to protect nests from predators, were splayed on the ground.

The fires changed my perception of what it means to 'do' conservation in today's world. For a long time, I ruminated on feelings of not being good enough. I left Kangaroo Island knowing that, although we can improve the trajectory of threatened species, intractable

Fig. 34.2. Daniella Teixeira in a once familiar landscape on Kangaroo Island, transformed by fire. (Photo: Nicolas Rakotopare/Threatened Species Recovery Hub)

Fig. 34.3. One of the lucky ones. A female glossy black-cockatoo flies out of her nest hollow in a sugar gum that escaped the wildfires. (Photo: Nicolas Rakotopare/Threatened Species Recovery Hub)

problems like climate change demand a new era of conservation. We need to be more creative in our thinking and bolder in our doing.

I'm often asked, 'Why should we save the glossy black-cockatoo?', as if their existence alone is not sufficient. My response is usually, 'It's not about the glossy black-cockatoo. It's about how we value life'. With every extinction, the world becomes less rich, less beautiful; and, in the words of poet Mark Tredinnick, 'the living world grows less Alive'. When we lose species, we lose billions of years of evolution; losing such complexity is a huge misfortune.

Evidently, my solastalgia extends beyond what was lost in the 2019–20 fires. It's about this whole, changing world. Knowing that countless species and ecosystems have been lost, and will be lost, in my lifetime, due to problems that I inevitably contribute to, is a difficult truth to accept. But this has only strengthened my resolve to contribute meaningfully to saving species.

I find hope in the birds' resilience. A month after the fires, glossy black-cockatoos were nesting in the habitat that remained (Fig. 34.3). We now know that over 450 birds survived by September 2020. In post-fire surveys in south-east Queensland, which also burnt, I found birds in areas that had experienced low severity fire. These observations certainly do not mean that everything is good, but they do remind me that there is resilience in nature. And for that I am hopeful.

Jenny Gray

Jenny is the Chief Executive Officer of Zoos Victoria.

Fighting extinctions

We all know where we were in December 2019. Smoke covered our skies and the daily news revealed an unfolding tragedy of unprecedented proportions. Every day we watched with dread as fires expanded across Australia.

Fig. 34.4. Providing food supplementation to help improve breeding success for the threatened mountain pygmy-possum. (Photo: Jenny Gray)

On 23 December 2019, I was in the Victorian High Country. Sitting on a balcony at Falls Creek, I saw that the sky was an unhealthy grey. With fires hundreds of kilometres away, the smoke served as a warning of the approaching destruction. I was on a call to Jon Paul Rodriguez, the Chair of the IUCN Species Survival Commission, talking about the partnership between zoos and aquariums and the IUCN Species Survival Commission to reverse Red List trends. There are days when our efforts to save every species from extinction seem impossible, days when the impending species catastrophe overcomes us like the grey smoke that filled the sky. And there are days when our resolve hardens and we commit to doing everything we can to fight for all the creatures that share this planet. Most of all, there are partners that remind us we can win this fight when we work together and use all the tools and resources available.

The 2019–20 wildfires served to engage and mobilise our local and international communities to act against the impending crisis of climate change and the threat of extreme weather and fire events on species. In addition, the fires brought together a wide range of professional conservation organisations, governments and the private sector to address the impacts. For a moment in time the attention of the world, from small children to billionaires, focused on the situation in Australia. The smoke of a thousand fires was experienced around the world. And the world reacted: fire fighters and vets arrived, and we mobilised professionals and volunteers. We could not be bystanders.

After the fires we assessed the loss, to humans and to biodiversity. We also assessed the need to help in the recovery of nature. Our ability to help the recovery after a catastrophic event is limited. It takes time, money and skill. At worst we can only count the cost; at best we can assist in removing compounding threats allowing species to bounce back.

Sometimes we just buy a little time, to allow species to survive long enough for their habitats to recover.

I was in the high country to visit and support the work of Zoos Victoria in supplementary feeding of mountain pygmy-possum (*Burramys parvus*). The lack of Bogong moths (*Agrotis infusa*) over the prior 3 years was impacting the breeding of this critically endangered marsupial and Zoos Victoria was trialling supplementary feeding during the critical period for female possums. The work is hard and tiring: crossing unstable boulder fields to fill purpose-built feeders and to change camera trap cards to secure evidence and knowledge of the access to food (Fig. 34.4). Little did we know that only 5 months later mountain pygmy possums in New South Wales, their habitat ravaged by the 2019–20 wildfires, would benefit from this trial: the food developed and trialled in Victoria would become the saviour of these small mammals.

Libby Rumpff

Libby is an ecologist and decision scientist who conducted many regional workshops after the fires to help coordinate responses for fire-affected species and ecological communities.

Clearing the smoke

There was claustrophobia, brought on by weeks of smoke and breathing in a heady mix of charred landscapes and substances I'd rather not think about. Smoke was the beast that fed anxiety and guilt. We were all pinned down, unable to escape, the once steady horizon no longer discernible. The soundtrack for the summer is still vivid, filled with ABC radio presenters working hard to pronounce the names of towns in the path of the fire, provide updates and give voice to stunned and grieving locals. And when the rainforests started to burn and the ecologists were projecting the number of animals impacted, it seemed there was a very public realisation that there was a loss of life that we don't often tally, that we don't understand well.

I knew I wasn't alone in feeling a sense of hopelessness. Pleas on social media, the international spotlight, governments announcing funding packages: the money was flooding in from everywhere. And this is where I got involved, as a member of an expert panel for the Australian Government, tasked with prioritising action for affected species and ecological communities. It was the collaboration with those involved in recovery efforts across the most heavily impacted regions in New South Wales, South Australia, Queensland, Victoria and the Australian Capital Territory that cleared the smoke. When the list of priorities is immense and the impacts are not yet understood, how do you make decisions about what to do, and where? How do you plan for a future where such fires are likely to be more commonplace? I asked questions, and I listened. These were people working together to create opportunities from the ashes, to plan for the future.

Alone, grief can floor me, but when I can see and talk to others, it unites. I found inspiration and hope in others. Yet there is no deluded sense of security; I see the past, the likely future, the tangle of threats and management applied to our landscape, and the shifting baselines for our ecosystems, plants, animals and fungi. It terrifies me. But if I don't keep moving, and trying, then the smoke descends and quashes hope. And I refuse that.

Dan Rogers

Dan is an ecologist with the South Australian Department for Environment and Water.

The power of community

Kangaroo Island is one of a small number of places in the world to which I am connected. Not by heritage, or even by the island's objectively spectacular wildlife and natural aesthetic. This connection was born of knowledge, recognition, familiarity. Intimacy. A place that, given time, grabs your soul and doesn't let go. Those of you who have deep connection with a place, as I'm sure most of you do, will know what I mean.

My first interaction with the Kangaroo Island wildfires was via a map, on a screen: an abstraction, of the smoke and heat, of the intense pyrrhic dance between air and ground and life. What happened next, however, shook us all. Ultimately, ~200 000 ha – half of the island's total area – burnt in December 2019 and January 2020. At least 80% of the known habitat for many species already threatened with extinction. We were facing the real possibility of watching a mammal species – the Kangaroo Island dunnart (*Sminthopsis fuliginosus aitkeni*) – go extinct before our eyes. On our watch.

Close to 2 years on, the visceral response to the memory of such devastation remains. Surprisingly, perhaps, what remains the most emotive memory for me, more than the impacts of the fires, relates to the response of friends and colleagues, from many walks of life, from across the country. A number are contributors to this volume. The generosity of people, to offer all the support they were able, at a time when they were often personally suffering through the worst fire seasons in modern Australian history, still takes me aback. Receiving phone calls, offering assistance, while the fires burnt. In the small town of American River, we hosted close to 100 people (Fig. 34.5), not more than a month after the

Fig. 34.5. Participants from the American River workshop visit a feral predator exclosure set up after the fires on Kangaroo Island. (Photo: Dan Rogers)

fires, there to collectively offer whatever they could, to ensure that we would not allow species to go extinct. Such passion, such intuitive generosity, will stay with me, and I will always be grateful to those who gave it.

Nature conservation, in a world of such overwhelming change, can be a tough game. But what the fires on Kangaroo Island taught me, more than anything, is that it's a game that we're not in alone. We belong to a community who have such diverse experience, and who are incredibly knowledgeable, ferociously passionate and deeply generous. The power of that collective is enormous and, particularly when I think about global extinction, I remind myself of that power. It gives me hope.

Kangaroo Island, its unique biodiversity and its community are recovering from the fires, more slowly for some than others. We now know that some species are far more resilient to such events than we gave them credit for, although the tireless recovery efforts of many people, over many years, have certainly helped. What I know is that the strength of the partnerships, developed over years leading up to these events, was fundamental to how we were able to respond, and how we were able to support each other, during and immediately after. Those partnerships will, no doubt, have a legacy well beyond.

Fern Hames

Fern is the Director of the Arthur Rylah Institute for Environmental Research

The human-nature recovery loop

One hundred and seventy-three.

That number, 173, is etched into my mind, my memory, my very being. One hundred and seventy-three is the number of people who died in the Victorian Black Saturday fires of February 2009. The scale of that human tragedy shaped our response: our personal response in our own lives, families and communities; and our response as ecologists, cradling an urgency to get out on the fireground and support affected biodiversity. In my own case, I gently wrapped up my own trauma, and put it very firmly in a quiet hidden space in the back of my head. I needed to focus on urgent recovery actions for two species of threatened fish: the Macquarie perch (*Macquaria australasica*) and the barred galaxias (*Galaxias fuscus*). Over the next few weeks, we collected 'insurance' populations of fish from battered streams about to be deluged with sediment sludge. Over the next 2 years we cared for those fish in our chilled aquaria, and checked on the slow recovery of their habitat, every month.

At the same time, because of the massive human impacts, we also worked very closely with the small communities who lived alongside these fishes. We connected them with the story of the little fish, patiently living in Melbourne while their streams recovered. We nurtured connections between local kids, Traditional Owners, Landcare champions, fish ecologists and artists. We created the opportunity for people to design a local interpretive bridge about the fish story. Kindergarten kids wrote a song about the fish. The arts group created a fishy parade and an ephemeral art installation in the river. Anglers planted trees to help the habitat along. Farmers helped find eggs to identify spawning habitats. We deliberately applied the principles of working with trauma-affected communities, generating a sense of calm, safety, hope, connectedness, and agency/control.

When the little fish were finally returned to their home streams, it coincided with many people returning to their newly rebuilt homes. The synergy of this shared story of return was powerful. It was a hugely celebratory moment, which cemented these recovery stories together. That connection drove a powerful recovery loop, in which people advocated for the fish, supported fish recovery, witnessed fish recovery, gained hope, connectedness and a sense of agency, and experienced recovery in themselves.

We saw the power in this nature–human recovery loop. And we recognised it would probably be a very useful thing, in the future. Especially in a future littered with more extreme events.

And, indeed, here we are again. The 2019–20 fires were another massive event in Australian history, with all kinds of associated traumas. Keenly mindful of the lessons of 2009, we argued that biodiversity recovery actions should be delivered alongside, or integrated with, an active nature-led community recovery program. Connect people with local nature recovery, and it will support their own personal and community recovery. The program was launched, but was then almost immediately overwhelmed by the restrictions of the COVID-19 pandemic. The restrictions hampered our ability to meet with fire-affected communities, apply the psychological principles of trauma-sensitive engagement, or share the stories of nature recovery.

Unsurprisingly, perhaps, the cycle happened anyway. Of course, many people already, inherently, intuitively, know these things about nature. Belatedly, now lockdowns are over, we are meeting local people and exploring streams, forests and wetlands across Victoria's North East and East Gippsland. The people we wander with are thrilled at the orchids coming up, delighted at the frogs and fish thriving in the wetlands, and in awe at the return of certain birds. They are painting the greening mountain beyond their doorstep, teaching each other new wildflowers, and hauling out willows along the river. They are monitoring their new nest boxes, photographing epicormic shoots, and loading data into citizen science apps. We are nourishing, amplifying and supporting their efforts. We are joining up local and agency-based experts, nudging emerging citizen scientists, and providing small grants to nurture local, place-based, community-led actions, which make a genuine difference for nature. We are also paying more attention to evaluation, with a dedicated analysis of the impact of such actions, to inform our future practice. But before we begin any formal analysis, we are confident this approach is powerful, important and meaningful. As the primary-school-aged kids from the 'Sarsfield Snaps' post-fire photography project (Fig. 34.6) said: 'If this happens to you, go and watch nature. You'll be OK'.

Bruce Pascoe

Bruce is a Yuin, Bunurong and Tasmanian Aboriginal man.

Spades and shovels

I could hear it coming up the hill. A dull roar like a truck on a bridge. I was alone and so I headed back to the ute and put all the fire kit in the tray. I looked back at the trees where I had been fighting spot fires. See you tomorrow, I told them. Because I knew I would.

I got back to the house and began dampening down as the fire continued to march up to the ridge. During the night it crept down into the paddocks, just fast enough to make you check the window every half-hour and to make sure that you had your car keys in your pocket. You had slept in your clothes for over a month. Hypervigilance. I saw fire from one window or another for 5 weeks. It wouldn't go away, like jackals stalking a zebra. Turn away and it's on you again. My nostrils have the smell of smoke as a default. Everything smells burnt.

But it was stopped short of the house on three occasions and I was able to go back to the ridge and see my trees. Large trees and, as I'd predicted, there they were, scorched half-way up the trunks but the crowns untouched. The rest of the forest, honestly, looked like a

Fig. 34.6. The 'Sarsfield Snaps' post-fire photography project group at Sarsfield Snaps exhibition launch at Melbourne Zoo. (Photo: Bob Carter)

bomb had gone off. Shattered branches, crowns obliterated, every small twig gone. But the big trees, still standing, still a haven for the birds that remained.

Too big to burn. Old trees. Saved from the timber industry because the iron barks weren't straight enough and the angophoras were perceived as useless timber. Our lesson is size. Old, old trees, the first limb so high off the ground that even catastrophic fires hadn't reached more than half-way up their trunks.

The bomb of gas-laden air that exploded on Mallacoota, taking over 140 houses, smashed everything, but here on the farm a month later a tree could survive fire simply by its size and the naturally dynamic distance from those around it. A sparse, massive forest as the old people had designed it. (A postscript on the Mallacoota fire: 40 families are still without a house 2 years later because they were renters, not owners, and capital only rewards the capitalist. Working families take pot luck.)

The design of that old forest had been regulated, legislated in fact, by people using cool fire as a tool. Those people wore no shoes. The rule was if you can't walk backwards and forwards through the fire it is too hot. My son and I learnt to light fires like that and were able to say hullo to parrots and possums as they passed us in leisurely fashion into the unburnt section. It was civilised; civil and civilised.

Most of my family are in the Country Fire Authority, but when we fight fires we dress as if to go to war. We are at war, at war with the land. The boots, helmets and heavy coats and trousers are designed to save our lives because we have designed a forest that has the potential to kill us.

When we knew the baleful fires of 2019–20 were about to descend on us, I counted the trees on several acres of land on the border of New South Wales and Victoria. Three hundred and thirty trees to the acre. A monoculture of same-sized silvertop ash destined for the chip mill. After seeing my daughter and family safely on the road to Canberra on New Year's day I came back along the highway, part of which was still burning where flaming trees had dropped across it. I looked for the trees I had counted and they were gone but for piles of 'snow' where they had been. The field of ash resembled Switzerland. But nobody was yodelling.

Colonialism insists that you have to decree that the invaded are not part of God's imagination. They have no culture, they have no economy, they left no mark on the earth. Therefore, it belongs to the invader. Sim sala bim, QED, all squared.

But they did leave a mark. They created a forest of huge, sparse trees, food crops beneath them and almost impossible to burn. Unless the Fat Controller gets a contract to supply timber to the world: cedar for the gentleman's sanctum, ironbark for the railways, stringybark for the soldiers' garrison.

Suddenly, and it was quite sudden, no more forest and because no barefoot fire fighters patrolled the country all the regrowth of wattle, gum, burgan and hop bush came back unimpeded because the designer was imprisoned by the missionary and their method impugned by the priest, pastoralist and politician. We allowed the bomb components to gather around us and when they burnt we blamed the country: poor old, patient, Mother Earth – we never blamed ourselves. We cursed the country and its snakes and spiders and sharks because it was not Shropshire. We lauded the writers who wrote of her terror.

She's not terrible, our country, our mother, she's just missing the care bestowed on her for 100 000 years. Argue amongst yourselves whether it's 65 000 years or 120 000 because I don't have the time or inclination and if, in a decade, white, and therefore acceptable, science reveals that, in fact it is 120 000 years, then you owe me a cup of tea. AND a biscuit.

Save yourself, cuddle your tea caddy close to your chest, because some archaeologists are already saying 120 000 years and all of them are young, except Jim Bowler, but he was always an iconoclast! And an excellent scientist. But Australia, will you pillory the science of the scientists: the young woman in Perth, the young men in Brisbane, the prime-of-life lady in Melbourne when they bring you this science or will you gargle bleach instead?

You may have noticed that I descended into sarcasm for a moment – forgive me, for I am old and sad and deeply offended by the careless and heartless attitude to our mother and the ignorance of Australian history beyond a rack of royal teaspoons and a Panama hat attendance at the Boxing Day Test.

Spades and shovels, Australia – it is time we started talking about what is really what and the words we use to describe it. It was our system that created that fire and then decided that not all would have their houses replaced, just the rich. Even the hard-working families who had just enough to buy a house and then insure it are still richer than those who remain without.

It might seem that the politics of financial security and forest design have nothing in common but both are about care, care of earth, care of people. If that is seen as mawkish heart on the sleeve pixieness, then I think our values have blinded us to decency.

Our values need reformation and if, in a rare moment of generosity towards the original forest designers, we deign to contemplate the Aboriginal economy and conservatism, let us do so without an excess of New Agey guilt and apology but instead the rock-solid foundation of common sense: care for our home. We have security cameras on our gated residences, so why not the same for our foreheads so that we are forced to bear witness to poor farming practice and misuse of water, both designed to favour those who need no favour.

If the flighty environmentalists can remove their gaze from the organic potatoes of the weekend market for long enough, they might realise that their most likely coalition partner is the farmer. Not the Akubra and rugby jersey-type nationalist, but the one who looks out over the land and worries how it will be for their children. How do we grow food in such a way that I hand the soil on in good condition to my children?

Plenty of those modest paddock thinkers are telling me that when they got so broke they had nothing left but debt, they divorced Coles and Woolies, and separated from the farm chemical companies. Result: reduced income but first profit in 30 years.

We don't need to abandon the market, just add back in the commonsense ingredients that make it sustainable and equitable. Of course, the incredibly wealthy will don a miner's helmet and tell us we can't afford change. Well, we can – we just can't afford them. They will tell us that if you remove incentive you end up with a lethargic Soviet tractor factory; but what if we just removed greed and corruption?

I am confident that the world is capable of this kind of contemplation, but we need a truly free press and a parliament where the first woman to declare that her first political motivation was kindness and justice is not laughed at in the chamber of green leather.

I am confident because of the children I talk to. The logic of their concern is palpable. Their ambition to ensure fairness and care is lambent, inspiring. They are young enough to realise that the people risking the health of the world for the sake of their pocket will be dead before the big death happens, counting their coins but leaving the kids to pay their taxes. It seems to me that the kids are ashamed of us. We could, of course, pull up our socks, but I suspect that the young are expecting us to have our stockings at our ankles while they attempt to clean up our mess.

That's my hope: that the young will turn their back on a generation that changed the dictionary so that indulgence was defined as self-expression, living for today.

When did we lose the most human of wishes, to ensure our children had a future? Swallows still do it but we have lost the ticker for it. It's a shame we don't watch the dignity of animals more closely.

John Woinarski

John is a conservation scientist motivated by the wonder of nature, feelings of despair, and hope.

A fragile hope

The Australian bush is resilient, long inured to adversity. We recognise such hardiness as a defining feature of our country, and indeed have long claimed it as a core of our national character. Much will come back after the fires; a blackened landscape will become green. Life will return from the ashes –or, in many cases, from unburnt patches or from beyond the periphery of burnt areas. Over time, the now silent bush will resound again with a familiar chorus of bird song. Such recovery is an antidote to grief, a future to hope for.

But recovery is a fragile process, and we can all help or subvert it, directly or indirectly. Hope alone is insufficient – we need also to act, else the green will wither and become black again, and nature's return will be transitory and superficial – less beautiful, less productive, less diverse, less resilient than it was before these fires.

There are signs of hope – that the many actions taken after the fires have helped recover nature. Hope comes to us with the return to the wild of fire-injured wildlife, after they have been lovingly and expertly nursed back to health. It comes with making and placing nest boxes in fire-degraded landscapes, remedying in part the losses to fire of resources vital to some species. Hope and satisfaction come when many years of painstaking research yield life-saving dividends in the aftermath of catastrophe: because we then know what to do to aid recovery. Hope and joy also come when – after dogged searches to seek any survivors of a species feared extinct because of the fires – some live individuals are at last found.

Many individuals, communities and organisations took actions after the 2019–20 fires to seek to heal country, in part because they recognised that we belong to this land and have responsibilities for it. In many cases, they were rewarded as their hopes for recovery have been incrementally realised. Local actions are necessary; they help restore the sense of place we value. But local actions alone are insufficient: there are more pervasive issues that must also be remedied.

At its core, hope is wanting for the future to be better than the present, or at least better than some other possible future. The 2019–20 wildfires provided a forewarning of a possible dystopian future, of our generation bequeathing a tarnished world to our descendants. If we are to avoid that unwanted future, our society needs to learn the lessons from these fires. There is hope that it has, evident in official and community recognition that the way we are managing country is not good enough, and that we need to better respect and apply the management knowledge held by Indigenous Australians. I see hope also in official and community recognition that the root cause of these fires was climate change, and mixed hope (we haven't yet committed to doing enough) that our nation and all nations need to do far more to constrain greenhouse gas emissions; else we will experience an ever-increasing frequency of environmental catastrophes.

Perversely, I see hope also in the magnitude of grief felt by our community to the loss of nature in these fires. Such a response demonstrates explicitly the value we place on our environment, and how much it informs and nestles our sense of place. This is our home. We are part of this country. Hope lies in such acknowledgement, and in recognising that it comes with responsibilities.

35

Recommendations

John C. Z. Woinarski, Sarah M. Legge and Libby Rumpff

Introduction

Heroic efforts helped to control the Australian wildfires of 2019–20. Extraordinary responses were also made during and after these fires to rescue wildlife and support the recovery of fire-affected biodiversity. Much of this highly collaborative effort was successful and will result in long-term benefits. However, the scale of these fires severely tested existing capabilities, planning, practices and policy. Such stress revealed some critical shortcomings, and those inadequacies contributed to the magnitude of the impacts of these fires on biodiversity and may constrain or compromise recovery. In a future likely to be characterised by an increasing frequency of comparable severe and extensive wildfires, these shortcomings need to be remedied – for otherwise there will be little long-term recovery of the biodiversity affected by the 2019–20 wildfires, and future fires will replay and magnify the losses of species and environments.

The official inquiries into the 2019–20 wildfires made many recommendations to address some of these shortcomings (Chapter 30). Less formally, Chapter 29 reports on some of the experiences of individuals closely involved in the management of the 2019–20 wildfires and the attempts to protect wildlife during them, highlighting their assessments of what worked and what didn't work before and during the fires. Most other chapters in this book also include recommendations that reflect a range of different perspectives and foci, all aiming to enhance preparedness for the future and to better secure recovery after these wildfires. In this chapter, we distil and coalesce these recommendations. Our focus is particularly on factors that affect biodiversity, and we do not presume here to venture into many of the other components of fire management and disaster response that were the subject of most recommendations in the official inquiries.

We structure the recommendations below to reflect the needs to make improvements before, during and after wildfires (with sets of recommendations broadly in that order), and to enhance legislation, policy, planning, knowledge, practice and management. The challenge is immense, and limited and piecemeal refinements will not suffice: many of the recommendations are interrelated (Fig. 35.1) and major broad-scale changes are needed.

Although these recommendations are targeted explicitly to wildfire in Australia, most are also applicable to comparable environmental catastrophes across the globe, and we would hope that lessons learnt in the 2019–20 wildfires will help decrease the risk of such catastrophes, constrain their impacts, and aid recovery after them.

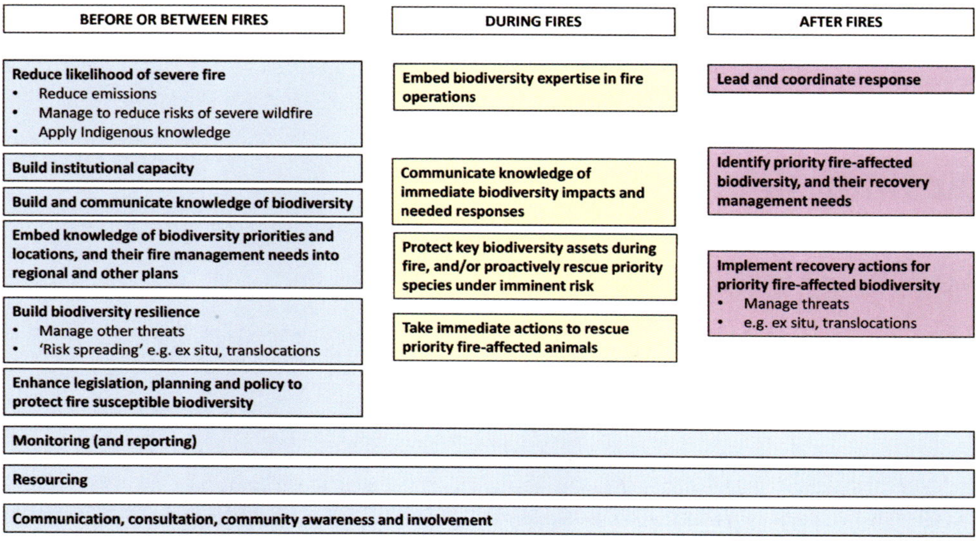

Fig. 35.1. Summary schematic diagram of main recommendations.

List of recommendations

Mitigate climate change

The magnitude of the 2019–20 wildfires, and consequent losses of biodiversity, was driven by climate change; and climate change will lead to a future marked by an increasing frequency of comparable catastrophes. To constrain and reduce such risks and compounding impacts on biodiversity, there is a foundational requirement to mitigate climate change.

- *Immediately implement stronger commitments to urgent reductions in greenhouse gas emissions.*

Support Indigenous land management

The 2019–20 wildfires showed that the management currently implemented in Australia may be inadequate to pre-empt and control the fires that are likely to characterise our future. As a community, we will need to also recognise and apply other fire knowledge systems and practices, in particular the long-honed and intricate knowledge held by First Nations people to care for Country. Limiting the spread of wildfire in the catastrophic fire weather conditions that will become increasingly common may remain a management challenge, but there is much to be gained from following Indigenous leadership to reset our relationship with fire, and its application for looking after and connecting with our country.

- *Enable the broader application of Indigenous-led wildfire planning and recovery, and Indigenous rights and authority to care for Country through fire and other management practices. This will require more support for increased capacity for planning, on-ground management and learning. It will also require the removal of barriers that currently impede such uptake, including current decision making frameworks and processes, insufficient resourcing, current regulatory and legal frameworks, disconnection with and lack of access to Country, conflicting views around fire and burning regimes, fragmented partnerships, and a lack of information about Indigenous wildfire management.*

Enhance capability to evaluate and mitigate severe fire risks and to suppress wildfire

Government inquiries examined in detail many operational matters relating to fire control and these are not revisited here. However, we note three recommendations that may reduce risks of future conflagration, thereby benefiting biodiversity:

- *Enhance modelling capability to evaluate risks to biodiversity from severe wildfire and fire regimes in different landscape contexts, and include such risk assessments in fire plans and operations.*
- *Improve capability and resourcing for early detection of wildfires such that there is more chance of suppression in conditions conducive to uncontrollable fire.*
- *Cease timber-harvesting in native forests, given the supporting evidence that logged landscapes increase fire risks, and the practice is detrimental for biodiversity as a whole.*

Build resilience

The period between fires is critical to build resilience in natural environments. Biodiversity is likely to fare better in future fires if – in the period before fire – ongoing targeted conservation management can increase the population size and number of locations of fire-susceptible species (especially threatened species) and can improve the condition of ecological communities. Furthermore, for the conservation of biodiversity, most other threats can be controlled more tractably than fire, so there is practical advantage in reducing the overall threat burden by controlling the more manageable threats.

- *Supported by increased long-term investment, implement more enduring and effective programs to control threats that are currently affecting biodiversity and likely to compound fire impacts.*
- *More specifically, identify and better manage areas of current and potential future critical habitat, including fire and climate change refuges.*
- *Develop and implement a risk-informed program of translocations for highly imperilled species that are likely to be especially susceptible to future wildfires, in order to establish these species across multiple locations, and thus reduce risks of extinction in single events.*

Risk of species' loss in future fires will be reduced, and capability to respond after fires will be increased, if strategic and prioritised *ex situ* programs are implemented, especially for threatened species whose fate may be most precarious in future fires.

- *Implement a program of priority* ex situ *measures for species whose existence is likely to be jeopardised by future fires. Such actions should build from the existing* ex situ *(e.g. seed) collections, enhance knowledge about germination and restoration, enhance knowledge of husbandry for at risk animals, develop appropriate holding facilities for any emergency collections of fire-affected species and, where appropriate, develop maintenance of insurance populations. Note that this recommendation links also to capability for translocations, as highlighted above.*

Enhance the evidence base

Biodiversity is more likely to be effectively managed, protected during fire and recovered successfully following fire if there is more knowledge about the distribution, baseline condition (including abundance), ecology (including traits associated with fire-susceptibility and potential for recovery), threats, responses to fires and fire regimes (including consideration of mortality rates during and in the aftermath of fires) and management requirements of species and ecological communities; spatial information on the age composition of vegetation types or plant populations relative to thresholds of potential concern (or 'immaturity risk'); and the effectiveness of threat management. Such information provides guidance on where management is critical, what actions are required, and when. The knowledge base is currently biased, being particularly poorly developed for some groups, notably invertebrates and fungi, for which there is also a need for much taxonomic resolution. Likewise, knowledge of suitable fire regimes (and responses to fire regimes generally) is generally better for plants and ecological communities than for most animals. The utility of such knowledge also depends on its accessibility, interpretability, currency and capability to provide information from local to national scales.

- *Implement a strategic research program to build the evidence base for the responses of biodiversity to fire regimes, and especially to fires of the intensity likely to characterise the future (noting that the research response to the 2019–20 wildfires will provide a major contribution to this goal). Such a research program should focus especially on understanding which species, communities and places of significance for biodiversity are highly susceptible to fire impacts, with poor opportunities for recovery. Filling major knowledge gaps that currently impede effective management is a priority, with much greater focus on relatively poorly known species and environments.*
- *Consolidate and enhance data and modelling for biodiversity (including distributions, baseline condition), particularly to help circumscribe and help prioritise sites of biodiversity significance, and for short-range endemic species that may be particularly susceptible to loss of entire populations during fire.*
- *Undertake research to build the evidence base for the responses to fire of other threat factors (e.g. introduced predators), their distribution in the landscape, compounding impacts with fire, and the effectiveness of management to control them in different landscape contexts (noting that the research response to the 2019–20 wildfires will provide a major contribution to this goal).*
- *Implement a program of taxonomic research to fill major knowledge gaps (e.g. for some invertebrate groups and fungi).*
- *Ensure that the resulting knowledge of biodiversity responses to fire is readily accessible and informs prioritisation to support management actions before, during and following wildfire events.*
- *Increase public awareness of fire impacts on biodiversity, and how fire affects ecological health and function, and hence our society as a whole.*

Enhance and implement long-term biodiversity monitoring programs

Monitoring provides a mechanism for enhancing the evidence base, particularly with respect to capability to assess impacts of wildfire (i.e. through before/after sampling), setting targets for and measuring the rate and success of recovery, and assessing and refining the effectiveness of threat management before and after fire.

- *Implement a comprehensive and enduring monitoring program for threatened and fire-susceptible species and ecological communities, sites of biodiversity significance, threats and ecological processes, and ensure that results from such monitoring are regularly reported and publicly accessible in a coordinated, regularly updated database. Ensure such monitoring data can and are used to assess and help continually refine management (i.e. adaptive management), and can measure progress towards post-fire recovery.*

Identify and protect significant sites for biodiversity

Some parts of the landscape are more critical for biodiversity conservation than others: such places include centres of endemism and genetic diversity, wetlands of international significance, World Heritage sites, sole locations and critical current and future habitat for threatened species, climate change refuges, longer unburnt patches and old-growth forests, and places of Indigenous cultural significance. Damage to such irreplaceable sites may be irreversible. Such sites are more likely to be afforded protection during fire if, beforehand, their value is explicitly recognised and accorded priority in policy and planning and fire management, their locations explicitly delineated, their responsiveness or susceptibility to fire and fire management (e.g. vulnerability to use of retardants) appropriately documented, and such information is routinely and rapidly available during fire control operations.

- *Appropriately delineate all sites of biodiversity significance with their importance and prioritisation documented, and with such information explicitly described in fire management plans and appropriately recognised in relevant policy and law.*
- *Appropriately manage all such sites of biodiversity significance to reduce fire risk (e.g. by prescription fire around their perimeter) and target for prioritised protection during fire control operations.*
- *Appropriately fund management of all such sites of biodiversity significance to reduce compounding threats.*
- *Amend environmental legislation to provide more explicit recognition of the need to protect sites of biodiversity significance (i.e. legislative compulsion), to clarify responsibility for such protection, and to provide greater accountability when such sites are not adequately protected.*
- *Evaluate and prioritise the significance of, and risks to, these sites in a manner that is fungible with respect to the value and risks to infrastructure and human life, such that the asset protection hierarchy in fire control operations is not inevitably ordered infrastructure > nature.*

Enhance fire management and planning for fire-susceptible and threatened species and ecological communities

Many threatened and fire-susceptible species and ecological communities have particular requirements for fire management, and particular management needs following wildfire. These management requirements should be clearly articulated and promulgated in relevant planning documents, and implemented.

- *Develop or update management plans (including recovery plans and conservation advices, where relevant) to include more specific information on responses to fire, preferred fire regime, need for protection from fire (e.g. through prescriptive burns around the periphery of important locations), constraints during fire operations (e.g. detrimental impacts of retardants) and post-fire management needs for all threatened species and ecological communities.*
- *Ensure such information informs regional fire management plans and is readily available during fire operations. Ensure that these plans provide explicit wildfire advice (including*

clarification of roles and responsibilities, priority locations and actions under different wildfire scenarios, and resources required) such that they can be implemented readily (i.e. plans and their actions are 'shovel ready').

- *Develop state/territory and regional biodiversity emergency response plans that chart priorities and provide the mechanism, and trigger the funding, required to implement post-fire urgent responses. Such plans should have explicit legislative support.*
- *Ensure currency of a Key Threatening Process under national legislation for 'fire regimes that cause biodiversity decline' and develop, resource and implement a threat abatement plan that provides a strategic, proactive and prioritised basis for regional and national management of fire to benefit biodiversity.*
- *Undertake scenario modelling and contingency planning to stress test existing planning and policy, to ensure readiness for future fires, and to ensure that planning and policy settings are optimally honed for the recovery of biodiversity after fire.*

Build and coordinate skill sets and capability for individuals and groups that have important roles in the protection of biodiversity during and after wildfire

The likelihood of success of actions taken to protect and recover wildlife during and after wildfire is much influenced by the capability, knowledge and coordination of respondents.

- *In preparedness for future wildfire, enhance the expertise and capacity for individuals and groups that have responsibilities or interests in wildlife care and biodiversity conservation during and after fires, and formalise communication pathways among such respondents.*
- *Provide appropriate training or knowledge about biodiversity to fire operations staff (at command centre and on-ground).*
- *Undertake scenario training for wildlife respondents to ensure adequate capability and preparation for future emergencies.*
- *Undertake scenario modelling and contingency planning to stress test capability and to ensure readiness of respondent groups for future fires.*
- *Help build governance and other capacity for conservation non-government organisations (NGOs) and other bodies to manage large donations flowing rapidly in response to catastrophic wildfire, including appropriate collaboration across groups to best harness skills, clarification of the distribution of donations, and a process for clear specification about the use of funds.*

Embed biodiversity expertise and responsibilities in disaster planning and fire control operations

In some jurisdictions, biodiversity agency representatives were embedded in the emergency management structure (e.g. in incident management teams). Experience in the 2019–20 wildfires demonstrated that this allowed for ready access and dissemination of biodiversity data, guided prioritisation of natural assets, and facilitated connections to a broader network of biodiversity respondents, and that this helped to provide protection of important biodiversity assets during the fires (Chapter 29).

- *Ensure legislation, planning and policy provide for representatives of conservation agencies to be embedded in the emergency management structure, with due authority. Such representation and liaison should be enacted at the central control level, but also at local on-ground operational level.*

Develop, maintain and resource capability for rapid response for biodiversity protection, during and immediately after fires and other catastrophes

Biodiversity may be at most risk during and immediately after wildfires, and the opportunity to prevent extinction and initiate recovery may then be contingent on rapid responses. Accordingly, there is a need to facilitate ready access to necessary resources in the aftermath of wildfire.

- *Where irreplaceable loss for high priority biodiversity assets is likely during fires or fire operations, implement emergency rescue, supported by 'shovel ready' plans.*
- *Ensure financial systems can readily provide emergency funding (e.g. pre-established disaster recovery fund) for wildlife response during and after catastrophic wildfire.*
- *Ensure regulatory provisions (e.g. permits for emergency collection of wildlife) are in place before fires, to allow for emergency response.*

Establish a permanent body to provide oversight of or advice on preparedness for fire, prioritisation of actions during fire, and recovery post-fire for biodiversity

The Australian Government established an expert panel to help prioritise responses for the recovery of biodiversity in the aftermath of the 2019–20 wildfires, but this panel was disbanded after 9 months. If made permanent, such a panel would help ensure better and more collaborative planning and management before fires, and rapid response during and after future fires. Note that the Royal Commission into National Natural Disaster Arrangements (see Chapter 30) recommended a suite of such permanent collaborative entities, including a national disaster advisory body, a 'resilience and recovery entity' and a 'national preparedness and response entity'. One option may be to fit a biodiversity component into these broader groups.

- *Establish a permanent collaborative national body focused on planning, policy and management for biodiversity in relation to fire (and other environmental catastrophes).*

Prioritise species, places and actions to support the recovery of biodiversity after fire

The responses made to the 2019–20 wildfires provided a sound building block for planning and implementing the suite of actions that will need to be taken to support recovery from future fires. Key lessons are listed as recommendations here:

- *Ensure that there is coordination and a clear demarcation of response roles amongst government agencies, conservation NGOs, animal welfare groups, Indigenous groups and other stakeholders, and national leadership that helps foster such collaboration.*
- *Coordinate responses and actions from national to local scales.*
- *Involve and empower local communities in the response to heal country and manage recovery.*
- *Use spatial analysis, on-ground sampling and expert opinion where required to assess the extent of species' overlap with fire and the likely proportional short- and long-term population loss (accounting for the likelihood of future recovery) for all components of biodiversity – including taxonomic groups that are generally poorly known and uncharismatic.*
- *Provide appropriate legislative protection (listing as threatened) for those species and ecological communities that have been most fire-affected.*

- *Undertake surveys to locate important surviving populations and key refuge areas following fire, for subsequent protection.*
- *Map and protect unburnt patches that may be essential sources for persistence in largely burnt landscapes, and recruitment to burnt areas.*
- *Rapidly undertake actions to control other threats that may compound the impacts of fire and/or would otherwise compromise recovery.*
- *Don't exacerbate the impacts of wildfire by undertaking post-fire actions that will further reduce resources or disturb affected areas (such as salvage logging).*
- *Communicate impacts, responses and results – including through collaborative regional workshops and accessible databases – to all communities and affected interests to ensure public awareness of and involvement in the recovery process.*

Recognise that recovery is a long-term proposition and requires long-term resourcing commitments

The impacts of catastrophic wildfire on biodiversity are likely to be long-lasting (Fig. 35.2), but funding for biodiversity recovery following fire tends to be short-term, targeting immediate post-fire needs. Also, it is likely to be more cost-effective to provide ongoing funding to effective fire management (including to build resilience) than to be required to episodically spend extraordinary funds in the aftermath of successive wildfires.

- *Ensure adequate enduring funding is available to support the long-term recovery of fire-affected biodiversity.*
- *Provide adequate funding in the intervals between fires to build resilience and establish appropriate fire plans, policies and management.*

Fig. 35.2. Nature returns after wildfire, but much investment, management, planning and action need to be done to support the recovery. (Photo: Alex Pike/Saving our Species/DPE)

36

Our future

John C. Z. Woinarski, Libby Rumpff, Sarah M. Legge, Stephen van Leeuwen and Brendan A. Wintle

Our present begets our future. Our generation gives hurt or love, blight or gifts, to the generations that follow us. This book was prompted by, and represents a response to, a single catastrophic event – the unprecedented series of Australian megafires of 2019–20. Our objectives were to review the many impacts that these fires had on biodiversity and the natural environment, to describe management responses, and to assess what we have learnt in our attempt to ameliorate impacts and support recovery. In this chapter, we use that assessment as a baseline and perspective for looking into the future, to attempt to discern what it may hold for Australian biodiversity and to consider the extent to which we may shape that future.

Our primary focus in this book on the conservation impacts of that single season of megafires matches a long-established societal perspective, to look immediately and retrospectively at a single catastrophe rather than the longer-term pattern, broader context and underlying causality. Our society concerns itself with wildfire mostly in its urgency and aftermath, when shocked by the damage a single fire or set of fires has caused, rather than on the recurrent pattern. For example, the historian Tom Griffiths noted community responses to the 2009 wildfires in Victoria:

> *The 2009 fires were 'unprecedented,' as many commentators have said ... But it is the recurrent realities that are more striking. For those of us who know the history, the most haunting aspect of this tragedy is its familiarity. The 2009 bushfires were 1939 all over again, laced with 1983. The same images, the same stories, the same words and phrases, and the same frightening and awesome natural force that we find so hard to remember and perhaps unconsciously strive to forget. It is a recurrent nightmare. (Griffiths 2009)*

We are readily mesmerised by catastrophe and acute impacts; and, of course, we need to pay attention to and understand the effects of events such as the 2019–20 wildfires. But it is at least as important to more broadly contextualise such events and impacts. Any single fire is part of a fire regime that extends dynamically across time and space, with a kaleidoscopic variety of impacts. The 2019–20 fires are part of a pattern that has been characterised by an increasing frequency of high severity fires in Australia and globally (Abram *et al.* 2021; Canadell *et al.* 2021). This patterning has important consequences that influenced the impacts of the 2019–20 wildfires. In some areas the 2019–20 wildfires burnt over country not yet recovered from previous fires (Lindenmayer and Taylor 2020); some habitats and places have now been irrevocably changed as a result of such recurrent fire (Fairman *et al.* 2016). Indeed, the now extinct *Banksia montana* mealybug (*Pseudococcus markharveyi*) might have survived the 2019–20 wildfires were it not for the fact that a fire a few years before caused the loss of a second population (see Chapter 11): fire impacts compound.

Future fires may compromise the management efforts directed to, and hence the recovery of, species and communities affected by the 2019–20 wildfires (Legge *et al.* in press): the time available for recovery is diminishing. Given the vast extent of the 2019–20 wildfires, there is a high risk that this subversion of recovery will occur over large areas, ratchetting up the rate of population loss in species (Wagner *et al.* 2020) and of ecosystem collapse (Bergstrom *et al.* 2021). But a future marked by more wildfires of the severity of the 2019–20 wildfires will also touch much more of the landscape, leading to far more widespread risks and impacts, including losses of some of those important refuge areas and culturally significant places that were unburnt in 2019–20.

These fires had severe impacts on Australian biodiversity, but that biodiversity is declining rapidly due to many other pervasive and ongoing threats (Woinarski *et al.* 2015, 2019; Bayraktarov *et al.* 2021) and to the inadequacy of policy, legislation, management and resourcing directed to the conservation of our nature (Woinarski *et al.* 2017a; Ward *et al.* 2019; Wintle *et al.* 2019). As an example, the vast toll of Australian wildlife affected by the 2019–20 wildfires (Chapter 12) galvanised much public concern in Australia and globally, yet for mammals and birds, the number of individuals involved (180 million birds and 143 million mammals 'affected' by fire) is appreciably less than the estimated numbers killed – far less conspicuously – each year in Australia by feral cats (272 million and 815 million, respectively) (Woinarski *et al.* 2017b; Murphy *et al.* 2019). Collectively, this medley of threat factors is likely to drive increasing rates of loss of Australian biodiversity in the future (Geyle *et al.* 2018). This ongoing, indeed escalating, biodiversity decline is also part of a global trend, driven historically by many of the same threats, and increasingly by the consequences of climate change (Collen *et al.* 2009; Ceballos *et al.* 2017; Johnson *et al.* 2017).

While much public attention is focused on declines in a few high-profile iconic species (The Senate Environment and Communications References Committee 2011), the decline of biodiversity is far more extensive and less known, and includes a wide range of ecosystems (Bergstrom *et al.* 2021). In decline are areas of old growth and other long-unburnt vegetation, which provide important resources and carbon storage (Lindenmayer and Taylor 2020), species that provide critical support for ecosystem processes and health, and poorly known species with low public profiles (Lintermans *et al.* 2020; Braby *et al.* 2021; Geyle *et al.* 2021). In many cases, the rate of loss and extent of imperilment in Australia is greatest among species and environments that are the most distinctive and of the most ancient lineages – as these prove least adaptable to the increasing incidence and severity of disturbance and rapidly changing climate (Woinarski and Recher 1997; Holz *et al.* 2020). The fabric of our land is fraying; our nature is in retreat.

Acute episodes of loss, such as those due to the 2019–20 wildfires, cause a step change in the underlying trajectory of this decline. Indeed, the chronic factors that are driving the momentum of biodiversity loss – introduced animals, land clearing, timber harvesting, modification and loss of water resources, weeds, disease – rendered environments and species less resilient to the 2019–20 wildfires. These threats will further compromise

Fig. 36.1. Our future risks being increasingly transformed by severe wildfires and other environmental catastrophes. Kangaroo Island after the 2019–20 wildfires. (Photo: © WWF-Australia/Sii Studio)

recovery following the fires (Chapters 17, 18 and 19). Furthermore, some of the factors that catalysed the 2019–20 wildfires – such as extensive drought, rising temperatures, and days of extreme heat – themselves directly drive both abrupt and gradational declines and losses of many components of biodiversity (Welbergen *et al.* 2008; Mac Nally *et al.* 2009; van der Zande *et al.* 2020; Williams and de la Fuente 2021). These factors are part of a syndrome of pressures that threaten biodiversity – and our lives and livelihoods – and they can't readily be unpicked.

Other chapters of this book have recommended responses that need to be made – in law, policy, management, and in our values – to help support the recovery of biodiversity and landscapes following the 2019–20 wildfires and to reduce the risk of comparable losses in future wildfires. For the conservation of fire-prone species and environments, it is vital to implement these recommendations at all scales (Chapter 35). But these alone will be insufficient: it is likely to be futile to try to recover biodiversity following the 2019–20 wildfires if the larger package of threats also affecting that biodiversity is not also addressed; and if the resources provided for conservation remain nugatory. Furthermore, these wildfires bore the fingerprint of climate change, where episodes of catastrophic environmental events are escalating globally (Abatzoglou *et al.* 2019; Fig. 36.1); constraining the impacts will require global solutions.

The 2019–20 wildfires forced us to imagine a possible – indeed, likely – future, terrifyingly close, in which catastrophe becomes commonplace (Fig. 36.1).

It doesn't have to happen like this. We can shape a different future.

The 2019–20 wildfires as catalyst for change

The 2019–20 wildfires showed that our governments and our community are deeply affected by the loss of biodiversity, and are committed to attempting to repair those losses. For many fire-impacted species and environments, this commitment is paying dividends, achieving some recovery. We want to indicate here that it is possible that the 2019–20 wildfires could also leave an enduring legacy of positive change, of a community unwilling to bequeath to future generations a land in continuing decline. The 2019–20 wildfires showed us that the rate of change is now so rapid we can recognise the fragility of the future, and that it is within our hands to shape it – or to do too little and allow that future to wither. The royal commission into the 2019–20 wildfires provides an important springboard: it recognised that we need a different way of managing and living with fire, to use the interfire periods to build resilience – else we will be overwhelmed (Royal Commission into National Natural Disaster Arrangements 2020). Though not explicitly stated, this is so also for biodiversity.

The Victorian response to the 2019–20 wildfires includes a 'nature-led community recovery' (see Chapter 23) that provides a blueprint for such holistic and enduring change (Inspector-General for Emergency Management 2021). This recognises that helping to recover nature is a vital component of restoring the economic and social fabric of rural communities affected by fire, and that our wellbeing cannot be divorced from that of nature (Fig. 36.2). From this viewpoint, healing involves far more than simply responding to wildfire and patching up its casualties. Instead, it has as its foundation an ethic of connection to and responsibility for the care of nature. This is in part because our lives are dependent upon and enriched by healthy and diverse natural environments: they provide the necessary foundation for our life.

Such appreciation of the need to care for Country is, of course, a central tenet of the culture of Indigenous Australians. The 2019–20 wildfires demonstrated many failings in the prevailing official methods of land management, and that we will need to do things

Fig. 36.2. The 2019–20 wildfires affected our society and natural environments; here, a kangaroo is framed by a burning house at Lake Conjola, December 2019. Nature-led community recovery offers a pathway for a future built on caring for country. (Photo: Bruce Detorres)

differently in order to reduce the risks of comparable future catastrophes. While the post-fire government inquiries recognised some of this needed change involves improved technology, more resourcing and better coordination (Government of NSW 2020; Inspector-General for Emergency Management 2020), they also recognised the urgent need for a cultural change that respects the long-established fire knowledge held by Indigenous Australians, and to implement such practice, led by Traditional Owners (Inspector-General for Emergency Management 2020; Royal Commission into National Natural Disaster Arrangements 2020; Inspector-General for Emergency Management 2021). Importantly, such Indigenous fire management is predicated not on the application of fire as an end in itself but rather as an integral part of nurturing and healing Country. We can hope that a legacy of the 2019–20 wildfires involves a much greater appreciation and respect by all Australians of such expertise, and of the culture on which it is based, and recognition that such knowledge and ethics will be needed to maintain and restore the health of our country.

An attitudinal shift to caring for country offers hope for recovery and for reversing the baleful trajectory of decline in our nature. But over generations we have sown the seeds for a perilous future. A recent assessment concluded 'the rate of change in fire risk delivered by climate change is outstripping the capacity of ecological and social systems to adapt' (Nolan *et al.* 2021). Climate change risks sabotaging any improvements we may make in the management of fire, and in actions taken to support the recovery of species and environments affected by the 2019–20 wildfires and by any comparable future fires. A fundamental lesson from the 2019–20 wildfires is that we must strive harder and more urgently to constrain the emissions that drive such change, else our nature and society will not cope. There is hope that this lesson has been learnt: a recent Australian public survey found that 52% of respondents 'take climate change more seriously now as a result of the bushfires' (morestrategic 2020). Climate change affects our nature, it affects us; it will affect even more our descendants.

These are difficult challenges to overcome. Climate change is now deeply embedded. It may take generations for our society to learn how to adapt to the changes in fire regimes that seem inevitable. An ethic of caring for Country may not yet fit readily into those long accustomed to exploiting the land. But we fail, as a society, to evaluate the true costs of inaction. In not planning for the very real future scenario where climate change exists, we are not even guaranteeing short-term economic gain. We are locking ourselves into a future of continually mopping up after natural disasters of our making, and locking ourselves into a future of social, ecological and economic decline. So our conclusion to this book reflects our hope, but also our plea – for more investment into mitigation, preparation and protection, aided by a better understanding of what biodiversity we have, and how it can be effectively conserved.

There are challenges and threats that we can manage now, with short-term benefits. Many of the factors that compound the impacts of fire (Chapters 17 and 18), or have their own impacts, can be more readily managed than fire, provided there is will, resources, capability and legislative support (Garnett *et al.* 2018). Our laws continue to allow land clearing at a profligate rate, even where it causes the loss of habitat for threatened species (Ward *et al.* 2019). Aquatic environments continue to be degraded and exploited (Kingsford *et al.* 2017). Many feral animals, weeds and diseases remain largely unchecked, driving ongoing declines in many native species and environments (Legge *et al.* 2017). Such enhancement in the control of factors affecting our environments requires strong will in the face of competing demands, resources, capability and legislative support (Garnett *et al.* 2018).

The sudden transformation of loved landscapes due to fire (Fig. 36.1) caused many in our society to see and be confronted by the loss of nature, and to recognise that such loss should not be acceptable. These fires showed that we value nature, are affronted by its loss, and are willing to invest in its recovery. The challenge now is to extend such concerns and actions beyond response to this single catastrophe to also encompass biodiversity loss attributable to all other factors. For the future of our nature, and hence our own future, depends upon such transformation.

References

Abatzoglou JT, Williams AP, Barbero R (2019) Global emergence of anthropogenic climate change in fire weather indices. *Geophysical Research Letters* **46**, 326–336. doi:10.1029/2018GL080959

Abram NJ, Henley BJ, Sen Gupta A, Lippmann TJR, Clarke H, *et al.* (2021) Connections of climate change and variability to large and extreme forest fires in southeast Australia. *Communications Earth & Environment* **2**. doi:10.1038/s43247-020-00065-8

Bayraktarov E, Ehmke G, Tulloch AIT, Chauvenet AL, Avery-Gomm S, *et al.* (2021) A threatened species index for Australian birds. *Conservation Science and Practice* **3**, e322. doi:10.1111/csp2.322

Bergstrom DM, Wienecke BC, van den Hoff J, Hughes L, Lindenmayer DB, *et al.* (2021) Combating ecosystem collapse from the tropics to the Antarctic. *Global Change Biology* **27**, 1692–1703. doi:10.1111/gcb.15539

Braby MF, Yeates DK, Taylor GS (2021) Population declines and the conservation of insects and other terrestrial invertebrates in Australia. *Austral Entomology* **60**, 3–8. doi:10.1111/aen.12519

Canadell JG, Meyer CP, Cook GD, Dowdy A, Briggs PR, *et al.* (2021) Multi-decadal increase of forest burned area in Australia is linked to climate change. *Nature Communications* **12**, 6921. doi:10.1038/s41467-021-27225-4

Ceballos G, Ehrlich PR, Dirzo R (2017) Biological annihilation via the ongoing sixth mass extinction signaled by vertebrate population losses and declines. *Proceedings of the National Academy of Sciences of the United States of America* **114**, E6089–E6096. doi:10.1073/pnas.1704949114

Collen B, Loh J, Whitmee S, McRae L, Amin R, *et al.* (2009) Monitoring change in vertebrate abundance: the Living Planet index. *Conservation Biology* **23**, 317–327. doi:10.1111/j.1523-1739.2008.01117.x

Fairman TA, Nitschke CR, Bennett LT (2016) Too much, too soon? A review of the effects of increasing wildfire frequency on tree mortality and regeneration in temperate eucalypt forests. *International Journal of Wildland Fire* **25**, 831–848. doi:10.1071/WF15010

Garnett ST, Latch P, Lindenmayer DB, Woinarski JCZ (Eds) (2018) *Recovering Australian Threatened Species: A Book of Hope.* CSIRO Publishing, Clayton.

Geyle HM, Woinarski JCZ, Baker GB, Dickman CR, Dutson G, *et al.* (2018) Quantifying extinction risk and forecasting the number of impending Australian bird and mammal extinctions. *Pacific Conservation Biology* **24**, 157–167. doi:10.1071/PC18006

Geyle HM, Braby MF, Andren M, Beaver E, Bell P, *et al.* (2021) Butterflies on the brink: identifying the Australian butterflies (Lepidoptera) most at risk of extinction. *Austral Entomology* **60**, 98–110. doi:10.1111/aen.12525

Government of NSW (2020) 'Final report of the NSW Bushfire Inquiry.' Government of NSW, Sydney.

Griffiths T (2009) We have still not lived long enough: testimony from the 1939 and 2009 fires reveals what we haven't learnt from history. *Inside Story*, 16 February, <https://insidestory.org.au/we-have-still-not-lived-long-enough/ https://insidestory.org.au/we-have-still-not-lived-long-enough/>.

Holz A, Wood SW, Ward C, Veblen TT, Bowman DMJS (2020) Population collapse and retreat to fire refugia of the Tasmanian endemic conifer *Athrotaxis selaginoides* following the transition from Aboriginal to European fire management. *Global Change Biology* **26**, 3108–3121. doi:10.1111/gcb.15031

Inspector-General for Emergency Management (2020) *Inquiry into the 2019–20 Victorian Fire Season. Phase 1. Community and Sector Preparedness For and Response To the 2019–20 Fire Season*. Government of Victoria, Melbourne.

Inspector-General for Emergency Management (2021) *Inquiry into the 2019–20 Victorian Fire Season. Phase 2. Progress and Effectiveness of Victoria's Immediate Relief and Recovery Arrangements*. Government of Victoria, Melbourne.

Johnson CN, Balmford A, Brook BW, Buettel JC, Galetti M, *et al.* (2017) Biodiversity losses and conservation responses in the Anthropocene. *Science* **356**, 270–275. doi:10.1126/science.aam9317

Kingsford RT, Bino G, Porter JL (2017) Continental impacts of water development on waterbirds, contrasting two Australian river basins: global implications for sustainable water use. *Global Change Biology* **23**, 4958–4969. doi:10.1111/gcb.13743

Legge S, Murphy BP, McGregor H, Woinarski JCZ, Augusteyn J, *et al.* (2017) Enumerating a continental-scale threat: how many feral cats are in Australia? *Biological Conservation* **206**, 293–303. doi:10.1016/j.biocon.2016.11.032

Legge SM, Rumpff L, Woinarski JCZ, Whiterod NS, Ward M, *et al.* (in press) The conservation impacts of ecological disturbance: Time-bound estimates of population loss and recovery for fauna affected by the 2019–2020 Australian megafires. *Global Ecology and Biogeography* doi:10.1111/geb.13473.

Lindenmayer DB, Taylor C (2020) New spatial analyses of Australian wildfires highlight the need for new fire, resource, and conservation policies. *Proceedings of the National Academy of Sciences of the United States of America* **117**, 12481–12485. doi:10.1073/pnas.2002269117

Lintermans M, Geyle HM, Beatty S, Brown C, Ebner B, *et al.* (2020) Big trouble for little fish: identifying Australian freshwater fishes in imminent risk of extinction. *Pacific Conservation Biology* **26**, 365–377. doi:10.1071/PC19053

Mac Nally R, Bennett AF, Thomson JR, Radford JQ, Unmack G, *et al.* (2009) Collapse of an avifauna: climate change appears to exacerbate habitat loss and degradation. *Diversity & Distributions* **15**, 720–730. doi:10.1111/j.1472-4642.2009.00578.x

morestrategic (2020) 'Public attitudes to bushfire fundraising: insights from national opinion poll 16–18 January 2020'. Fundraising Institute Australia, Sydney.

Murphy BP, Woolley LA, Geyle HM, Legge SM, Palmer R, *et al.* (2019) Introduced cats (*Felis catus*) eating a continental mammal fauna: the number of individuals killed in Australia. *Biological Conservation* **237**, 28–40. doi:10.1016/j.biocon.2019.06.013

Nolan RH, Bowman DMJS, Clarke H, Haynes K, Ooi MKJ, *et al.* (2021) What do the Australian Black Summer fires signify for the global fire crisis? *Fire (Basel, Switzerland)* **4**, 97. doi:10.3390/fire4040097

Royal Commission into National Natural Disaster Arrangements (2020) 'Royal Commission into National Natural Disaster Arrangements report.' Commonwealth of Australia, Canberra.

The Senate Environment and Communications References Committee (2011) *The Koala – Saving Our National Icon*. Commonwealth of Australia, Canberra.

van der Zande RM, Achlatis M, Bender-Champ D, Kubicek A, Dove S, *et al.* (2020) Paradise lost: end-of-century warming and acidification under business-as-usual emissions have severe consequences for symbiotic corals. *Global Change Biology* **26**, 2203–2219. doi:10.1111/gcb.14998

Wagner B, Baker PJ, Stewart SB, Lumsden LF, Nelson JL, *et al.* (2020) Climate change drives habitat contraction of a nocturnal arboreal marsupial at its physiological limits. *Ecosphere* **11**, e03262. doi:10.1002/ecs2.3262

Ward MS, Simmonds JS, Reside AE, Watson JEM, Rhodes JR, *et al.* (2019) Lots of loss with little scrutiny: the attrition of habitat critical for threatened species in Australia. *Conservation Science and Practice* **1**, e117. doi:10.1111/csp2.117

Welbergen JA, Klose SM, Markus N, Eby P (2008) Climate change and the effects of temperature extremes on Australian flying-foxes. *Proceedings of the Royal Society of London. Series B, Biological Sciences* **275**, 419–425.

Williams SE, de la Fuente A (2021) Long-term changes in populations of rainforest birds in the Australia Wet Tropics bioregion: a climate-driven biodiversity emergency. *PLoS One* **16**, e0254307. doi:10.1371/journal.pone.0254307

Wintle BA, Cadenhead NCR, Morgain RA, Legge SM, Bekessy SA, *et al.* (2019) Spending to save: what will it cost to halt Australia's extinction crisis? *Conservation Letters* **12**, e12682. doi:10.1111/conl.12682

Woinarski JCZ, Recher HF (1997) Impact and response: a review of the effects of fire on the Australian avifauna. *Pacific Conservation Biology* **3**, 183–205. doi:10.1071/PC970183

Woinarski JCZ, Burbidge AA, Harrison PL (2015) The ongoing unravelling of a continental fauna: decline and extinction of Australian mammals since European settlement. *Proceedings of the National Academy of Sciences of the United States of America* **112**, 4531–4540. doi:10.1073/pnas.1417301112

Woinarski JCZ, Garnett ST, Legge SM, Lindenmayer DB (2017a) The contribution of policy, law, management, research, and advocacy failings to the recent extinctions of three Australian vertebrate species. *Conservation Biology* **31**, 13–23. doi:10.1111/cobi.12852

Woinarski JCZ, Murphy BP, Legge SM, Garnett ST, Lawes MJ, *et al.* (2017b) How many birds are killed by cats in Australia? *Biological Conservation* **214**, 76–87. doi:10.1016/j.biocon.2017.08.006

Woinarski JCZ, Braby MF, Burbidge AA, Coates D, Garnett ST, *et al.* (2019) Reading the black book: the number, timing, distribution and causes of listed extinctions in Australia. *Biological Conservation* **239**, 108261. doi:10.1016/j.biocon.2019.108261

Index